The Cosmic Century

A History of Astrophysics and Cosmology

The twentieth century witnessed the emergence of the disciplines of astrophysics and cosmology, from subjects which scarcely existed to two of the most exciting and demanding areas of contemporary scientific inquiry. There has never been a century in which fundamental ideas about the nature of our Universe and its contents have changed so dramatically. This book reviews the historical development of all the key areas of modern astrophysics, linking the strands together to show how advances have led to the extraordinarily rich panorama of modern astrophysics and cosmology. While many of the great discoveries were derived from pioneering observations, the emphasis is upon the development of theoretical concepts and how they came to be accepted. These advances have led astrophysicists and cosmologists to ask some of the deepest questions about the nature of our Universe and to stretch our ability to address them by advanced observation to the very limit. This is a fantastic story, and one which would have defied the imaginations of even the greatest story-tellers.

MALCOLM LONGAIR completed his Ph.D. in the Radio Astronomy Group of the Cavendish Laboratory, University of Cambridge, in 1967. From 1968 to 1969 he was a Royal Society Exchange Visitor to the Lebedev Institute, Moscow. He has been an exchange visitor to the USSR Space Research Institute on six subsequent occasions and has held visiting professorships at institutes and observatories throughout the USA. From 1980 to 1990, he held the joint posts of Astronomer Royal for Scotland, Regius Professor of Astronomy at the University of Edinburgh and Director of the Royal Observatory, Edinburgh. He has been Head of the Cavendish Laboratory, Cambridge since 1997, and was made a CBE in the 2000 Millennium Honours list. Longair's primary research interests are in the fields of high-energy astrophysics and astrophysical cosmology. He has published 15 books and over 250 journal articles on his research work.

**For
Deborah**

The Cosmic Century

A History of Astrophysics and Cosmology

MALCOLM S. LONGAIR

CAMBRIDGE
UNIVERSITY PRESS

CAMBRIDGE UNIVERSITY PRESS
Cambridge, New York, Melbourne, Madrid, Cape Town, Singapore, São Paulo

Cambridge University Press
The Edinburgh Building, Cambridge CB2 2RU, UK

Published in the United States of America by Cambridge University Press, New York

www.cambridge.org
Information on this title: www.cambridge.org/9780521474368

First published 2006

Printed in the United Kingdom at the University Press, Cambridge

A catalogue record for this publication is available from the British Library

ISBN-13   978-0-521-47436-8 hardback
ISBN-10   0-521-47436-1 hardback

Contents

Preface

How this book came about

The origin of this book was a request by Brian Pippard to contribute a survey of astrophysics and cosmology in the twentieth century to the three-volume work that he edited with Laurie Brown and the late Abraham Pais, *Twentieth Century Physics* (Bristol: Institute of Physics Publishing and New York: American Institute of Physics Press, 1995). This turned out to be a considerable undertaking, my first draft far exceeding the required page limit. By drastic editing, I reduced the text to about half its original length and the survey appeared in that form as Chapter 23 of the third volume.

I was reluctant to abandon all the important material which had to be excised from the published survey and was delighted that the Institute of Physics agreed to my approaching Cambridge University Press about publishing the full version. The Press were keen to take on the project, with some further expansion of the text and, in particular, with a number of explanatory supplements to chapters where a little simple mathematics can make the arguments more convincing for the enthusiast. I have also made liberal use of references to my other books, where I have already given treatments of topics covered in this book. The result has been a complete rethink of the whole project and an expansion of the text by a factor of five as compared with the original published version.

As when I was writing my book, *Theoretical Concepts in Physics*, 2nd edn (Cambridge: Cambridge University Press, 2003), I have learned so much during the preparation of this book that I wish I had known when I was learning these subjects. The historical material provides real *physical* insight into the intellectual infrastructure of astrophysics and cosmology, and it is saddening that it is not more easily accessible to the student, researcher and lecturer. Even worse, in many cases, the folk-tales of astrophysics and cosmology have acquired mythical status, which do not necessarily coincide with how many of the great insights came about.

The original sub-title of the book was to be *A History of Twentieth Century Astrophysics and Cosmology*, reflecting its origin as a chapter of *Twentieth Century Physics*. As pointed out by the CUP editors, the story runs right up to 2005 and furthermore astrophysics and cosmology in their modern physics-related guises scarcely existed before 1900. Therefore, it seemed much more appropriate to drop the words 'Twentieth Century' from the subtitle.

Warnings and apologies

The magnitude of the task I set myself only became apparent once I was well into the writing of the final text. It is folly to pretend to completeness, or to hope to make reference to all the important contributions of so many distinguished colleagues. Therefore, I have had to be selective and am only too aware of the limitations of what is published here. Even worse, I do not believe it is possible to write a wholly objective history of as complex a field as the development of astrophysical and cosmological understanding over the twentieth century. I have tried to be fair in my assessments of what is of lasting importance, but this is bound to be a subjective process.

Equally significant is the fact that I am one of the lucky generation who began research in the early 1960s when the whole astrophysical and cosmological landscape changed forever from one dominated by optical astronomy to one of multi-wavelength astronomy in which quite different types of astrophysics began to dominate much of the scene. The influx of physicists into astrophysics from that time onwards has been one of the most important features of this story and I write from that perspective. One of the most revealing aspects of the story told in this book is the close link between developments in physics and their impact upon astrophysics and cosmology, and *vice versa*, and the fact that this symbiosis has been at the heart of these disciplines from the beginning. I find it revealing that the author index of this book includes references to large numbers of physicists as well as to astrophysicists and cosmologists.

Although the author index includes about 1000 individuals, I am aware that it omits many who have made important contributions, sometimes simply because they were not the first author on the paper. The problem of attributing credit to individuals has become very much more difficult during the last few decades of the twentieth century when many of the key papers can involve tens or hundreds of authors. This reflects the fact that many of the large space- and ground-based projects can now involve very large numbers of individuals, and so the credit should go to the project team rather than to individual scientists. I have made value judgements about whom to credit in these cases, often giving up and simply giving the detailed authorship in the bibliography. I hope my colleagues will understand the impossibility of doing justice to everyone involved.

I am bound to repeat the disclaimer that I am not a professional historian, and far less a philosopher, of science. My objectives in this book are astrophysical and cosmological, specifically to track the intellectual history of the development of astrophysics and cosmology through what has been one of the most extraordinary centuries in the history of scientific endeavour. Therefore, this is not a history of astronomy *per se*, but astronomy viewed through the mirror of physical understanding. Numerous controversial topics will be treated in this history, but my approach has been to concentrate upon the astrophysical and cosmological issues rather than the more sensational aspects of the story.

Secondary literature

There is an enormous wealth of fascinating material on the history of twentieth-century astrophysics and cosmology which I have had to condense into a modest space. In the references, I have given complete bibliographical citations to all the original articles discussed. In preparing this book, I have found the following volumes particularly helpful:

Bernstein, J. and Feinberg, G. (1986). *Cosmological Constants: Papers in Modern Cosmology* (New York: Columbia University Press). This volume includes translations of many of the seminal papers in cosmology.

Bertotti, B., Balbinot, R., Bergia, S. and Messina, A., eds (1990). *Modern Cosmology in Retrospect* (Cambridge: Cambridge University Press).

Bondi, H. (1960). *Cosmology*, 2nd edn (Cambridge: Cambridge University Press).

Gillespie, C. C., ed. (1981). *Dictionary of Scientific Biography* (New York: Charles Scribner's Sons).

Gingerich, O., ed. (1984). *The General History of Astronomy, Vol. 4. Astrophysics and Twentieth-Century Astronomy to 1950: Part A* (Cambridge: Cambridge University Press).

Harrison, E. (2001). *Cosmology: The Science of the Universe* (Cambridge: Cambridge University Press).

Hearnshaw, J. B. (1986). *The Analysis of Starlight: One Hundred and Fifty Years of Astronomical Spectroscopy* (Cambridge: Cambridge University Press).

Hearnshaw, J. B. (1996). *The Measurement of Starlight: Two Centuries of Astronomical Photometry* (Cambridge: Cambridge University Press).

Kragh, H. (1996). *Cosmology and Controversy: The Historical Development of Two Theories of the Universe* (Princeton: Princeton University Press).

Lang, K. R. and Gingerich, O., eds (1979). *A Source Book in Astronomy and Astrophysics, 1900–1975* (Cambridge, Massachusetts: Harvard University Press). This volume contains reprints of and brief historical introductions to many of the original articles published between 1900 and 1975 referred to in this survey. All the articles are translated into English.

Learner, R. (1981). *Astronomy through the Telescope* (London: Evans Brothers Limited).

Leverington, D. (1996). *A History of Astronomy from 1890 to the Present* (Berlin: Springer-Verlag).

Martinez, V. J., Trimble, V. and Pons-Bordeía, M. J., eds (2001). *Historical Development of Modern Cosmology*, ASP Conference Series, vol. 252 (San Francisco: ASP).

North, J. D. (1965). *The Measure of the Universe* (Oxford: Clarendon Press).

A key resource for all aspects of astrophysics and cosmology is the series entitled *Annual Review of Astronomy and Astrophysics*, which first appeared in 1963. These reviews are authoritative and represent understanding at the year of the review. The more recent volumes include autobiographical essays by a number of the key personalities who appear in this book. For many topics, I have given references to authoritative books and reviews in the Notes to each chapter.

I have also assumed some familiarity with astronomical terminology. For more details of the terminology and reviews of many areas of astronomy, the following can be recommended:

Nicholson, I. (1999). *Unfolding our Universe* (Cambridge: Cambridge University Press). This is an elementary text, but it includes a large amount of useful background material on all aspects of astronomy.

Maran, S. P., ed. (1992). *The Astronomy and Astrophysics Encyclopedia* (New York: Van Nostrand Reinhold, and Cambridge: Cambridge University Press).

Murdin, P., ed. (2001). *Encyclopaedia of Astronomy and Astrophysics* (4 vols) (Bristol and Philadelphia: Institute of Physics Publishing, and London, New York and Tokyo: Nature Publishing Group).

Acknowledgements

My thanks are warmly accorded to the many friends and colleagues who have helped in numerous ways in bringing this book into being. Clearly, the first set of thanks goes to Brian Pippard who started the whole project off. Many colleagues provided advice about my chapter 'Astrophysics and cosmology' in *Twentieth Century Physics*. They included Tony Hewish, David Dewhirst and the late Peter Scheuer. Once the book project was under way, John Hearnshaw kindly read the whole long first draft of my chapter and made many key corrections and observations about what I had assembled. I also thank him for his excellent hospitality in Christchurch, New Zealand, where the final proofreading and finishing touches took place. The late Sir William McCrea also kindly read the first draft of this book and made many valuable comments about the history as he had experienced it. I have quoted from the wonderful letter he wrote to me on 11 December 1993 in the text.

The stimulus for looking deeper into the history of the technology of modern astrophysics and cosmology was provided by the invitation to participate in the excellent Valencia conference organised by Vicent Martínez, Virginia Trimble and Maria-Jesus Pons-Bordeía entitled *Historical Development of Modern Cosmology*. Michael Hoskin kindly reviewed the contents of my paper for that meeting. The invitation by Wendy Freedman to celebrate the centenary of the foundation of the Carnegie Observatories by providing a brief history of twentieth-century cosmology for the symposium *Measuring and Modelling the Universe* also contributed to the enrichment of the present text.

Special thanks are due to Leon Mestel and John Faulkner for their help with the history of the understanding of stellar evolution. I have picked the brains of countless colleagues on the contents of this book – my apologies if I cannot record them all. I also thank many colleagues for allowing me to quote their birth years in the main text – I hope colleagues of all ages will take encouragement from the wide range of ages at which individuals have made seminal contributions to astrophysics and cosmology.

Special thanks are also due to David Green for his help in customising the CUP LaTeX macros to format the book just as I wanted it to appear in its published form.

Particular thanks are due to the following whose help was invaluable in tracking down many of the obscurer references which have been consulted: Judith Andrews, for her help in tracking down many of the old references referred to in the bibliography, and the birth and death years of the individuals mentioned in the text; Gillian Wotherspoon and Nevenka Huntic of the Rayleigh Library of the Cavendish Laboratory, for help in finding old books and journals; Mark Hurn, librarian at the Institute of Astronomy, for his help in tracking down many old astronomical journals held in the Institute's splendid library; the librarians at the Gordon and Betty Moore Library, the Cambridge University Library and the Royal Observatory, Edinburgh, for their assistance. Judith Andrews also deserves special thanks for acting as my secretary for the last eight years and for defending me from the excessive demands of management of the Laboratory so that this book could be completed in a reasonable time.

As ever, it is an enormous pleasure to dedicate this book to Deborah, whose love and support mean so much more than can be adequately expressed in words.

Picture acknowledgements

I am most grateful to the following publishers and organisations for permission to reproduce the diagrams and pictures which appear in this book.

Addison-Wesley Publishing Company (Fig. 15.4)

American Astronomical Society – *Astronomical Journal* (Figs 10.2, 10.3, 10.7, 14.9, 15.14)

American Astronomical Society – *Astrophysical Journal* (Figs 3.6, 5.2, 5.4, 5.6, 5.7, 5.8, 6.1(b), 7.4, 7.12, 7.14, 7.15(a) & (b), 8.10, 9.5, 9.6, 9.10, 9.15, 10.1, 10.5, 11.1(b), 11.3, 11.5, 11.7, 12.1, 12.4, 12.5, 12.6, 13.6, 13.7, 14.13, 14.17, 15.9, 15.17)

American Astronomical Society – *Astrophysical Journal Supplement Series* (Figs 7.8, 9.8, 15.12)

American Physical Society – *Physical Review D* (Fig. 15.13)

American Physical Society – *Physical Review Letters* (Figs 7.7, 8.4)

American Physical Society – *Reviews of Modern Physics* (Figs 8.1, 12.2)

American Science and Engineering (AS&E) (Fig. 7.6)

Anglo-Australian Observatory (Fig. 10.9)

Annual Review of Astronomy and Astrophysics (Figs 9.2, 9.7, 11.12, 11.20)

Annual Review of Nuclear and Particle Physics (Fig. 9.12)

Astronomical Society of Japan (Fig. 9.4)

Astronomical Society of the Pacific (Figs 7.11, 13.4)

Astronomical Society of the Pacific Conference Series (Fig. 11.14)

Astronomy and Astrophysics (Figs 7.19(a) & (b), 8.5(b), 14.5(b), 15.11)

Astrophysics and Space Science (Fig. 15.2)

AT&T (Fig. 7.17)

Birr Scientific and Heritage Foundation (Figs 1.4(a) & (b))

CalTech Submillimetre Observatory (Fig. 7.16)

Cambridge University Press (Figs 6.3, 6.4(a) & (b), 7.1(a) & (b), 8.3, 8.11, 10.11(b), 15.1, A15.1, A15.2, 16.1, 16.2)

Deutsches Museum, Munich (Fig. 1.2)

Edition Frontières (Fig. 13.8)

Elsevier Publishers (Figs 15.15, 15.16)

European Space Agency (Figs 3.4, 8.5(a), 8.6, 8.7, 15.10)

Harvard College Observatory (Figs 2.1, 5.3)

Harvard-Smithsonian Astrophysical Observatory (Fig. 10.8)

Harvard University Press (Fig. 7.9)

Huntingdon Library and the Observatories of the Carnegie Institution (Fig. 1.6)

International Gemini Observatory (Fig. 7.18)

Living Review in Relativity (Fig. 8.12)

Los Angeles Times (Fig. 4.2)

Mary Lea Shane Archives of the Lick Observatory (Fig. 1.5)

Max Planck Institute for Extraterrestrial Physics (Fig. 14.5(a))

National Academy of Sciences of the USA (Fig. 6.1(a))

National Aeronautics and Space Administration (NASA) (Figs 5.8, 7.10, 7.13, 8.15, 9.11, 11.2, 11.6, 11.19, 11.21, 12.4, 14.15(a), 15.9)

National Radio Astronomy Observatory of the USA (Figs 7.3, 10.11(a), 11.1(b), 11.13(a) & (b))

Nature (Figs 3.1, 3.3(a) & (b), 7.5(a) & (b), 8.2, 8.14, 9.9, 10.4, 11.1(a), 11.8, 11.9(a) & (b), 11.10, 11.11, 11.15, 11.16(a) & (b), 13.9(b), 14.8)

The Observatory (Figs 12.3, 13.1)

Physica Scripta (Fig. 10.12)

Potsdam Astrophysical Observatory (Fig. 3.2)

Royal Astronomical Society – *Memoirs* (Fig. 9.3)

Royal Astronomical Society – *Monthly Notices* (Figs 3.7, 9.1, 10.6, 11.4, 11.13(b), 11.17, 13.5, 13.10, A13.1, 14.3, 14.4, 14.6(a) & (b), 14.7, 14.10, 14.12, 15.7, 15.8)

Royal Astronomical Society – *Quarterly Journal* (Fig. 8.9)

Royal Society of London – *Proceedings* (Figs 5.1, 8.13, 15.5)

Science Museum of London/Science and Society Picture Library (Fig. 1.3)

Space Telescope Science Institute (Figs 7.13, 8.15, 9.11, 11.2, 11.6, 14.15(b))

Springer-Verlag (Figs 8.8, A9.1)

Springer-Verlag – D. Reidel Publishing Company (Figs 7.2, 9.13, 9.14, 13.9(a))

Springer-Verlag – Kluwer Academic Publishers (Figs 10.10, 14.2, 14.14)

SUSSP Publications (Fig. 15.6)

Swiss Society of Astronomy and Astrophysics (Fig. 14.1)

Tartu University Astronomical Observatory (Fig. 4.1)

W. H. Freeman and Company (Fig. 13.2)

World Scientific Publishers (Figs 14.11, 14.15(a), 14.16)

Yale University Press (Figs 5.5, 6.2)

Zeitschrift für Astrophysik (Fig. 3.5)

Part I

Stars and stellar evolution up to the Second World War

1 The legacy of the nineteenth century

1.1 Introduction

The great revolutions in physics of the early years of the twentieth century have their exact counterparts in the birth of astrophysics and astrophysical cosmology – these astronomical disciplines scarcely existed before 1900.

The history of the interaction between astronomy and fundamental physics is long and distinguished. From the birth of modern science, astronomy has provided scientific information on scales and under physical conditions which cannot be obtained in laboratory or terrestrial experiments. There is no better example than the history of the discovery of Newton's law of gravity, which provides a model for the process by which astronomical discovery is absorbed into the infrastructure of physics.[1] The technological and managerial genius of the great Danish astronomer Tycho Brahe (1546–1601) and his magnificent achievements in positional astronomy during the period 1575 to 1595 provided the data which led to the discovery of the three laws of planetary motion of Johannes Kepler (1571–1630) during the first two decades of the seventeenth century. The technical skill of Galileo Galilei (1564–1642) in telescope construction resulted in his discovery in 1610 of the satellites of Jupiter, which were recognised as a scale-model for the Copernican System of the World. Finally, in an extraordinary burst of scientific creativity, Isaac Newton (1643–1727) used Kepler's laws to discover the inverse square law of gravity and synthesised the laws of mechanics and dynamics into his three laws of motion. This story is too well known to need further comment, except to emphasise its astronomical roots – Newton had unified the laws of celestial mechanics with those of free fall on Earth. It is difficult to top this achievement in any branch of the physical sciences, but it illustrates beautifully the intimate relation between the astronomical and physical sciences – this is a theme which will be emphasised throughout this book.

Until the late nineteenth century, astrophysics as such did not exist. Astronomy meant positional astronomy, and the techniques of accurate observation had improved steadily since the time of Tycho Brahe. The accurate measurement of the motions of the Sun, Moon and planets against the background of the fixed stars had a practical application as a means of keeping track of time and of measuring position at sea. One of the early by-products of accurate time keeping was the first reasonably accurate measurement of the speed of light in 1676 by the Danish astronomer Ole Rømer (1644–1710), who observed that the interval between eclipses of Jupiter's innermost satellite, Io, by the planet was greater when the Earth moved away from the planet and was shorter when the Earth moved towards it. Interpreting these differences as resulting from the changing distance between the Earth and Jupiter, Rømer found a value for the speed of light of $c = 225\,000$ km s^{-1}.

All observations were made by eye using telescopes as large as the astronomers could afford. The revolution which was to take place at the beginning of the twentieth century can be traced to three important technical developments during the nineteenth century – the invention of astronomical spectroscopy, the first measurements of astronomical parallaxes for nearby stars and the invention of photography. To take full advantage of these developments, telescope design and operation had to be substantially improved, and the resulting instruments were to dominate the astronomy of the first half of the twentieth century. Let us review briefly these technical developments, since they were to provide the observational foundations for the great revolutions in astrophysics and cosmology in the first decades of the twentieth century.

1.2   From Joseph Fraunhofer to Gustav Kirchhoff

The first decades of the nineteenth century marked the beginnings of quantitative experimental spectroscopy. The breakthrough resulted from the pioneering experiments and theoretical understanding of the laws of interference and diffraction of waves by Thomas Young (1773–1829). It is said that his ideas on the interference of light waves were stimulated by observing the patterns of radiating ripples in the pond in the Paddock at Emmanuel College, Cambridge, where he was a Fellow Commoner. In his Bakerian Lecture of 1801 to the Royal Society of London, 'On the theory of light and colours', he used the wave theory of light of Christian Huyghens (1629–1695) to account for the results of interference experiments, such as his famous double-slit experiment (Young, 1802). In the same lecture, Young introduced the tri-chromatic theory of colour vision in its modern form. Among the most striking achievements of this paper was the measurement of the wavelengths of light of different colours using a diffraction grating with 500 grooves per inch. From this time onwards, wavelengths were used to characterise the colours in the spectrum.

In 1802, William Wollaston (1766–1828) made spectroscopic observations of sunlight and discovered five strong dark lines, as well as two fainter lines.[2] He interpreted the dark lines as delineating the four primary colours of sunlight, rather than the seven colours of the rainbow of Newton or the three colours of the tri-colour theory of colour vision (Wollaston, 1802).

The full significance of these observations only began to be appreciated following the remarkable experiments of Joseph Fraunhofer (1787–1826). Fraunhofer was the son of a glazier and he became one of the two directors of the Benediktbraun glassworks in Bavaria in 1814. The firm manufactured high-quality optical glass for military and surveying instruments. Fraunhofer's motivation for studying the solar spectrum was his realisation that accurate measurements of the refractive indices of glasses should be made using monochromatic light. In his spectroscopic observations of the Sun, he rediscovered the narrow dark lines which would provide precisely defined wavelength standards. His visual

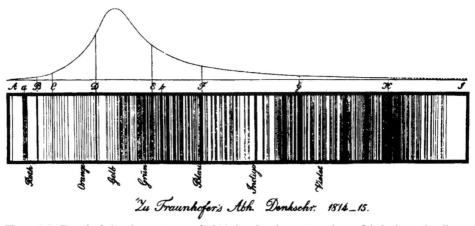

Figure 1.1: Fraunhofer's solar spectrum of 1814 showing the vast numbers of dark absorption lines. The colours of the various regions of the spectrum are labelled, as are the letters A, a, B, C, D, E, b, F, G and H, indicating the most prominent absorption lines. The continuous line above the spectrum shows the approximate solar continuum intensity, as estimated by Fraunhofer (Fraunhofer 1817a,b).

observations were made by placing a prism in front of a 25 mm aperture telescope. In his words,

I wanted to find out whether in the colour-image (that is, spectrum) of sunlight, a similar bright stripe was to be seen, as in the colour-image of lamplight. But instead of this, I found with the telescope almost countless strong and weak vertical lines, which however are darker than the remaining part of the colour-image; some seem to be completely black.

He labelled the ten strongest lines in the solar spectrum by the letters A, a, B, C, D, E, b, F, G and H, and he recorded 574 fainter lines between the B and H lines (Figure 1.1); see Fraunhofer (1817a,b).[3] This notation is still used to describe the prominent absorption lines in the spectra of the Sun and stars.

From the technical point of view, a major advance was the invention of the spectroscope with which the deflection of light passing through the prism could be measured precisely. To achieve this, Fraunhofer placed a theodolite on its side and observed the spectrum through a telescope mounted on the rotating ring (Figure 1.2).

In a second paper, Fraunhofer measured the wavelengths of what are now referred to as the *Fraunhofer lines* in the Solar spectrum using a diffraction grating, which consisted of a large number of equally spaced thin wires (Fraunhofer, 1821) – he was one of the early pioneers in the production of diffraction gratings. He found that the wavelengths of these lines were stable and so provided accurate wavelength standards. In addition to his observations of the Sun, Fraunhofer was the first to make spectroscopic observations of the planets and the stars. In his papers of 1817, he reported the observation of Fraunhofer lines in the spectrum of Venus, inferring that the spectrum was the same as sunlight. In the case of the first magnitude star Sirius, he found, to his surprise,

. . . three broad bands which appear to have no connection with those of sunlight.

Figure 1.2: A portrait of Fraunhofer with his spectroscope (Courtesy of the Deutsches Museum, Munich). This portrait is located in the Hall of Fame of the museum.

In 1823, Fraunhofer made further observations of the spectra of the planets and the brightest stars, anticipating by about 40 years the next serious attempts to measure the spectra of the stars (Fraunhofer, 1823). He concluded that the stars have dark lines in their spectra similar to those seen in the Sun, but that the lines present differ from star to star.

From the perspective of the glass industry, Fraunhofer was then able to characterise the chromatic properties of glasses and lenses quantitatively and precisely. These developments led to much superior glasses, as well as to much improved polishing and testing methods for glasses and lenses. These technical improvements also resulted in the best astronomical telescopes then available. Fraunhofer's masterpiece was the 24-cm Dorpat Telescope built for Wilhelm Struve at the Dorpat, now Tartu, Observatory in Estonia. In addition, he built a heliometer for Friedrich Bessel at Königsberg, to which we will return in Section 1.3.

The understanding of the dark lines in the solar spectrum had to await developments in laboratory spectroscopy. In his first report of the multitude of lines in the solar spectrum, Fraunhofer had noted that the dark D lines coincided with the bright double line seen in lamplight. In 1849, Léon Foucault (1819–1868) performed a key experiment in which sunlight was passed through a sodium arc so that the two spectra could be compared precisely. To his surprise, the solar spectrum displayed even darker D lines when passed through the arc than without the arc present (Foucault, 1849). He followed up this observation with an experiment in which the continuum spectrum of light from glowing charcoal was passed through the arc, and the dark D lines of sodium were found to be imprinted on the transmitted spectrum.[4]

Ten years later, the experiment was repeated by Gustav Kirchhoff (1824–1887), who made the further crucial observation that, to observe an absorption feature, the source of the light had to be hotter than the absorbing flame. From these considerations, Kirchhoff concluded that sodium was present in the solar atmosphere. These results were immediately followed up in 1859 by his understanding of the relation between the emissive and absorptive properties of any substance, now known as *Kirchhoff's law* of emission and absorption of radiation (Kirchhoff, 1859). This states that, in thermal equilibrium, the radiant energy emitted by a body at any frequency is precisely equal to the radiant energy absorbed at the same wavelength. From thermodynamics arguments, he was able to show that there must be a unique spectrum of radiation in thermal equilibrium, which depended only upon temperature and frequency.[5] This profound insight was the beginning of the long and tortuous story which was to lead to Planck's discovery of the formula for black-body radiation and the inevitability of the concept of quantisation over 40 years later.

Throughout the 1850s, there was considerable effort in Europe and in the USA aimed at identifying the emission lines produced by different substances in flame, spark and arc spectra. The fact that different elements and compounds possessed distinctive patterns of spectral lines was established, and attempts were made to relate these to the lines observed in the solar spectrum. In 1859, for example, Julius Plücker (1801–1868) identified the Fraunhofer F line with the bright Hβ line of hydrogen, and the C line was more or less coincident with Hα, demonstrating the presence of hydrogen in the solar atmosphere. The most important work, however, resulted from the studies of Robert Bunsen (1811–1899) and Kirchhoff. In Kirchhoff's great papers of 1861 to 1863 entitled 'Investigations of the solar spectrum and the spectra of the chemical elements', the solar spectrum was compared with the spark spectra of 30 elements using a four-prism arrangement with which it was possible to view the spectrum of the element and the solar spectrum simultaneously (Kirchhoff, 1861, 1862, 1863). He concluded that the cool, outer regions of the solar atmosphere contained iron, calcium, magnesium, sodium, nickel and chromium and probably cobalt, barium, copper and zinc as well.

1.3 The first stellar parallaxes

From the seventeenth century onwards, most astronomers assumed that the stars were objects similar to the Sun, but at vastly greater distances.[6] The method of distance determination used by Newton and others involved assuming that the Sun and stars have the same intrinsic luminosities, a procedure known as the method of *photometric parallaxes*. Then, the inverse square law can be used to measure the relative distances of the Sun and the stars. The major technical problem was that the Sun is so much brighter than the brightest stars that it was difficult to obtain good estimates of the ratio of their observed flux densities, or apparent magnitudes. An ingenious solution was discovered in 1668 by James Gregory (1638–1675), who used Jupiter as an intermediate luminosity calibrator, assuming that its light was entirely composed of sunlight reflected from the disc of the planet and that its surface was a perfect reflector. Then, the apparent magnitudes of Jupiter and the bright star Sirius could be compared, and the distance of Sirius from the Earth was found to be 83 190 astronomical units (Gregory, 1668). The same method was used by John Michell (1724–1793) in 1767 to estimate a distance of 460 000 astronomical units for Vega, or α Lyrae, from the Earth (Michell, 1767).[7] This distance was about a factor of 4 smaller than that found in 1838 by Wilhelm Struve, who used the method of trigonometric parallax. The problem with this approach is that it depends upon the assumption that the intrinsic luminosities of the Sun and the stars are the same.

Direct evidence for the large distances of the stars came from James Bradley's first definitive measurements of the effects of the aberration of light caused by the Earth's motion about the Sun in 1728 (Bradley, 1728). Ever since the time of Copernicus (1473–1543) it had been realised that a test of the hypothesis that the Earth moved about the Sun would be the observation of the annual parallax of the stars. Attempts to measure these small movements of the stars had been subject to a variety of insidious systematic errors. Instead of the expected effect, Bradley (1693–1762) discovered the phenomenon of the aberration of light due to the motion of the Earth, the effect amounting to about ± 20 arcsec for the star γ Draconis. A consequence of this remarkable result was that an upper limit could be derived for the annual parallax of γ Draconis and hence a lower limit to its distance of 400 000 astronomical units,[8] a figure consistent with Newton's estimate using the method of photometric parallax published in the same year. Bradley's pioneering observations ushered in a new epoch of precision astrometry.

The first definitive distance measurements were made in the 1830s by the method of *trigonometric parallax*, the apparent motion of nearby stars against the background of the distant stars due to the Earth's motion about the Sun. Priority for the first trigonometric parallax is accorded to Friedrich Bessel (1784–1846) at Königsberg. The instrument he used was a 16-cm heliometer custom-built by Fraunhofer. The heliometer consisted of a lens cut in half to form two D-shapes, after a design by John Dollond (1706–1761). The images of separated stars could be brought together and their separation measured by the reading on a micrometer screw. Bessel used this telescope to measure the movement of the high proper motion star 61 Cygni relative to distant background stars, and he announced its parallax in 1838 (Bessel, 1839). The parallax amounted to only about one-third of an arcsec, corresponding to a distance of 10.3 light-years. Three months later, Thomas Henderson (1798–1844) published a parallax of 1.16 arcsec for the southern star α Centauri (α Cen) (Henderson, 1840), and almost contemporaneously Wilhelm Struve (1793–1864) measured the parallax of α Lyrae to be 0.12 arcsec (Struve, 1840). Henderson was unlucky not to publish the first parallax – he had measured a parallax of 1 arcsec in declination a few years earlier, but delayed publication until he had reduced his data in right ascension as well. These observations set the scale of the Universe of stars and showed unambiguously that the stars are objects similar to our Sun.

One of the key programmes for the development of astrophysics in the late nineteenth century and the early years of the twentieth century was the gradual accumulation of trigonometric parallaxes for nearby stars, but it was a difficult and demanding task. By 1900, less than 100 parallaxes for nearby stars had been measured with any accuracy.[9] The measurement of parallaxes is still the only direct method of measuring astronomical distances for stars and it remains one of the great challenges of observational astronomy. Matters improved

dramatically in the final decade of the twentieth century with the magnificent set of parallaxes measured by the *Hipparcos* satellite of the European Space Agency, which has
measured precision parallaxes for many thousands of stars (see Figure 3.3).

1.4 The invention of photography

The third major contribution to the development of astrophysics was the invention of the
photographic process by Louis-Jacques-Mandé Daguerre (1789–1851) and William Henry
Fox Talbot (1800–1877). Daguerre began life as an inland revenue official and then became
a scene painter at the opera. The search for methods of recording images by what was to
become the photographic process began with the discovery that some natural compounds
are rendered insoluble when they are exposed to light. In the course of his experiments,
Daguerre discovered that iodine-treated silver paper was also sensitive to light. By 1835,
he had made the important discovery of the *latent image* which was recorded on sensitised
paper, even if the light was not intense enough to darken the paper. The latent image could
then be developed by exposure to mercury vapour and fixed by a strong salt solution. The use
of the latent image meant that exposures could be reduced to 20 to 30 minutes. Interestingly,
the announcement of the discovery of what was called the *daguerreotype* process was made
by François Arago (1786–1853), the director of the Paris Observatory, on 7 January 1839.

A similar announcement was made almost simultaneously by Fox Talbot in England. One
of the earliest, and for me most moving, images is the picture taken in February 1839 by John
Herschel (1792–1871) of his father's 40-foot telescope. In a two-hour exposure, the support
for the large tube of the telescope can be clearly seen – the telescope was dismantled in
the following year (Figure 1.3). John Herschel had a passionate interest in photography and
invented much of its terminology, including the terms 'photography', 'positive', 'negative'
and so on.[10]

The first astronomical images were taken in the succeeding years, but the process was
slow. Isolated examples of successful daguerreotype images of astronomical objects were
reported over the following decade and included the Moon, a solar eclipse and the Sun.
Among the most significant images of these early years of photography was the first
daguerreotype spectrum of the Sun obtained by Edmond Becquerel (1820–1891) in 1842
which showed the complete spectrum of Fraunhofer lines as well as many lines in the
visually unobservable ultraviolet region of the spectrum (Becquerel, 1842). The problem
with the daguerreotype process was that, even for terrestrial objects, the typical exposure
times were about 30 minutes. This was greatly reduced by the invention of the wet collodion
process by Frederick Scott Archer (1813–1857) in 1851 (Archer, 1851). This process produced finely detailed negatives, and typical terrestrial exposures were reduced to 10 seconds.
Astronomical exposures were limited to 10 to 15 minutes because the plates had to remain
wet during exposure.[11] The net result was faster, fine-grained plates which quickly superseded the daguerreotype process. These inventions sparked an enormous popular interest
in photography in the 1850s and many commercial photographic studios were set up. The
wet collodion process was used by Julia Margaret Cameron (1815–1879) in her spectacular
portraits of great nineteenth-century figures, including her famous images of the aged John
Herschel.

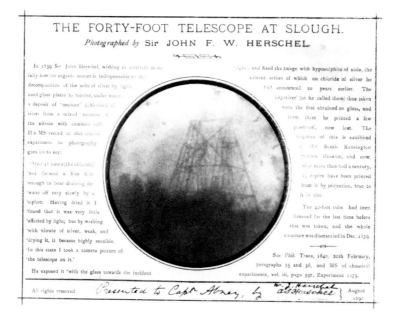

Figure 1.3: John Herschel's photograph of 1839 of part of the support structure of his father's 40-foot telescope just before the telescope was dismantled. The details of the photographic process are described in the text. (Courtesy of the Science Museum/Science and Society Picture Library.)

The story now diverges in two directions. Firstly, the wet collodion process was sufficiently fast for astronomical images and spectra to be recorded, and the search for improved photographic materials continued throughout the remaining years of the century. The boom in photography meant that there was no lack of plates for astronomical use. Secondly, telescope design had to be considerably improved. To take advantage of the use of photographic plates, it had to be possible to track and guide the telescope with very much improved precision as compared with a telescope used visually. In the latter case, the length of the exposure is determined by the response time of the eye, which is only about one-tenth of a second. Let us first complete the story of the development of photographic techniques.

The development of photographic astronomy was largely in the hands of inspired amateurs. Warren de la Rue (1815–1889) in England designed and built a photographic camera for taking daily images of the Sun from Kew Gardens in London using very short exposures. The result was a remarkably complete set of daily sunspot records for the period 1858 to 1872. The first photographic spectrum using the wet collodion process was obtained for the bright star Vega by Henry Draper (1837–1882) in 1872 (Draper, 1879). The spectrum showed the $H\gamma$ and $H\delta$ lines of hydrogen, as well as the first detections of the next seven ultraviolet lines in this hydrogen series. These ultraviolet lines were discovered by astronomical spectroscopy seven years before they were measured in the laboratory. Subsequent observations of the spectra of Vega and Sirius by William Huggins were used by the Swiss schoolmaster Johann Jakob Balmer (1825–1898) in his remarkable papers of 1885 on the *Balmer formula*, which describes the wavelengths of these lines in the spectrum of hydrogen.

Balmer wrote the formula as follows:

$$\text{wavelength} = \frac{m^2}{m^2 - 4} h, \tag{1.1}$$

where $m = 3, 4, 5, \ldots$ and $h = 3645$ Å. Using Huggins' spectral data, Balmer was able to test his formula up to $m = 16$ (Balmer, 1885).[12] This was the first quantum mechanical formula to be discovered. Sadly, Balmer died 15 years before the deep significance of his numerological discovery was appreciated by Niels Bohr.

A key development for astronomical photography was the invention of dry collodion plates, which were much easier to use than wet plates. The speed of the dry plates was similar to that of the wet plates but they allowed much longer exposure times to be used. The search for improved materials continued and culminated in the discovery of emulsions consisting of silver salts suspended in gelatin by Richard L. Maddox (1816–1902) and Charles Bennett (1840–1927) in 1879. It was soon found that the speed of the gelatin emulsions could be vastly increased by prolonged exposure to heat, or by the addition of ammonia. This was the beginning of the dark art of hypersensitising photographic plates to increase their quantum efficiencies.[13] As a result of these developments, the typical exposure time for terrestrial photography was reduced to about 1/15 second. Over the next few years, some superb astronomical images were taken of star clusters and nebulae, revealing unambiguously the remarkable power of photography for astronomy.

It is striking that the photographic pioneers developed their techniques on small telescopes – the larger telescopes were still used for the traditional pursuits of astronomers, the accurate measurement of time and stellar positions. As expressed by Richard Learner,[14]

The lessons of photography and spectroscopy, where astronomy of the highest class had been carried out by observers with very modest telescopes ... were not learned by the astronomical establishment. To them, Urania, the muse of astronomy, was cold and distant, concerned with the smooth and silent motions of the stars, not a grubby figure in an apron, standing at the laboratory sink and doing the washing up.

1.5 The new generation of telescopes

The need to be able to track and guide the telescope accurately for long exposures required major improvements in telescope design. The early pioneers of this story were Lewis Morris Rutherfurd (1816–1892), also famous for his pioneering spectroscopic observations of bright stars, and John Draper (1811–1882), the father of Henry Draper. The key developments concerned the tracking and guiding of the telescope, as well as the continued improvement in the quality of the lenses and mirrors. Rutherfurd invented a clockwork drive for his photographic telescope and, during the 1850s and 1860s, he produced some excellent astronomical images. Besides obtaining photographic images of star fields, Rutherfurd obtained detailed photographic spectra for the Sun. In his observations made in the 1870s, the solar spectrum consisted of 28 overlapping plates totalling about 3 metres in length.

John Draper devoted huge efforts to optimising telescopes for photographic purposes. Over a three-year period, he devised a series of seven grinding and polishing machines

(a)   (b)

Figure 1.4: (a) Lord Rosse's 72-inch telescope at Birr Castle in Central Ireland following refurbishment of the instrument in the 1990s. (b) A drawing of the nebula M51 and its nearby dusty companion made by Lord Rosse from visual observations with the 72-inch telescope, showing clear evidence for spiral structure in the galaxy. (Birr Scientific and Heritage Foundation, courtesy of the Earl of Rosse.)

and produced over 100 mirrors ranging up to 19 inches in diameter. In the last year of his life, 1882, he succeeded in obtaining the spectra of 10th-magnitude stars in the region of M42. Draper's legacy, in a financial as well as a technical sense, was to be crucial for the development of astrophysics over the succeeding years. These endeavours, all carried out on small telescopes, were to pave the way for the spectacular burst of telescope construction in the late nineteenth and the beginning of the twentieth centuries.

Refracting telescopes had been the preferred choice for astrometric applications, but this development reached the end of the line with the completion of the 1-metre (40-inch) refractor at the Yerkes Observatory of the University of Chicago located at Williams Bay, Wisconsin. The refractors had outstanding capabilities for the visual determination of parallaxes and for the detection of double stars. In the latter case, the observer simply waited until a period of good seeing occurred and then, by direct observation, noted whether the stellar image was single or double.

Several large reflecting telescopes had been constructed earlier in the century. The largest of these was built by William Parsons, the third Earl of Rosse, a 1.8-metre (72-inch) reflector known as the 'Leviathan' at his home at Birr Castle in Ireland (Figure 1.4(a)). Despite almost insuperable problems, Rosse (1800–1867) was able to make good visual observations of diffuse nebulae, perhaps his greatest achievement being the observation of spiral arms in nebulae such as M51 (Figure 1.4(b)). It was, however, a struggle, not only against the weather, but also with the materials of the telescope itself.

The biggest problem lay with the large reflector. Domestic flat mirrors had been produced for many years, the reflection being produced by depositing tin compounds on the back surface of a sheet of flat glass. The technical problem of producing parabolic mirrors, which were silvered on the back surface, had not been solved. In consequence, telescope builders, including Newton, used metal mirrors. In Lord Rosse's telescope, the mirror was made of speculum metal, an alloy of tin and copper with a pinch of arsenic, which is 50% reflective. The problem was that speculum metal is a very brittle material and consequently it is very difficult to work with. When the mirror tarnished, the mirror had to be repolished, a hazardous procedure which could potentially destroy the mirror. In fact, Rosse had two speculum mirrors so that one could be installed on the telescope while the other was being repolished.

The solution to the problem of producing large silvered telescope mirrors was discovered by Justus von Liebig (1803–1873) who, in 1835, showed how metallic silver could be deposited by reducing silver nitrate chemically. At the Great Exhibition of 1851, glass-makers had on show decorative items in which silver had been chemically deposited on glass. The telescope builders realised that this was the solution to the problem. The film of silver could be deposited on the front surface of the mirror and could be made thin and uniform, with the result that, when the silver tarnished, rather than having to repolish the mirror, the layer of silver could be removed chemically and a new surface laid down. Foucault used this process to silver the rapidly rotating mirror in his famous speed-of-light experiment in 1850. The first reflecting telescopes using silvered mirrors were built by Karl Steinheil (1801–1870) in 1856, a 10-cm reflector, and by Foucault, who constructed successively larger telescopes, culminating in his 80-cm reflector which was housed at the Marseilles Observatory in 1864.[15]

The problems of constructing larger reflectors were considerable, not least because the reflector design is much more susceptible to flexure and to vibrational and temperature effects. The challenge was taken up by Andrew A. Common (1841–1903), the telescope designer and astronomer, and George Calver (1834–1927), the mirror-maker. During the 1870s, they made a major effort to overcome the inherent problems of the design of reflecting telescopes and introduced a number of innovations which were to be incorporated into the next generation of instruments. The principal innovations involved in constructing their 91-cm reflector were to relieve the weight on the bearings, by submerging a hollow steel float in mercury, and the introduction of an adjustable plate-holder. The result was that the tracking and guiding of the telescope were very smooth – the adjustable plate-holder had the great advantage that a guide-star could be selected outside the field of view of the photographic plate and continuously monitored to ensure that precisely the same field was exposed on the photographic plate within the limits of the seeing disc. Their 90-minute exposure of the Orion Nebula won the Gold Medal of the Royal Astronomical Society in 1884.

The next advance came through the generosity of the English amateur astronomer Edward Crossley (1841–1905). In 1895, he presented his 91-inch reflector, built to the design of the Calver–Common telescope, to the Lick Observatory of the University of California at Santa Cruz (Figure 1.5(a)). An important development was that the observatory was located on an excellent Californian mountain site at Mount Hamilton, where the transparency and stability of the atmosphere were very good and there was a large percentage of clear nights. The mirror was repolished by Howard Grubb (1849–1931) and the mounting of the telescope was stiffened by James E. Keeler (1857–1900). During the commissioning of the Crossley

(a)   (b)

Figure 1.5: (a) The Crossley 91-cm reflector at the Lick Observatory on Mount Hamilton. (b) A photograph of the galaxy M51 taken by Keeler and his colleagues during the commissioning of the telescope in 1900. (Courtesy of the Mary Lea Shane Archives of the Lick Observatory, University of California at Santa Cruz.)

reflector in 1900, Keeler obtained spectacular images of spiral nebulae, including his famous image of M51 (Figure 1.5(b)). Not only were the details of its spiral structure observed in unprecedented detail, but there were also large numbers of fainter spiral nebulae of smaller angular size. If these were objects similar to the Andromeda Nebula, M31, they must lie at very great distances from our Solar System. Tragically, just as this new era of astronomy was dawning, Keeler died of a stroke later in 1900 at the early age of 42.[16]

The next step in increased aperture followed the appointment of George Ellery Hale (1868–1938) as founding Director of the Mount Wilson Observatory in 1904. He persuaded his father to buy the 1.5-metre blank for a 60-inch reflecting telescope. The design was to be an enlarged version of the Calver–Common design for the 91-cm reflector at the Lick Observatory. Before the 60-inch telescope was completed, however, he persuaded John D. Hooker (c.1838 – 1911), an elderly Los Angeles businessman with a passionate interest in astronomy, to fund an even bigger telescope, the 100-inch telescope, to be built on Mount Wilson. In 1906, the American philanthropist Andrew Carnegie visited the fledgling Mount Wilson Observatory and pledged an additional $10 million to the endowment of the Carnegie Institution, specifically requesting that the benefaction be used to enable the work of the Observatory to proceed as rapidly as possible.

The technological challenges presented by the 100-inch were proportionally greater, the mass of the telescope being 100 tonnes, but the basic Calver–Common design was retained. The tracking was provided by a large 2-ton weight, very much like the mechanism of a grandfather clock, which had to be wound up at the beginning of each night's observing.

Figure 1.6: The 100-inch Hooker Telescope at the Mount Wilson Observatory. (Courtesy of the Observatories of the Carnegie Institution of Washington and the Huntingdon Library, Pasadena.)

The optics were the responsibility of George Ritchey, an optical designer of genius, who invented the ingenious optical configuration known as the Ritchey–Chrétien design, which enabled excellent imaging to be achieved over a wide field of view. This was the telescope which was to be at the heart of observational cosmology through the key years from 1918 until 1950 when the 200-inch telescope was commissioned.

1.6 The prehistory concluded

Thus, by the first decades of the twentieth century the tools and techniques which were to provide the foundations for the revolutions in astrophysics and cosmology that were about to take place were well developed. The number of professional astronomers was, however, still very small. The ability to carry out large surveys of the sky with advanced facilities and more complex procedures and instruments needed a new generation of professionals. It is noteworthy that Hale had the foresight to hire Harlow Shapley and Edwin Hubble as staff astronomers for the new Mount Wilson Observatory – they were to play central roles in the history of astrophysics and cosmology in the first half of the twentieth century.

Notes to Chapter 1

1 I have given an account of the achievements of Tycho Brahe, Johannes Kepler, Galileo Galilei and Isaac Newton in Case Study 1 of Malcolm Longair, *Theoretical Concepts in Physics* (Cambridge: Cambridge University Press, 2003).

2 John Hearnshaw makes the interesting point that Isaac Newton narrowly missed discovering the dark lines in the solar spectrum. The principal reason was that he used a small circular aperture 1/4 inch (0.64 cm) in diameter, whereas Wollaston used a slit 1/20 inch (0.13 cm) wide. Wollaston's discovery of the dark lines in the solar spectrum was essentially a footnote to his paper, which was principally concerned with the refractive indices of a wide range of different substances.

3 Fraunhofer's discoveries were first reported in lectures to the Munich Academy of Sciences in 1814 and 1815 and printed in the *Denkschriften der München Akademie der Wissenschaften* and *Gilbert's Annalen der Physik* in 1817. This paper was published in English in the *Edinburgh Philosophical Journal* in two parts; see Fraunhofer (1917a).

4 The principal artificial sources of light for laboratory experiments during the early nineteenth century were flame, arc and spark spectra. Flame spectra, obtained by burning gas in air as in a Bunsen burner, had a typical temperature of about 2000 K. The hotter arc spectra could have temperatures between about 3000 and 6000 K. The hottest sources were the spark spectra, which, on average, had temperatures similar to arc spectra but, because of the presence of hot-spots, small regions of very much higher temperature gas were produced. Thus, in terms of the ionisation state of the material under investigation, spark spectra contained the lines of the highest excitation and flame spectra the lowest.

5 I have given a simple derivation of Kirchhoff's laws in Section 11.2 of Longair, *Theoretical Concepts in Physics*. Case Study 5 of that book describes in some detail the subsequent history which led to Planck's and Einstein's discoveries of quantisation and quanta.

6 An excellent review of early estimates of stellar distances is contained in M.A. Hoskin, *Stellar Astronomy* (Chalfont St Giles, Buckinghamshire: Science History Publications, 1982), Section A.

7 Michell noted that Vega and Saturn have the same brightness when Saturn is in opposition, that is in the direction away from the Sun. Therefore, since he knew the angular diameter of Saturn, he could work out how much of the Sun's light was intercepted by the planet and, assuming that all Saturn's light was reflected sunlight, he could use the inverse square law to estimate how far away Vega must be. Specifically, the angular size of Saturn as observed from the Sun is 17 arcsec and so the illuminated circular hemisphere of Saturn intercepts only $(17/3600)^2 \times (\pi/720)^2$ of the Sun's light. If Vega has the same intrinsic luminosity as the Sun, Vega must be $(3600/17) \times (720/\pi)$ = 48 500 times further away than Saturn, that is, 460 000 astronomical units. See Z. Kopal, in *Dictionary of Scientific Biography*, Vol. 9, ed. C. C. Gillespie (New York: Charles Scribner's Sons, 1981), pp. 370–371.

8 See the article entitled 'Hooke, Bradley and aberration' in M. A. Hoskin *Stellar Astronomy*.

9 A contemporary account of the state of stellar distance measurements in 1900 is contained in Chapter 20 of the remarkable book by Agnes M. Clerke, *The System of the Stars* (London: MacMillan and Company, 1890; 2nd edn, 1905).

10 John Herschel's long-standing interest in photography predated Daguerre's announcement. As a result, within weeks of the announcement, Herschel was able to produce his own images.

11 According to Learner, the wet collodion process involved the following procedure.
   - Cover a clean glass plate with a mixture of collodion (gun-cotton or cellulose nitrate) and potassium iodide dissolved in ether.
   - Allow the ether to evaporate and, while still tacky, immerse in a solution of silver nitrate, later to be improved by mixing with silver bromide.
   - The silver nitrate reacted with the potassium iodide to precipitate insoluble silver iodide.
   - Expose the plate, but do not let it dry out. Once exposed, developed and dried, a permanent negative image is created.
   - Then, the positive could be printed at leisure on albumen coated paper.

12 Balmer's paper of 1885 published in the *Annalen der Physik und Chemie* was a synthesis of two papers originally published in the *Verhandlungen der Naturforschenden Gesellschaft in Basel* **7**, pp. 548–560, 750–752.

13 Eventually, in the 1970s, the hypersensitised IIIaJ plates developed by the Kodak company reached a quantum efficiency of about 1–2%, which has proved to be the effective limit for the photographic process. These quantum efficiencies should be compared with those of current CCD detectors, which can reach 80% or even greater in the red region of the spectrum. As a result, for most astronomical applications, CCDs have replaced the photographic plate as the preferred detector for astronomy, the exception being for very wide field astronomy with large Schmidt telescopes.

14 See R. Learner, *Astronomy Through the Telescope* (London: Evans Brothers, 1981.)

15 A splendid account of the contributions of Léon Foucault to these and many other areas of physics and astronomy is contained in the book by William Tobin, *The Life and Science of Léon Foucault* (Cambridge: Cambridge University Press, 2003).

16 An excellent biography of Keeler and his pioneering contributions to astrophysics has been written by Donald E. Osterbrock, *James E. Keeler: Pioneer American Astrophysicist* (Cambridge: Cambridge University Press, 1984).

2 The classification of stellar spectra

Somewhat surprisingly, Fraunhofer's great discoveries in astronomical spectroscopy were not followed up in any detail until 1863, almost 40 years later, when a number of independent investigators, Giovanni Donati (1826–1873) in Florence, Rutherfurd in New York, George Airy (1801–1892) at the Royal Greenwich Observatory, Huggins in London and Secchi in Rome, began the systematic study of the spectra of the stars and nebulae.[1]

2.1 William Huggins – the founder of stellar astrophysics

William Huggins (1824–1910) was inspired to take up astronomical spectroscopy on reading Kirchhoff's great papers of 1861 to 1863 on the chemical composition of the solar atmosphere. In his words,[2]

This news came to me like the coming upon a spring of water in a dry and thirsty land. Here, at last presented itself the very order of work for which in an indefinite way I was looking for – namely, to extend his novel methods of research upon the Sun to the other heavenly bodies.

Huggins was an inspired amateur astronomer who had no formal university training in the sciences, but from 1856 until his death in 1910 he supported himself by his private income and dedicated his efforts to the advance of astrophysics. Much of his early work was carried out in collaboration with William Miller (1817–1870), who was professor of chemistry at King's College London and an expert on spectral analysis, as well as being his friend and neighbour at Tulse Hill in London. Together, Huggins and Miller immediately began a programme of stellar spectroscopy, the distinctive feature of their observations being that they were carried out with good spectral resolution. In 1864, they published the first results of these studies, those for the brightest stars being of particular importance (Huggins and Miller, 1864a). For Aldebaran, for example, 70 lines were recorded, and for Betelgeuse about 80 lines could be measured. About a dozen spectra were described in detail and the common elements found in all of them. Sodium, magnesium and iron lines were very common, while hydrogen lines were observed in some stars but not in others. Huggins' conclusion is best summarised in his own words, written many years later:[3]

One important object of this original spectroscopic investigation of the light of the stars and other celestial bodies, namely to discover whether the same chemical elements as those of our Earth are present throughout the Universe, was most satisfactorily settled in the affirmative; a common chemistry, it was shown, exists throughout the Universe.

At the time he concluded (Huggins and Miller, 1864a):

It is remarkable that the elements most widely diffused through the host of the stars are some of those most closely connected with the constitution of living organisms on our globe, including hydrogen, sodium, magnesium and iron. . . These forms of elementary matter, when influenced by heat light and chemical force, all of which we have certain knowledge are radiated from the stars, afford some of the most important conditions which we know to be indispensable to the existence of living organisms such as those with which we are acquainted.

It is probably not coincidental that *The Origin of Species* by Charles Darwin (1809–1882) was published in 1859.

In the same year, 1864, Huggins turned his attention to the nebulae, the nature of which was uncertain. The common view was that they consisted of associations of unresolved stars, in which case their spectra would be expected to display the common stellar absorption features. While some nebulae displayed the expected absorption features, in eight of them there were prominent bright emission lines, quite unlike those of any stellar spectrum (Huggins and Miller, 1864b). The four most common emission lines were the $H\beta$ and $H\gamma$ lines of hydrogen and two strong unidentified lines at wavelengths of 500.7 and 495.9 nm. Precise measurements of the wavelengths of the latter lines showed that they could not be associated with any of the lines found in absorption in typical stellar spectra, and they became known as the 'nebulium' lines. On the basis of the observations of the strong emission lines of hydrogen observed in some nebulae, Huggins and Miller correctly concluded that these objects were not associations of unresolved stars but rather

. . .must be regarded as enormous masses of luminous gas or vapour.

By 1868, they had observed some 70 nebulae, about one-third of them displaying strong emission-line spectra, while the others possessed continuous, star-like spectra with prominent absorption lines (Huggins, 1868).

The solutions to the problem of the nebulium lines and the nature of the star-like nebulae had to await the 1920s. In 1927, Ira S. Bowen showed that the nebulium lines were the forbidden lines of doubly ionised oxygen, which can be emitted by an ionised gas at low densities (Bowen, 1927). The problem of the nature of the nebulae with essentially stellar spectra was conclusively resolved by Hubble in 1925 when he showed that the sample of nebulae consisted of a mixture of diffuse gas clouds belonging to our own Galaxy, star clusters in our Galaxy and nearby galaxies, the light of which is the integrated emission of millions of stars.

The first photographic spectra using wet collodion plates were recorded by Henry Draper in 1873, the spectrum of Vega showing the $H\gamma$ and $H\delta$ lines of hydrogen as well as the next seven members of the series (Draper, 1879). Huggins took up stellar spectroscopy again in 1876. He was the first to use the new dry collodion plates for spectroscopy, but he was soon converted to the use of dry gelatin plates which had greater sensitivity. By 1880, he had obtained excellent photographic spectra of about a dozen of the brightest stars. Of particular significance for atomic spectroscopy were the spectra of the 'white stars' in his sample, which extended into the ultraviolet region of the spectrum. In these, he found 12 strong absorption lines of the hydrogen series extending from Hγ into the ultraviolet region of the spectrum. He noted that these were all likely to be associated with hydrogen. For the first four lines in the series, these identifications were confirmed by Hermann W. Vogel (1834–1898) in his laboratory studies in Berlin. As noted in Section 1.4, Balmer used these observations to demonstrate the accuracy of his formula for the Balmer series of hydrogen up to transitions originating from the principal quantum number $m = 16$ (see equation (1.1)).

2.2   The first spectral classification systems

Huggins' brilliant analyses involved high-spectral-resolution studies of small numbers of the brightest stars and demonstrated the power of spectroscopy in understanding their nature. At the same time, much effort was devoted to the classification of the spectra of much larger samples of stars in an attempt to bring some order to the diverse features which they exhibited.

Although Rutherfurd had made the first attempt to place stellar spectra into different classes, the most influential of the pioneers of stellar classification was the Italian Jesuit priest Father Angelo Secchi (1818–1878), who founded the Roman College Observatory, the Collegio Romano, in 1852 with the generous support of Pope Pius IX (1792–1878). At the observatory, the principal instrument was a 24-cm refractor equipped with a direct-vision spectroscope. Secchi was a prolific observer whose 700 publications appeared over a period of 30 years up to the time of his death in 1878. The definitive version of his classification system was completed by 1868 on the basis of spectroscopic observations of about 500 stars (Secchi, 1866, 1868). He placed the stars into four classes.

- *Class I* consisted of white or blue stars, such as Sirius, which exhibit hydrogen absorption lines. This class included those with the hydrogen lines in emission.
- *Class II* consisted of slightly coloured, yellow or solar-type stars, which displayed the principal Fraunhofer lines.
- *Class III* were red stars with wide absorption bands, an example of which was Betelgeuse.
- *Class IV* were the 'carbon stars', as they are now known, which have 'luminous bands separated by dark intervals'. These were only identified after the other classes among samples of faint red stars.

The spectra of stars in each of these classes have similar patterns of spectral lines, for example the Class II stars resembling the spectrum of the Sun, although there were considerable variations within each class. Secchi continued his spectroscopic observations and, by the time of his death, had classified over 4000 stellar spectra, including most of the stars visible to the naked eye in the northern hemisphere.

Many of the pioneers of stellar spectroscopy and the classification of stellar spectra died in the period 1870 to 1880, leaving Huggins as the sole survivor and the father figure of stellar spectroscopy. Of the succeeding generation, major contributions were made by Hermann C. Vogel (1841–1907), who improved greatly the techniques of precision spectroscopy, in particular in the precise measurement of stellar velocities through their Doppler shifts. He also devised his own system of spectral classification, which was similar to Secchi's (Vogel, 1874). It excluded the Class IV spectra, but included the subdivision of the three classes into subclasses.

As the techniques of photographic spectroscopy developed, the objectivity of the classification procedures improved, but the full complexities of stellar spectra also began to be appreciated. The basic problem with the classification schemes was that there was a lack of understanding of the physical basis for the classification procedures. It was not clear the extent to which the colours and properties of the stars were affected by the presence of different elements in their atmospheres. There was some evidence that the blue stars were hotter than the red stars, but it was not clear that the colour of a star was an indicator of its temperature – Huggins argued that the redness of some stars could simply be due to the presence of large numbers of absorption lines towards the blue end of the spectrum.

David DeVorkin estimates that 23 different spectral classification systems had been proposed by 1900. While a number of groups studied the problems of stellar classification, the whole enterprise was overtaken by the mammoth surveys of stellar spectra which were undertaken at Harvard under the direction of Edward C. Pickering. The ultimate results of these efforts were the Harvard system of spectral classification and the *Henry Draper Catalogue*. These endeavours were to lead to the physical understanding of stellar spectra and also to new aspects of atomic physics.

2.3   The Harvard classification of stellar spectra

Henry Draper was trained in medicine but, after a visit in 1858 to William Parsons, the third Earl of Rosse, at Parsonstown in Ireland, he devoted all his energies to pioneering the application of photography to astronomical observation. As discussed in Section 1.4, he obtained the first photographic spectrum using the wet collodion process in August 1872. Following a visit to Huggins in 1879, he was converted to the use of dry photographic plates, which were to revolutionise astronomical spectroscopy. Among his technical innovations, he built an excellent clockwork drive for the telescope which enabled him to obtain long exposure spectra, his longest exposure being for 140 minutes.

Following his untimely death at the age of 45 in 1882, his widow, Mrs Anna Palmer Draper (1839–1914), established the Henry Draper Fund. She provided funds to the Harvard Observatory for Edward Pickering and his assistants to photograph, measure and classify the spectra of stars and to publish the resulting catalogue in the *Annals of the Harvard Observatory* as a memorial to Draper.

Edward Pickering (1846–1919) entered the Engineering Department at Harvard University and graduated *summa cum laude* on his nineteenth birthday. Two years later, he was appointed assistant professor of physics at the newly founded Massachusetts Institute of Technology. He revolutionised the teaching of physics by instituting a carefully designed practical course in physics, subsequently published as *Elements of Physical Manipulations*.[4]

In 1876, Pickering was appointed director of the Harvard College Observatory, a controversial appointment since he was not an observational astronomer. It was, however, an appointment of great foresight since the nature of astronomical research was changing. The introduction of spectroscopy as a tool for astronomical research was the springboard for the new science of 'astro-physics'. Symbolic of this new direction of astronomical research was the foundation of the *Astrophysical Journal* by George Ellery Hale and James Keeler in 1895, which included the interpretative subtitle *An International Review of Spectroscopy and Astronomical Physics*.[5]

Another feature of the revolution which was overtaking the nature of astronomical research was the increase in funding necessary to carry out large programmes in astronomy. The Henry Draper Fund provided several hundred thousand dollars over a number of years to support the Harvard programme. The ambitious programme which Pickering was to establish received other major benefactions. The Paine Fund donated about $400 000 in 1886 and the Boyden Fund gave $230 000 in 1887. In addition, the Bruce Fund contributed $50 000, and Pickering himself provided more than $100 000 from his own resources. This private sponsorship was to enable Pickering to carry out his huge programme to completion. Already, at the very birth of the science of astrophysics, astronomy was big science, needing substantial ongoing funding for the construction of state-of-the-art telescopes and their long-term operation.

Pickering fully realised the great scientific potential of astrophysics, as opposed to positional astronomy, which remained the principal concern of the national observatories. He pioneered three fields: visual photometry, stellar spectroscopy and stellar photography. The visual photometry was undertaken by means of a meridian photometer, in which the brightness of a star on the meridian is compared with the brightness of the pole star, Polaris. Over 1.5 million visual photometer readings were made, most of them by Pickering himself. In 1908, these studies culminated in the publication of the 'Revised Harvard photometry', which was crucial for the study of variable stars and which was to prove to be of special importance for astrophysics and cosmology (Pickering, 1908).

Stellar spectroscopy was funded by the Henry Draper Fund. Pickering decided that the most effective means of undertaking spectral studies of very large numbers of stars was to equip the survey telescope with an objective prism to disperse the images of all the stars in the region of sky under observation.[6] The first experiments were carried out in May 1885, and regular observing began in October of that year. The Bache 8-inch telescope at the Harvard College Observatory had a field of view of 10° and, with an objective prism of angle 13°, the spectra of 6th-magnitude stars could be recorded on photographic plates in 5 minutes – for comparison, the very faintest stars visible to the unaided eye are about 5th magnitude. The programme was in three parts: a general survey of stellar spectra for all stars north of declination −25° brighter than 6th magnitude; a study of the spectra of fainter stars; and a detailed investigation of the spectra of the brighter stars. The principal investigators were Williamina P. Fleming (1857–1911), Annie Jump Cannon (1863–1941) and Antonia C. Maury (1866–1952). They were supported by a large corps of women 'computers', the team being jokingly referred to as 'Pickering and his harem' (Figure 2.1).

The first part of the observing programme was complete by January 1889 and consisted of 633 plates containing the spectra of 10 351 stars. The tasks of examining and classifying the spectra, as well as estimating their magnitudes, were carried out by Fleming. The *Draper Memorial Catalogue* describing the spectra and properties of these stars was published in 1890[7] and provided by far the largest and most systematic classification completed in the

Figure 2.1: Pickering and his team of 'computers' in 1913. Annie Cannon is the second to the right of Professor Pickering. (Courtesy of Harvard College Observatory)

nineteenth century (Pickering, 1890). The spectral classification was based upon Secchi's four classes, but they were now divided into further subclasses. Class I was divided into four subclasses A, B, C and D, Class II into seven subclasses, E, F, G, H, I, K, L, and Classes III and IV were renamed M and N. Special letters were reserved for particular classes of object. The designation O was used to describe the class of star discovered by Charles J. F. Wolf (1827–1918) and Georges A. P. Rayet (1839–1906), which displayed prominent very broad emission lines in the blue spectral region on a continuous background and which are now known as *Wolf–Rayet stars*. The planetary nebulae were designated P and Q for stars otherwise unclassifiable through the sequence A to P. The intention was to provide uniformity of spectral type within each of the subclasses and continuity along the sequence so that, for example, progressing from E, F, G, through H, the prominent Fraunhofer lines seen in solar-type stars become more prominent, while the spectra became weaker at wavelengths shorter than 431 nm.

The task of analysing the spectra of the bright stars was undertaken by Antonia Maury, who graduated from Harvard in 1887 and who was a niece of Henry Draper. The observations for this programme were made with the 11-inch Draper telescope at the Harvard College Observatory with much higher spectral resolution than the general surveys, a maximum dispersion of $11\,\text{Å}\,\text{mm}^{-1}$ being available. A total of 4800 plates of the 681 stars in the survey were analysed by Maury, and Pickering allowed her to develop her own system of spectral classification. The classification scheme was similar to that devised by Mrs Fleming but consisted of 22 classes. One important difference was that she placed the B stars, which are similar to the blue stars in Orion, earlier in the sequence than the A stars because of their simpler spectra. The work was complete by 1895 and published in 1897 (Maury and Pickering, 1897).

Maury's most important contribution, however, lay in her further subdivision of the spectra on the basis of the appearance of the spectral lines. She described three different types of line, the bulk of the stars belonging to class a in which the lines were clearly defined and of 'average' width. In class b, the lines were much broader and hazy, while in class c the lines were unusually narrow and sharp. Of the 681 stars in her sample, 355 were class a, only 18 were of class c and 17 were classed as ac intermediate between classes a and c. The class b stars were mostly rapid rotators or double-lined spectroscopic binaries, although this was not understood at the time. The division of the stars into classes a and c was to lead to the discovery of the giant stars by Hertzsprung in 1905.

There was one further wrinkle in the story before the definitive Harvard system of classification was established. During the solar eclipse of 18 August 1868, spectroscopic observations were made of the emission from solar prominences, and Rayet discovered an intense emission line which he identified with the sodium D lines. The line was reobserved by Norman Lockyer (1836–1920) in October 1868 by placing the spectroscopic slit tangential to the limb of the Sun. He established that its wavelength did not correspond to either of the strong sodium D lines, but had a wavelength of 587.6 nm. There was no corresponding line in the Fraunhofer spectrum of the Sun, nor was there any corresponding feature in the spectra of the known elements. It was inferred that the line was due to some new element which had not been isolated in the laboratory and he named it 'helium', after Helios, the Greek god of the Sun. Helium was only discovered in the mineral cleveite by William Ramsay (1852–1916) in 1895. When the spectrum of the gas was observed in a discharge tube, the line at 587.6 nm was observed along with five other lines. Lockyer showed that some of these lines were also present in chromospheric spectra and, in particular, that they were present in some of the Orion stars which had been placed in spectral class B. He recognised that stars which exhibited helium absorption lines had to be hotter than stars such as Vega on the basis of his laboratory observations of arc and the hotter spark spectra. Lockyer's theoretical ideas were controversial, to say the least, but they foreshadowed future developments. He concluded that as the temperature increases, the elements are dissociated into 'proto-metals'. In his words (Lockyer, 1900),

We have then to face the fact that on the dissociation hypothesis, as the metals which exist at the temperature of the arc are broken up into finer forms, which I have termed proto-metals, at the fourth stage of heat (that of the high tension spark) which gives us the enhanced spectrum; so the proto-metals are themselves broken up at some temperature which we cannot reach in our laboratories into other simpler gaseous forms, the cleveite gases, oxygen, nitrogen and carbon being among them.

Does the story end here? No there is still a higher stage; as the cleveite gases have disappeared as the arc lines and enhanced lines did at the lower stages; the raw form of hydrogen to which I have before drawn attention and which we may think of as 'proto-hydrogen', makes its appearance.

What might have been 'proto-hydrogen' was discovered by Pickering in 1896 in the star ζ Puppis. He discovered a sequence of absorption lines resembling the Balmer series, which became known as the *Pickering series* (Pickering, 1896). He showed that the lines could be described by Balmer's formula provided half-integral values of the principal quantum number m were used (Pickering, 1897).[8] As a result, Pickering considered that the lines were associated with hydrogen under conditions of density and temperature not accessible in the laboratory. In 1912, however, Alfred Fowler (1868–1940) showed that the Pickering series could be observed in laboratory experiments in which the spectra of mixtures of hydrogen and helium were measured. In 1913, in his first great paper on the quantum theory of the hydrogen atom, Niels Bohr (1885–1962) showed that the lines of the Pickering series were not associated with half-quantum numbers, but rather were due to singly ionised helium atoms which have twice the nuclear electric charge[9] (Bohr, 1913a) – the Pickering series resulted from transitions from energy levels with principal quantum numbers $n > 4$ into the $n = 4$ level.

The most famous work which led to the standard Harvard classification was carried out by Annie Cannon. She attended Wellesley College in Norfolk County, Massachusetts, in 1884, one of the first girls of her native state of Delaware to go away to college, and joined the staff of the Harvard College Observatory in 1896. The Boyden Fund donation of \$230 000 was used by Pickering to establish a southern station of the Harvard College Observatory at Arequipa in Peru in 1887. In addition to the 8-inch telescope, which was no longer needed for observations in the northern hemisphere, a 13-inch refractor was also purchased from the Boyden Fund. The objective was to complete the survey of the little-known southern sky. The first plates of the survey were taken in 1891 and, by 1899, 5961 plates had been taken of the spectra of 1122 stars. Cannon was assigned the task of classifying 813 of these stars.

In classifying the stellar spectra, Cannon did not adopt Maury's scheme, but reverted to Fleming's original Harvard scheme with some important amendments. She adopted Maury's proposal that the O stars are the hottest classes of star and that the B stars should precede the A stars in the stellar sequence. She dropped a number of the classes introduced in the original Harvard sequence so that the basic sequence now became O, B, A, F, G, K, M. In addition, there was a class P for planetary nebulae and Q for peculiar stars. Another innovation was the introduction of a decimal notation to represent the spectra of stars intermediate between the main classes. Thus, B stars were renamed B0 stars and A stars as A0. Stars intermediate between them could be assigned classes B1, B2, B3, etc. Stars were placed along this linear sequence on the basis of the presence or absence of different spectral lines, the intention being that the progression of spectral features should be continuous along the sequence. Unlike in Maury's classification system, there was no distinction between lines of different widths. Although there were subsequent enhancements to Cannon's system of spectral types, this is the origin of the basic Harvard classification system which was published in the *Harvard Annals* in 1901 (Cannon and Pickering, 1901).

It was already apparent from the investigations of Lockyer that the sequence was basically a temperature sequence, but it would take a great deal more work before the precise relation between spectral type and temperature was established through the work of Saha, Fowler and Milne.[†] A simplified version of the modern classification system, which is known as the MK classification, is given in Table 2.1.[10]

[†] See Section 3.3.

Table 2.1. *The principal features of the modern system of stellar spectra*

The MK system does not include the classes R, N, S (Morgan, Keenan and Kellman, 1943; Johnson and Morgan 1953)

Class	Class characteristics	Type	Effective temperature (T_{eff}/K)
O	hot stars with He II absorption lines; strong ultraviolet continuum	O5	40 000
B	He I lines attain maximum strength; no He II lines; H developing later	B0 B5	28 000 15 000
A	H lines attain maximum strength at A0, decreasing later; Ca II increasing	A0 A5	9900 8500
F	Ca II stronger; Fe and other metal lines appear	F0 F5	6030 6500
G	Ca II very strong; Fe and other metals strong; H weaker; solar-type spectrum	G0 G5	6030 5520
K	neutral metallic lines dominate; CH and CN bands developing; continuum weak in blue	K0 K5	4900 4130
M	very red; TiO_2 bands developing strongly	M0 M5 M8	3480 2800 2400
R	strong CN bands and C_2 bands increasing		
N	C_2 bands; CN bands decreasing		
S	heavy metal stars; ZrO bands		

Cannon went on to classify a further 1477 northern stars according to the 1901 system and then began the classification of a further 1688 southern stars to fainter magnitudes. These were published in the *Harvard Annals* in 1912. The result of the efforts of Pickering, Maury and Cannon was that, by 1912, almost 5000 stars had been classified over the whole sky with classifications far superior to those previously available. The quality of the data was such that in 1912 Pickering was able to begin the analysis of the distribution of stars of different spectral type throughout the Galaxy. He concluded (Pickering, 1912):

These figures show very clearly that the spectra of Classes A and B are more numerous in the Milky Way than outside it, and that the maximum point is a little south of the Galactic Equator.

This analysis marks the beginning of the delineation of Galactic structure using stars of different spectral type.

In addition to these heroic efforts by Pickering, Cannon and Maury, Williamina Fleming carried out detailed studies of all the stellar spectra which did not fit naturally into what was to become the standard Harvard classification scheme. From 1899 to 1911, when she died, she was curator of astronomical photographs for the Observatory and devoted her astronomical energies to the stars with peculiar spectra, including meticulously describing the spectra, cataloguing the discoverers of the different types of star and the literature references. These stars included novae, gaseous nebulae, peculiar O stars, emission line A and B stars, spectroscopic binaries, variable stars, N and R stars, Oe5 stars and other anomalous stars. This list gives some impression of the enormous advances in spectroscopic knowledge of the stars since the inception of the Henry Draper Memorial project.

At the time of Fleming's death, Pickering was planning an even more ambitious programme, which was to become the *Henry Draper* (or *HD*) *Catalogue*. On 11 October 1911, Cannon began the classification of 225 300 stars and completed the task just under four years later. In an extraordinary feat of concentrated effort, she was able to classify spectra at a rate of about of three per minute, and her classifications were repeatable over the years of the survey. The huge sample of stars comprised all those brighter than 8th magnitude in the northern hemisphere and about 9th magnitude in the south. The *HD Catalogue* was published between 1918 and 1924.[11] Pickering died in 1919, having lived to see the first three sections of the catalogue published – he had been director of the Harvard College Observatory for 42 years. He showed no special interest in interpreting his results but was content to be, in his words, 'a collector of astronomical facts'. Under Pickering, the Harvard College Observatory became the worldwide distribution centre for astronomical information.

Cannon supervised the publication of the other volumes and continued to classify spectra under Pickering's successor, Harlow Shapley. An extension of the *HD Catalogue* was prepared to extend the northern survey to the same magnitude limit as the southern survey. By the time of her death in 1941, Cannon had classified almost 400 000 spectra. According to Cecilia Payne-Gaposchkin[12]

Miss Cannon was not given to theorising; it is probable that she never published a controversial word or a speculative thought. That was the strength of her scientific work – her classification was dispassionate and unbiassed.

Cannon was almost completely deaf throughout her career. She received many international honours. In 1938, she was appointed William Cranch Bond Astronomer, one of the first women to receive an appointment from the Harvard Corporation, and she was the first woman to be awarded an honorary degree by Oxford University.

Notes to Chapter 2

1 Many more details on the history of astronomical spectroscopy are contained in John Hearnshaw's outstanding book *The Analysis of Starlight: One Hundred and Fifty Years of Astronomical Spectroscopy* (Cambridge: Cambridge University Press, 1986, 1990). In addition to the achievements of the pioneers of astronomical spectroscopy, Hearnshaw provides many details of the instrumental innovations which made the astrophysical advances possible.

2 This is a quotation from Huggins' autobiographical eassy which appeared in the *Nineteenth Century Review*, June 1897.

3 This remark appears as a footnote to the 1909 reprint of the 1864 paper by Huggins and Miller in Huggins' collected scientific papers in Huggins, Sir W. and Huggins, Lady M., eds, *Publications of Sir William Huggins Observatory, Vol. II* (London: W. Wesley and Son).

4 The *Elements of Physical Manipulations* was published in two volumes by Macmillan and Co. in the period 1873 to 1876. The second volume contains a major section on practical astronomy. Pickering wrote in the preface to this volume, 'One of the most important features of this volume

is the introduction of a chapter on Astronomy... A careful examination of the subject seems to show that the laboratory method may be used to teach Astronomy as successfully as Chemistry and Physics.'

5 The history of the founding of the *Astrophysical Journal* was recounted in 1995 by Donald E. Osterbrock in the centenary edition of the journal (see *Astrophysical Journal* **438**, 1995, 1–8).

6 The many surveys undertaken by Pickering and his team and the details of the different classification schemes are described in detail by Hearnshaw, *The Analysis of Starlight*, Chapter 5.

7 This catalogue was based on the work of Williamina Fleming and was entitled *The Draper Memorial Catalogue of Stellar Spectra*.

8 In his paper of 1896, Pickering suggested a modified version of Balmer's formula, which had only been discovered 11 years earlier. In this paper, his proposed formula was given by

$$\lambda = \left(4650 \frac{m^2}{m^2 - 4} - 1032 \right) \text{ Å}.$$

Note the different 'Rydberg' constant as compared with the value appearing in equation (1.1) and the subtraction of the wavelength 1032 Å. In his paper of 1897, however, he found that the formula

$$\lambda = 3646.1 \frac{m^2}{m^2 - 16} \text{ Å} \tag{1}$$

gave an excellent fit to the data and had the advantage of using the same Rydberg constant as that of the Balmer series of hydrogen (see equation (1.1)). Pickering suggested that both the Balmer lines and the lines of the Pickering series could be explained by the original Balmer formula, provided half-integral values of m were allowed. Thus, replacing m in equation (1.1) by $m/2$ enables equation (1) to be recovered. Note that, once the Bohr model of the atom was applied to singly ionised helium, the Pickering series results from transitions into the $n = 4$ level from higher principal quantum numbers. The result is that the lines resulting from transitions from even principal quantum numbers are coincident with those of the Balmer series of hydrogen. The odd transitions were members of the series described by Pickering in his 1896 paper and which led to the idea that the lines originated in an even more primitive form of hydrogen, what Lockyer termed 'proto-hydrogen'.

9 There is a delightful story concerning Bohr's identification of the Pickering series with the Balmer series of singly ionised helium. Bohr argued that singly ionised helium atoms would have exactly the same spectrum as hydrogen, but the wavelengths of the corresponding lines would be four times shorter, as observed in the Pickering series. Fowler objected, however, that the ratio of the Rydberg constants for singly ionised helium and hydrogen was not 4, but 4.00163. Bohr realised that the problem arose from neglecting the contribution of the mass of the nucleus to the computation of the moments of inertia of the hydrogen atom and the helium ion. If the angular velocity of the electron and the nucleus about their centre of mass is ω, the condition for the quantisation of angular momentum is given by

$$\frac{nh}{2\pi} = \mu \omega R^2,$$

where $\mu = m_e m_N / (m_e + m_N)$ is the *reduced mass* of the atom, or ion, which takes account of the contributions of both the electron and the nucleus to the angular momentum; R is their separation. Therefore, the ratio of Rydberg constants for ionised helium and hydrogen should be

$$\frac{R_{\text{He}^+}}{R_{\text{H}}} = 4 \left(\frac{1 + \dfrac{m_e}{M}}{1 + \dfrac{m_e}{4M}} \right) = 4.00160,$$

where M is the mass of the hydrogen atom. Thus, precise agreement was found between the theoretical and laboratory estimates of the ratio of Rydberg constants for hydrogen and ionised helium.

In his biography of Bohr, *Niels Bohr's Times, in Physics, Philosophy, and Polity* (Oxford: Oxford University Press, 1991), Pais tells the story of the encounter of George Hevesy (1885–1966) with Einstein in September 1913. When Einstein heard of Bohr's analysis of the Balmer series of hydrogen, he remarked cautiously that Bohr's work was very interesting, and important if right. When Hevesy told him about the helium results, Einstein responded, 'This is an enormous achievement. The theory of Bohr must then be right.'

10 The Yerkes system of spectral classification was published by W. W. Morgan, P. C. Keenan, and Edith Kellman (b. 1911) in 1943 (Morgan, Keenan and Kellman, 1943), and was known as the MKK system. Some further refinements were made by Johnson and Morgan in 1953 (Johnson and Morgan, 1953) and this revised Yerkes system is known as the MK system.

11 *The Henry Draper (HD) Catalogue* was published in the *Harvard Annals*, vols **51** and **55–62** between 1918 and 1924.

12 This quotation is included in Cecilia Payne-Gaposchkin's obituary of Annie Cannon published in *The Telescope*, **8**, 1941, 62–63.

3 Stellar structure and evolution

3.1 Early theories of stellar structure and evolution

The origin of the theory of stellar structure and evolution can be traced to the understanding of the first law of thermodynamics. As a result of the experimental ingenuity of Julius Mayer (1814–1878) and, particularly, of James Prescott Joule (1818–1889), and the deep theoretical insights of Rudolph Clausius (1822–1888) and William Thomson, later Lord Kelvin (1824–1907), the two laws of thermodynamics were established in the early 1850s.[1] In popular terms, they can be stated as follows.

(1) Energy is conserved when heat is taken into account.
(2) The entropy of any isolated system can only increase.

Applying the first law to the stars, the source of energy could be attributed to the heat liberated when matter is accreted onto their surfaces. The kinetic energy of infall from infinity, which is equal to the gravitational binding energy of the material at the surface, is converted into heat when the matter hits the surface. A popular version of the theory involved meteoritic bombardment of stars as the means of providing the necessary energy release. This proposal contained, however, the serious flaw that the necessary flux of meteoroids would perturb the orbits of the inner planets and would also have resulted in a quite unacceptably high rate of meteoroid bombardment of the Earth. Hermann von Helmholtz (1821–1894) and Kelvin proposed that, rather than the gravitational potential energy of meteoroids, the contraction of the Sun itself provided an enormous reservoir of energy (Helmholtz, 1854; Thomson, 1854a,b).[2] The Sun was conceived of as a liquid sphere which gradually contracted and cooled. Energy transport through the Sun was assumed to be by convection.

Kelvin and Helmholtz realised that they could estimate the age of the Sun since its present luminosity is known and its gravitational potential energy can be estimated. This timescale, nowadays referred to as the *Kelvin–Helmholtz timescale*, t_{KH}, for cooling of the Sun, or any star, can be estimated by dividing its gravitational potential energy, $\sim GM^2/R$, by its present luminosity, L, as follows:

$$t_{KH} \approx \frac{E}{L} \approx \frac{GM^2}{LR}. \qquad (3.1)$$

For the Sun, this timescale is only $\sim 10^7$ years. Kelvin used a similar argument to estimate the age of the Earth. He knew the temperature gradient in the outer layers of the Earth and so he could estimate how long it would take the Earth to cool by thermal conduction through its surface. This age turned out to be about 20–40 million years, not so different from the Kelvin–Helmholtz timescale for the Sun.

These estimates were considerably shorter than those favoured by the geologists, who suggested that the age of the Earth was greater than 100 million years from stratigraphic analyses; but these estimates were subject to some uncertainty. Kelvin argued forcibly and successfully against such long geological timescales, much to the chagrin of the geologists. The first reliable estimates of the age of the Earth were made in 1904 by Ernest Rutherford (1871–1937) using the relative abundances of radioactive and stable isotopes of very heavy elements such as uranium. His age for the Earth was at least 700 million years. Rutherford's announcement of this result in Kelvin's presence is delightfully told in his own words.[3]

I came into the room which was half dark and presently spotted Lord Kelvin in the audience and realised that I was in for trouble with the last part of my speech dealing with the age of the Earth where my views conflicted with his. To my relief, Kelvin fell asleep, but as I came to the important point, I saw the old bird sit up, open an eye and cock a baleful glance at me!

Then a sudden inspiration came and I said Lord Kelvin has limited the age of the Earth, provided no new source was discovered. The prophetic utterance refers to what we are now considering tonight, radium! Behold! the old boy beamed upon me.

Rutherford elaborated and refined these techniques over the following years (Rutherford, 1907). The Kelvin–Helmholtz theory of the energy source of the Sun had remained unchallenged for almost 50 years, but now there was a major problem concerning the origin of its luminosity and that of the stars. The origin of the Sun's energy remained a thorny issue for the pioneers of stellar structure over the next two decades. As late as 1932, when Chandrasekhar visited Copenhagen to work with Niels Bohr and his associates, Kameshwar C. Wali records Bohr's attitude to the problems of understanding the physics of the stars:[4]

I cannot be really sympathetic to work in astrophysics because the first question I want to ask when I think of the Sun is where does the energy come from. You cannot tell me where the energy comes from, so how can I believe all the other things?

By then, however, the feasibility of nuclear energy as the source of the Sun's luminosity was well on the way to becoming established, as we will see. Let us first return to the early pioneers of stellar structure and evolution.

The first person to investigate the internal structure of the Sun as a gaseous, rather than a liquid, body was the American J. Homer Lane (1819–1880), who was employed by the US Patent Office in Washington as an expert examiner from 1847 to 1857, and then by the Office of Weights and Measures from 1869. He was highly regarded by Joseph Henry (1797–1878) as a mathematical physicist. During the period of the American Civil War, he carried out the first calculations to determine whether or not the Sun's surface properties were consistent with it being considered to be a sphere of a perfect gas. In 1869, he reversed the calculations and, assuming the material of the Sun to be a perfect gas, attempted to reproduce its surface properties and to determine the variation of its density, temperature and pressure with radius (Lane, 1870). This programme was not particularly successful. Nonetheless, he was the first person to adopt the correct equations of hydrostatic equilibrium and mass conservation for the stars. Although not included in his 1870 paper, he was the first person to derive the important and somewhat non-intuitive result that, if a star loses energy by radiation and contracts, the temperature *increases* rather than decreases.[5] This occurs because, as the star contracts through a series of quasi-equilibrium states, the negative gravitational potential energy becomes more negative and since, according to the virial theorem the internal thermal energy must be minus one half of the gravitational potential energy, the internal energy of the gas must increase, raising its temperature.[†]

In the late 1870s, similar calculations were carried out independently by Augustus Ritter (1826–1908), Professor of Mechanics at the Polytechnical School at Aachen, who identified the initial phase of evolution of the star as the contraction of a perfect gas sphere, which then cooled according to the Kelvin prescription (Ritter, 1883a,b, 1898). The culmination of these early physical models for stars was the treatise by Robert Emden (1862–1940), *Gaskugeln*, published in 1907 (Emden, 1907). Emden was at that time an assistant professor of physics and meteorology at the Technische Hochshule in Munich, and, delightfully, the intention of the treatise was to attract students into theoretical physics by giving as 'practical examples' the internal structure of the stars.

Like Lane and Ritter, Emden assumed that his stellar models were in convective equilibrium, but he went further than Lane and Ritter by introducing *polytropic solutions*, which allowed a much wider range of stellar models to be constructed. In particular, they allowed different variations of the density and pressure of the gas with radius consistent with the laws of physics. The set of equations used by Lane, Ritter and Emden are very powerful indeed. The equations of hydrostatic equilibrium and mass conservation are given by

$$\frac{dp}{dr} = -\frac{GM\rho}{r^2} \quad \text{and} \quad \frac{dM}{dr} = 4\pi r^2 \rho, \tag{3.2}$$

where p and ρ are the pressure and density of the stellar material, respectively, and M is the mass of the star within radius r, $M = M(\leq r)$. The equation of state of the gas was assumed to be of power-law form, $p = \kappa \rho^\gamma$. To simplify equations (3.2), it is convenient to introduce a dimensionless measure of density, w, through the relation $\rho = \rho_c w^n$, where $n = (\gamma - 1)^{-1}$ is the polytropic index and ρ_c is the central density of the star. Writing the distance r from the centre in terms of the dimensionless radius z, $r = az$, the *Lane–Emden equation* can be written as a second-order differential equation:

$$\frac{1}{z^2} \left[\frac{d}{dz} \left(z^2 \frac{dw}{dz} \right) \right] + w^n = 0, \tag{3.3}$$

with

$$a = \left[\frac{(n+1)\kappa \rho_c^{\frac{1}{n}-1}}{4\pi G} \right]^{1/2}.$$

This equation determines the dependence of w upon radius and so provides solutions for the structure of the star for different values of n.

Emden showed that the solutions of equation (3.3) result in stars which have a boundary at a finite radius. As discussed elegantly by Kippenhahn and Weigert in their classic text *Stellar Structure and Evolution*, many important aspects of the astrophysics of stars of different

---

[†]   A demonstration of this result is presented in Section A3.1.

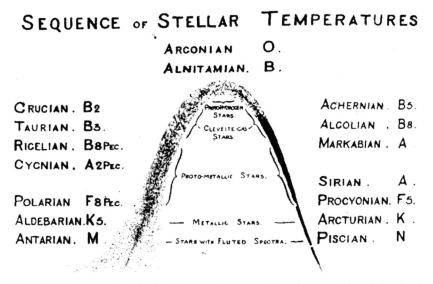

SEQUENCE of STELLAR TEMPERATURES

ARCONIAN   O.
ALNITAMIAN.   B.

CRUCIAN . B2
TAURIAN . B3.
RICELIAN . B8Pec.
CYCNIAN . A2Pec.

POLARIAN   F8Pec.
ALDEBARIAN.K5.
ANTARIAN .   M

PROTOHYDROGEN STARS.
CLEVEITE-GAS STARS.
PROTO-METALLIC STARS.
—— METALLIC STARS. ——
— STARS WITH FLUTED SPECTRA. —

ACHERNIAN . B5.
ALCOLIAN . B8.
MARKABIAN . A

SIRIAN .   A .
PROCYONIAN. F5.
ARCTURIAN . K .
PISCIAN .   N

Figure 3.1: An example of Lockyer's temperature curve for stellar spectral evolution (Lockyer, 1914). His assignments of the spectral types to different parts of the arch are indicated.

types can be understood on the basis of this equation alone.[6] These theories did not gain the recognition they deserved because astronomers argued that the physical conditions inside the stars were unknown and there was no understanding of the process of energy generation.

While these physical models were being developed, the commonly held view was that the various schemes of spectral classification must have some evolutionary significance. Among the most vocal advocates of this picture, Norman Lockyer played an important role as a populariser of astronomy and of science in general. He was an amateur enthusiast of enormous energy, who, following his employment by the British War Office, became Director of the Solar Physics Observatory, set up in South Kensington, London, in 1879. In 1869, he founded the weekly science magazine *Nature*, which he edited for the next 50 years, and which provided a ready-made forum for promoting his opinions. In the 1880s, Lockyer attempted to identify the various spectral classes of star with an evolutionary scheme based upon the meteoritic hypothesis. Figure 3.1 shows a sketch of his theory of stellar evolution.[7] The evolutionary 'temperature arch' begins at the bottom left and shows a cloud of meteoroids colliding and vaporising, thus giving rise to gaseous nebulae and comets. The nebulae condense and contract to form stars, at which point the star attains its maximum temperature at the peak of the arch. The star was then assumed to cool to become a compact red star. Lockyer assigned spectral classes to different parts of this evolutionary arch, although the reasons for the assignments to the ascending and descending branches were often rhetorical rather than physical. Nonetheless, it was implicit in his scheme that there should exist large- and small-diameter stars at the same temperature. There are similarities to Ritter's more physical picture of stellar evolution.

3.2   The origin of the Hertzsprung–Russell diagram

With the publication of the *Draper Memorial Catalogue of Stellar Spectra* in 1890 (Pickering, 1890), it became possible to make the first tests of the hypothesis that the stars evolved down the spectral sequence from hot A and B stars to cool K stars.[8] By 1893, William Monck (1839–1915) and Jacobus Kapteyn had independently come to the conclusion that something must be wrong with this simple picture (Kapteyn, 1892; Monck, 1895). Although parallaxes are the best distance measures for stars, cruder information can be obtained from their proper motions, meaning their apparent drift motions on the sky relative to very distant background stars. Generally, nearby stars can have appreciable proper motions while distant stars have none. Monck and Kapteyn discovered that, among the bright stars, those with the greatest proper motions, and hence nearby, lowest-luminosity objects, were not the K and M but the F and G stars. This was scarcely consistent with a scheme in which stars cooled and grew fainter along the spectral sequence from O to M.

The breakthrough was made by the Danish astronomer Ejnar Hertzsprung (1873–1967). Hertzsprung had trained as a chemical engineer in Copenhagen and St Petersburg and then studied electrochemistry with Wilhelm Ostwald (1853–1932) in Leipzig. He returned to Denmark in 1901 and, as an amateur, began his serious study of astronomy. He was strongly influenced by the discovery by Max Planck (1858–1947) in 1900 of the formula for black-body radiation and realised that, if it were assumed that the stars radiated like black bodies and their distances were known, it is a straightforward calculation to work out their physical sizes. In 1906 he showed that the diameter of Arcturus is roughly the same as the diameter of the orbit of Mars – he immediately inferred that some very large stars must exist (Hertzsprung, 1906).

Hertzsprung had already deduced that there must be a wide range of luminosity among the stars. From the proper-motion data of Monck and Kapteyn, he inferred that statistically the A and Orion-type B stars must be of high luminosity. In addition, he related these data to the information on the distinction between the *a* and *c* class spectra, which was part of Maury's classification scheme (Maury and Pickering, 1897). It was immediately apparent that the red *c* stars were distant luminous stars, with luminosities similar to that of the A and B stars, while the non-*c* stars were low-luminosity nearby objects. As he stated (Hertzsprung, 1905):

The result confirms the assumption of Antonia C. Maury that the *c*-stars show some intrinsic characteristic.

The distinction between what became known as the *dwarf* and the *giant* stars was confirmed by parallax studies (Hertzsprung, 1905, 1907). It turned out that, among the brightest stars in the sky, there are more giants than dwarfs, and this accounted for Monck and Kapteyn's strange result that the red stars are more luminous than the yellow stars.

To Hertzsprung's distress, when he received the *Revised Harvard Photometry*, published in 1908, Cannon's new spectral classifications were included, but Maury's classifications, which were based on the appearance of the spectral lines, were not mentioned. He wrote to Pickering on 22 July 1908:

In my opinion the separation by Antonia C. Maury of the *c-* and *ac-* stars is the most important advancement in stellar classification since the trials by Vogel and Secchi ... To neglect the *c*-properties in classifying stellar spectra, I think, is nearly the same thing as if the zoologist, who detected the deciding differences between a whale and a fish, would continue in classifying them together.

Pickering was not convinced. He attached little significance to Miss Maury's classification, believing that the characteristics she had identified were too subtle to be real.

In 1909 Hertzsprung was invited by Karl Schwarzschild (1873–1916) to visit Göttingen, where he was appointed associate professor. In the same year, Schwarzschild was appointed to the directorship of the Potsdam Astrophysical Observatory and Hertzsprung joined him there. In 1907, he had turned his attention to star clusters for which it can be safely assumed that the stars are all at the same distance. As early as 1900, Schwarzschild had recognised the importance of stellar colours in estimating their spectral types and defined the concept of *colour index* as the difference between a star's photographic and visual brightness (Schwarzschild, 1900b).

A closely related concept was that of the *effective wavelength*, which described the mean wavelength of the spectral energy distribution of the star and so was also a measure of its colour. It had been shown by George Comstock (1855–1934), Director of the Washburn Observatory in Wisconsin, that the effective wavelength was strongly correlated with spectral type (Comstock, 1897), and this was the approach adopted by Hertzsprung to study the colours of the stars. In 1911, he published the first luminosity–colour diagrams for the Pleiades and Hyades star clusters (Hertzsprung, 1911). In these diagrams there was a prominent continuous sequence of stars which he named the *main sequence*, but there was also a very wide range of luminosity among the red stars (Figure 3.2). If attention is restricted only to the main sequence, which is composed of dwarf stars, the red stars are indeed less luminous than the yellow F and G stars and so the systematic trend of stars becoming intrinsically fainter and redder along the spectral sequence was correct. These were the first published colour–magnitude diagrams, which have dominated studies of stellar evolution ever since. Notice that Hertzsprung's research was based upon photometric studies of the stars.

Independently, Henry Norris Russell (1877–1957) arrived at the same diagram by a rather different route. Russell had graduated from Princeton in 1897 and completed his doctorate in 1900. From 1902 to 1905, he worked at the Observatories at Cambridge, England, and began one of the first photographic parallax programmes for stars. With Arthur Hinks (1873–1945), the chief assistant at the University Observatory, Russell perfected the procedures for measuring stellar parallaxes photographically. He returned to Princeton in 1905, and the reduction of the parallax data was completed by 1910. In 1908, Russell made contact with Pickering, who agreed to provide magnitudes and spectra for the 300 stars in the parallax programme. The data were supplied by September 1909, and it was immediately apparent that high- and low-luminosity red stars were present in the sample, a result similar to Hertzsprung's. Russell's famous luminosity–spectral class diagram (Figure 3.3(a)) was first published simultaneously in *Nature* and *Popular Astronomy* in 1914 (Russell, 1914a–e). The scale on the ordinate is in absolute magnitudes, M ($M = \text{constant} - 2.5 \log L$, where L is the intrinsic luminosity of the star), and the spectral class is plotted along the abscissa. The

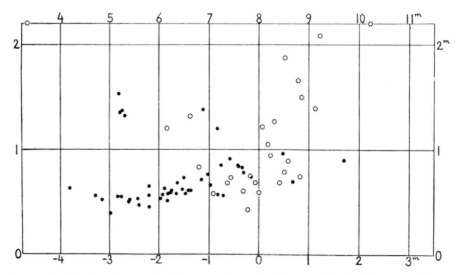

Figure 3.2: The effective wavelength–apparent magnitude diagram for the Hyades star cluster published by Hertzsprung in 1911. The vertical axis is the effective wavelength, which is proportional to the colour, or colour index, of the star. The horizontal axis is the magnitude of the star. This was the first published colour–magnitude diagram. The main sequence runs along the lower part of the diagram and then moves to longer effective wavelengths at faint magnitudes (Hertzsprung, 1911). The giant stars lie in the area above the main sequence in this form of colour–magnitude diagram.

correlation between spectral type and luminosity indicated by the bounding diagonal lines in Figure 3.3(a) is apparent and corresponds to the *main sequence* described by Hertzsprung. In addition, there are red stars above the main sequence, in a region which became known as the *giant branch*. It can be seen that the luminosities of the K and M stars span about 10 magnitudes, corresponding to a factor of 10 000 in luminosity. It is intriguing that the diagram includes one low-luminosity A star well below the main sequence. This is the star 40 Eridani B, which was the first white dwarf to be identified (see Section 4.2). Russell's famous papers also include the luminosity–spectral class diagram for four star clusters in which the giant branch is more clearly defined (Figure 3.3(b)).

The luminosity–spectral class (or colour–magnitude) diagram was known as the 'Russell diagram' until 1933 when Strömgren introduced the term Hertzsprung–Russell diagram, recognising Hertzsprung's key contributions. It is intriguing to compare Figures 3.2 and 3.3 with the Hertzsprung–Russell diagram, or H–R diagram, derived from observations made by the *Hipparcos* astrometric satellite which includes 2927 stars for which parallaxes have been measured to better than 5% accuracy (Figure 3.4).

Hertzsprung's pioneering studies had shown how features in stellar spectra could be used to determine whether stars are dwarfs or giants. Independently, Walter Adams (1876–1956) and Arnold Kohlschütter (1883–1969) discovered other spectral features which could be used as luminosity indicators (Adams and Kohlschütter, 1914b). By combining data on the parallaxes, proper motions and spectral types, they discovered that, within a given spectral class, certain spectral features were sensitive luminosity indicators. Specifically, for stars

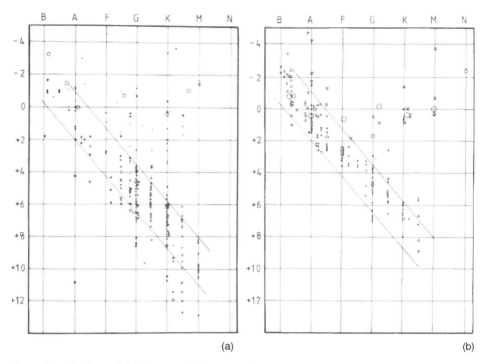

(a) (b)

Figure 3.3: The first published 'Russell diagrams' showing the relation between absolute magnitude and spectral type. (a) The relation for all nearby stars. (b) The relation derived from studies of four star clusters.

of the same spectral type, the low-luminosity stars had weaker ultraviolet continua than the luminous stars, the hydrogen absorption lines were much stronger in the luminous stars and certain metallic line ratios were shown to be sensitive luminosity indicators. In their paper of 1914, they showed that, using these criteria, the absolute magnitudes of stars could be estimated with an accuracy of about 1.5 magnitudes, corresponding to a factor of 4 in intrinsic luminosity. In consequence, the distances of stars could be roughly estimated from the characteristics of their spectra alone. Although the distance estimates would be uncertain by a factor of 2, this represented an enormous advance because, without this additional information, a star's absolute magnitude could be uncertain by more than 10 magnitudes, a factor of 10 000 in luminosity. The procedure was entirely empirical and relied upon calibration of the relations using stars of known parallax. Distances estimated in this way were referred to as *spectroscopic parallaxes*.

The luminosity indicators were built into the system of spectral classification which eventually superseded the Harvard classification system. The Morgan, Keenan and Kellman (MKK), or Yerkes, system was published in 1943 in their *Atlas of Stellar Spectra, with an Outline of Spectral Classification*. A two-dimensional classification system was introduced in which, in addition to the basic spectral types listed in Table 2.1, stars were assigned to five luminosity classes from type I, the supergiant stars, to type V, main-sequence stars (Morgan,

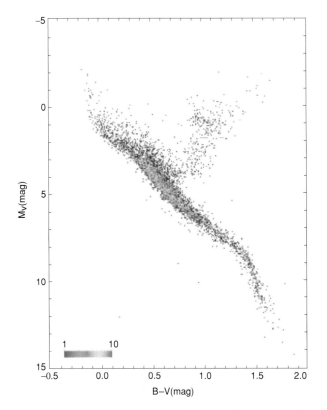

Figure 3.4: The Hertzsprung–Russell diagram for 2927 nearby stars for which parallaxes have been determined with an accuracy of better than 5% by the *Hipparcos* astrometric satellite of the European Space Agency. (M. A. C. Perryman, *The Hipparcos and Tycho Catalogues*, vol. 1 (Noorwijk, The Netherlands: ESA Publication Division, ESTEC, ESA SP-1200, 1997).)

Keenan and Kellman, 1943). The modern MK system was described in 1953 and involves only minor changes as compared with the MKK system (Johnson and Morgan, 1953).

Another of the great problems of stellar astronomy was the determination of stellar masses, which could only be found for stars which are members of binary star systems. Pioneering analyses for the determination of stellar sizes and masses were carried out by Russell using the light curves of eclipsing binary star systems. In 1912, these techniques were elaborated in collaboration with Russell's first graduate student, Harlow Shapley (1885–1972) (Russell, 1912a,b; Russell and Shapley, 1912a,b). Shapley went on to make a detailed study of the light curves of 90 eclipsing binary stars, and from these he was able to demonstrate that there is a wide range of diameters among these stars. The classical techniques of dynamical astronomy for binary star systems could be used in conjunction with accurate radial velocities to make estimates of the masses of the stars. Shapley found that some of them had mean densities similar to the Sun, while the very brightest yellow and red stars were giant stars of much lower mean density (Shapley, 1915). It soon became apparent that the range of luminosities was enormous compared with the range of masses. Russell found at best weak evidence for a correlation between luminosity and mass for the stars in his samples.

This evidence was consistent with the prevailing view, favoured by Russell, that the red giant stars represented the earliest phases of stars, which then contracted and heated up to join the upper end of the main sequence. The main sequence then represented a cooling sequence for stars as they grew older. Any weak correlation of mass with luminosity could be attributed to mass loss. A relic of these early, and quite incorrect, theories remains in the use of the term *early-type stars* to mean stars on the upper part of the main sequence and *late-type stars* for those on the lower main sequence.

3.3   The impact of the new physics

Within ten years, the picture would change dramatically, many pieces of evidence contributing to these profound changes. Continuing the theme of the mass–luminosity relation for dwarf stars, contrary to Russell's assertion that the dwarf stars on the main sequence had similar masses, first Jacob Halm (1866–1944) in 1911 and then Hertzsprung in 1915 showed that there *is* a correlation between mass and luminosity along the main sequence (Halm, 1911). By 1919, Hertzsprung had derived an empirical mass–luminosity relation for main-sequence stars, $L \propto M^x$ with $x \approx 7$ (Hertzsprung, 1919), somewhat greater than present best values, which are closer to $x = 4$ for stars with mass roughly that of the Sun. This finding ran contrary to the expectations of the Russell–Lockyer theory, according to which stars on the main sequence have the same mass. To rescue the standard theory, the stars would have to lose mass.

Another important development was the idea that energy could be transported through the gaseous envelope of the star by radiation rather than by convection. This concept had first been discussed by Ralph Sampson (1866–1939) in 1894 (Sampson, 1895) and studied in more detail by Arthur Schuster (1851–1934) and Schwarzschild in the early 1900s (Schuster, 1902, 1905; Schwarzschild, 1906). Specifically, Schwarzschild showed that convection would only occur if the temperature gradient exceeded the adiabatic gradient for the gas in the envelope. The radiative transport of energy did not, however, find favour with the majority of astronomers, but in 1916 Arthur Eddington revived the idea and applied it to radiative transfer in the envelopes of giant stars (Eddington, 1916b).[†]

Bohr's theory of atomic structure, published in 1913, had an immediate impact upon astrophysics (Bohr, 1913a–c). It enabled the energy levels of elements in different states of ionisation to be determined, and this had important implications for the measurement of the temperatures of stellar atmospheres. The earliest attempts to measure the surface temperatures of the stars had assumed that they emitted like black bodies, the technique which had been used by Hertzsprung to measure the diameter of Arcturus (Hertzsprung, 1907). This technique had been used to measure temperatures by a number of workers, but it suffered from the problem of taking proper account of the presence of absorption lines in the stellar spectra. To estimate the continuum intensity, observations had to be made between the prominent absorption lines. There remained the problem, however, of the unknown

---

[†] The radiative transport of energy in stars and a derivation of the fourth equation of stellar structure are presented in Section A3.2.2.

extent to which weak absorption lines depressed the continuum, a phenomenon known as *line blanketing*, as well as the problem of the *Balmer jump*, the discontinuity which takes place at the limit of the Balmer series, for late B and A type stars.

The first astronomer to apply the idea of using the state of ionisation of atoms in stellar atmospheres as a means of measuring temperatures was the Indian astrophysicist Megh Nad Saha (1893–1956). In 1919, Saha visited the German physical chemist Walther Nernst (1864–1941), who was studying the thermodynamic theory of the equilibrium state of chemical reactions. Saha acknowledged that this work was the inspiration for his formulation of equilibrium ionisation states. In his own words (Saha, 1920), he described ionisation as

a sort of chemical reaction, in which we have to substitute ionisation for chemical decomposition.

John Eggert (1891–1973), a pupil of Nernst, had already calculated the equilibrium state for eight-times ionised iron in stellar interiors (Eggert, 1919), and Saha applied the same formalism to studies of the solar atmosphere. These considerations led to the *Saha equation*, which describes the state of ionisation of a gas in thermal equilibrium at a given temperature (Saha, 1920). Saha combined Boltzmann's equation with the equations of ionisation equilibrium and so determined how the state of ionisation depends upon both the density and the temperature of the gas. To estimate temperatures, he used the method of 'marginal appearances' of lines based upon the first appearance or disappearance of the different spectral lines employed in the Harvard sequence of stellar types listed in Table 2.1. He concluded his important paper of 1921 with the remark that (Saha, 1921)

It will be admitted from what has gone before that the temperature plays the leading role in determining the nature of the stellar spectrum. Too much importance must not be attached to the figures given, for the theory is only a first attempt for quantitatively estimating the physical processes taking place at high temperature. We have practically no laboratory data to guide us, but the stellar spectra may be regarded as unfolding to us, in an unbroken sequence, the physical processes succeeding each other as the temperature is continually varied from 3000 K to 40 000 K .

These concepts were developed by Ralph Fowler (1889–1944) and Edward A. Milne (1896–1950), who provided a much more complete description of the equilibrium ionisation states, including the effects of excited states of atoms and ions (Fowler and Milne, 1923, 1924). Rather than simply use the first appearance or disappearance of different ions and atoms, they worked out the conditions under which the absorption lines would have maximum strength. The way was opened up for determining the abundances of the elements in detail, and this task was undertaken by Cecilia Payne (1900–1979), a pupil of Milne's, who carried out these studies at Harvard under the supervision of Shapley. Her doctoral degree was the first awarded by Harvard in astronomy, and it was published in 1925 as a monograph with the title *Stellar Atmospheres* (Payne, 1925). According to Otto Struve (1897–1963),[9]

It is undoubtedly the most brilliant Ph.D. thesis ever written in astronomy.

Payne's dissertation concerned the application of the Saha–Fowler–Milne theory to stellar atmospheres and summarised everything that was known about laboratory and stellar spectra at that time. The most famous aspect of her work was the demonstration that, although the spectra of stars can vary widely, they all have remarkably similar chemical compositions, the principal cause of the observed differences being the surface temperature of the star. In her monograph, she stated that

the uniformity of composition of stellar atmospheres appears to be an established fact.

She further showed that these abundances were similar to the terrestrial abundances, with the exception of the elements hydrogen and helium, which she found to be vastly more abundant in the stars than on Earth. Although she had obtained the correct answer, she did not believe it. She wrote:

Although hydrogen and helium are manifestly very abundant in stellar atmospheres, the actual values derived from the estimates of their marginal appearances are regarded as spurious.

This conclusion simply reflected the prevailing prejudice. Three years later, in 1928, Albrecht Unsöld (1905–1995) showed that the abundance of hydrogen was indeed very much greater than all the other elements (Unsöld, 1928). This was confirmed by William McCrea (1904–1999), who used the relative intensities of flash spectra to show that the number density of hydrogen atoms at the base of the chromosphere was the same as Unsöld's value (McCrea, 1929).

Payne's brilliant analyses indicated the power of spectroscopy in determining the chemical abundances and physical conditions in stellar atmospheres. The story was taken up by Russell, who had been one of the pioneers in testing Saha's theory by comparing the relative intensities of the spectral lines of potassium and rubidium in the solar atmosphere and in sunspots – he found excellent agreement with the theory (Russell, 1922). In 1925, he investigated the problem of understanding the anomalous triplet terms of the alkaline earth metals, calcium, scandium and barium. In collaboration with Frederick Saunders (1875–1963), he developed the vector model of the atom of Alfred Landé (1888–1976) to account for what became known as Russell–Saunders or L–S coupling (Russell and Saunders, 1925). With this new understanding of atomic spectra, Russell, Walter Adams and Charlotte Moore (1898–1990) began a detailed study of the chemical abundances in the solar atmosphere (Russell, Adams and Moore, 1928). They used 1288 absorption lines in 228 different multiplets to find the relation between the strengths of the absorption features and the number of absorbing atoms or ions. In agreement with the work of Unsöld and McCrea, they found that hydrogen was by far the most abundant of the elements. In Russell's analysis of 1929, the solar abundances for 56 different elements and 6 diatomic molecules were determined (Russell, 1929). These abundances were all within a factor of 2 of present estimates.

These procedures were adapted for the determination of the abundances of elements in the stars by Marcel Minnaert (1893–1970) and Gerard Mulders (1908–1993) in 1930 (Minnaert and Mulders, 1930). They introduced the concept of the *equivalent width* of the spectral line, meaning the waveband of continuum radiation of the star which would correspond to the same amount of radiation removed from the continuum by integrating over the observed line profile. They developed the procedures for relating the equivalent widths of lines in the spectrum to the number of absorbing atoms, taking account of the different processes which broaden the absorption lines – radiation damping, natural damping and thermal broadening. This led to what Minnaert called the *curve of growth* technique for

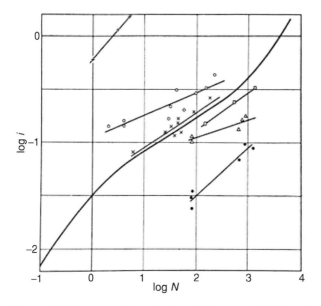

Figure 3.5: The curve of growth due to Minnaert and Mulders showing the dependence of the abundance of an element upon its equivalent width (Minnaert and Mulders, 1930). The continuous curve shows a mean curve of growth for a wide range of equivalent widths.

relating the equivalent width of the line to the number of absorbing atoms (Figure 3.5). The same type of procedure was developed independently by Donald Menzel (1901–1976) in 1930 for emission lines (Menzel, 1931). These became the standard techniques for analysing the abundances of the elements in stars, in particular for investigating chemical differences between stars as a means of determining their evolutionary status.

3.4 Eddington and the theory of stellar structure and evolution

Arthur Stanley Eddington (1882–1944) was the central figure in the development of the theory of the internal structure and evolution of the stars. After a very distinguished undergraduate career in mathematics at Cambridge University, where he became senior wrangler in only his second year, in 1906 he became a senior assistant at the Royal Greenwich Observatory, where he obtained valuable experience of practical astronomy. In 1913, he returned to Cambridge as the Plumian Professor of Astronomy, where he was to remain for the rest of his career.[10]

Between 1916 and 1924, Eddington published over a dozen papers, which were collected and extended in his great book *The Internal Constitution of the Stars* (Eddington, 1926b). According to Henry Norris Russell (Russell, 1925), whose theory of stellar evolution was comprehensively demolished by Eddington,

Several investigators – Jeans, Kramers, Eggert – have contributed to this field, but much the largest share is Eddington's.

In a perceptive letter to me in 1993, William McCrea wrote

[People] don't realise that before, say, 1916 astronomers simply had no idea what the inside of a star was like, and had no idea how to find out anything about this. The speed at which Eddington transformed the sitution was incredible.

There is no simpler way of describing Eddington's achievement than to quote Chandrasekhar's assessment:[11]

In the domain of the internal constitution of the stars, Eddington recognised and established the following basic elements of our present understanding.

(i) Radiation pressure must play an increasingly important role in maintaining the equilibrium of stars of increasing mass.
(ii) In parts of the star in which radiative equilibrium, as distinct from convective equilibrium, obtains, the temperature gradient is determined jointly by the distribution of the energy sources and of the opacity of the matter to the prevailing radiation field. Precisely,

$$\frac{\mathrm{d}p_{\mathrm{r}}}{\mathrm{d}r} = -\kappa \frac{L(r)}{4\pi cr^2} \rho, \qquad p_{\mathrm{r}} = \frac{1}{3}aT^4 \qquad (3.4)$$

and

$$L(r) = 4\pi \int_0^r \epsilon \rho r^2 \, \mathrm{d}r, \qquad (3.5)$$

where $p_{\mathrm{r}}, \kappa, \epsilon$ and ρ denote, respectively, the radiation pressure, the coefficient of stellar opacity, the rate of energy generation per gram of stellar material, and the density.[†] Also a is Stefan's constant and c is the velocity of light.
(iii) The principal physical processes contributing to the opacity, κ, is determined by the photo-electric absorption coefficient in the soft X-ray region, i.e., by the ionisation of the innermost K- and L-shells of highly ionised atoms.
(iv) With electron scattering as the ultimate source of stellar opacity, there is an upper limit to the luminosity, L, that can support a given mass M. The maximum luminosity, set by the inequality

$$L < \frac{4\pi cGM}{\sigma_{\mathrm{e}}}, \qquad (3.6)$$

where σ_{e} denotes the Thomson scattering-coefficient, is now generally referred to as the *Eddington limit*. This Eddington limit plays an important role in current investigations relating to X-ray sources and the luminosity of accretion discs around black holes.[‡]
(v) In a first approximation, in normal stars (that is in stars along the main sequence), the (mass, luminosity, effective temperature)-relation is not very sensitive to the distribution of the energy sources through the star. Therefore, a relation is available for comparison with observations even in the absence of a detailed knowledge of the energy sources of the star.
(vi) The burning of hydrogen into helium is the most likely source of stellar energy.
(vii) The phenomenon of Cepheid variability is due to the adiabatic radial pulsations of these stars.

These great insights were not gained without a considerable struggle and, in particular, there were continuing disputes with James Jeans (1877–1946) and others about many of the fundamental issues concerning the internal structure of the stars.[12] Many of these heated discussions are faithfully recorded in the Reports of the meetings of the Royal Astronomical

---

[†] The origin of the third and fourth equations of stellar structure are described in Section A3.2.
[‡] This result is derived in the context of the physics of active galactic nuclei in Section A11.1.1.

Society published in *The Observatory* through this remarkable era. Some appreciation of the heat which was generated may be gained from this extract from a letter which Jeans published in *The Observatory* in November 1926 (Jeans, 1926):

May I conclude by assuring Prof. Eddington it would give me great pleasure if he could remove a long-standing source of friction between us by abstaining in future from making wild attacks on my work which he cannot substantiate, and by making the usual acknowledgements whenever he finds that my previous work is of use to him? I attach all the more importance to the second part of the request because I find that some of the most fruitful ideas which I have introduced into astronomical physics – e.g., the annihilation of matter as the source of stellar energy, and highly dissociated atoms and free electrons as the substance of the stars – are by now fairly generally attributed to Prof. Eddington.

The problem was that, to determine theoretically the internal structure of a star, the four equations of stellar structure, equations (A3.1), (A3.2), (A3.13) and (A3.19) had to be supplemented by knowledge of the equation of state of the stellar material, as well as the density and temperature dependence of the energy generation rate, $\varepsilon(\rho, T)$, and the opacity of the stellar material, $\kappa(\rho, T)$. This was far beyond what was feasible in 1916. Some sweeping approximations were needed to make progress, and Eddington had exactly the right qualities of fearless imagination and technical skill to make the problem tractable. For example, in order to simplify the mathematics of his 'standard model', Eddington made what Leon Mestel (b. 1927) refers to as the 'hair-raising approximation' that the radiation pressure is a constant fraction of the total pressure throughout the star. It is no surprise that the subject of stellar structure provoked heated debate.

In his first paper on the internal structure of the stars, Eddington assumed that the mean atomic mass of the particles was 54, meaning that the star was predominantly composed of iron atoms (Eddington, 1916b). As soon as his paper was read at the Meeting of the Royal Astronomical Society on 8 December 1916, Jeans pointed out in the discussion that (Jeans, 1917)

For these temperatures and energy, we have very hard Roentgen radiation, and so the atoms in the gas will be smashed up.

In the opening paragraph of his next paper, Eddington acknowledged this important contribution (Eddington, 1917), stating that

Jeans has convinced me that a rather extreme state of disintegration is possible, and indeed seems more plausible.

Eddington adopted a mean atomic weight of 2, corresponding to the complete ionisation of the atoms, assuming that there is a negligible fraction of hydrogen present. As noted in Section 3.3, it was only in the early 1930s that the high cosmic abundance of hydrogen was established. The adoption of fully ionised gas as the material of the star was an important change of perspective in that, if all the atoms of the stellar material are fully ionised, the perfect gas law could be applied at very much higher densities and temperatures than are found in terrestrial environments. As a bonus, this change of the mean atomic weight enabled Eddington to obtain better agreement between his theory and the observed masses of the stars.

Eddington still adhered to the prevailing Russell–Lockyer picture and so the theory could not be applied to main-sequence stars. The analyses of the properties of red giants by

Hertzsprung and Russell had indicated, however, that they were of enormous size and so their very-low-density envelopes were likely to be gaseous. Eddington was inspired to apply his theory of radiative transfer to the envelopes of the red giant stars. In his paper of 1917, he showed that the luminosity is predominantly determined by the mass of the star and that the meagre observational data on their mass–luminosity relation was in good agreement with the theory. Furthermore, if the release of gravitational energy was the source of the luminosity of the giant stars, they could not radiate for more than 100 000 years, which is very much less than the age of the Earth (Eddington, 1917). To avoid the short timescales, another source of energy had to be found. At this time, his view was that

Probably the simplest hypothesis ... is that there may be a slow process of annihilation of matter (through positive and negative electrons occasionally annulling one another).

Eddington had little doubt about the correctness of his theory of red giants, and in 1919 the opportunity arose of testing directly that they have the large diameters inferred from theory. Albert A. Michelson (1852–1931) had been developing the techniques of optical interferometry for almost 30 years (Michelson, 1890) and George Ellery Hale, director of the Mount Wilson Observatories, decided that the 100-inch Hooker telescope should be equipped with a Michelson interferometer to determine the separations of close binary stars and, potentially, the diameters of stars. Michelson originally thought that impractically long interferometer baselines would have to be used to measure the diameters of red giants but, being aware that the instrument was in the process of construction, Eddington used his theory of the structure of red giant envelopes to predict the angular size of Betelgeuse. In the light of this prediction, Michelson built a 6-metre interferometer, which was mounted on the top ring of the 100-inch telescope by Francis Pease (1881–1938) and John A. Anderson (1876–1959) (Figure 3.6). On the night of 13 December 1920, they measured the angular diameter of Betelgeuse to be 0.047 arcsec, just slightly less than Eddington's prediction (Michelson and Pease, 1921). This observation confirmed beyond any doubt the large diameters of the red giants, that of Betelgeuse being greater than the diameter of the Earth's orbit about the Sun. Michelson and his colleagues went on to measure the diameters of four other red giants (Pease, 1921).

In 1919, Eddington discovered to his surprise that his theory of the structure of red giants could also account for the observed mass–luminosity relation for main-sequence stars (Figure 3.7) (Eddington, 1924). The implications were profound – the main-sequence stars were not slowly contracting incompressible liquid spheres, but rather gaseous spheres. This cut the foundation from under the standard Russell picture. The conclusion was vigorously opposed by Jeans, who believed the result was spurious since Eddington's standard model avoided addressing the problem of the processes of energy generation inside the stars. Jeans proposed that the source of energy in the Sun was radioactive decay.

In his early papers in this series, Eddington advocated the annihilation of matter as an inexhaustible source of energy for the stars. In 1920, he realised that, although there was no known mechanism by which nuclear energy could be released, at least energetically this provided a very attractive means of powering the stars. In a remarkably prescient paragraph of his Presidential Address to the Mathematical and Physics Section of the British Association for the Advancement of Science at its Annual Meeting, held in Cardiff, he stated (Eddington, 1920)

Figure 3.6: Michelson's interferometer mounted on the top ring of the Hooker 100-inch telescope at Mount Wilson. This instrument was used to measure the angular diameter of Betelgeuse in 1919 (Michelson and Pease, 1921).

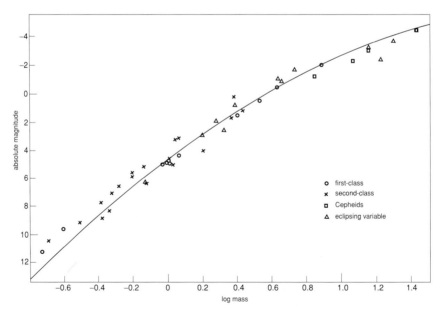

Figure 3.7: The observed mass–luminosity relation for stars compared with Eddington's theoretical mass–luminosity relation (Eddington, 1924). The different symbols refer to different classes of star for which Eddington had made estimates of luminosities and masses. More than half the stars are main-sequence stars.

Certain physical investigations in the past year ... make it probable to my mind that some portion of this sub-atomic energy is actually being set free in the stars. F. W. Aston's experiments seem to leave no room for doubt that all the elements are constituted out of hydrogen atoms bound together with negative electrons. The nucleus of the helium atom, for example, consists of 4 hydrogen atoms bound with two electrons. But Aston has further shown conclusively that the mass of the helium atom is less than the sum of the masses of the 4 hydrogen atoms which enter into it; and in this at any rate the chemists agree with him. There is a loss of mass in the synthesis amounting to about 1 part in 120, the atomic weight of hydrogen being 1.008 and that of helium 4. ... Now mass cannot be annihilated, and the deficit can only represent the mass of the electrical energy set free in the transmutation. We can therefore at once calculate the quantity of energy liberated when helium is made out of hydrogen. If 5 per cent of the star's mass consists initially of hydrogen atoms, which are gradually being combined to form more complex elements, the total heat liberated will more than suffice for our demands, and we need look no further for the source of a star's energy.

Eddington had the good fortune to be working at the Observatories at Cambridge University, only a 20 minute walk from the Cavendish Laboratory, where Francis Aston (1877–1945) was carrying out his precise measurements of atomic and isotopic masses. At that time, this could be no more than a hypothesis, but Eddington had indeed hit upon the correct solution for the energy source of the Sun. Note that Eddington's insight was based upon the prevailing view at the time that the nucleus consisted of protons and electrons. Although Rutherford had postulated as early as 1920 the existence of neutrons as the particles which, along with the protons, made up the total mass of the nucleus, there was little enthusiasm, and even less evidence, for their existence (Rutherford, 1920). The beauty of Eddington's argument was that it did not depend upon the precise nature of the nucleus, but only upon the conservation of energy and the mass–energy relation $E = mc^2$.

It is hardly surprising that the unravelling of the internal structure of the stars resulted in heated debate. In Leon Mestel's critique[13] of Eddington's *The Internal Constitution of the Stars*, he shows that Eddington made a number of rough approximations and used quite a bit of sleight of hand in order to find tractable solutions of the equations. In fact, Eddington was lucky in that, as noted by Chandrasekhar and Mestel, the form of the mass–luminosity relation is remarkably independent of the precise process of energy production and the opacity law.[†] As Mestel remarks, even in an extreme model in which the energy is assumed to originate in a point source at the centre of the star, Thomas Cowling (1906–1990) found that the resulting variation of density with radius was not so different from Eddington's standard model (Cowling, 1935).

The scene was set for probing much more deeply into the physical processes that determine the opacity of stellar material and its energy generation rate. In 1923, Hendrik Kramers (1894–1945) used classical arguments to work out the opacity of a fully ionised plasma for bremsstrahlung, what become known as free–free radiation in the post-quantum era, and which is a good approximation for the central regions of stars like the Sun (Kramers, 1923). To obtain the average energy flux transmitted through the star, the absorption coefficient has to be averaged over all frequencies; the procedure for doing this was worked out by Svein Rosseland (1894–1985) in 1924, the resulting opacities being known as *Rosseland mean*

[†] These weak dependences of the surface properties of main-sequence stars upon the energy generation rate and the opacity law can be appreciated from the analyses given in Section A3.3.

opacities (Rosseland, 1924) (see Section A3.3.3). The strong dependence of nuclear energy generation rates upon temperature was established in the 1930s once the great discoveries of quantum mechanics had been assimilated into the toolkit of the astrophysicist (see Section 3.5).

3.5   The impact of quantum mechanics and the discovery of new particles

The solution of the problem of energy generation in the Sun was one of the fruits of the remarkable theoretical developments which led to the discoveries of quantum mechanics and Dirac's theory of the electron. Specifically, the key developments were the following.

- The discovery of Fermi–Dirac statistics, which found immediate application in the equation of state of dense matter in stars (Fermi, 1926).
- The phenomenon of quantum mechanical tunnelling, discovered by George Gamow (1904–1968), and its application to the inelastic scattering of α-particles by nuclei (Gamow, 1928).
- The formulation of the theory of β-decay by Enrico Fermi (1901–1954), which included the proposal of the existence of the neutrino (Fermi, 1934a,b).

Equally important were the experimental discoveries of this golden age of physics.

- Nuclear transmutations were discovered by Ernest Rutherford in collisions between fast α-particles and nitrogen nuclei (Rutherford, 1919; Rutherford and Chadwick, 1921), and these were photographed in the remarkable automatic cloud chamber experiments of Patrick Blackett (1897–1974) (Blackett, 1925).
- The discovery of the positron in cosmic-ray cloud-chamber experiments was announced by Carl Anderson (1905–1991) in 1931 (Anderson, 1932).
- In the same year, Harold Urey (1893–1981) and his colleagues discovered deuterium in spectroscopic studies of 'distilled' liquid hydrogen (Urey, Brickwedde and Murphy, 1932).
- Also, in 1932, the neutron was discovered by James Chadwick (1891–1974) (Chadwick, 1932).
- Finally, in 1932, the experiments of John Cockcroft (1897–1967) and Ernest Walton (1903–1995) not only provided the first destruction of lithium nuclei by artificially accelerated fast protons, but also provided direct experimental confirmation of the correctness of the mass–energy relation $E = mc^2$ (Cockcroft and Watson, 1932).

These discoveries were quickly assimilated into astrophysics.

From the perspective of astrophysics, a key development was the much improved understanding of nuclear structure. Eddington's proposal that nuclear energy could provide the luminosity of the Sun could now be placed on a proper physical basis.

The problem was that, even at the high temperatures of stellar interiors, the Coulomb repulsion between protons and nuclei is so great that, according to classical physics, protons could not pentrate the nucleus and so this energy source could not be tapped. The solution of this problem had to await Gamow's theory of quantum mechanical tunnelling of 1928 (Gamow, 1928). One year later, Robert Atkinson (1893–1981) and Fritz Houtermans (1903–1966) applied Gamow's theory to the physics of nuclear reactions in the hot central regions of stars (Atkinson and Houtermans, 1929). By considering the process of barrier penetration by a Maxwellian distribution of protons, they established two key features of the process of nuclear energy generation in stars. Firstly, the most effective energy sources involve interactions with nuclei of small electric charge since the Coulomb barriers are smaller than for nuclei with large charges. Secondly, the particles which can penetrate the Coulomb barriers are those few particles in the high-energy tail of the Maxwellian distribution. As a result, nuclear reactions can take place at temperatures which are considerably smaller than might have been expected. These ideas also suggested why the luminosity of the stars should be a sensitive function of temperature. As the temperature increases, the rate of barrier penetration increases exponentially and so hotter stars should be more luminous than less massive stars.

By 1931, the evidence was accumulating that hydrogen is by far the most abundant element in the stars. In addition to the studies of stellar atmospheres discussed above, Bengt Strömgren had shown that the mass–luminosity relation is sensitive to the hydrogen abundance and, using opacities in which hydrogen constituted about 35% by mass, much improved agreement with the observations was obtained (Strömgren, 1932, 1933). Reluctantly, Eddington agreed with Strömgren's conclusion, but the adopted abundances of the heavy elements was still large. This was the abundance used by Chandrasekhar in his influential book *An Introduction to the Study of Stellar Structure* (Chandrasekhar, 1939). In the 1930s, it was realised that even lower abundances of the heavy elements would also be consistent with the data, but it was only in the 1940s that the heavy-element abundance was further revised downwards, close to the values used today, as a result of rather general astrophysical arguments by Fred Hoyle (Hoyle, 1946).

Atkinson's objective was to account for the origin of the chemical elements by the successive addition of protons to nuclei. He argued that the process of forming helium by the combination of four protons was very unlikely and proposed instead that helium could be formed by the successive addition of protons to heavier nuclei which, when they became too massive for nuclear stability, would eject α-particles and so create helium (Atkinson, 1931a,b). This proposal was the precursor of the carbon–nitrogen–oxygen (CNO) cycle, which was discovered independently by Carl von Weizsäcker (b. 1912) and Hans Bethe (1906–2005) in 1938 (Weizsäcker, 1937, 1938; Bethe, 1939). In this cycle, carbon acts as a catalyst for the formation of helium through the successive addition of protons accompanied by two β^+ decays as follows:

$$^{12}C + p \rightarrow \ ^{13}N + \gamma \ ; \ ^{13}N \rightarrow \ ^{13}C + e^+ + \nu_e \ ; \ ^{13}C + p \rightarrow \ ^{14}N + \gamma;$$
$$^{14}N + p \rightarrow \ ^{15}O + \gamma \ ; \ ^{15}O \rightarrow \ ^{15}N + e^+ + \nu_e \ ; \ ^{15}N + p \rightarrow \ ^4He + \ ^{12}C.$$

In the meantime, it had become possible to make estimates of the reaction rates for the simplest nuclear reaction, the combination of pairs of protons to form deuterium nuclei which can then combine with other deuterons to form ^{3}He and ^{4}He. The first calculations were carried out by Atkinson in 1936 (Atkinson, 1936) and were much refined in 1938 by Bethe and Charles L. Critchfield (1910–1994), who combined Fermi's theory of weak interactions with Gamow's theory of barrier penetration (Bethe and Critchfield, 1938). The principal series of reactions in the proton–proton (or p–p) chain are as follows:[†]

$$p + p \rightarrow\ ^2H + e^+ + \nu_e; \qquad ^2H + p \rightarrow\ ^3He + \gamma;$$
$$^3He +\ ^3He \rightarrow\ ^4He + 2p.$$

The crucial first reaction in the chain involves a weak interaction, in which a positron and neutrino are released in what may be thought of as the transformation of one of the protons into a neutron. This reaction accounts for most of the energy release in the p–p chain but it has never been measured experimentally at the energies of interest for nucleosynthesis in the Sun. Bethe and Critchfield showed that this series of reactions could account for the luminosity of the Sun. In addition, they found that the rate of energy production, ε, of the p–p chain depends upon the central temperature of the star as $\varepsilon \propto T^4$. In 1939, Bethe worked out the corresponding energy production rate for the CNO cycle and found a very much stronger dependence, $\varepsilon \propto T^{17}$ (Bethe, 1939). He concluded that the CNO cycle was dominant in massive stars, whereas the p–p chain was the principal energy source for stars with mass $M \leq M_\odot$. These conclusions were confirmed by the much more detailed models of stellar structure which became available after the Second World War and, in particular, with the development of computer codes, which have converted the study of stellar structure into one of the most precise of the astrophysical sciences.

The laws of energy generation, $\varepsilon(\rho, T)$, and the dependence of the opacity of stellar material upon density and temperature, $\kappa(\rho, T)$, were the keys to understanding the theory of stellar structure and evolution in much more detail. Using simple power-law approximations for the ε and κ, Fred Hoyle (1915–2001) and Raymond Lyttleton (1911–1995) were able to derive homology relations, which illustrated clearly how the properties of stars depended upon the different processes of energy generation and upon the opacity law (Hoyle and Lyttleton, 1942). A simple illustration of the power of these methods is given in Section A3.3.

Thus, by the time of the outbreak of the Second World War, much of the basic physics of main-sequence stars was beginning to be understood.

Notes to Chapter 3

1 I have given an account of the origins of the two laws of thermodynamics in Case Study IV of Malcolm Longair, *Theoretical Concepts in Physics* (Cambridge: Cambridge University Press, 2003).
2 The same suggestion had been made earlier by Mayer and John James Waterston (1811–1883) in two independent unpublished papers (Virginia Trimble, personal communication).
3 This quotation is contained in the biography of Lord Kelvin by C. W. Smith and M. N. Wise, *Energy and Empire: A Biographical Study of Lord Kelvin* (Cambridge: Cambridge University Press, 1989).

[†]   More details of these reactions are given in Section 8.2 in the context of the generation of the flux of solar neutrinos.

4 Chandrasekhar, S., as quoted by Wali, K. C. *Chandra: A Biography of S. Chandrasekhar* (Chicago: University of Chicago Press, 1991).

5 These results are contained in Lane's unpublished notes in the US National Archives.

6 My favorite recommendations as introductions to the theory of the structure and evolution of the stars are R. J. Tayler, *The Stars: Their Structure and Evolution* (Cambridge: Cambridge University Press, 1994) at the elementary level and R. Kippenhahn and A. Weigert, *Stellar Structure and Evolution* (Berlin: Springer-Verlag, 1990) at a more advanced level.

7 Lockyer's first version of his temperature arch was published in 1887 in his paper 'Researches on the spectra of meteorites. A report to the Solar Physics Committee', published in *Proceedings of the Royal Society of London*, **43**, 1887, 117–156. He continued to publish various versions of the temperature arch; that shown in Figure 3.1 dates from 1914.

8 An excellent account of the early history of the theory of stellar structure and evolution is given by David DeVorkin, Stellar evolution and the origin of the Hertzsprung–Russell Diagram, in *The General History of Astronomy, Vol. 4. Astrophysics and Twentieth-Century Astronomy to 1950: Part A*, ed. O. Gingerich (Cambridge: Cambridge University Press, 1984), pp. 90–108.

9 This remark appears on p. 220 of the book by O. Struve and V. Zebergs, *Astronomy of the 20th Century* (New York: Macmillan and Company, 1962).

10 Eddington has been the subject of a number of biographical studies, including A. Vibert Douglas, *Arthur Stanley Eddington* (London: Thomas Nelson and Sons Ltd, 1956), S. Chandrasekhar, *Eddington: The Most Distinguished Astrophysicist of His Time* (Cambridge: Cambridge University Press, 1983) and D. S. Evans, *The Eddington Enigma* (Princeton, New Jersey: Xlibris Corporation, 1998).

11 Chandrasekhar's short book *Eddington: The Most Distinguished Astrophysicist of His Time* is essential reading for anyone wishing to understand Eddington's thinking on astrophysical and cosmological problems. The list of six insights into the physics of the stars in that book was expanded to seven in the Preface, which Chandrasekhar wrote as an introduction to the 1988 reprint of Eddington's *The Internal Constitution of the Stars* (Cambridge: Cambridge University Press). I have included the list of seven in the present text.

12 It is interesting that Eddington acknowledged that Jeans had adopted different positions on many of the key issues in his bibliographical references at the end of *The Internal Constitution of the Stars*, p. 402, but left it to the reader to judge their importance.

13 Mestel's penetrating critique was presented at a meeting of the Royal Astronomical Society on 12 March 2004 to celebrate the 60th anniversary of Eddington's death. It is published as L. Mestel, *Stellar Structure and Stellar Atmospheres* (Cambridge: Cambridge University Press, 2005).

A3 Explanatory supplement to Chapter 3

A3.1 The virial theorem and Homer Lane's insights

We can demonstrate Lane's important result quantitatively from the first two equations of stellar structure. The first is the *equation of hydrostatic equilibrium*:

$$\frac{dp}{dr} = -\frac{GM\rho}{r^2}, \tag{A3.1}$$

which must apply at all points in a quasi-static star. By quasi-static, we mean that, although the star is losing energy and so changing slowly, the star can be considered to be in hydrostatic equilibrium to a very high degree of accuracy so far as its structure is concerned. This equation describes the balance between the gravitational attraction of the mass $M = M(\leq r)$ within radius r and the pressure gradient of the hot gas pushing outwards.

The second is the *equation of conservation of mass*, namely that the mass in the spherical shell of thickness dr is $dM = 4\pi r^2 \rho \, dr$ and so

$$\frac{dM}{dr} = 4\pi r^2 \rho. \tag{A3.2}$$

Note that in this analysis M is a variable, the mass within radius r inside the star.

We first use these equations to derive a form of the *virial theorem* as applied to stars in hydrostatic equilibrium. Dividing equation (A3.1) by equation (A3.2), we find

$$\frac{dp}{dM} = -\frac{GM}{4\pi r^4}. \tag{A3.3}$$

Reorganising equation (A3.3) and integrating from the centre to the surface of the star, we find

$$4\pi r^3 \, dp = 3V \, dp = -\left(\frac{GM}{r}\right) dM, \tag{A3.4}$$

$$\int_{p_c}^{p_s} 3V \, dp = -\int_0^{M_s} \left(\frac{GM}{r}\right) dM. \tag{A3.5}$$

We recognise that the quantity on the right-hand side of equation (A3.5) is the total gravitational energy of the star, which we write as Ω, noting that Ω is a *negative* quantity. Assuming that the surface pressure, p_s, is negligible compared with the central pressure p_c, we integrate the left-hand side by parts to find

$$-3 \int_0^{V_s} p \, dV = \Omega. \tag{A3.6}$$

Finally, we write dV in terms of the corresponding mass element dM, $dM = \rho \, dV$, and so

$$3 \int_0^{M_s} \frac{p}{\rho} \, dM + \Omega = 0. \tag{A3.7}$$

This is the *virial theorem* for stars. Many important general results can be derived from the virial theorem.

For our present purposes, let us relate the integral on the left-hand side of equation (A3.7) to the internal thermal energy of the star. The energy density per unit mass is given by the general relation,

$$u = \frac{p}{(\gamma - 1)\rho}, \tag{A3.8}$$

where γ is the ratio of specific heats of the material of the star. Therefore, the integral on the left-hand side of equation (A3.7) becomes

$$3 \int_0^{M_s} \frac{p}{\rho} \, dM = 3 \int_0^{M_s} (\gamma - 1)u \, dM = 3(\gamma - 1)U, \tag{A3.9}$$

where U is the total internal thermal energy of the star. For a monatomic gas, which is an excellent approximation for a fully ionised gas, $\gamma = 5/3$, and so

$$2U + \Omega = 0. \tag{A3.10}$$

Thus, the magnitude of the gravitational potential energy is twice the internal thermal energy of the star. This explains why the *Kelvin–Helmholtz* timescale is often referred to as the *thermal* timescale of the star. The thermal energy is half the gravitational energy, and so the time it takes the star to radiate away its internal thermal energy is the Kelvin–Helmholtz timescale.

We can now return to the apparent paradox that, as stars radiate away their energy, they heat up. The total energy of the star is the sum of its thermal and gravitational potential energies, $E = U + \Omega$. But the virial theorem tells us that $U = -\Omega/2$, and so the total energy is given by

$$E = \frac{\Omega}{2} = -U; \tag{A3.11}$$

in other words, a negative quantity. Thus, as the star loses energy, the total energy must become even more negative and so U must increase; in other words, the star becomes hotter. This was Homer Lane's great insight, and it is entirely associated with the fact that the gravitational potential energy is a negative quantity. This insight also influenced early theories of the evolution of the stars.

A3.2   The third and fourth equations of stellar structure

A3.2.1   Energy generation

The third equation of stellar structure describes the energy generation rate within the star. The energy generated within the star diffuses outwards and so the contribution to the outflow of energy from the shell of radius r and thickness dr is given by

$$dL = 4\pi r^2 \rho \varepsilon \, dr, \tag{A3.12}$$

where ε is the energy generation rate *per unit mass* and is a function of the local temperature and density conditions. Note that L is the rate of flow of energy, or the power, passing through the spherical surface at radius r. Hence, the differential equation for L is given by

$$\frac{dL}{dr} = 4\pi r^2 \rho \varepsilon. \tag{A3.13}$$

A3.2.2   Radiative transport of energy through a star

The fourth equation describes how radiation diffuses through the star. There are two principal mechanisms for the transport of energy through stars: by *radiation* and *convection*. Schwarzschild, in his important paper of 1906 (Schwarzschild, 1906), derived the conditions under which energy would be transported by convection. If the temperature gradient in the star exceeds the adiabatic gradient, that is, it is superadiabatic, convective motions stabilise the energy transport so that the variation of temperature with pressure, or density, is limited to the adiabatic gradient. Specifically, the condition is given by

$$\frac{d \ln T}{d \ln p} \geq \frac{\gamma - 1}{\gamma},$$

where γ is the ratio of specific heats of the material of the star. In practice, what is done is to work out the structure of the star and then test whether or not there are regions which are superadiabatic and in which convective transport of energy should be adopted.

Radiative transport of energy is much more important than thermal conduction because the mean free path for photons, although small, is still very much greater than the mean free path for electrons. The standard form of the heat diffusion equation is given by

$$F = -\lambda \frac{\mathrm{d}T}{\mathrm{d}r}, \tag{A3.14}$$

where F is the power per unit area parallel to the direction of the temperature gradient and λ is the heat diffusion coefficient. Therefore, the total rate of flow of energy through the spherical surface at radius r is $L = 4\pi r^2 F$.

In the radiative transport of energy within stars, the radiation is scattered many times, because of the very high density of the material and the large cross-section for scattering. Because of the very large numbers of scatterings, it is safe to assume that the radiation at any point inside the star is almost precisely isotropic and has a black-body spectrum at the local temperature of the material of the star. The diffusion of energy takes place through the very gradual decrease in temperature with increasing radius.

Rather than work with a heat diffusion coefficient, astrophysicists work with the quantity κ, which is known as the *opacity* of the stellar material. It is defined in terms of the fraction of the flux density of radiation which is absorbed or scattered per unit mass per unit path-length. Thus, if the increment of flux density, $\mathrm{d}F$, is scattered by the material of the star on traversing a distance $\mathrm{d}r$, κ is defined by

$$\mathrm{d}F = -\kappa\rho F \,\mathrm{d}r. \tag{A3.15}$$

Astrophysicists make use of the fact that the spectrum of the radiation inside the stars is very close indeed to a black-body spectrum at the local temperature to rewrite the equation of radiative transfer in an alternative form that is more directly related to local physical conditions in the star. The flux density decrease corresponds to a decrease in radiation pressure with radius through the star. The energy loss per second from the increment of path length $\mathrm{d}r$, is $-\kappa\rho F \,\mathrm{d}r$, and hence the corresponding change in momentum per unit area per unit time, that is, the change of radiation pressure, is given by

$$\mathrm{d}p = -\frac{\kappa\rho F}{c} \,\mathrm{d}r. \tag{A3.16}$$

The radiation is locally black-body radiation at temperature T, however, and so, according to the Stefan–Boltzmann law, $p = \frac{1}{3}aT^4$. Therefore,

$$\frac{\mathrm{d}p}{\mathrm{d}T} = \frac{4}{3}aT^3 = \frac{\mathrm{d}p}{\mathrm{d}r}\frac{\mathrm{d}r}{\mathrm{d}T}. \tag{A3.17}$$

We have derived an expression for $\mathrm{d}p/\mathrm{d}r$ from equation (A3.16), however, which involves the flux density of radiation, F. Therefore, finally we obtain

$$F = -\frac{4}{3}\frac{acT^3}{\kappa\rho}\frac{\mathrm{d}T}{\mathrm{d}r}, \tag{A3.18}$$

or, in terms of the luminosity passing through the sphere at radius r,

$$L = -\frac{16\pi acr^2 T^3}{3\kappa\rho}\frac{dT}{dr}. \tag{A3.19}$$

This is the standard form of the *fourth equation of stellar structure*. The opacity, κ, is determined by the most important processes which impede the escape of radiation from the star and involves a great deal of detailed atomic physics.

A3.3 The origin of the main sequence

At first sight, finding solutions of the equations of stellar structure appears to be a formidable task, but Hoyle and Lyttleton realised that tractable power-law solutions could be found which provide important insights into the relevant physics which determine the observed properties of stars (Hoyle and Lyttleton, 1942). Let us recall the four equations of stellar structure:

$$\frac{dp}{dr} = -\frac{GM\rho}{r^2}, \qquad\qquad \text{hydrostatic equilibrium,} \tag{A3.20}$$

$$\frac{dM}{dr} = 4\pi r^2\rho, \qquad\qquad \text{conservation of mass,} \tag{A3.21}$$

$$\frac{dL}{dr} = 4\pi r^2\rho\varepsilon, \qquad\qquad \text{energy generation,} \tag{A3.22}$$

$$\frac{dT}{dr} = -\frac{3\kappa\rho}{16\pi acr^2 T^3}L, \qquad \text{energy transport.} \tag{A3.23}$$

These equations and the notation were discussed in Sections A3.1 and A3.2. It is possible to make estimates of the central pressure of the star and a lower limit to its central temperature from these equations, without knowledge of the detailed physics.

A3.3.1 The central pressure and temperature of the Sun

Let us write $dp/dr \sim p/R$. Then, equation (A3.20) becomes

$$p \sim \frac{GM^2}{4\pi R^4}, \tag{A3.24}$$

where we have used the approximation $M \sim \rho R^3$. Inserting the values for the Sun, $M_\odot = 2 \times 10^{30}$ kg and $r = 7 \times 10^8$ m, we find $p \sim 2 \times 10^{13}$ N m$^{-2} \approx 2 \times 10^8$ atmospheres. This is the typical pressure inside the Sun. Integrating equation (A3.3) from the centre to the surface of the star shows that the central pressure is twice this value. Thus, the central pressure in the Sun is enormous.

A lower limit to the central temperature of the Sun is found from the *virial theorem*, equation (A3.7), derived in Section A3.1:

$$3\int_0^{M_s} \frac{p}{\rho}\,dM + \Omega = 0. \tag{A3.25}$$

The gravitational potential energy Ω is given by

$$-\Omega = \int_0^{M_s} \frac{GM\,dM}{r}.$$

(A3.26)

We really need to know the variation of mass with radius, but we can obtain a firm lower limit to Ω if we replace r by r_s, the radius of the star. Hence,

$$-\Omega > \int_0^{M_s} \frac{GM\,dM}{r_s} = \frac{GM_s^2}{2r_s}.$$

(A3.27)

For a perfect gas, $p = nkT$, and so the internal thermal energy of the star can be written as

$$U = 3 \int_0^{M_s} \frac{p}{\rho}\,dM = \frac{3k}{m} \int_0^{M_s} T\,dM = \frac{3k}{m}\overline{T}M_s,$$

(A3.28)

where m is the average mass of the particles which contribute to the pressure and $\overline{T}$ is the mass-weighted average of the temperature inside the Sun. Thus, since $-\Omega = 2U$,

$$\overline{T} > \frac{GM_s m}{6kr_s}.$$

(A3.29)

Anticipating the answer we are about to obtain, the temperature is greater than the ionisation potential of hydrogen, and so it is a good approximation to assume that the gas is fully ionised. Since hydrogen is the most abundant element, the mean mass of the particles contributing to the pressure is the average of the proton and electron mass, $m = m_p/2$. Hence,

$$\overline{T} > \frac{GM_s m_p}{12kr_s} = 2 \times 10^6 \text{ K},$$

(A3.30)

for the Sun, confirming that it is a good approximation to assume that the material of the star is fully ionised. This argument also tells us that we need to understand radiative transfer processes for radiation in the far-ultraviolet and X-ray wavebands since $k\overline{T} > 200$ eV. In fact, we now know that the central temperature of the Sun is about 1.5×10^7 K, and so we need to know the opacity at energies $kT \sim 2$ keV. Eddington had the important insight that the opacity of stellar material at X-ray wavelengths was the key to understanding stellar structure.

A3.3.2 Homologous stars

One of the most useful approximations for understanding the physics of the stars is to assume that the material of the star has the same composition at all radii and that the same properties of energy generation and transport apply throughout the star. These stellar models are known as *homologous stars*. Then, the equations of stellar structure can be rewritten in such a way that they depend only upon the mass of the star. The variation of quantities such as the pressure, temperature and luminosity with radius within the star all follow the same relations which scale as different powers of the total mass of the star.[1] If, in addition, we assume that we can represent the opacity, κ, and the energy generation rate, ε, by power laws over appropriate ranges of temperature and density, we can find analytic expressions for relations between the mass of the star and its luminosity, radius and effective temperature. The essence of these arguments can be appreciated from dimensional and order-of-magnitude calculations.

Let us replace all the derivatives in expressions (A3.20) to (A3.23) by their order-of-magnitude values, that is, $dp/dr \sim p/R$, $dM/dr \sim M/R$ and so on. Then, we obtain the following four relations:

$$p \propto \frac{M^2}{R^4}, \qquad \text{hydrostatic equilibrium,} \qquad \text{(A3.31)}$$

$$M \propto R^3 \rho, \qquad \text{conservation of mass,} \qquad \text{(A3.32)}$$

$$L \propto \rho \varepsilon R^3, \qquad \text{energy generation,} \qquad \text{(A3.33)}$$

$$L \propto \frac{RT^4}{\kappa \rho}, \qquad \text{energy transport,} \qquad \text{(A3.34)}$$

where we have used the approximation $M \sim \rho R^3$. These equations need to be supplemented by the equation of state of the material of the star, which we take to be that of a perfect gas:

$$p = nkT, \qquad \text{that is} \qquad p \propto \frac{MT}{R^3}. \qquad \text{(A3.35)}$$

Combining equations (A3.31) and (A3.35),

$$T \propto \frac{M}{R}. \qquad \text{(A3.36)}$$

We can see immediately that, if the radius of the star is only a weak function of its mass, $T \propto M$ and so $L \propto T^4 \propto M^4$. It will turn out that this is a surprisingly good approximation for stars of mass $M \sim M_\odot$.

A3.3.3 Energy generation rates and stellar opacities

Finally, we need to know the density and temperature dependence of the opacity, κ, and the energy generation rate, ε, upon temperature and density. In the case of the energy generation rate, the p–p chain and the CNO cycle for the conversion of hydrogen into helium can be described by $\varepsilon \propto \rho T^\alpha$, where α takes the values 4 and 17, respectively.

The opacity, κ, is a more complex function of temperature. It is most convenient to describe κ by power-law relations of the form $\kappa \propto \rho^\beta T^\gamma$, where the values of β and γ take different values in different temperature ranges. The values quoted by Tayler are shown in Table A3.1.

It is interesting that the most important of these processes can be understood in terms of the classical processes of the emission and absorption of radiation. At the very highest temperatures, the plasma is fully ionised and the dominant scattering process is *Thomson scattering* for which the Thomson cross-section, $\sigma_T = e^4 / 6\pi \epsilon_0^2 m_e^2 c^4 = 6.653 \times 10^{-29} \, \text{m}^{-2}$, is independent of frequency.

In the intermediate temperature range, one of the dominant processes is *free–free* or *bremsstrahlung absorption*. The appropriate values of β and γ can be found from semi-classical arguments (Kramers, 1923) and results in the formula known as the *Kramers opacity*. In outline, the calculation proceeds as follows.

Table A3.1. *The approximate variations of the opacity, κ, with temperature and density in different temperature ranges for typical densities and temperatures found in main-sequence stars*

Temperature	Temperature range (K)	Physical processes	β	γ
Low	10^4–$10^{4.5}$	atomic and molecular absorption	0.5	4
Medium	$10^{4.5}$–10^7	bound–free and free–free absorption	1	−3.5
High	$>10^7$	electron scattering	0	0

The radiation spectrum of a free electron moving through a fully ionised plasma of ions with charge Z and number density N_i can be worked out by semi-classical methods and is found to have the following form:

$$j_v \propto Z^2 N_i T^{-1/2} g(v, T) e^{-hv/kT},$$

where $g(v, T)$ is a slowly varying function of v and T known as the *Gaunt factor*.[2] The corresponding absorption coefficient can be found from Kirchhoff's law, $\kappa_v = j_v/B(v)$, where $B(v)$ is the black-body spectrum. Finally, we need to sum over all the contributions of the different frequencies to the average opacity κ. As shown by Kippenhahn and Weigert,[3] the correct weighting is given by

$$\frac{1}{\kappa} = \frac{\pi}{ac T^3} \int_0^\infty \frac{1}{\kappa_v} \frac{\partial B}{\partial T} dv.$$

This expression is known as the *Rosseland mean opacity*, first derived by Svein Rosseland in 1924 (Rosseland, 1924). It is then straightforward to work out the dependence of κ upon the temperature and density of the plasma, $\kappa \propto \rho T^{-7/2}$, where ρ is the density of the plasma.

A3.3.4 The solutions

Given the power-law dependences discussed in Section 3.3.3, equations (A3.33) and (A3.34) become

$$L \propto \frac{M^3}{R^3} T^\alpha, \qquad \text{energy generation,} \qquad (A3.37)$$

$$L \propto \frac{T^{4-\beta} R^{4+3\gamma}}{M^{1+\gamma}}, \qquad \text{energy transport.} \qquad (A3.38)$$

We now have sufficient relations to determine the dependence of L, T and R upon the mass of the star, M. Firstly, we can use equations (A3.37) and (A3.38) to eliminate L:

$$T^{4-\beta-\alpha} R^{7+3\gamma} \propto M^{3+\gamma}. \qquad (A3.39)$$

Finally, combining equations (A3.36) and (A3.39), we find

$$R \propto M^{-(*)} \qquad \text{where} \quad * = \frac{1-\alpha-\beta-\gamma}{3+\alpha+\beta+3\gamma}. \qquad (A3.40)$$

Table A3.2. *The approximate variations of the opacity, κ, with temperature and density in different temperature ranges for typical densities and temperatures found in main-sequence stars*

α	β	γ	$R \propto M^a$ a	$L \propto M^b$ b	$L \propto T_{\mathrm{eff}}^c$ c
4	0	0	3/7	3	$28/5 = 5.6$
17	0	0	4/5	3	$60/7 \approx 8.6$
4	1	3.5	1/13	$71/13 \approx 5.5$	$284/69 \approx 4.1$
17	1	3.5	9/13	$67/13 \approx 5.2$	$268/49 \approx 4.5$

All the other relations between observables and the mass of the star follow immediately.

Let us first work out the relation between R and M for stars like the Sun, a medium-temperature star, from the above discussion. Inserting $\alpha = 4$ for the p–p chain, $\beta = -3.5$ and $\gamma = 1$, we find

$$R \propto M^{1/13}. \tag{A3.41}$$

This is a very weak dependence upon the mass of the star. We can immediately use equations (A3.36) and (A3.39) to find the temperature– and luminosity–mass relations:

$$T \propto \frac{M}{R} \propto M^{12/13}, \qquad L \propto \frac{M^3}{R^3} T^\alpha \propto M^{71/13}. \tag{A3.42}$$

Finally, we need to determine the dependence of the *effective temperature*, T_{eff}, of the star upon mass. It might seem as though equations (A3.42) provide the answer, but that is not correct since the above temperature relation refers to the central temperature of the star, not to the emission through its surface. The complication is that, for the homologous stars, the temperature changes from T_c to $T = 0$ at the surface and so the homology relations cannot be used. Instead, we know that the luminosity and radius attain finite values at the surface of the star and so we can use the Stefan–Boltzmann law to define an *effective temperature*, T_{eff}, through the relation

$$L = 4\pi R^2 a T_{\mathrm{eff}}^4. \tag{A3.43}$$

Hence,

$$T_{\mathrm{eff}} \propto \left(\frac{L}{R^2}\right)^{1/4} \propto \left(M^{71/13} M^{-2/13}\right)^{1/4}, \tag{A3.44}$$

$$T_{\mathrm{eff}} \propto M^{69/52}. \tag{A3.45}$$

Hence, from equations (A3.42),

$$L \propto M^{71/13} \propto T_{\mathrm{eff}}^{284/69}. \tag{A3.46}$$

This is the *main sequence* for stars of mass roughly that of the Sun in terms of their luminosities and effective temperatures and is roughly $L \propto T_{\mathrm{eff}}^{4.1}$.

Similar calculations can be carried out for the different expressions for the opacity and energy generation rates; some of those tabulated by Tayler are given in Table A3.2. Thus, for a wide range of different assumptions about the opacity of the stellar material and the energy generation rate, there is a power-law relation of the form $L \propto M^b$, where b lies in the range 3 to 5.5. For the physical conditions appropriate for main-sequence stars with mass roughly that of the Sun, the model in the third row of the table with $b \approx 5.5$ is a good approximation to what is observed among the stars. For high-mass main-sequence stars, the second row is a good approximation with $b \approx 3$.

These models are illustrative, but demonstrate how many features of the properties of stars can be explained in physical terms. In particular, they show that the properties of the stars are relatively insensitive to the details of the energy generation mechanism, a point fully appreciated by Eddington. It can be seen from Table A3.2 that changing the dependence of the energy-generation rate upon temperature by a huge amount makes remarkably little difference to the mass–luminosity relation.

Notes to Section A3

1 This approach is beautifully explained by Roger J. Tayler in his book *The Stars: Their Structure and Evolution* (Cambridge: Cambridge University Press, 1994).
2 See, for example, my version of the calculation in Malcolm Longair, *High Energy Astrophysics*, vol. 1 (Cambridge: Cambridge University Press, 1997), Chapter 3.
3 See R. Kippenhahn and A. Weigert *Stellar Structure and Evolution* (Berlin: Springer-Verlag, 1990), Section 5.1.

4 The end points of stellar evolution

4.1 The red giant problem

While the understanding of main-sequence stars proceeded apace through the 1920s and 1930s, there remained the problem of accounting for the red giant stars, which are very much more luminous than main-sequence stars at the same effective temperatures. Russell adopted the position that matter existed in different states in the dwarf and giant stars, what he termed 'giant stuff' and 'dwarf stuff'. Atkinson assumed that different nuclear processes were responsible for the luminosities of the giant stars.

The stellar models of Eddington are homogeneous, and it was assumed that homogeneity was maintained, probably by large-scale meridional circulation driven by the internal rotation of the star. It was only in the early 1950s, that a number of astrophysicists, Peter Sweet (1921–2005), Martin Schwarzschild, Ernst Öpik and Leon Mestel, showed that the mixing assumption was highly implausible.[1]

The solution to the red giant problem was discovered in 1938 by the Estonian astrophysicist Ernst Öpik (1893–1985), then working at the University of Tartu (Öpik, 1938). Öpik realised that if the stars are not well mixed, it is inevitable that they become inhomogeneous. Within the central core of the star, nuclear burning of hydrogen into helium leads to the depletion of the nuclear fuel in the core. In Öpik's model it was assumed that the central core of the star was maintained in convective equilibrium, resulting in a uniform depletion of hydrogen in this region. Once core burning had exhausted all the available hydrogen, nuclear burning would continue in a shell about an inert, isothermal core, while the core itself would begin to collapse, releasing gravitational potential energy. Öpik argued that the rapid release of energy during these phases would lead to the expansion and cooling of the envelope, resulting in the red giant phase of a star's evolution. His picture of the structure of a red giant is shown in Figure 4.1.

As a result of Öpik's work, there was no need to seek separate physical processes to account for the giant stars – they form naturally at the end of the phase of hydrogen burning on the main sequence. The giant phase could only last a short time compared with the age of stars on the main sequence because they must burn the available nuclear fuel at thousands of times the rate at which it is consumed on the main sequence to account for the huge luminosities of the red giants. Thus, the giant phase is a brief final fling before the star settles down to some form of dead star. As Öpik pointed out, this picture is entirely

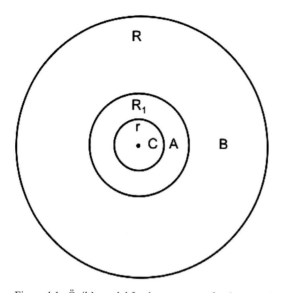

Figure 4.1: Öpik's model for the structure of a giant star. Hydrogen has been exhausted in the contracting convective core, C, which has radius r. Hydrogen is converted into helium in the hydrogen-burning shell, A, which has outer radius R_1. The extensive envelope of the giant star, B, was assumed to be in radiative equilibrium (Öpik, 1938).

consistent with the observation that the red giant stars are very much rarer per unit volume of space than the dwarf stars. The quantitative theory of red giant stars needed a more complete theory of the nuclear reactions involved in post-main-sequence evolution. Specifically, as the core continued to collapse and heat up, helium burning in the core would result in the formation of carbon, and this was first discussed by Öpik and Edwin E. Salpeter (b. 1924) after the Second World War (Öpik, 1951; Salpeter, 1952) (see Section 8.2).

An important link in the chain was provided in 1942 by the Brazilian astrophysicist Mario Schönberg (1914–1990) and Subrahmanyan Chandrasekhar (1910–1995). Their studies concerned the stability of stellar models with inert isothermal cores, which are expected even if the central regions are in radiative, rather than convective, equilibrium. In the radiative case, the hydrogen fuel is depleted first in the hottest central regions and the size of the inert core grows with time. They found the important result that there do not exist stable stellar models in which the inert stellar core contains more than about 10% of the mass of the star (Schönberg and Chandrasekhar, 1942). Physically, the pressure at the base of the hydrogen-burning shell becomes too great and causes the inner regions to collapse. As shown by Kippenhahn and Weigert, the key quantity is the ratio of the mean molecular weights in the core and the envelope.[2]

This result, known as the *Schönberg–Chandrasekhar limit*, explained the formation of red giant stars during the course of stellar evolution. Stars spend most of their lifetimes on the main sequence, the energy source being the conversion of hydrogen into helium, by the p–p chain for stars with masses less than about $1.5 M_\odot$ and by the CNO cycle for stars with greater masses. Hydrogen is depleted in the central regions, resulting in the formation of an inert core. When the core grows to about 10% of the mass of the star, core collapse and the formation of the red giant envelope ensues.[3] This result also enables good estimates for the lifetimes of the Sun and main-sequence stars to be made (see Section A4.1).

4.2   White dwarfs

The discovery of white dwarfs is charmingly told in a reminiscence of Henry Norris Russell delivered at a Princeton colloquium in 1954. In 1910, Russell suggested to Pickering that it would be useful to obtain the spectra of stars for which parallaxes had been measured. Russell's reminiscence continues:[4]

Pickering said 'Well, name one of these stars.' Well, said I, for example, the faint component of Omicron Eridani. So Pickering said, 'Well, we make rather a specialty of being able to answer questions like that'. And so we telephoned down to the office of Mrs. Fleming and Mrs. Fleming said, yes, she'd look it up. In half an hour she came up and said, I've got it here, unquestionably spectral type A. I knew enough, even then, to know what that meant. I was flabbergasted. I was really baffled trying to make out what it meant. Then Pickering thought for a moment and then said with a kindly smile, 'I wouldn't worry. It's just these things which we can't explain that lead to advances in our knowledge.' Well, at that moment, Pickering, Mrs. Fleming and I were the only people in the world who knew of the existence of white dwarfs.

The remarkable feature of the faint companion of o-Eridani was that it was a very-low-luminosity star and yet it had the type of spectrum associated with hot stars on the upper part of the main sequence. Russell included it without comment in his first 'Russell' diagram (Figure 3.3(a)), the single A star lying roughly 10 magnitudes (a factor of 10 000 in luminosity) below typical main-sequence A stars. Adams drew attention to its remarkable properties in 1914 (Adams, 1914) and discovered another example in the following year, Sirius B, the faint companion of Sirius A (Adams, 1915).

Eddington realised that these observations immediately implied that white dwarf stars had to be very dense indeed. Their masses could be determined from the fact that they were members of binary star systems, and their radii could be estimated using Planck's radiation formula and their observed luminosities. Their mean densities had to be about 10^8 kg m^{-3}, but Eddington argued that there was nothing inherently implausible about such large densities (Eddington, 1924). Matter at such high temperatures would be completely ionised and so there was no reason at that time why matter could not be compressed to much higher densities than typical terrestrial densities. In fact, he argued that even nuclear densities were quite conceivable. In his paper of 1924, he also worked out the magnitude of the gravitational redshift which would be expected from such a compact star according to general relativity and found that it corresponded to a Doppler shift of the spectral lines to longer wavelengths of about 20 km s^{-1}. Adams made careful spectroscopic observations of Sirius B with the 100-inch telescope in 1925 and, once account was taken of the orbital motions of the binary stars, a shift of 19 km s^{-1} was measured (Adams, 1925a,b).[5] Eddington was jubilant (Eddington, 1926a):

Prof. Adams has killed two birds with one stone; he has carried out a new test of Einstein's theory of general relativity and he has confirmed our suspicion that matter 2000 times denser than platinum is not only possible, but is actually present in the universe.

The theory of white dwarfs was one of the first triumphs of the new quantum theory of statistical mechanics as applied to astrophysics. Wolfgang Pauli (1900–1958) enunciated the exclusion principle in 1922 (Pauli, 1925), and this led to Fermi–Dirac statistics and the concept of degeneracy pressure. In 1926, Fowler used these concepts to derive the equation of state of a cold degenerate electron gas (Fowler, 1926) and found the important result

$$p = \frac{(3\pi^2)^{2/3}}{5} \frac{\hbar^2}{m_e} \left(\frac{\rho}{\mu_e m_u}\right)^{5/3},$$ (4.1)

where μ_e is the mean molecular weight of the material of the star per electron and m_u is the unified atomic mass constant.[†] The important aspect of this equation of state is that it is independent of temperature and so the structure of white dwarfs can be derived directly from the Lane–Emden equation (3.3).[6] Unlike main-sequence stars, in which pressure support is provided by the thermal pressure of hot gas, the white dwarfs are supported by electron degeneracy pressure. The source of their luminosity is the internal thermal energy with which they were endowed on formation. According to Fowler's picture, the white dwarfs simply radiate away their internal thermal energies and end up as inert cold stars with all the nuclei and electrons in their lowest ground states.

In 1929, Wilhelm Anderson (1880–1940) showed that the degenerate electrons in the centres of white dwarfs with mass roughly that of the Sun become relativistic (Anderson, 1929). In the extreme relativistic limit, the equation of state of the degenerate electron gas becomes

$$p = \frac{(3\pi^2)^{1/3} \hbar c}{4} \left(\frac{\rho}{\mu_e m_u}\right)^{4/3}.$$ (4.2)

Once again, the result is independent of temperature, but the change in the dependence of pressure upon density from $p \propto \rho^{5/3}$ to $p \propto \rho^{4/3}$ has profound implications. Anderson and Edmund Stoner (1899–1968) realised that the consequence was that there do not exist equilibrium configurations for degenerate stars with mass greater than about the mass of the Sun (Anderson, 1929; Stoner, 1929). The most famous analysis of this result was carried out by Chandrasekhar, who had begun working on this problem before he arrived to take up a fellowship at Trinity College, Cambridge, in 1930. According to Wali's biography,[7] he derived the key result while on board the ship *Lloyd Triestino* which was taking him as a 19-year-old from Bombay to London. He found the crucial result that, in the extreme relativistic limit, there is an upper limit to the mass of stable white dwarfs,

$$M_{\text{Ch}} = \frac{(3\pi)^{3/2}}{2} \left(\frac{\hbar c}{G}\right)^{3/2} \times \frac{2.01824}{(\mu_e m_u)^2} = \frac{5.836}{\mu_e^2} M_\odot.$$ (4.3)

This mass is known as the *Chandrasekhar mass*[‡] (Chandrasekhar, 1931). The critical mass depends upon the chemical composition of the material of the star through the value of μ_e, the mean molecular weight of the stellar material per electron. Other than that, the

---

[†]   Order-of-magnitude calculations demonstrating the origins of the forms of the non-relativistic and relativistic equations of state for degenerate matter are given in Section A4.2.
[‡]   Order-of-magnitude calculations illustrating the origins of the Chandrasekhar mass are given in Section A4.3.

Chandrasekhar mass only depends upon fundamental constants. Since $\mu_e \approx 2$ for the material of compact stars, the Chandrasekhar mass is usually quoted as $M_{Ch} = 1.46 M_\odot$.

The cause of the instability is that, in the extreme relativistic limit, both the internal thermal energy, U_{th}, and the gravitational potential energy, U_{grav}, of the star depend upon radius in the same way, $U_{th} = (1/2)U_{grav} \propto R^{-1}$. Now, the gravitational potential energy is proportional to M^2, whereas the thermal energy is proportional to the mass of the star and so, for massive enough stars, the gravitational energy term dominates, causing collapse, which cannot be stabilised by the pressure of the degenerate gas since the two energies always depend upon radius in the same way. The inference is that there is nothing to prevent degenerate stars more massive than M_{Ch} from collapsing to very high densities indeed and possibly to a state of complete gravitational collapse.

This conclusion was vigorously challenged by Eddington and led to the famous dispute with Chandrasekhar. Eddington found the idea of complete gravitational collapse unacceptable, believing that there must be some new unspecified physical process which prevented its occurrence. Chandrasekhar's work was publicly repudiated by Eddington, employing arguments which were more polemical than physical, at the meeting of the Royal Astronomical Society on January 11 1935 (Eddington, 1935):

Dr. Chandrasekhar has got this result before, but he has rubbed it in in his last paper; and, when discussing it with him, I felt driven to the conclusion that this was almost a *reductio ad absurdum* of the relativistic degeneracy formula. Various accidents may intervene to save the star, but I want more protection than that. I think there should be a law of Nature to prevent a star behaving in this absurd way! ... The formula is based upon a combination of relativity mechanics and non-relativistic quantum theory, and I do not regard the offspring of such a union as born in lawful wedlock.

Eddington fully realised that what we would now call a black hole was the natural outcome of gravitational collapse. In his words (Eddington, 1941)[8]

If the star is symmetrical and not in rotation, it would contract to a diameter of a few kilometres, until according to the theory of relativity, gravitation becomes too great for the radiation to escape.

Eddington objected instinctively, however, to what he called elsewhere 'this stellar buffoonery'. His physical concerns centred upon the use of the Anderson–Stoner relativistic equation of state for a degenerate gas, which is the ultimate cause of the gravitational collapse of the star. While the vast majority of physicists and astrophysicists agreed with Chandrasekhar's analysis, what Leon Mestel calls a 'ding-dong dispute' continued in the literature for a number of years.[9]

Chandrasekhar, then a brilliant young mathematical physicist recently arrived from India to an alien environment, took the rebuff badly and it rankled for many years, despite his lasting respect for, and friendship with, the older man. In his address to the General Assembly of the International Astronomical Union in Montreal in 1979, he stated (Chandrasekhar, 1980):

It is difficult to understand why Eddington, who was one of the early enthusiasts and staunchest advocates of general relativity, should have found the conclusion that black holes may be formed during the course of the evolution of stars so unacceptable. But the fact is that Eddington's supreme authority in those years effectively delayed the development of fruitful ideas along these lines for some thirty years.

Quite independently, Lev Landau (1908–1968) had come to the conclusion in 1932 that gravitational collapse to a singularity should be taken seriously (Landau, 1932), and in 1938 Robert Oppenheimer (1904–1967) and Hartland Snyder (1913–1962) gave the first general relativistic analysis of what would be observed in the final stages of gravitational collapse of a pressureless sphere (Oppenheimer and Snyder, 1939). In their paper they described the key observed features of what are now termed black holes.

4.3   Supernovae and neutron stars

The neutron was discovered in 1932 by Chadwick (Chadwick, 1932), and the model of the nucleus consisting of neutrons and protons was quickly adopted, although the problem of how the nucleus could be held together remained to be resolved. The first mention of the possibility of neutron stars appears as the famous 'Additional remark' to a paper by Walter Baade (1893–1960) and Fritz Zwicky (1898–1974) of 1934 (Baade and Zwicky, 1934b). In that year, they published two papers on the energetics of what they termed 'super-novae'.

The extragalactic nature of the spiral nebulae had been established beyond doubt by Hubble in 1926. Among the objects which played a part in that debate were the novae, or 'new stars', which increase rapidly in brightness and fade away again. As will be discussed in Section 5.3, the nova of 1885 which exploded in the Andromeda Nebula appeared to be exceptional, being about 100 times more luminous than the more common nova. Lundmark had suggested that there were two classes of novae, that of 1885 belonging to the upper class. In their first paper, Baade and Zwicky proposed that the population of novae consists of two types: the ordinary novae, which are relatively common phenomena and which had been used by Lundmark as distance indicators for spiral nebulae (see Section 5.3), and the supernovae, which are very rare but very energetic indeed (Baade and Zwicky, 1934a). They identified the bright nova observed in the Andromeda Nebula in 1885 as the archetype of this class of extremely violent explosion – they suggested that Tycho Brahe's nova of 1572 was another example of this class. The frequency of occurrence of these events was estimated to be only about once per 1000 years per galaxy but, when they occurred, an enormous amount of energy was released, corresponding to a significant fraction of the rest mass energy of the precursor star. In their second paper, they suggested that such events might be the sources of the cosmic rays which had been discovered by Victor Hess in 1912 (Hess, 1912). Both proposals are remarkably close to the truth. As an addendum to the second paper (Baade and Zwicky, 1934b), they wrote

With all reserve we advance the view that a super-nova represents the transition of an ordinary star into a *neutron star*, consisting mainly of neutrons. Such a star may possess a very small radius and an extremely high density. As neutrons can be packed much more closely than ordinary nuclei and electrons, the 'gravitational packing' energy in a *cold* neutron star may become very large, and under certain circumstances, may far exceed the ordinary nuclear packing fractions. A neutron star would therefore represent the most stable configuration of matter as such. The consequences of this hypothesis will be developed in another place, where also will be mentioned some observations that tend to support the idea of stellar bodies made up mainly of neutrons.

Be Scientific with OL' DOC DABBLE.

Figure 4.2: The cartoon which appeared in the *Los Angeles Times* of 19 January 1934 in the comic strip entitled 'Be Scientific with Ol' Doc Dabble'.

It is best to allow Zwicky to describe how these ideas were received in a quotation from the extraordinary preface to his *Catalogue of Selected Compact Galaxies and of Post-eruptive Galaxies* of 1968 (Zwicky, 1968),

In the *Los Angeles Times* of January 19, 1934, there appeared an insert in one of the comic strips, entitled 'Be Scientific with Ol' Doc Dabble' quoting me as having stated 'Cosmic rays are caused by exploding stars which burn with a fire equal to 100 million suns and then shrivel from $\frac{1}{2}$ million miles diameter to little spheres 14 miles thick', says Prof. Fritz Zwicky, Swiss Physicist.' This, in all modesty, I claim to be one of the most concise triple predictions ever made in science. More than 30 years were to pass before this statement was proved to be true in every respect.

(See Figure 4.2.) Baade and Zwicky's idea that a neutron star might be the remnant left behind after a supernova explosion was proved correct 33 years later with the discovery of pulsars by Antony Hewish, Jocelyn Bell and their colleagues in 1987 (Hewish *et al.*, 1968).

In the meantime, Gamow showed in 1937 that a gas of neutrons could be compressed to a much higher density than a gas of nuclei and electrons and estimated the probable densities of such stars to be about 10^{17} kg m^{-3} (Gamow, 1937, 1939). The issue of the maximum mass of neutron stars was discussed by Landau in 1938 (Landau, 1938) and in much greater detail by Oppenheimer and George Volkoff (1914–2000) in the following year (Oppenheimer and Volkoff, 1939). The result they found is not so different from expression (4.3) if we set $\mu_e = 1$ and $m_u = m_n$. The physics is the same as in the case of the white dwarfs, but now neutron degeneracy pressure holds up the star. Complications arise because it is necessary to take into account the details of the equation of state of neutron matter at nuclear densities, and the effects of general relativity can no longer be neglected. They found an upper mass limit of about $0.7M_\odot$. This result is not so different from the best modern estimates, which correspond to about $2–3M_\odot$.

This is a very much more serious situation than the case of the white dwarfs. The neutron stars are so compact that general relativity is no longer a small correction but is central to the stability of the star. Typically, for neutron stars, the general relativistic parameter $2GM/Rc^2 = R_g \sim 0.3$, and so they have radii which are only about three times the Schwarzschild radius, R_g, of a spherically symmetric black hole of the same mass.

This work created some theoretical interest but little enthusiasm from the observers. The radii of typical neutron stars were expected to be about 10 km and so there was no prospect of detecting significant fluxes of thermal radiation from such tiny stars. Nonetheless, many of the objects which were to play a leading role in the development of high-energy astrophysics in the years following the Second World War were already in place in the literature, even if there was not a great deal that the astronomers could do about them at that time.

Notes to Chapter 4

1 I am most grateful to Leon Mestel for his deep insights into the history of the astrophysics of stellar structure and evolution.

2 The result quoted by Kippenhahn and Weigert is that the fraction of the mass of the star in the core should not exceed $(\mu_{core}/\mu_{env})^2$, where the μs are the mean molecular weight of the material per electron. For the case of a helium core surrounded by an envelope with normal cosmic abundance, the limit corresponds to about 10% of the mass of the star being in the form of an inert core.

3 One of the more controversial issues concerned the precise cause of the initiation of the red giant phase. Every numerical calculation of the evolution of stars shows that, as hydrogen burning in the core is exhausted, the core collapses and the envelope expands, but the dramatic changes in the star's structure are associated with a number of different processes taking place almost simultaneously – the collapse of the central regions, changes in chemical composition of the stellar material with radius, in particular the discontinuity in atomic weight at the core–envelope boundary, and consequently changes in its opacity, the development of extensive convective zones in the stellar envelope and so on. The most detailed discussion of the likely cause of the formation of red giants has been presented by John Faulkner (see, for example, J. Faulkner, Fred Hoyle, Red giants and beyond, *Astrophysics and Space Science*, **285** (2003), 339, and in the memorial volume to celebrate the life and work of Fred Hoyle, Red giants, then and now, in D. O. Gough, ed., *New Frontiers of Astronomy* (Cambridge: Cambridge University Press, 2005)).

4 See A. G. Davis Philip and D. H. DeVorkin, eds, *In Memory of Henry Norris Russell* (Dudley Observatory Report no. 13, 1977), pp. 90–107.

5 See also A. V. Douglas, *The Life of Arthur Stanley Eddington* (London: Thomas Nelson and Sons Ltd, 1956), pp. 75–78.

6 I have demonstrated how this analysis can be carried out in Section 15.3 of Malcolm Longair, *High Energy Astrophysics*, vol. 2 (Cambridge: Cambridge University Press, 1997). The section also includes a derivation of the formula for the Chandrasekhar mass. An order of magnitude analysis is given in Section A4.3.

7 See endnote 4, Chapter 3.

8 This quotation is taken from the discussion of Eddington's paper, The theory of white dwarf stars (See A. S. Eddington, The theory of white dwarf stars, in *Novae and White Dwarf Stars*, ed. A. J. Shaler (Paris: Herrmann et Cie), pp. 249–262), which immediately preceded the discussion. A more extended quote from the discussion recorded by the editors is of interest. 'Sir Arthur Eddington replies that in stars of mass greater than the critical masses mentioned by Dr. Chandrasekhar there is no limit to the contraction, so that if the star is symmetrical and not in rotation, it would contract to a diameter of a few kilometers, until, according to the theory of relativity, gravitation becomes too great for the radiation to escape. This is not a fatal difficulty, but it is nevertheless surprising; and, being somewhat shocked by this conclusion, Sir Arthur was led to reexamine the physical theory and so finally to reject it.'

9 See, for example, S. Chandrasekhar, *Eddington: The Most Distinguished Astrophysicist of his Time* (Cambridge: Cambridge University Press, 1983), p. 47 *et seq.* Leon Mestel has written: 'Chandra was anxious to get backing from leading physicists if only because astronomers were overawed by Eddington's reputation. Bohr, Rosenfeld, Dirac, Peierls, Pauli, Fowler all supported Chandra against Eddington, at least in private, though there seems to have been some reluctance to stand up and be counted.' This account agrees with the views expressed to me by William McCrea and Lyman Spitzer, who were present during many of these debates. Both agreed that everyone thought that Chandrasekhar was right. McCrea wrote to me in 1993 as follows: 'No one who knew Eddington . . . would remotely imagine his being unpleasant. The last few years have seen most unfortunate, completely misleading accounts of this matter. I was present at the core episode. Unfortunately misunderstandings arose, mainly on Chandra's part. Unhappily he has allowed them to prey upon him, which I deeply regret. Eddington did use some sloppy arguments. But, if Chandra was so certain that he was right, why did he not pursue the consequences at the time?' Mestel provides detailed references to the published articles involved in the dispute (see note 13 to Chapter 3).

A4   Explanatory supplement to Chapter 4

A4.1   The lifetimes of the Sun and the stars

The Schönberg–Chandrasekhar limit enables simple estimates of the lifetimes of main-sequence stars to be made. Most of the lifetime of a main-sequence star is spent burning hydrogen to helium in its core. This nuclear fusion reaction releases by far the largest fraction of the nuclear energy available to the star, about 0.7% of the rest mass of the hydrogen nuclei being liberated in forming helium. Stellar evolution models show that, once the star has settled onto the main sequence, its luminosity changes very little until it begins to move off the main sequence when the core contracts and the red-giant phase begins. The subsequent phases of stellar evolution are all short compared with the main-sequence lifetime.

These considerations enable a simple estimate of the main-sequence lifetime of the star to be made. The star moves off the main sequence when the central 10% of its mass has been converted into helium. The energy released in this process is given by $E = 0.007 (0.1 \times M)c^2$. Therefore, since the luminosity of the star is L, its main-sequence lifetime is given by

$$T_{MS} = \frac{E}{L} = \frac{0.007(0.1 \times M)c^2}{L}.$$

Inserting the values for the Sun, $L = 3.9 \times 10^{26}$ W and $M = M_\odot = 2 \times 10^{30}$ kg, we find $T_\odot = 10^{10}$ years.

We can use this result to find the main-sequence lifetimes of main-sequence stars of different masses. If the mass–luminosity relation has the form $L \propto M^x$, where $x \sim 4$ for stars with $M \sim M_\odot$, then, by exactly the same argument, the lifetime of the star is given by

$$T(M) = 10^{10} \left(\frac{M}{M_\odot} \right)^{-(x-1)} \text{years.} \tag{A4.1}$$

A4.2   The equation of state for degenerate matter

In the cases of both white dwarfs and neutron stars, there is no internal heat source – the stars are held up by degeneracy pressure. The significance of degeneracy pressure comes about naturally because, in the centres of stars at an advanced stage of evolution, the central densities become high and the use of the pressure formulae for a classical gas is inappropriate. The Heisenberg uncertainty principle ensures that, at very high densities, when the inter-particle spacing becomes small, the particles of the gas must possess large momenta according to the relation $\Delta p \, \Delta x \approx \hbar$. These large quantum mechanical momenta provide the pressure of the degenerate gas.

First of all, we work out the physical conditions under which degeneracy pressure is important. If the electron–proton plasma is in thermal equilibrium at temperature T, the root-mean-square velocity of the particles, $\langle v^2 \rangle$, is given by $\frac{1}{2}m\langle v^2 \rangle = \frac{3}{2}kT$ and hence the typical momenta of the particles are $p = mv \approx (3mkT)^{1/2}$. Because the electrons are much lighter than the protons and neutrons, they become degenerate at much larger inter-particle spacings than the protons and neutrons. According to the Heisenberg uncertainty principle, the inter-electron spacing at which quantum mechanical effects become important is $\Delta x \approx \hbar/\Delta p$ and hence, setting $\Delta p = p$, the density of a hydrogen plasma, which is mostly contributed by the protons, is given by

$$\rho \approx m_p/(\Delta x)^3 \approx m_p \left(\frac{3m_e kT}{\hbar} \right)^{3/2}, \tag{A4.2}$$

where m_p is the mass of a proton. Thus, the density at which degeneracy sets in for electrons in the non-relativistic limit is proportional to $T^{3/2}$. A better estimate of the critical density can be found by equating the degeneracy pressure of a non-relativistic gas, (equation (A4.9) below) to the pressure of a classical gas. Performing this sum, we find that the critical density is given by

$$\rho \approx 3.3 \times 10^{-4} T^{3/2} \, \text{kg m}^{-3}, \tag{A4.3}$$

where T is the temperature in kelvin. Hence, for stars like the Sun with central temperature about 1.6×10^7 K and central density $\rho \approx 1.5 \times 10^5$ kg m^{-3}, the equation of state of the gas can always be taken to be that of a classical gas. When the star moves off the main sequence, however, the central regions contract and, although there is a modest increase in temperature, the matter in the core becomes degenerate, and this plays a crucial role in the evolution of stars on the giant branch. Ultimately, in the white dwarfs, the densities are typically about 10^9 kg m^{-3} and so they are degenerate stars.

The next consideration is whether or not the electrons are relativistic. To order of magnitude, we can find the condition for the electrons to become relativistic by setting $\Delta p \approx m_e c$ in Heisenberg's uncertainty principle; then, by the same arguments as above, we find that the density is given by

$$\rho \sim \frac{m_p}{(\Delta x)^3} \sim m_p \left(\frac{m_e c}{\hbar}\right)^3 \sim 3 \times 10^{10} \text{ kg m}^{-3}. \tag{A4.4}$$

A better calculation, with exactly the same physics but expressed in a slightly different way,[1] is to require the Fermi momentum of a degenerate Fermi gas in the zero-temperature limit to be $m_e c$. In this case, the density at which the electrons become relativistic is given by

$$\rho = \frac{m_u}{3\pi^2} \left(\frac{m_e c}{\hbar}\right)^3 \mu_e = 9.74 \times 10^8 \mu_e \text{ kg m}^{-3}. \tag{A4.5}$$

In this expression, m_u is the atomic mass unit and $\mu_e = m_B/(m_u Y_e)$ is the mean molecular weight of the material per electron, m_B being the mean baryon rest mass of the material, and Y_e is the mean number of electrons per baryon.[2] In the centres of the most massive white dwarfs, the densities attain these values, and so the equation of state for a relativistic degenerate electron gas has to be used. It is this feature which determines the upper mass limit for white dwarfs and neutron stars.

Next, we can work out by these rough methods the equations of state for degenerate matter in the non-relativistic and relativistic regimes. In general, the relation between pressure and energy density can be written as $p = (\gamma - 1)u$, where p is the pressure, u is the energy density of the matter or radiation which provides the pressure and γ is the ratio of specific heat capacities.

In the non-relativistic regime, the energy of an electron in the degenerate limit is given by

$$E = \tfrac{1}{2}m_e v^2 = \frac{p^2}{2m_e} = \frac{\hbar^2}{2m_e a^2}, \tag{A4.6}$$

where $a = \Delta x$ is the inter-electron spacing. Therefore, to order of magnitude, the energy density of the material is given by $u \approx E/a^3 = \hbar^2/2m_e a^5$. Since the density of matter is $\rho \sim m_p/a^3$, it follows that $p \propto \rho^{5/3}$, and hence the ratio of specific heat capacities is given by $\gamma = 5/3$. The pressure of the gas is therefore roughly given by

$$p \approx \frac{\hbar^2}{3m_e a^5} \approx \frac{\hbar^2}{3m_e} \left(\frac{\rho}{m_p}\right)^{5/3}. \tag{A4.7}$$

We can repeat this calculation for a relativistic electron gas, in which case $E \approx pc \approx \hbar c/a$ and hence $u \approx E/a^3 \approx \hbar c/a^4$. Since $\rho \sim m_p/a^3$, $p \propto \rho^{4/3}$ and $\gamma = 4/3$. The pressure of the gas is roughly given by

$$p \approx \frac{\hbar c}{3a^4} \approx \frac{\hbar c}{3} \left(\frac{\rho}{m_\mathrm{p}} \right)^{4/3}. \tag{A4.8}$$

The exact results found from the application of statistical mechanics to a Fermi–Dirac distribution in its ground state are as follows:

non-relativistic $$p = \frac{(3\pi^2)^{2/3}}{5} \frac{\hbar^2}{m_\mathrm{e}} \left(\frac{\rho}{m_\mathrm{u}\mu_\mathrm{e}} \right)^{5/3}; \tag{A4.9}$$

relativistic $$p = \frac{(3\pi^2)^{2/3} \hbar c}{4} \left(\frac{\rho}{m_\mathrm{u}\mu_\mathrm{e}} \right)^{4/3}. \tag{A4.10}$$

We obtain the corresponding results for degenerate neutrons if we substitute neutrons for electrons in the above expressions and set $\mu_\mathrm{e} = 1$. Then, the expressions for the pressure of the neutron gas in the two limits are as follows:

non-relativistic $$p = \frac{(3\pi^2)^{2/3}}{5} \frac{\hbar^2}{m_\mathrm{n}} \left(\frac{\rho}{m_\mathrm{n}} \right)^{5/3}; \tag{A4.11}$$

relativistic $$p = \frac{(3\pi^2)^{1/3} \hbar c}{4} \left(\frac{\rho}{m_\mathrm{n}} \right)^{4/3}. \tag{A4.12}$$

A4.3 The Chandrasekhar mass

We can use these techniques to illustrate the physical origin of the Chandrasekhar mass. The total internal energy of the star can be found from our order-of-magnitude derivation of the equation of state of a relativistic degenerate gas, equation (A4.8). Since $p = \frac{1}{3}u$,

$$U = Vu = 3Vp \approx V\hbar c(\rho/m_\mathrm{p})^{4/3}. \tag{A4.13}$$

In Section A3.1, we derived the *virial theorem* for stars, according to which the total internal energy, U, is related to the total gravitational potential energy, $|\Omega|$, by $3(\gamma - 1)U = |\Omega|$. Setting $\gamma = 4/3$ for a relativistic gas, $U = |\Omega|$, and so

$$V\hbar c \left(\frac{\rho}{m_\mathrm{p}} \right)^{4/3} \approx \frac{GM^2}{R}. \tag{A4.14}$$

Now, $V \approx R^3$ and $\rho V = M$. Therefore, the left-hand side of equation (A4.14) becomes

$$\frac{\hbar c}{R} \left(\frac{M}{m_\mathrm{p}} \right)^{4/3}. \tag{A4.15}$$

Note the key point that, because we have used the relativistic equation of state, the left-hand side of the equation depends upon radius as R^{-1}, exactly the same dependence as the gravitational potential energy. Thus, the mass of the star does not depend upon its radius. From equations (A4.13) and (A4.14), we find

$$M \approx \frac{1}{m_\mathrm{p}^2} \left(\frac{\hbar c}{G} \right)^{3/2} \approx 2M_\odot, \tag{A4.16}$$

dropping constants of order 1. This is an order-of-magnitude derivation of the *Chandrasekhar mass.*[3]

Note the physical meaning of the result given in equation (A4.16). For lower-mass stars, the non-relativistic equation of state, $p \propto \rho^{5/3}$, should be used, and then equating the gravitational potential energy and the internal thermal energy results in stars with a definite radius. The origin of the collapse of relativistic degenerate stars can be understood as follows. For masses less than the Chandrasekhar mass, inspection of equations (A4.14) and (A4.15) shows that the internal energy exceeds the gravitational energy and so collapse does not occur. However, for masses greater than the Chandrasekhar mass, the gravitational potential energy exceeds the internal thermal energy and so the star collapses. Since both energies depend upon radius R in the same way, if the gravitational energy once dominates, it will always dominate. Hence, there is nothing to prevent collapse to a black hole. This conclusion was the source of contention between Chandrasekhar and Eddington.

Note that the Chandrasekhar mass depends only upon fundamental constants. One of the more intriguing ways of rewriting the expression (A4.16) is in terms of a 'gravitational fine-structure constant', α_G. The standard fine-structure constant is given by $\alpha = e^2/4\pi\epsilon_0\hbar c$. The equivalent formula for gravitational forces can be found by replacing $e^2/4\pi\epsilon_0$ in the inverse square law of electrostatics by GM^2 from Newton's law of gravity, where m_p is the mass of the proton. Thus, $\alpha_G = Gm_p^2/\hbar c$. Putting in the values of the constants, we find $\alpha^{-1} = 137.04$ and $\alpha_G = 5.6 \times 10^{-39}$, the ratio of these constants being $\alpha_G/\alpha = 2.32 \times 10^{40}$, reflecting the enormous difference in the strengths of the electrostatic and gravitational forces. Therefore, the Chandrasekhar mass is roughly given by

$$M \approx m_p\alpha_G^{-3/2}.$$

In other words, in terms of the basic constants of physics, stars are typically objects with 10^{60} protons. Note also that the calculation applies equally to white dwarfs and neutron stars, the only difference being that the neutron stars are very much denser than the white dwarfs.

Notes to Section A4

1 R. Kippenhahn and A. Weigert's book, *Stellar Structure and Evolution* (Berlin: Springer-Verlag, 1990), deals with these and many other aspects of stellar evolution, bringing out the physical principles very clearly.
2 S. I. Shapiro and S. A. Teukolsky's book, *Black Holes, Neutrons Stars and White Dwarfs: The Physics of Compact Objects* (New York: Wiley Interscience, 1983), can be thoroughly recommended for a detailed treatment of degenerate stars.
3 I have given a more detailed derivation of the Chandrasekhar mass in Malcolm Longair, *High Energy Astrophysics*, vol. 2 (Cambridge: Cambridge University Press, 1997).

Part II

The large-scale structure of the Universe, 1900–1939

5 The Galaxy and the nature of the spiral nebulae

The second part of our history concerns the understanding of the large-scale distribution of matter in the Universe. At the beginning of the period 1900 to 1939, little was known even about the structure of our own Galaxy; by the end of it, the Universe of galaxies was established, the system was known to be expanding and general relativity provided a theory capable of describing the distribution of matter in the Universe on the very largest scales.

5.1 'Island universes' and the cataloguing of the nebulae

The earliest cosmologies of the modern era were speculative conjectures. The 'island universe' model of René Descartes (1596–1650), published in *The World* of 1636, involved an interlocking jigsaw puzzle of solar systems. In 1750, Thomas Wright of Durham (1711–1786) published *An Original Theory or New Hypothesis of the Universe*, in which the Sun was one of many stars which orbit about the 'Divine Centre' of the star system. Immanuel Kant (1724–1804) in 1755 and Johann Lambert (1728–1777) in 1761 took these ideas further and developed the first hierarchical, or fractal, models of the Universe.[1] Kant made the prescient suggestion that the flattening of these 'island universes' was due to their rotation. The problem with these early cosmologies was that they lacked observational validation, in particular because of the lack of information on the distances of astronomical objects.

Towards the end of the eighteenth century, William Herschel (1738–1822) was one of the first astronomers to attempt to define the distribution of stars in the Universe in some detail on the basis of careful observation. To determine the structure of the Milky Way, he counted the number of stars in different directions, assuming they all have the same intrinsic luminosities. In this way, he derived his famous picture for the structure of our Galaxy, consisting of a flattened disc of stars with diameter about five times its thickness, the Sun being located close to its centre (Figure 5.1) (Herschel, 1785).

Herschel inferred that the nebulae were island universes similar to our Galaxy. A test of this picture was to show that the nebulae could be resolved into stars, and he believed this had been achieved in a number of cases. In others, he assumed that the nebulae were too distant to be resolved into individual stars. This model came into question, however, when Herschel discovered that among the nebulae were the planetary nebulae, which consist of a central star surrounded by a shell of gas. Herschel recognised that these nebulae were

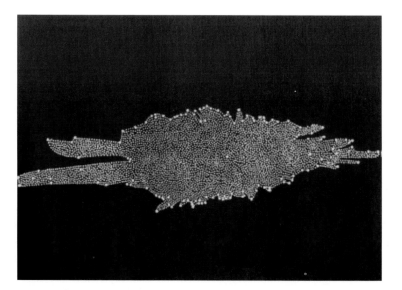

Figure 5.1: William Herschel's model of the Galaxy based upon star counts in different directions. The Sun is located close to the centre of the disc of stars (Herschel, 1785).

unlikely to be resolved into stars but rather consisted of 'luminous fluid' surrounding the central star.

John Michell (1734–1793) had already warned Herschel that the assumption that the stars have a fixed luminosity was a poor approximation. This is the same John Michell who was Woodwardian Professor of Geology at Queens' College, Cambridge, before becoming the rector of Thornhill in Yorkshire in 1767. He designed and built what we now know as the Cavendish experiment to measure the mean density of the Earth. Nowadays, he is also remembered as the first person to realise that light could not escape from the surface of a massive enough body, what would now be called a black hole (see Section 11.3) (Michell, 1784). In his remarkable pioneering paper of 1767, he introduced statistical methods into astronomy in order to show that binary stars and star clusters must be physical associations and not random associations of stars on the sky (Michell, 1767). Consequently, there must be a dispersion in the absolute luminosities of the stars from the observed range of apparent magnitudes in bright star clusters, such as the Pleiades. Despite this warning, Herschel proceeded to produce a number of different versions of his model for the structure of our Galaxy, adding appendages to account for various features of the star counts in different directions.

In 1802, Herschel measured the magnitudes of visual binary stars and was forced to agreed with Michell's conclusion about the wide dispersion in the luminosities of the stars (Herschel, 1802). Equally troubling was the fact that observations with his magnificent 40-foot telescope showed that as he studied fainter samples of stars, the more he continued to find. Evidently, the stellar system was unbounded – there was no edge to the Galaxy. Eventually, Herschel lost faith in his model of the Galaxy. On top of all these problems, the importance of extinction by interstellar dust was not appreciated – it was only in the

1930s that its central importance for studies of our Galaxy was finally established (see Section 5.6).

Meanwhile, the cataloguing of the nebulae was progressing steadily. Among the first lists of bright nebulae to be published was that of Charles Messier (1730–1817), whose catalogue of 103 objects was compiled during the years 1771–1784 (Messier, 1784).[2] The list was not intended primarily as a catalogue of interesting nebulae, but rather as a list of objects to be avoided by comet-hunters, of whom Messier was a leading exponent. Many of the bright nebulae are still referred to by their Messier numbers, for example, the Andromeda Nebula, being the 31st entry in the catalogue, is M31; the Orion Nebula is M42 and the Crab Nebula is M1.

The systematic cataloguing of the nebulae was begun by William Herschel, assisted by his sister Caroline (1750–1848), using his 20-foot reflector at Slough. The first catalogue was published in 1786 and consisted of 1000 nebulae. This was followed by a further 1000 entries in 1789 and 500 more in 1802. This work was continued by William's son John Herschel (1792–1871), who took the 20-foot telescope to the recently completed Royal Observatory at the Cape of Good Hope in South Africa where he surveyed the Southern Sky for nebulae. In 1864, John Herschel published the *General Catalogue of Nebulae* containing 5079 objects, of which all but 449 were discovered by the Herschels (Herschel, 1864). The catalogue was compiled entirely from visual observations, using the 20-foot telescope as a transit instrument, before the use of photographic methods became practicable for these studies. This catalogue provided a large fraction of the entries in the *New General Catalogue of Nebulae and Clusters of Stars* published by John Dreyer (1852–1926) in 1888 (Dreyer, 1888). This catalogue, the *NGC Catalogue*, is still the fundamental catalogue of bright nebulae and contains positions and descriptions of the catalogued nebulae. Dreyer produced two supplements to the NGC catalogues known as the *Index Catalogues* and the objects therein are referred to by their IC numbers (Dreyer, 1895, 1908). In all, these catalogues contained some 15 000 nebulous objects. The process of cataloguing bright nebulae was completed by 1908.[3]

Among the nebulae there were undoubtedly numerous star clusters, and a common view was that many of the diffuse nebulae were simply too distant to be resolved into stars. Some of them were certainly 'enormous masses of luminous gas or vapour', as had been convincingly demonstrated by Huggins and Miller's important spectroscopic observations of the 1860s (see Section 2.1). There remained, however, the issue of the nature of the spiral nebulae – the definitive solution of this problem had to await the 1920s (see Section 5.3).

5.2   The structure of our Galaxy

The determination of the large-scale distribution of stars in our Galaxy and the understanding of its internal dynamics became feasible once measurements of stellar motions and distances became available. One of the objectives of these studies was the measurement of the motion of the Sun relative to the nearby stars. This was accomplished by measuring the proper motions of the stars, that is, their apparent angular motions on the sky relative to very distant stars. In 1718, Edmund Halley (1656–1742) had noted that the positions of Aldebaran,

Betelgeuse and Sirius differed from the positions listed in the *Almagest* of Claudius Ptolemy (second century AD), but it was not clear whether these differences were due to the motion of the Sun, to the motions of the stars or to some combination of both (Halley, 1718). By 1783, William Herschel had measured the mean motion of the Sun relative to 13 bright stars (Herschel, 1783), and in 1837 Friedrich Argelander (1799–1875) used the proper motions of 330 stars to find a mean motion of the Sun similar to Herschel's estimate (Argelander, 1838). When Hermann Kobold (1858–1942) extended the analysis to a fainter sample of over 1000 stars in 1895, however, a different answer was found – these observations showed that the motions of the stars were not random but contained a systematic component (Kobold, 1895).

This was the situation when Jacobus Kapteyn (1851–1922) began his studies of the proper motions of 2400 stars. By 1904, he had confirmed Kobold's result that the stars did not move randomly but rather that, when averages were taken within 28 separate areas, there were systematic motions (Kapteyn, 1905). He found that the stars tended to move in two preferred opposite directions. This result was confirmed by Eddington in 1908, who suggested that there were two interpenetrating streams of stars (Eddington, 1908).

Just as the Harvard Observatory had become pre-eminent in the classification of stellar spectra, so the Lick Observatory, first under James E. Keeler and then William Campbell (1862–1938), became the leading observatory for the measurement of the radial velocities of stars through spectroscopic measurements of their Doppler shifts. Although Huggins had claimed to have measured the Doppler shift of Sirius A as early as 1868, the errors were large, so large in fact that he claimed a positive radial velocity when it should have been negative. Keeler, at the Lick Observatory, and Herman C. Vogel and Julius Scheiner (1858–1913) at Potsdam understood the necessity of taking great care to eliminate the many systematic errors which can enter into radial velocity measurements and reduced the typical error to a few kilometres per second.

In 1896, the Lick Observatory received a benefaction from Mr D. O. Mills (1825–1910) for the construction of a spectrograph, which Campbell optimised for radial velocity measurements. Campbell already had ambitious plans for a very large survey of the radial velocities of stars in the northern and southern hemispheres. When he became Director of the Lick Observatory following Keeler's death in 1900, he obtained a further major grant from Mills for the construction and maintenance of a southern observing station at Cerro San Cristobal, near Santiago in Chile, to undertake the southern part of the survey. The initial part of the southern survey, consisting of 899 spectrograms, was completed by 1906 and the results for 150 stars were published in 1911 (Wright *et al.*, 1911). The radial velocity programme continued for many years, and it was 1928 before the catalogue of 2771 radial velocities for stars with magnitudes brighter than 5.51 was completed (Campbell and Moore, 1928).

One of Campbell's main interests was the determination of the solar motion, and as early as 1901 he had derived a value of 19.89 ± 1.52 km s^{-1}, as well as determining the direction of the apex of the solar motion. His estimates improved steadily over the years of the survey, and by 1910 he had shown that the relative velocity of the two streams discovered by Kapteyn was 40 km s^{-1} (Campbell, 1910). In 1907, Karl Schwarzschild suggested that it was not necessary to think in terms of two star streams but rather that the velocity distribution of the stars could be described in terms of a *velocity ellipsoid*, meaning that the local velocity

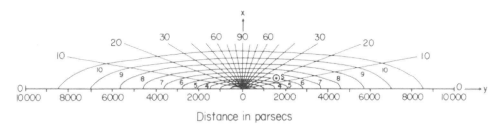

Figure 5.2: Kapteyn's model for the distribution of stars in the Galaxy (Kapteyn, 1922). The diagram shows the distribution of stars in a plane perpendicular to the Galactic plane. The curves are lines of constant number density of stars and are in equal logarithmic steps. The Sun, S, is slightly displaced from the centre of the system.

dispersion of the stars had different values along three orthogonal directions,

$$p(v_x, v_y, v_z) \propto \exp\left[-\frac{1}{2}\left(\frac{v_x^2}{\sigma_x^2} + \frac{v_y^2}{\sigma_y^2} + \frac{v_z^2}{\sigma_z^2}\right)\right],$$

in which $\sigma_x \neq \sigma_y \neq \sigma_z$ (Schwarzschild, 1907). The longest axis of the ellipsoid lay along the direction of Kapteyn's two star streams.

To determine the scale and structure of the system of stars in the Galaxy, Kapteyn drew up a plan of 206 Selected Areas in which deep star counts and proper motions would be measured (Kapteyn, 1906). By this time, it was well known that there is a very wide range of intrinsic luminosities among random samples of stars and so, to interpret the star counts, it was necessary to work out the distribution of their luminosities in a typical volume of space – this distribution is known as the *luminosity function* of the stars. By 1920, Kapteyn and Pieter van Rhijn (1886–1960) had determined the luminosity function of stars near the Sun and found that it could be approximated by a Gaussian distribution with mean absolute magnitude $\overline{M} = 7.7$ and half-width a few magnitudes (Kapteyn and van Rhijn, 1920). Assuming this luminosity function applied throughout the Galaxy, they were able to work out the space distribution of stars from star counts in different directions (Kapteyn, 1922). They found that the Galaxy was highly flattened, with dimensions 1500 pc perpendicular to the plane and about eight times that size in the Galactic plane (Figure 5.2).

Kapteyn used this model of the Galaxy to work out the gravitational acceleration which binds the stars to the plane of the Milky Way (Kapteyn, 1922). The distribution of stars perpendicular to the Galactic plane can be taken to be a plane parallel atmosphere in which the number density of stars is described by a Boltzmann distribution perpendicular to the plane, $n = n_0 \exp -(|z|/z_0)$, where z_0 is the scale height of the distribution. In a simple picture, the scale height, z_0, can be related to the gravitational acceleration, g, perpendicular to the Galactic plane, since $z_0 \approx \frac{1}{2} m \langle v_z^2 \rangle / g$, and hence to the mass distribution once the velocity dispersion of stars, $\langle v_z^2 \rangle$, perpendicular to the plane is known. The mass density in the Galactic plane was found to be 10^{-20} kg m^{-3} or $0.15 M_\odot$ pc^{-3}, very close to modern values. This density is often referred to as the *Oort limit* and provides an upper limit to the total mass of stars, interstellar matter and dark objects of all types in the plane of the Galaxy (Oort, 1932).

Meanwhile, Harlow Shapley had adopted a different approach to the determination of Galactic structure. One of the most important methods for measuring astronomical distances was discovered as a result of the systematic studies carried out at the Harvard Observatory's Arequipa observing station in Peru. From 1893 to 1906, the nearby companions of our own Galaxy, the Large and Small Magellanic Clouds, were systematically surveyed photographically by the 24-inch telescope. At Harvard, Henrietta Leavitt (1868–1921), who, like Annie Cannon, was extremely deaf, was assigned the task of finding variable stars in the Magellanic Clouds. She had graduated from Radcliffe College in 1892 and after 1902 became the head of the photographic photometry department of the observatory. Whilst she is best remembered for her work on the Cepheid variables, her main work was the establishment of the North Polar Sequence, the accurate determination of the magnitude scale for stars in a region of sky which would always be accessible to observers in the northern hemisphere. By the time of her death, in 1921, she had extended the North Polar Sequence from 2.7 to 21st magnitude with errors less than 0.1 magnitudes. To achieve this, she used observations from 13 telescopes ranging from 0.5 to 60 inches in diameter and compared her scale using 5 different photographic photometric techniques. From 1915 to 1940, these pioneering efforts were refined and developed almost single-handedly by Frederick Seares (1873–1964) at the Mount Wilson Observatory.

The advantage of studying systems such as the Magellanic Clouds is that, although their absolute distances may not be known, it is safe to assume that all the stars are at the same distances, and hence that the *relative* luminosities of the stars can be found. Leavitt's technique for measuring small variations in the brightnesses of stars was to make a positive plate of the star field; by overlaying this plate over negative plates taken at later epochs, small changes in brightness could be measured rather precisely. Among the 1777 variable stars which she discovered in the Clouds were a number of Cepheid variable stars. These stars were named after the variable star δ Cephei in which the periodic light curve has a distinctive temporal behaviour, the brightness of the star increasing rapidly and then decaying slowly to minimum light. By 1908, she had identified eight Cepheid variables in the Small Magellanic Cloud and noted that the long-period variables had greater luminosities than the short-period variables. In her famous paper of 1912 (Leavitt, 1912), the periods and apparent magnitudes of 25 Cepheid variables were reported and their remarkable period–luminosity relation, which has played such a prominent role in twentieth-century astronomy, was displayed for the first time (Figure 5.3).

This discovery provided a powerful means for measuring astronomical distances because the Cepheid variables are intrinsically luminous stars and their distinctive light curves can be recognised in stars in distant systems. Once the absolute luminosities of the Cepheid variables had been determined from studies of nearby examples, the period–luminosity relation could be calibrated. Therefore, by measuring the period of a Cepheid variable, its absolute luminosity could be found, from which the distance can be estimated from its apparent magnitude using the inverse square law. This procedure was first carried out in 1913 by Hertzsprung, who derived a distance of 10 kpc for the Small Magellanic Cloud, the greatest distance for any astronomical object measured at that time (Hertzsprung, 1913).[4] In fact, this value is five times smaller than the present best estimate for the distance of the Cloud.

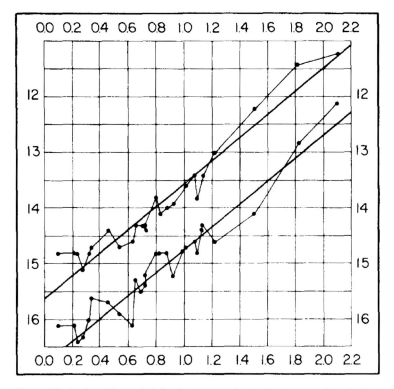

Figure 5.3: A plot of the period–luminosity relation for the 25 Cepheid variables discovered by Leavitt in the Small Magellanic Cloud (Leavitt, 1912). The upper locus is found for the maximum light of the Cepheid variables and the lower line for their minimum brightnesses.

The Cepheid variables were the tools used by Harlow Shapley to determine the structure of the Galaxy through his studies of globular clusters. Globular clusters were among the objects classed among the 'nebulae', but they have a spherically symmetric appearance and can be clearly resolved into individual stars. In contrast to most of the components of the Milky Way, they extend to high Galactic latitudes. By 1918, Shapley had realised that the system of globular clusters provided a means of determining the scale of our Galaxy. Among the stars which could be distinguished in the globular clusters were the Cepheid variables, which he used to establish their distances. The distances found in this way were entirely consistent with estimates based on observations of giant stars and other characteristic stars found in these clusters. The scale of the system of globular clusters was found to be enormous, the most distant globular cluster having a distance of 67 kpc. Furthermore, the globular cluster system was not centred upon the Solar System, but rather most of the globular clusters were found in a direction centred upon the constellation of Sagittarius. Shapley plotted a map of the globular clusters and found that the Solar System is located towards one edge of the globular cluster system (Figure 5.4) (Shapley, 1918). He estimated the distance to the Galactic Centre to be about 20 kpc. Shapley's picture of the Galaxy differed radically from Kapteyn's Sun-centred Universe, Shapley arguing that Kapteyn's studies referred only to the nearby part of the Galactic system.

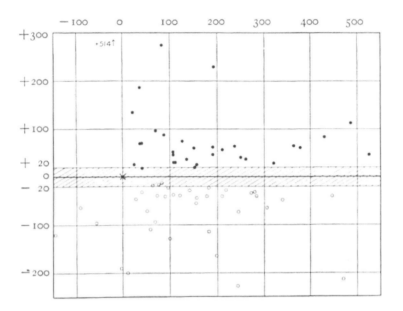

Figure 5.4: The distribution of globular clusters in the Galaxy according to Shapley's distance measurements (Shapley, 1918). The scales on the abscissa and ordinate are in units of 100 pc and correspond to distances in and perpendicular to the Galactic plane, respectively. The Sun, located at zero coordinates on the abscissa and ordinate, lies towards one edge of the globular cluster system.

5.3   The Great Debate

These different approaches to the determination of the size and structure of our Galaxy and the nature of the spiral nebulae led to what came to be known as the *Great Debate*.[5] There were two separate questions to be resolved, the first concerning the scale and structure of our Galaxy and the second concerning the nature of the spiral nebulae. The first question concerned the contrast between Shapley's model of the Galaxy, which had dimensions of at least 60 kpc and in which the Sun was located towards the outer boundary of the system, and the Sun-centred model of Kapteyn, in which the Galaxy has dimensions of 10 kpc. The second question concerned the issue of whether the spiral nebulae were 'island universes', or whether they were constituent members of our own Galaxy.

   In 1917, George W. Ritchey (1864–1945), the brilliant optician of the 60-inch and 100-inch telescopes at Mount Wilson, discovered by chance a nova[6] in the spiral nebula NGC 6946 (Ritchey, 1917). This led to searches through the plate archives of the major observatories for further examples of novae in spiral nebulae. Heber D. Curtis (1872–1942) and Shapley announced the discovery of several other novae, so that, by the end of 1917, 11 novae were known to have taken place in 7 spiral nebulae, 4 of them having been observed in the Andromeda Nebula (Curtis, 1917; Shapley, 1917). Curtis noted that, at maximum, the novae in our Galaxy typically have apparent magnitudes about 5.5, whereas those in the spiral nebulae were about 10 magnitudes fainter. In consequence, if they were the same types of object, the spiral nebulae would have to be 100 times more distant than their Galactic counterparts. Shapley drew the same conclusion, estimating the distance to the Andromeda Nebula to be 50 times the distance of the nearby novae, that is, about 300 kpc.

There were, however, two big problems. The first was that an extraordinarily bright nova had been observed in 1885 in the Andromeda Nebula, M31. It was 6 magnitudes brighter than the novae which had been used to measure the distance to the Andromeda Nebula and so, if it really were at a distance of 300 kpc, the nova of 1885 would have been more than 100 times brighter than the typical nearby novae. If the nova of 1885 were regarded as a typical nova, the distance of the nebula would have been 10 times smaller.

The second problem was that Adriaan van Maanen (1884–1946) claimed to have measured proper motions in the arms of the bright spiral nebula M101 (van Maanen, 1916). Similar results were reported for the galaxies M33, M51 and M81 in 1921 (van Maanen, 1921).[7] The motions seemed to correspond to both rotational and radial motions, van Maanen favouring motions along the spiral arms. His measurements of the rotational components of these motions corresponded to rotation periods about the centres of the nebulae of between 45 000 and 160 000 years. If these spiral nebulae had sizes similar to that estimated by Shapley for our own Galaxy, the speed of rotation would exceed the speed of light. As Shapley remarked,

Measurable internal proper motions, therefore, can not well be harmonised with 'island universes' of whatever size, if they are composed of normal stars.

This was the background to the Great Debate between Shapley and Curtis, which took place at the National Academy of Sciences in Washington on 20 April 1920 (Curtis, 1921; Shapley, 1921).The course of the debate was complex, and the two separate issues identified at the beginning of this section became interwoven. Shapley took the 'scale of the Universe' to mean the size of the globular cluster system, which he found to have dimensions of about 30 kpc. He accepted van Maanen's observations of proper motions in the arms of spiral nebulae, which he assumed must form part of an extensive halo about the Galaxy. He also pointed out that the surface brightnesses of the spiral nebulae are very much greater than the surface brightness of the plane of our Galaxy in the vicinity of the Sun and so it was not evident that the spiral nebulae were the same class of object as our own Galaxy. There was also the question of whether or not individual stars had been resolved in the spiral nebulae. Furthermore, if Shapley's large dimensions for our Galaxy were adopted, then, even if a distance as large as 300 kpc were adopted for the spiral nebulae, our own Galaxy would have been very much larger than the typical spiral nebula and so retain a unique position in the Universe.

Curtis defended the smaller distances inferred from Kapteyn's statistical studies and the 'island universe' picture. He made what turned out to be the correct inference that van Maanen's reported proper motions of the spiral arms of nebulae were spurious[8] and placed considerable weight upon the use of the novae as distance indicators, regarding the nova of 1885 in the Andromeda Nebula as an abnormality. He remarked further:

With one, and only one, exception, all known genera of celestial objects show such a distribution with respect to the plane of our Milky Way, that there can be no reasonable doubt that all classes, save this one, are integral members of our galaxy. We see that all the stars, whether typical, binary, variable, or temporary, even the rarer types, show this unmistakable concentration towards the galactic plane. So also for the diffuse and the planetary nebulae and, though somewhat less definitely, for the globular star clusters.

The one exception is formed by the spirals; grouped about the poles of our galaxy, they appear to abhor the regions of greatest star density. They seem clearly a class apart. *Never* found in the Milky Way, there is no other class of celestial objects with their distinctive characteristics of form, distribution, and velocity in space.

This was the origin of the term *zone of avoidence*, coined by Hubble. As Virginia Trimble has pointed out, the arguments used by Shapley and Curtis were sound, so long as they were discussing the areas in which they were experts (see endnote 5).

There were two problems in reconciling these different pictures. Most serious was the neglect of interstellar extinction, that is, absorption and scattering of light by interstellar dust, which affected Shapley's and Kapteyn's analyses in different ways. Curtis was well aware of the importance of obscuring matter in the discs of spiral nebulae, as revealed by his images of 'a band of absorbing or occulting matter' observed in those spiral nebulae observed edge-on (Curtis, 1918b). Interstellar dust absorption in the plane of the Galaxy was indeed responsible for the observation that the spiral nebulae avoid the Milky Way. The central regions of our Galaxy in fact have very similar surface brightness to those of the spiral nebulae, but interstellar extinction prevents us observing these regions directly in the optical waveband. The second problem was that it cannot be assumed that the local luminosity function of stars necessarily applies throughout the Galaxy.

Gradually, the discrepancies between the two pictures were resolved. Between 1917 and 1919, the Swedish astronomer Knut Lundmark (1889–1958) discovered 22 novae in the Andromeda Nebulae and, if these were assumed to be similar to Galactic novae, a distance of 200 kpc was found (Lundmark, 1920). Lundmark made the distinction between two classes of novae, those used to make the distance measurements belonging to the 'lower class', while novae such as that of 1885 were assigned to the 'upper class' – these were to be identified with 'super-novae' by Baade and Zwicky in 1934 (see Section 4.3).

In 1899, Julius Scheiner had obtained a spectrum of the central regions of M31 in a $7^1/_2$ hour exposure and found that its spectrum was similar to that of the Sun (Scheiner, 1899). In 1921, Lundmark extended this observation, making a detailed spectroscopic study of the spiral arms of M33, as well as some of its brightest stars, and found them to be typical of luminous stars in our Galaxy. If he assumed the brightest stars in the spiral nebulae had absolute magnitude $M = -6$, the distance of M33 would be about 300 kpc. In 1921, he wrote (Lundmark, 1921):

Some objects [in the arms] have a nebular spectrum but most of the objects belonging to the spiral show a strong continuous spectrum without bright lines. It is of course hard to give an accurate spectral type but a solar or somewhat earlier type seems to be predominant. From the spectral evidence, it seems probable that the spiral nebula consists of ordinary stars, clusters of stars, and some nebular [i.e. gaseous] material.

Further evidence for the extragalactic nature of M31 was provided by Ernst Öpik, who used measurements of its rotational velocity by Francis Pease to show that if its mass-to-luminosity ratio were similar to that of stars in our Galaxy, its distance would have to be about 480 kpc, in fact a more accurate estimate than that found subsequently by Hubble (Öpik, 1922).

The conclusive proof of the extragalactic nature of the spiral nebulae was provided by Edwin Hubble (1889–1953) in 1925 (Hubble, 1925). Using the Hooker 100-inch telescope, he discovered 22 Cepheid variables in M33 and 12 in M31. These displayed exactly the same form of period–luminosity relation found for Cepheids in the Magellanic Clouds. He was therefore able to make good distance estimates for the spiral nebulae, which he found to be 285 kpc, much greater than Shapley's largest estimate for the size of our Galaxy.

5.4   Hubble and the Universe of galaxies

The extragalactic nature of the spiral galaxies was established and Hubble immediately began to use the galaxies as tools for studying the large-scale structure of the Universe. He realised that the galaxies provided the means by which fundamental cosmological problems could be addressed by astronomical observation. In the next year, 1926, he published a major study of galaxies which begins with his famous classification scheme, distinguishing between the main classes of galaxies – the ellipticals, normal spirals, barred spirals and irregulars (Hubble, 1926). Elliptical galaxies were ordered according to the ellipticity of their images, and the spirals and barred spirals were divided into subclasses labelled a, b and c according to the tightness of the winding of the spiral structure and the relative importance of the disc and bulge in the distribution of stars in the galaxy. This classification scheme was eventually presented in the form of a 'tuning-fork' diagram published in 1936 (Figure 5.5) (Hubble, 1936). Hubble interpreted the diagram as an evolutionary sequence in which the galaxies were supposed to evolve from spherical elliptical galaxies at the left of the diagram through the sequence of spiral galaxies. This speculation proved to be wholly incorrect, but the terms 'early-type' galaxy and 'late-type' galaxy are still used, reflecting Hubble's original prejudice.

Of particular significance for cosmology was his realisation that the number counts of galaxies brighter than a given apparent magnitude provide a test of the homogeneity of the distribution of galaxies in the Universe. It is a simple calculation to show that, if the galaxies are distributed uniformly in local Euclidean space, the number brighter than limiting apparent magnitude, m, is expected to be $\log N = 0.6m + $ (constant), independent of the luminosity function of the galaxies (see Section A5.1). In 1926, Hubble's galaxy counts extended to 16.7 magnitude, and he found that the number of galaxies increased with increasing apparent magnitude exactly as expected for a uniform distribution. This result was to have profound implications for the construction of cosmological models because it meant that, as a first approximation, the Universe could be taken to be homogeneous on the large scale.

Next, Hubble worked out the typical masses of galaxies, and from this he estimated the mean mass density in the Universe. The value he found was $\rho = 1.5 \times 10^{-28}$ kg m^{-3}. Already in this paper of 1926, Hubble recognised that this figure had cosmological

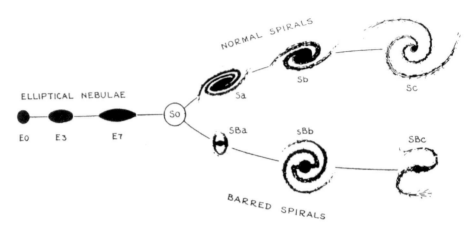

Figure 5.5: Hubble's 'tuning-fork' diagram illustrating the sequence of nebular types. As Hubble noted in the caption to this diagram, which appears in *The Realm of the Nebulae* (Hubble, 1936), 'The diagram is a schematic representation of the sequences of classification. A few nebulae of mixed types are found between the two sequences of spirals. The transitional stage, S0, is more or less hypothetical. The transition between E7 and SBa is smooth and continuous. Between E7 and Sa, no nebulae are definitely recognised.' The S0 galaxies were later recognised in photographic surveys of nearby galaxies and may be thought of as disc galaxies with central bulges but without spiral arms.

significance. Adopting Einstein's static model for the Universe (Einstein, 1917) (see Section 6.2.2), he found that the radius of curvature of the spherical geometry was 27 000 Mpc and that the number of galaxies in this closed Universe was 3.5×10^{15}. In the last paragraph of this 1926 paper, he noted that the 100-inch telescope could observe typical galaxies to about 1/600 of the radius of the Einstein Universe and bright galaxies such as M31 to several times this distance. He concluded this remarkable paper by remarking that (Hubble, 1926)

... with reasonable increases in the speed of plates and sizes of telescopes it may become possible to observe an appreciable fraction of the Einstein Universe.

Thus, by 1926, the first application of the ideas of relativistic cosmology to the Universe of galaxies had been made. It comes as no surprise that in 1928 George Ellery Hale, Director of the Mount Wilson Observatory, began his campaign to raise funds for the construction of the Palomar 200-inch telescope – the study of the Universe of distant galaxies needed the largest telescopes that could be built (Hale, 1928). In the great American tradition of private sponsorship of observational astrophysics, in which the USA had taken a decisive lead, Hale was successful in obtaining a grant of $6 000 000 from the Rockefeller Foundation for the telescope before the year was out.

Before tackling the remarkable story of the discovery of the expanding Universe and the development of theoretical cosmology, let us complete the story of the understanding of Galactic structure and the key role of interstellar extinction.

5.5 The discovery of Galactic rotation

The first clues which were to lead to the discovery of Galactic rotation came from the radial velocity programmes established at the Lick and Mount Wilson Observatories. Most of the radial velocities of the stars were less than about 50 km s^{-1}, but, by 1901, Campbell at the Lick Observatory had noted seven stars with velocities greater than 76 km s^{-1} (Campbell, 1901). In 1914, Adams and Kohlschütter found that among the high-velocity stars with velocities greater than 50 km s^{-1}, those approaching the Solar System outnumbered those moving away from it by a ratio of three to one (Adams and Kohlschütter, 1914a). Their sample included two stars with radial velocities of -325 and -242 km s^{-1}. In 1918, Benjamin Boss (1880–1970) showed that all the stars with negative velocities greater than 75 km s^{-1} lay in the range of Galactic longitude 140° to 340° (Boss, 1918). These results were soon confirmed by Adams and Alfred Joy (1882–1973) and by Gustaf Strömberg (1882–1962), who by 1924 had discovered about 100 high-velocity stars, all within a range of Galactic longitude 143° to 334°, indicating that this system of stars had a highly asymmetric velocity distribution relative to the Sun and the bulk of the nearby stars (Adams and Joy, 1919; Strömberg, 1924). On average, the high-velocity stars moved at about 300 km s^{-1} relative to the Sun. Strömberg pointed out that a possible interpretation of these data was that the system of high-velocity stars was in fact at rest and that the Sun and the nearby stars were moving through it at high velocities.

In 1925, Bertil Lindblad (1895–1965) introduced the idea that the local system of stars is rotating about the Galactic centre (Lindblad, 1925). His model of the Galaxy was similar to Shapley's, but he divided the system of stars into a number of separate subsystems which rotated at different velocities about the Galactic centre, which he identified with the centre of the distribution of globular clusters. In this picture, the high-velocity stars define the local standard of rest for the Galaxy as a whole and the local stellar distribution drifts through it. The nearby nebulae partake in the same motion as the local stars. He deduced a local rotational velocity of 350 km s^{-1}, a Sun-centre distance of 12 kpc and a Galactic mass of $1.8 \times 10^{11} M_\odot$. In a paper of 1927, he went on to show that, if the orbits of stars moving in purely circular orbits about the centre are perturbed, the result is epicyclic motion about the circular orbit (Lindblad, 1927). The epicycles are elongated along the circular orbits, but the major axis of the velocity ellipsoid associated with a system of stars in orbit about the Galactic centre is perpendicular to this direction. The magnitude of the perturbing velocities associated with the epicyclic motion was of the same order as the amplitude of the observed velocity ellipsoid. The beautiful result of this analysis was that it predicted that the major axis of the velocity ellipsoid should be perpendicular to the streaming velocity of the high-velocity stars, which is what was observed.

Following Lindblad's work, Jan Oort (1900–1992) realised that a direct way of testing the hypothesis of Galactic rotation was to look for the effects of differential rotation in the distribution of local stellar velocities. In a model such as Lindblad's, it is expected that the stars should not rotate as a solid body but rather that the effects of differential rotation as a function of distance from the Galactic centre should be detectable. Oort first showed

Table 5.1. *The residual radial velocities of c-stars from data by Schilt used by Oort (1927) to demonstrate the local differential rotation of stars in our Galaxy*

To match the observed sinusoidal variation of the radial velocity residuals, it had to be assumed that the direction of the centre of rotation was in the direction $l = 325°$. If the numbers in the final column are multiplied by 10, good agreement within the limits of the experimental errors is obtained.

Average longitude	Average peculiar velocity (km s^{-1})	Mean error (km s^{-1})	$\sin 2(l - 325°)$
30	+8	±3.5	+0.77
90	−8	±2.7	−0.94
150	0	±3.6	+0.17
210	+10	±3.9	+0.77
270	−7	±4.3	−0.94
330	0	±3.5	+0.17

by a simple geometrical argument that the variation of radial and tangential velocities, v_r and v_t, respectively, observed from the Sun at distance r and Galactic longitude l are given by

$$v_r = Ar \sin 2l, \qquad v_t = Ar \cos 2l + Br, \tag{5.1}$$

where

$$A = \frac{1}{2} \left[\frac{V_0}{R_0} - \left(\frac{dV}{dR} \right)_{R=R_0} \right] \quad \text{and} \quad B = -\frac{1}{2} \left[\frac{V_0}{R_0} + \left(\frac{dV}{dR} \right)_{R=R_0} \right] \tag{5.2}$$

(Oort, 1927).[†] Note that $V(R)$ is the circular velocity at radial distance R from the Centre and that the subscript 0 refers to the orbit of the local standard of rest at the Solar System. Oort's A and B constants contain information about the velocity of the local standard of rest about the Galactic centre, the distance of the Sun from the centre and the local variation of the rotational velocity with distance from the centre.

In his paper of 1927, Oort found clear evidence for this sinusoidal variation of the radial velocities with Galactic longitude in all the sets of data in which it might be expected to be found. The sinusoidal effect was particularly noticeable for the c-stars (Table 5.1). Taking the rotational velocity of the Sun about the Galactic centre to be 272 km s^{-1} from the radial velocities of globular clusters, a new distance estimate for the Galactic centre of 5.9 kpc was found, which was subsequently revised to 5.1 kpc. Note that this analysis gave an independent estimate of the direction of the centre of the Galaxy, which agreed with Shapley's estimate for the centre of the distribution of the globular clusters. Another attractive feature of this picture is that the epicyclic angular frequency, κ, is related to Oort's B constant by $\kappa^2 = -4B\omega_0$, where $\omega_0 = V_0/R_0$ is the angular velocity of the Sun about the Galactic centre. Oort recognised that this model of the Galaxy differed from that

---

[†]   A simple proof of Oort's equations is given in Section A5.2.

deduced by Kapteyn, but he proposed that the discrepancy was likely to be due to interstellar extinction.

5.6 Interstellar matter and extinction by dust

In 1904, Johannes Hartmann (1865–1936) reported the observation that the narrow H and K lines of ionised calcium do not share the periodic displacement of the lines seen in the double star δ Orionis (Hartmann, 1904). He concluded that the absorption must occur along the line of sight to the star. In 1909, Edwin Frost (1866–1935) detected similar stationary lines in several bright, young stars, and Vesto Slipher, in the same year, confirmed Hartmann's findings for binary stars (Frost, 1909; Slipher, 1909). These observations provided the first evidence for diffuse absorbing gas along the line of sight to the stars.

In 1923, John Plaskett (1865–1941) of the Dominion Astrophysical Observatory in Victoria, British Columbia, showed that the velocities of the calcium absorption lines differ from those of their background stars by up to 50 km s^{-1} (Plaskett, 1923). In 1933, he and Joseph Pearce (1893–1988) completed an exhaustive study of the radial velocities of O and B stars (Plaskett and Pearce, 1933). The stars exhibited the expected sinusoidal distribution of radial velocities with Galactic longitude due to differential rotation (equations (5.1)), but in addition the interstellar absorption lines also showed a sinusoidal behaviour with amplitude half that of the stellar absorption lines. Plaskett and Pearce inferred correctly that the interstellar material also partakes in the Galactic differential rotation and is, on average, smoothly distributed along the line of sight to the stars.

It had been suspected that interstellar absorption by dust was an important influence upon the magnitudes and spectra of distant stars, but most observers up till about 1930 preferred to assume that interstellar space was transparent. Kapteyn, for example, could find no definite evidence of extinction in his data. The first conclusive evidence that interstellar extinction could not be ignored came from the analysis of Robert Trumpler (1886–1956) of the properties of open clusters. In a paper published in 1930, he determined the absolute magnitudes of the stars in the clusters from their spectral properties and so was able to measure their distances, knowing their apparent magnitudes (Trumpler, 1930). Employing this procedure, he found that the clusters were systematically larger in physical size with increasing distance as compared with the nearby clusters. Assuming the clusters all had the same physical size, he estimated that the extinction along the line of sight to the clusters amounted to 0.67 mag kpc^{-1}.

An even more compelling estimate of the importance of interstellar extinction was made by Alfred Joy, who published the results of his study of the radial velocities of Cepheid variables in 1939 (Joy, 1939). For these stars absolute magnitudes could be estimated from the period–luminosity relation. He found that the more distant Cepheids were fainter than expected and demonstrated that a result consistent with Galactic rotation could be found, provided the light from the distant stars is attenuated by interstellar dust extinction amounting to 0.85 mag kpc^{-1} (Figure 5.6).

By 1940, Joel Stebbins (1878–1966), Charles Hufford (1894–1981) and Albert Whitford (1905–2002) had used photoelectric photometric techniques to show that

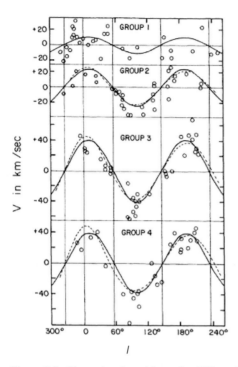

Figure 5.6: Illustrating the evidence for differential rotation within the disc of the Galaxy from Joy's data of 1939 for 156 Cepheid variables (Joy, 1939). The radial velocities, V, are plotted against Galactic longitude, l, for four groups of stars at progressively greater distances from the Sun. The solid lines show the predicted sinusoidal variation of the radial velocities with Galactic longitude assuming Oort's formula, $V = Ar \sin 2l$. A mean value of $A = 21$ km s^{-1} kpc^{-1} was found. Joy used these data to show that the interstellar extinction amounts to 0.85 mag kpc^{-1}.

interstellar extinction was sufficient to reconcile the dimensions of the Galaxy derived from the globular clusters and from the distribution of nearby stars (Stebbins, Hufford and Whitford, 1940). They also showed, however, that the distribution of absorbing dust was patchy and that the standard extinction quoted in the literature of 1 mag kpc^{-1} is only a global average. For precise work, it is necessary to evaluate the extinction separately along each line of sight.

5.7 The Galaxy as a spiral galaxy

The result of these studies was to demonstrate that our Galaxy, although among the more luminous and massive of the spiral galaxies, was by no means exceptional. The obvious question to ask was whether or not our Galaxy possesses spiral arms similar to those observed in the spiral galaxies. The means for studying this question was provided by Walter Baade, who undertook a magnificent set of observations of the Andromeda Nebula and its companion galaxies M32 and NGC 205 in 1943. At that time, Los Angeles and Hollywood were subject to a wartime black-out, which enabled plates of quite exceptional quality to be taken. The acquisition of these plates was a virtuoso performance of observing

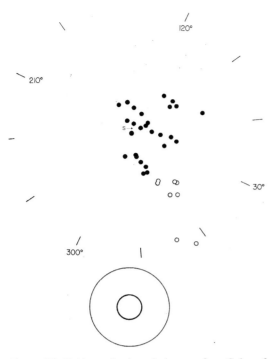

Figure 5.7: Evidence for the spiral arms of our Galaxy from the spatial distribution of associations of bright O and B stars. The position of the Sun is indicated by the letter S and the centre of the Galaxy is shown by the double circle at the bottom of the diagram. This diagram is taken from the paper of Morgan, Whitford and Arthur Code (b.1923) (Morgan, Whitford and Code, 1953), which included better distance measurements for the OB associations than the earlier abstract by Morgan, Sharpless and Osterbrock (1951).

technique, since the focus of the telescope changed during the exposure and Baade worked out how to guide and correct the focus from the comatic image of an off-axis star observed at a magnification of 2800. The importance of these observations was that he was able to resolve not only the brightest blue stars but also the fainter red stars.

In his famous paper of 1944, Baade reported the important discovery that different spectral classes of stars formed different populations within spiral galaxies (Baade, 1944). He divided the stars into two classes. Population I consisted of open star clusters and highly luminous O and B stars which were found exclusively in the regions of spiral arms. In a subsequent paper, he noted that gas clouds and dense regions of interstellar dust were also characteristic of Population I objects and made the correct inference that the young O and B stars had recently been formed in these dusty clouds (Baade, 1951). In contrast, Population II objects comprised the bulk of the old red population of stars which define the central bulge of the Galaxy and the old disc population. The globular clusters and the high-velocity stars are typical members of Population II. The differences between the populations are summarised in Table 5.2.[9] One of the important consequences of these studies was that he also discovered a difference in absolute magnitude between the Cepheid variables belonging to the two populations (Baade, 1952). This was to have important consequences for the cosmological distance scale.

Table 5.2. *These data, taken from Allen (1973), give some impression of the differences in properties of stars belonging to different stellar populations in our Galaxy*

	Population I			Population II	
	Extreme	Older	Older disc	Intermediate	Extreme or halo
General description	very hot blue stars; often associated with HII regions and dust				halo stars, old red stars; globular clusters belong to this population
	spiral arm population	disc population	old disc population		
Properties of stars	←metal-rich stars→			←metal-poor stars→	extremely metal-poor
Heavy-element abundance	0.04	0.02	0.01	0.004	0.001
Ages of stars in units of 10^9 years	≤0.1	0.1–1.5	1.5–5	5–6	>6
Spatial distribution					
Extent perpendicular to Galactic plane (kpc)	120	160	400	700	2000
Mean speed perpendicular to Galactic plane (kpc)	8	10	16	25	75
Axial ratio of stellar distribution	100	50	20	5	2
Smoothness of distribution	very patchy	patchy	smooth	smooth	smooth
Concentration to Galactic centre	little	little	strong	strong	strong

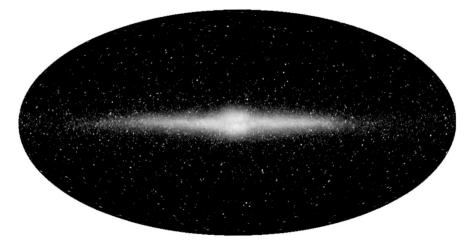

Figure 5.8: The COBE composite infrared image of the Galaxy in an Aitoff projection made from observations at 1.25, 2.2, 3.5 and 4.9μ, showing clearly its disc–bulge structure. Analysis of the photometry of the central bulge has suggested that it is elongated, as expected in a barred spiral galaxy (Dwek *et al.*, 1995).

In 1951, William Morgan (1906–1994), Stewart Sharpless (b. 1926) and Donald Osterbrock (b. 1924) published their study of the distribution of O and B stars within 2 kpc of the Sun and found clear evidence for 'spiral arms', the Sun lying on the inner edge of the local spiral arm (Figure 5.7) (Morgan, Sharpless and Osterbrock, 1951).

The final question in this saga is whether our Galaxy is a normal or a barred spiral. This is not a trivial question because of the effects of interstellar absorption and the fact that our Solar System is located within the disc of the Galaxy. The most recent infrared images of the Galaxy obtained by the COBE satellite, which are free from the effects of interstellar extinction, show clearly the disc and bulge structure of the old stars in our Galaxy (Figure 5.8). The same observations have suggested that the central bulge of old stars may be somewhat ellipsoidal, and so it may well be that we actually live in a barred spiral galaxy (Dwek *et al.*, 1995).

Notes to Chapter 5

1 Edward Harrison gives a delightful brief survey of these ideas in his text *Cosmology* (Cambridge: Cambridge University Press, 2001). Many of these early ideas are also reviewed by John North in his important book *The Measure of the Universe* (Oxford: Clarendon Press, 1965). Detailed studies of the contributions of Wright, Kant and Lambert are included in Section B, A century of speculative cosmologies, of Michael Hoskin's *Stellar Astronomy* (Chalfont St Giles, Buckinghamshire: Science History Publications, 1982).

2 The compilation of Messier's catalogue and how amateur astronomers can observe the objects listed are included in S. J. O'Meara's book, *Deep Sky Companions: The Messier Objects* (Cambridge: Cambridge University Press, 1998). Note that further objects were added to Messier's original catalogue, the list finally totalling 110 objects.

3 The most up-to-date catalogue of the brightest galaxies in the sky was published by the de Vau-couleurs and their colleagues: G. de Vaucouleurs, A. Vaucouleurs, H. G. Corwin, Jr., R. J. Buta, G. Paturel and P. Fouque, *Third Reference Catalogue of Bright Galaxies: Containing Informa-tion on 23,024 Galaxies With Reference to Papers Published Between 1913 and 1988* (Berlin: Springer-Verlag, 1991).

4 There was a numerical error in Hertzsprung's paper in which a distance of only 3000 light years, about 1 kpc, for the Magellanic Clouds was quoted. This was still a large enough distance to place the Clouds well above the plane of the Galaxy.

5 This story is told in R. Berendzen, R. Hart and D. Seeley, *Man Discovers the Galaxies* (New York: Science History Publications, 1976) and R. W. Smith, *The Expanding Universe: Astronomy's 'Great Debate' 1900–1931* (Cambridge: Cambridge University Press, 1982). As a number of authors have noted, the Great Debate is a somewhat exaggerated title for the two half-hour talks presented at the National Academy of Sciences (see M. A. Hoskin, The 'Great Debate': what really Happened, *Journal of the History of Astronomy*, **7**, 1976, 169–182 and V. Trimble, The 1920 Shapley–Curtis discussion: background, issues, and aftermath, *Publications of the Astronomical Society of the Pacific*, **107**, 1995, 1133–1144).

6 Novae, or 'new stars', are stellar explosions in which the luminosity of the star increases suddenly by a factor of hundreds to a million. It remains at about this luminosity for about three days to several months, after which it returns to its pre-nova luminosity. Novae in our Galaxy occur about four times per year. The explosions are associated with mass transfer in binary systems which contain a white dwarf. As the mass transfer onto the white dwarf continues, the surface temperature increases until hydrogen burning takes place in a thermonuclear runaway explosion.

7 Van Maanen continued to publish a series of papers describing these motions in a number of galaxies during the 1920s.

8 In fact, it took quite some time and effort to demonstrate that van Maanen's results were incorrect. The problem was that he was attempting to measure the proper motions of diffuse structures, which are notoriously difficult observations and very sensitive to the observing conditions. Van Maanen's proper motions were only finally laid to rest by Hubble's observations of 1935 (Hubble, 1935). Van Maanen retracted his claims as overestimates of the magnitudes of the proper motions in the same year (van Maanen, 1935).

9 Allen's compendium of astronomical data was a valuable resource and represented the state of knowledge in the 1960s and 1970s: C. W. Allen, *Astrophysical Quantities*, 3rd edn (London: Athlone Press, 1973).

A5   Explanatory supplement to Chapter 5

A5.1   Euclidean number counts of galaxies

In his famous monograph *The Realm of the Nebulae*, Edwin Hubble used counts of galax-ies to the limit of the Mount Wilson 100-inch telescope to demonstrate that, overall, the distribution of galaxies is homogeneous on the large scale (Hubble, 1936). The argument goes as follows.

Suppose the galaxies have a luminosity function $n(L)\,dL$ and that they are uniformly distributed in Euclidean space. The numbers of galaxies with flux densities greater than different limiting values, S, in a particular solid angle Ω on the sky is denoted $N(\geq S)$. Consider first galaxies with luminosities in the range L to $L + dL$. In a survey to a limiting flux density S, these galaxies can be observed out to some limiting distance r, given by the inverse square law, $r = (L/4\pi S)^{1/2}$. The number of galaxies brighter than S is therefore the number of galaxies within distance r in the solid angle Ω:

$$N(\geq S, L)\, dL = \frac{\Omega}{3} r^3 n(L)\, dL.$$

Therefore, substituting for r, the number of galaxies brighter than S is given by

$$N(\geq S, L)\, dL = \frac{\Omega}{3}\left(\frac{L}{4\pi S}\right)^{3/2} n(L)\, dL.$$

Integrating over the luminosity function of the galaxies, we obtain

$$N(\geq S) = \frac{\Omega}{3(4\pi)^{3/2}} S^{-3/2} \int L^{3/2} n(L)\, dL,$$

that is, $N(\geq S) \propto S^{-3/2}$, independent of the luminosity function $n(L)$. The result $N(\geq S) \propto S^{-3/2}$ is known as the *Euclidean number counts* for any class of extragalactic object. In terms of apparent magnitudes, $m = \text{constant} - 2.5 \log_{10} S$, the Euclidean number counts become

$$N(\leq m) \propto 10^{0.6m}, \quad \text{that is,} \quad \log N(\leq m) = 0.6m + \text{constant}.$$

This was the homogeneity test carried out by Hubble with the results shown in Figure 6.2.

A5.2 Oort's A and B constants

Oort's famous paper of 1927 includes an elegant demonstration of how the *rotation curve* for our Galaxy in the vicinity of the Sun can be determined (Oort, 1927). By rotation curve, we mean the variation of the circular velocity, $V(R)$, of stars about the centre of the Galaxy as a function of radial distance, R, from the centre. Oort's argument demonstrated conclusively that the disc of our Galaxy is in a state of differential rotation, provided a revised estimate of the direction of the Galactic centre and enabled an estimate of the distance of the Sun from the Galactic centre to be made.

The geometry of a differentially rotating disc is shown in Figure A5.1. Oort's analysis involved determining the radial and azimuthal components of the velocities of nearby stars as observed from the frame of reference of the Sun at O, which itself is rotating about the Galactic centre. Consider first the radial component, v_r, of stars at S, as observed from O. The projected velocities of the Sun and the stars at S along the direction OS are $V_0 \cos(\pi/2 - l)$ and $V(R) \cos \alpha$, and so, subtracting, yields

$$v_r = V(R)\cos\alpha - V_0 \cos\left(\frac{\pi}{2} - l\right) \tag{A5.1}$$

$$= R\omega(R)\cos\alpha - R_0\omega_0 \sin l, \tag{A5.2}$$

where ω and ω_0 are the angular velocities of the rotating disc about the centre at R and R_0, respectively. Similarly, the azimuthal velocity, v_θ, in the direction of increasing l is the difference of the projected velocities in the direction perpendicular to OS as shown:

$$v_\theta = V(R)\cos(\pi/2 - \alpha) - V_0 \cos l \tag{A5.3}$$

$$= R\omega(R)\sin\alpha - R_0\omega_0 \cos l. \tag{A5.4}$$

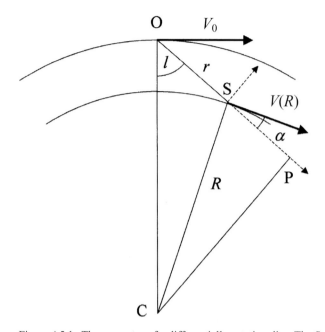

Figure A5.1: The geometry of a differentially rotating disc. The Sun is located at O and the Galactic latitude, l, of a star at distance r from O is indicated. Differential rotation means that the disc of stars does not rotate as a solid body, for which the angular velocity, ω, would be a constant for all radii r. The dashed arrows show the radial and azimuthal directions as observed from O.

Now apply the sine rule to the sides OC and SC of the triangle OCS:

$$\frac{\sin(\pi/2 + \alpha)}{R_0} = \frac{\sin l}{R}; \tag{A5.5}$$

$$R \cos \alpha = R_0 \sin l. \tag{A5.6}$$

Therefore, equation (A5.3) becomes

$$v_r = R_0(\omega - \omega_0) \sin l. \tag{A5.7}$$

Now consider the triangle OCP, where CP is the perpendicular to the radial direction from O, OS. The length OP is given by $R_0 \cos l$ and also $r + R \cos(\pi/2 - \alpha)$. Therefore,

$$R_0 \cos l = r + R \sin \alpha. \tag{A5.8}$$

Therefore, substituting for $R \sin \alpha$ in equation (A5.4), the azimuthal component of the velocity of the stars at S as observed from O is given by

$$v_\theta = R_0(\omega - \omega_0) \cos l - \omega r. \tag{A5.9}$$

So far, the analysis is exact. Oort noted that, for small distances, such that $r \ll R_0$, a simpler relation is found for the expected motions of the stars. Performing a Taylor expansion to first order in r/R_0, we obtain

$$\omega = \omega_0 + \left(\frac{d\omega}{dR}\right)_{R_0} (R - R_0). \tag{A5.10}$$

Therefore, writing $\omega = V/R$, equation (A5.10) becomes

$$\omega - \omega_0 = \frac{1}{R_0^2}\left[R_0\left(\frac{\mathrm{d}V(R)}{\mathrm{d}R}\right)_{R_0} - V_0\right](R - R_0).$$

(A5.11)

For small values of r, $(R_0 - R) \approx r\sin l$, and so we find

$$v_{\mathrm{r}} = \left[\frac{V_0}{R_0} - \left(\frac{\mathrm{d}V(R)}{\mathrm{d}R}\right)_{R_0}\right]r\sin l\cos l,$$

(A5.12)

or

$$v_{\mathrm{r}} = Ar\sin 2l,$$

(A5.13)

where Oort's A constant is given by

$$A = \frac{1}{2}\left[\frac{V_0}{R_0} - \left(\frac{\mathrm{d}V(R)}{\mathrm{d}R}\right)_{R_0}\right].$$

(A5.14)

Similarly, for the azimuthal component of apparent motion,

$$v_\theta = \left[\frac{V_0}{R_0} - \left(\frac{\mathrm{d}V(R)}{\mathrm{d}R}\right)_{R_0}\right]r\cos^2 l - \omega_0 r,$$

(A5.15)

or

$$v_\theta = Ar\cos 2l + Br,$$

(A5.16)

where Oort's B constant is given by

$$B = \frac{1}{2}\left[\frac{V_0}{R_0} + \left(\frac{\mathrm{d}V(R)}{\mathrm{d}R}\right)_{R_0}\right].$$

(A5.17)

Thus, the combination of the radial velocity residuals, v_{r}, and the proper motion residuals, v_θ, enables A and B to be found and hence V_0/R_0 and $(\mathrm{d}V/\mathrm{d}R)_{R_0}$ to be determined. Once V_0 has been determined, the distance of the Galactic centre, R_0, can be found.

6 The origins of astrophysical cosmology

6.1 Physical cosmology up to the time of Einstein

Gravity is the one long-range force which acts upon all matter. Soon after Isaac Newton had completed the unification of the laws of gravity and celestial physics through his discovery of the inverse square law of gravity, he appreciated that the unique form of this law has important consequences for the large-scale distribution of matter in the Universe.[1] In 1692–1693, the cosmological problem was addressed in a remarkable exchange of letters between Newton and the young clergyman Richard Bentley (1662–1742), later to become master of Trinity College, Cambridge. The correspondence concerned the stability of a Universe uniformly filled with stars under Newton's law of gravity.[2] The attractive nature of the force of gravity meant that matter tends to fall together, and Newton was well aware of this problem. His first solution was to suppose that the distribution of stars extends to infinity in all directions so that the net gravitational attraction on any star in the uniform distribution is zero. As he wrote,[3]

The fixt Stars, everywhere promiscuously dispers'd in the heavens, by their contrary attractions destroy their mutual actions.

Newton made star counts to test the hypothesis that the stars are uniformly distributed in space and found that the numbers increased more or less as expected with increasing apparent magnitude. The problem, which was fully understood by Newton and Bentley, was that a uniform distribution of stars is dynamically unstable. If any star is slightly perturbed from its equilibrium position, the attractive force of gravity causes the star to continue to fall in that direction. Newton had to adopt the unsatisfactory assumption that the Universe had been set up and remained in a perfectly balanced state.

During the late eighteenth century, non-Euclidean geometries began to be taken seriously by mathematicians who realised that Euclid's fifth postulate, that parallel lines meet only at infinity, might not be essential for the construction of a self-consistent geometry.[4] The first proposals that the global geometry of space might not be Euclidean were discussed by Girolamo Saccheri (1667–1733) and Johann Lambert. In 1766, Lambert noted that, if space were hyperbolic rather than flat, the radius of curvature of the space could be used as an absolute measure of distance.[5] In 1816, Carl Friedrich Gauss (1777–1855) repeated this proposal in a letter to Christian Gerling (1788–1864) and was aware of the fact that a test of the local geometry of space could be carried out by measuring the sum of the angles of a triangle between three high peaks in the Harz mountains: the Brocken, Hohenhagen and Inselberg. In 1818, Gauss was asked to carry out a geodetic survey of the state of Hanover and he devoted a large effort to carrying out and reducing the data himself. He was certainly aware of the fact that the sum of the angles of the triangle formed by the three Harz Mountains was 180° within the limits of geodetic measurements.[6]

The fathers of non-Euclidean geometry were Nicolai Ivanovich Lobachevsky (1792–1856), who became rector of Kazan University in Russia in 1827, and János Bolyai (1802–1860) in Transylvania, then part of Hungary. In the 1820s, they independently solved the problem of the existence of non-Euclidean geometries and showed that Euclid's fifth postulate could not be deduced from the other postulates (Bolyai, 1832; Lobachevsky, 1829, 1830). In his papers entitled 'On the principles of geometry', Lobachevsky also proposed an astronomical test of the geometry of space. If the geometry were hyperbolic, the minimum parallax of any object would be given by

$$\theta = \arctan\left(\frac{a}{\mathcal{R}}\right), \tag{6.1}$$

where a is the radius of the Earth's orbit and $\mathcal{R}$ is the radius of curvature of the geometry.[7] He found a minimum value of $\mathcal{R} \geq 1.66 \times 10^5$ AU $= 2.6$ light-years. It is intriguing that this estimate was made eight years before Bessel's announcement of the first successful parallax measurement of 61 Cygni. In making this estimate, Lobachevsky used the observational upper limit of 1 arcsec for the parallax of bright stars. In a statement which will warm the heart of observational astronomers, he remarked

There is no means other than astronomical observations for judging the exactness which attaches to the calculations of ordinary geometry.

Non-Euclidean geometry was placed on a firm theoretical basis by Bernhard Riemann (1826–1866) (Riemann, 1854), and the English-speaking world was introduced to these ideas through the works of William Kingdon Clifford (1845–1879) and Arthur Cayley (1821–1895). In 1900, Karl Schwarzschild returned to the problem of the geometry of space and was able to set more stringent limits to its radius of curvature. Repeating Lobachevsky's argument, he found $\mathcal{R} \geq 60$ light-years if space were hyperbolic (Schwarzschild, 1900a). If space were closed, he could set limits to the radius of curvature of the closed geometry because the total volume of the closed space is $V = 2\pi^2 \mathcal{R}^3$. Since there were only 100 stars with measurable parallaxes and at least 10^8 for which no parallax could be measured, he concluded that $\mathcal{R} \geq 2500$ light-years. He also noted that, if space were spherical, it should be possible to observe an image of the Sun in the direction precisely 180° away from its direction on the sky at any time.

Until Einstein's discovery of the general theory of relativity, considerations of the geometry of space and the role of gravity in defining the large-scale structure of the Universe were separate questions. After 1915, they were inextricably linked.

6.2   General relativity and Einstein's Universe

The history of the discovery of general relativity is admirably told by Abraham Pais in his scientific biography of Albert Einstein (1878–1955), *Subtle is the Lord ... The Science and Life of Albert Einstein*,[8] where many of the technical details of the papers published in the period 1907 to 1915 are discussed. Equally recommendable is the survey by John Stachel of the history of the discovery of both theories of relativity.[9] In seeking a fully self-consistent relativistic theory of gravity, Einstein was entering uncharted territory, and for many years he ploughed a lone furrow, making the ultimate spectacular success of the theory in 1915 all the more remarkable.

6.2.1   *Einstein's route to general relativity*

It is simplest to quote Einstein's words from his Kyoto address of December 1922 (Einstein, 1922b).

In 1907, while I was writing a review of the consequences of special relativity, ... I realised that all the natural phenomena could be discussed in terms of special relativity except for the law of gravitation. I felt a deep desire to understand the reason behind this ... It was most unsatisfactory to me that, although the relation between inertia and energy is so beautifully derived [in special relativity], there is no relation between inertia and weight. I suspected that this relationship was inexplicable by means of special relativity.

In the same lecture, he remarked

I was sitting in a chair in the patent office in Bern when all of a sudden a thought occurred to me: 'If a person falls freely he will not feel his own weight.' I was startled. This simple thought made a deep impression upon me. It impelled me towards a theory of gravitation.

In his comprehensive review of the special theory of relativity published in 1907, Einstein devoted the whole of the last section, Section V, to 'The principle of relativity and gravitation' (Einstein, 1907). In the very first paragraph, he raised the following question:

Is it conceivable that the principle of relativity also applies to systems that are accelerated relative to one another?

He had no doubt about the answer and stated the *principle of equivalence* explicitly for the first time:

... in the discussion that follows, we shall therefore assume the complete physical equivalence of a gravitational field and a corresponding acceleration of the reference system.

From this postulate, he derived the time-dilation formula in a gravitational field:

$$\mathrm{d}t = \mathrm{d}\tau \left(1 + \frac{\Phi}{c^2} \right), \tag{6.2}$$

where Φ is the gravitational potential, recalling that Φ is always negative, τ is proper time and t is the time measured at zero potential. Then, applying Maxwell's equations to the propagation of light in a gravitational potential, he found that the equations are form-invariant, provided the speed of light varies as

$$c(r) = c \left[1 + \frac{\Phi(r)}{c^2} \right],$$

(6.3)

according to an observer at zero potential. Einstein realised that, as a result of Huyghens' principle, or equivalently Fermat's principle of least time, light rays are bent in a non-uniform gravitational field. He was disappointed to find that the effect was too small to be detected in any terrestrial experiment.

Einstein published nothing on gravity and relativity until 1911, although he was undoubtedly wrestling with these problems through the intervening period. In his paper of that year, he reviewed his earlier ideas, but noted that the gravitational dependence of the speed of light would result in the deflection of the light of background stars by the Sun (Einstein, 1911). Applying Huyghens' principle to the propagation of light rays with a variable speed of light, he found the standard 'Newtonian' result that the angular deflection of light by a mass M would amount to

$$\Delta\theta = \frac{2GM}{pc^2},$$

(6.4)

where p is the collision parameter of the light ray, which in this case is just the radius of the Sun. The deflection amounts to 0.87 arcsec, although Einstein estimated 0.83 arcsec. Einstein urged astronomers to attempt to measure this deflection. Intriguingly, equation (6.4) had been derived by Johann von Soldner (1776–1833) in 1801 on the basis of the Newtonian corpuscular theory of light (Soldner, 1804).[10]

Following the famous Solvay conference of 1911, Einstein returned to the problem of incorporating gravity into the theory of relativity, and, from 1912 to 1915, his efforts were principally devoted to formulating the relativistic theory of gravity. It was to prove to be a titanic struggle. In summary, his thinking was guided by four ideas.

- The influence of gravity on light.
- The principle of equivalence.
- Riemannian space-time.
- The principle of covariance.

During 1912, Einstein realised that he needed more general space-time transformations than those of special relatively. Two quotations illustrate the evolution of his thought. The first is from Einstein (1912):

The simple physical interpretation of the space-time coordinates will have to be forfeited, and it cannot yet be grasped what form the general space-time transformations could have.

The second is from Einstein (1922b):

If all accelerated systems are equivalent, then Euclidean geometry cannot hold in all of them.

Towards the end of 1912, Einstein realised that what was needed was non-Euclidean geometry. From his student days, he vaguely remembered Gauss's theory of surfaces, which had been taught to him by Karl Friedrich Geiser (1843–1934). Einstein consulted his old school friend, the mathematician Marcel Grossmann (1878–1936), about the most general forms of transformation between frames of reference for metrics of the form

$$ds^2 = g_{\mu\nu}\, dx^\mu\, dx^\nu. \tag{6.5}$$

Although outside Grossmann's field of expertise, he soon came back with the answer that the most general transformation formulae were the Riemannian geometries, but that they had the 'bad feature' that they are non-linear. Einstein instantly recognised that, on the contrary, this was a great advantage, since any satisfactory theory of relativistic gravity must be non-linear.

The collaboration between Einstein and Grossmann was crucial in elucidating the features of Riemannian geometry, which were essential for the development of the general theory of relativity, Einstein fully acknowledging the central role which Grossmann had played. At the end of the introduction to his first monograph on general relativity, Einstein wrote (Einstein, 1916a)

Finally, grateful thoughts go at this place to my friend the mathematician Grossmann, who by his help not only saved me the study of the relevant mathematical literature but also supported me in the search for the field equations of gravitation.

The Einstein–Grossmann paper of 1913 was the first exposition of the role of Riemannian geometry in the search for a relativistic theory of gravity (Einstein and Grossmann, 1913a,b). The details of Einstein's struggles over the next three years are fully recounted by Pais. It was a huge and exhausting intellectual endeavour which culminated in the presentation of the theory in its full glory in November 1915 (Einstein, 1915, 1916b). In that month, Einstein discovered that he could account precisely for the perihelion shift of the planet Mercury.

In 1859, Urbain Le Verrier (1811–1877) had discovered that, once account was taken of the influence of the planets, there remained an unexplained component of the advance of the perihelion of Mercury's elliptical orbit about the Sun, amounting to about 40 arcsec per century (Le Verrier, 1859). In a feat of extraordinary technical virtuosity, Einstein showed in November 1915 that the advance of the perihelion of Mercury expected according to the general theory of relativity amounted to 43 arcsec per century, a value in excellent agreement with the present best estimates. He knew he must be right.

The theory also predicted the deflection of light by massive bodies because of the curvature of space-time in their vicinity. For the Sun, the predicted deflection of light rays from stars just grazing the limb of the Sun amounted to 1.75 arcsec. This deflection is a factor of two greater than that expected according to a Newtonian calculation (see above and endnote 10). This prediction resulted in the famous eclipse expeditions of 1919 led by Arthur Eddington and Andrew Crommelin (1865–1939) (Dyson, Eddington and Davidson, 1920). The Astronomer Royal, Frank Dyson (1868–1939), had long realised that the eclipse of 1919 would take place under the most advantageous conditions and had begun planning accordingly. Not only was the totality of the eclipse unusually long, about six minutes, but also the Sun would then be observed against the background of the Hyades star cluster, providing many bright target stars for the deflection experiments.

The eclipse of 29 May 1919 passed over Northern Brazil, across the Atlantic Ocean through the island of Principe and then across Africa. The British Government awarded a grant of £1100 to enable two expeditions to be made to photograph the eclipse, one to Sobral in Northern Brazil, led by Crommelin, and one to Principe, led by Eddington. The results were in agreement with Einstein's prediction, the Sobral result being 1.98 ± 0.16 arcsec and the Principe result 1.61 ± 0.4 arcsec.[11] These results were widely publicised, and Einstein's reputation was established in the public mind as the epitome of scientific genius.

The theory also predicted the gravitational redshift of light originating close to massive compact objects. As already described in Section 4.2, Adams' careful observations of the spectrum of the white dwarf Sirius B in 1925 showed a gravitational redshift amounting to a Doppler shift of 19 km s^{-1}, in precise agreement with the expectations of general relativity (Adams, 1925a). Thus, by the mid 1920s, the theory had triumphantly passed the three tests proposed by Einstein.[12]

6.2.2 Einstein's Universe

In 1916, the year after the discovery of the general theory of relativity was announced, Willem de Sitter (1872–1934) and Paul Ehrenfest (1880–1933) suggested in correspondence that a spherical four-dimensional space-time would eliminate the problem of the boundary conditions at infinity, which pose insuperable problems for Newtonian cosmological models.[13] In 1917, Einstein published his famous paper in which he derived a static closed model for the Universe which seemed to resolve the problems inherent in Newtonian cosmological models (Einstein, 1917).

Einstein's standard field equations can be written in the form

$$R_{mn} - \tfrac{1}{2} g_{mn} R = -\kappa T_{mn}, \tag{6.6}$$

where R_{mn} is the Ricci tensor, g_{mn} is the metric tensor, T_{mn} is the energy-momentum tensor, R is the contracted Ricci tensor and $\kappa = 8\pi G/c^2$. A strict relativist would adopt the point of view that the concept of gravitational force is unnecessary and the full content of the theory can only be appreciated in terms of the bending of the Riemannian geometry of space-time throughout the Universe. For illustrative purposes, we can work with the Newtonian analogue of equation (6.6) which becomes Poisson's equation,

$$\nabla f = -4\pi G\rho, \tag{6.7}$$

where f is the body force per unit mass due to gravity. The equivalence with the general relativistic formulation can be seen by writing $f = -\nabla\Phi$, where Φ is the (negative) gravitational potential, so that $\nabla^2\Phi = 4\pi G\rho$. The metric coefficients, g_{mn}, thus play the role of 'gravitational potentials' in general relativity.

Einstein realised that, in general relativity, he had a theory which could be used to construct models of the Universe as a whole. His motivation for taking this problem seriously was his objective of incorporating what he designated *Mach's principle* into the structure of general relativity. By Mach's principle, he meant that the local inertial frame of reference should be determined by the frame of reference of the distant stars. There were two obstacles to constructing self-consistent physical models. The first was that the static Newtonian model was unstable, in the sense that, even in the case of an infinite distribution of stars, local regions would collapse under gravity. The second problem concerned the boundary conditions at infinity. Einstein proposed to solve all these problems at one fell swoop by introducing an additional term into the field equations (6.6), the famous *cosmological constant*,[14] λ. The equations become

$$R_{mn} - \tfrac{1}{2} g_{mn} R - \lambda g_{mn} = -\kappa T_{mn}. \tag{6.8}$$

The corresponding modification to Poisson's equation would be given by

$$\nabla f = -4\pi G\rho + \lambda. \tag{6.9}$$

Note the key point that the gravitational force depends upon the density of the medium but the cosmological term is independent of density and is proportional to distance, $f = \tfrac{1}{3}\lambda r$. Inspecting equation (6.9), it can be seen that a static solution exists with constant gravitational potential Φ, $f = -\nabla \Phi = 0$, and

$$\lambda = 4\pi G\rho_0, \tag{6.10}$$

where ρ_0 is the density of the static Universe. Since λ is positive, the geometry of the Universe is closed and the radius of curvature of the geometrical sections is $\mathcal{R} = c/(4\pi G\rho_0)^{1/2}$. This geometry eliminated the problem of the boundary conditions at infinity since this model Universe is finite and closed. The volume of the spherical geometry is given by $V = 2\pi^2 \mathcal{R}^3$ and there is a finite number of galaxies in the Universe. Furthermore, Einstein believed he had incorporated Mach's principle into general relativity. The essence of the argument was that static solutions of the field equations did not exist in the absence of matter. In other words, according to equation (6.10), if $\lambda = 0$, $\rho_0 = 0$ and so, in the absence of the stabilising term λ, the only solution is a completely empty Universe. The cosmological constant was essential in creating a static closed model of the Universe with finite density.

This was the first fully self-consistent cosmological model, but it had been achieved at the cost of introducing the cosmological constant. This was to remain a thorn in the flesh of cosmologists from the time of its introduction in 1917, until the last few years of the twentieth century, when it came into its own. Einstein was somewhat uncomfortable about its introduction, acknowledging that the term was 'not justified by our actual knowledge of gravitation' but was merely 'logically consistent'.

In 1919, Einstein realised that a term involving the cosmological constant would appear in the field equations of general relativity, quite independent of its cosmological significance (Einstein, 1919). In the derivation of the field equations, the λ term appears as a constant of integration, which is normally set equal to zero in the development of standard general relativity. The significance of the cosmological constant can be appreciated by inspection of equation (6.9), which shows that, even if there is no matter present in the Universe, $\rho = 0$, there is still a repulsive force acting on a test particle. As Yakov Borisovich Zeldovich (1914–1987) remarked, the term corresponds to the 'repulsive effect of a vacuum' (Zeldovich, 1968). This type of force has no physical meaning according to classical physics. With the development of quantum field theory, the concept of the vacuum changed dramatically, but this is running far ahead of our story.

For most of the twentieth century, cosmologists adopted ambivalent views about the λ term. In 1919, Einstein was not enthusiastic about the term, remarking that it 'detracts from the formal beauty of the theory'. Willem de Sitter had similar views (de Sitter, 1917b) and wrote in 1919 that the term

detracts from the symmetry and elegance of Einstein's original theory, one of whose chief attractions was that it explained so much without introducing any new hypotheses or empirical constant.

Others regarded it as a constant which appears in the development of general relativity and its value should be determined by astronomical observation. Throughout the twentieth century, the cosmological constant made regular reappearances in the literature in response to various cosmological problems, and these will be recounted in the course of this chapter and in Part V. Suffice to say that none of these arguments withstood detailed scrutiny until, in the last decade of the twentieth century, compelling evidence for a non-zero value of the cosmological constant was found (see Chapters 13 and 15).

6.3   De Sitter, Friedman and Lemaître

In the same year that Einstein's first paper on cosmology was published, de Sitter showed that one of Einstein's objectives had not been achieved (de Sitter, 1917a). He found solutions of Einstein's field equations in the absence of matter, $\rho = p = 0$, and derived the following metric for isotropic world models with constant space curvature $\kappa = \mathcal{R}^{-2}$:

$$ds^2 = -dr^2 - R^2 \sin^2 \left(\frac{r}{\mathcal{R}}\right)(d\psi^2 + \sin^2 \psi \, d\theta^2) + \cos^2 \left(\frac{r}{\mathcal{R}}\right) c^2 \, dt^2. \tag{6.11}$$

Thus, although there is no matter present in the Universe, a test particle still has a perfectly well defined geodesic along which it can travel. As de Sitter asked, 'If no matter exists apart from the test body, has this inertia?' At that time, the principal issues at stake were the origin of inertia and Mach's principle, rather than any thought that these considerations might be of relevance to astronomical observation.

It was soon discovered that this solution could be written in terms of an expanding metric. In 1922, Cornelius Lanczos (1893–1974) showed that the de Sitter solution could be written alternatively in the form of a metric in which the test particles move apart at an exponentially increasing rate (Lanczos, 1922). To achieve this he separated the spatial and time components of the metric (6.11), so that it became

$$ds^2 = -dt^2 + \cosh^2 t [dr^2 + \cos^2 r (d\theta^2 + \cos^2 \theta \, d\phi^2)]. \tag{6.12}$$

Lanczos added the remark which has since become a platitude:

It is interesting to observe how one and the same geometry can appear with quite different physical interpretations ... according to the interpretation placed upon the particular coordinates.

At almost exactly the same time, Alexander Alexandrovich Friedman (1888–1925) published the first of two classic papers for both static and expanding world models (Friedman, 1922, 1924). Friedman[15] was a brilliant mathematician whose principal interests were the application of fluid and gas dynamics to meteorology. In 1922, he was employed by the Main Geophysical Observatory in Petrograd and led the theoretical division in studies of the physics of the atmosphere. As a result of the Soviet Revolution of 1917, the Civil War and the subsequent blockade of the Soviet Union, there was considerable delay before Soviet scientists became aware of Einstein's general relativity. Friedman was one of the first to appreciate fully the significance of the theory and sent his book *The World as Space and Time* to the publishers in 1922 (Friedman, 1923).

In the first of his classic papers published in 1922, Friedman wrote down explicitly the general equations for the dynamics of homogeneous isotropic world models. These can be written in an exactly equivalent form in terms of the *scale factor R* which, in the notation I will use, is the function which describes how the distance between any two points in the expanding Universe changes with time, R being normalised to the value unity at the present epoch. For the case of a uniform isotropic model, the field equations reduce to the following two equations:

$$\ddot{R} = -\frac{4\pi G R}{3}\left(\rho + \frac{3p}{c^2}\right) + \frac{1}{3}\lambda R; \tag{6.13}$$

$$\dot{R}^2 = \frac{8\pi G\rho}{3}R^2 - \frac{c^2}{\mathcal{R}^2} + \frac{1}{3}\lambda R^2. \tag{6.14}$$

In these equations, ρ is the mean density of the Universe, p is the pressure and $\mathcal{R}$ is the radius of curvature of the geometry of space at the present epoch. Note that the pressure term appears as a relativistic correction to the inertial mass density of the Universe.[16] To recover Einstein's Universe, we set $p = 0$, $\ddot{R} = \dot{R} = 0$ and $R = 1$. We can find Lanczos's solution by setting $p = \rho = 0$ and then, in the limit of large times, the scale factor changes as $R(t) = R_0 \exp[(\lambda/3)^{1/2}t]$, the exact analogue of the cosh t time dependence of the coefficient in front of the spatial part of the metric, equation (6.12).

Friedman explored the solutions of equations (6.13) and (6.14) for a variety of special cases, including the closed-world model in which the Universe eventually collapses back to a singular state. In 1925 Friedman died of typhoid in Leningrad before the fundamental significance of his work was appreciated. The neglect of Friedman's work in these early days is somewhat surprising since his papers appeared in the authoritative journal *Zeitschrift für Physik*. The problem may have been associated with a brief note by Einstein criticising a step in Friedman's paper of 1922 (Einstein, 1922a). Friedman wrote to Einstein pointing out that the criticism was incorrect and Einstein immediately published a brief note accepting that Friedman had not made an error (Einstein, 1923). It was not until Abbé Georges Lemaître (1894–1966) independently discovered the same solutions in 1927, and then became aware of Friedman's contributions, that the pioneering nature of these papers was appreciated. The standard world models of general relativity, with or without the cosmological constant, are nowadays usually referred to as the *Friedman world models*.

The significance of Lemaître's work was that he was seeking solutions of the field equations which avoided the problems which afflicted Einstein's Universe, which was of finite density, closed and static, and de Sitter's Universe, which was open, empty and expanding (Lemaître, 1927). In his independent discovery of the expanding solutions in his important paper of 1927, he ended by remarking, 'We still have to explain the cause of the expansion of the Universe.'

One of the problems facing the pioneers of relativistic cosmology was the interpretation of the space and time coordinates used in their calculations. De Sitter's solution could be written in apparent stationary form or as an exponentially expanding solution. From the metric, de Sitter had shown that a distance–redshift relation must exist for his empty-world model, but it was not clear whether or not this was relevant to the observable Universe. The answer came resoundingly in the affirmative with Hubble's discovery of the velocity–distance relation for galaxies in 1929, which ushered in a new epoch in astrophysical cosmology.

6.4 The recession of the nebulae

In 1917, Vesto Slipher (1874–1969) published a paper in which he reported heroic spectroscopic observations of 25 spiral galaxies made with the Lowell Observatory's 24-inch telescope (Slipher, 1917). He realised that, for the spectroscopy of low-surface-brightness objects such as the spiral nebulae, the crucial factor was the f-ratio, or speed, of the spectrograph camera, not the size of the telescope. Exposures of 20, 40 and even 80 hours were made to secure these spectra. He found that the velocities of the galaxies inferred from the Doppler shifts of their absorption lines were typically about 570 km s^{-1}, far in excess of the velocity of any known Galactic object. Furthermore, most of the velocities corresponded to the galaxies moving away from the Solar System, that is, the absorption lines were shifted to longer (red) wavelengths. This phenomenon became known as the *redshift, z*, of the galaxies, and it is defined by the relation

$$z = \frac{\lambda_o - \lambda_e}{\lambda_e},$$ (6.15)

where λ_e is the emitted wavelength of some spectral feature and λ_o is its observed wavelength.[17] Slipher noted that

This might suggest that the spiral nebulae are scattering but their distribution on the sky is not in accord with this since they are inclined to cluster.

In 1921, Carl Wilhelm Wirtz (1876–1939) searched for correlations between the velocities of the spiral galaxies and other observable properties (Wirtz, 1922) and concluded that, when the data were averaged in a suitable way,

an approximate linear dependence of velocity upon apparent magnitude is visible. This dependence is in the sense that the nearby nebulae tend to approach our galaxy whereas the distant ones move away ... The dependence of the magnitudes indicates that the spiral nebulae nearest to us have a lower outward velocity than the distant ones.

By 1925, Hubble had established the extragalactic nature of the spiral nebulae and, as discussed in Section 5.3, in his remarkable paper of 1926, he set out the basic types of galaxies, their masses and the contribution they made to the mass density of the Universe. He was fully aware of the importance of his estimate of the mean mass density of the Universe since, according to Einstein's Universe, once this was prescribed, all the other properties of the Universe followed immediately.

By 1929, he had assembled data on the distances of galaxies for which velocities had been measured (Hubble, 1929). It is interesting to note the methods used to estimate the distances of the 24 galaxies. The distances of the nearest seven objects, all within 500 kpc, were the best determined and used the Cepheid variable technique; the distances for the next 13 objects were found by assuming that the most luminous stars in the galaxies had an upper limit of absolute magnitude $M = -6.3$; the last four objects, believed to be members of the Virgo cluster, had distances assigned on the basis of the mean luminosities of the nebulae in the cluster. As Hubble acknowledged in *The Realm of the Nebulae*, most of the velocities used in his 1929 paper were due to Slipher. In fact, of the 44 galaxy redshifts known in 1925, 39 of them had been measured by Slipher. From these meagre data, Hubble's famous velocity–distance relation was derived (Figure 6.1(a)). It is intriguing that the main objective of Hubble's paper was not to derive the velocity–distance relation, but rather to use the velocities of the galaxies to derive the velocity of the local standard of rest of the Solar System relative to the extragalactic nebulae.

With hindsight, it is remarkable that Hubble found the redshift–distance relation from such a nearby sample of galaxies, but there was other evidence even at that time. He noted in his brief paper that Milton Humason (1891–1972) had measured a velocity of 3910 km s^{-1} for the galaxy NGC 7619, the brightest galaxy in a cluster. If the velocity–distance relation were correct, the absolute magnitude of this galaxy would be of the same order as those of the brightest galaxies in nearby clusters.

Although Hubble did not write down what is often referred to as the Hubble relation $v = H_0 r$ in this paper, he noted that 'the velocity–distance relation may represent the de Sitter effect'. De Sitter had shown that, according to his static metric, equation (6.11), there would be a redshift of spectral lines which increases with distance. It had already been appreciated, however, that the velocity–distance relation is a natural outcome of uniformly expanding world models. Both Lemaître and Howard Robertson (1903–1961) were aware of the fact that the Friedman solutions result locally in a velocity–distance relation (Lemaître, 1927; Robertson, 1928). Lemaître derived what he termed the 'apparent Doppler effect', in which 'the receding velocities of extragalactic nebulae are a cosmical effect of the expansion of the Universe' with $v \propto r$. Robertson found a similar result stating that 'we should expect . . . a correlation $v \approx (cl/R)$', where l is distance and v the recession velocity. From nearby galaxies, he found a value for what we now know as Hubble's constant, H_0, of 500 km s^{-1} Mpc^{-1}.

The subsequent story is told in Hubble's Silliman Lectures given at Yale University in 1935 and published as the famous and influential monograph *The Realm of the Nebulae* in the following year (Hubble, 1936). The task of extending the measurement of the radial velocities of galaxies to much greater distances was undertaken by Humason using the 100-inch Hooker Telescope at Mount Wilson. By 1935, he had measured the velocities of almost 150 further galaxies out to distances inferred to be 35 times greater than the distance of the Virgo cluster and to radial velocities of 42 000 km s^{-1}, roughly one-seventh of the speed of light. Although distances could not be measured directly, Hubble and Humason found that the luminosity functions of the galaxies in clusters are remarkably similar, and so they used the fifth brightest member of each cluster as a measure of its relative distance (Hubble and Humason, 1934). The resulting redshift–apparent magnitude relation is expected to follow the relation $\log v = 0.2m + \text{constant}$ if the galaxies follow a velocity–distance relation,

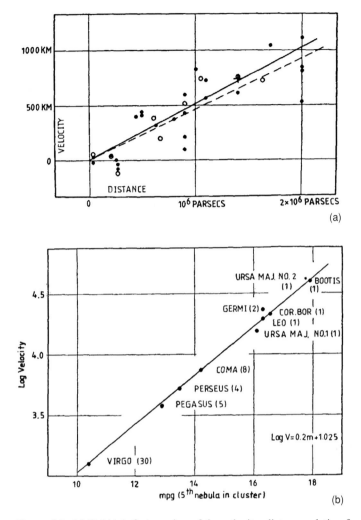

Figure 6.1: (a) Hubble's first version of the velocity–distance relation for nearby galaxies (Hubble, 1929). The filled circles and the full line represent a solution for the solar motion using the nebulae individually; the open circles and the dashed line represent a solution combining the nebulae into groups. The cross is an estimate of the mean distance of the other 20 galaxies for which radial velocities were available. (b) The velocity–apparent magnitude relation for the fifth brightest member of clusters of galaxies, corrected for galactic obscuration (Hubble and Humason, 1934). Each cluster velocity is the mean of the various individual velocities observed in the cluster, the number being indicated by the figure in brackets.

$v \propto r$. The results of these arduous programmes of observation are shown in Figure 6.1(b) and are in excellent agreement with a linear velocity–distance relation. Even today, the apparent magnitude–redshift relation for the brightest galaxies in clusters remains among the most convicting evidence for the extension of Hubble's law to significant cosmological distances.

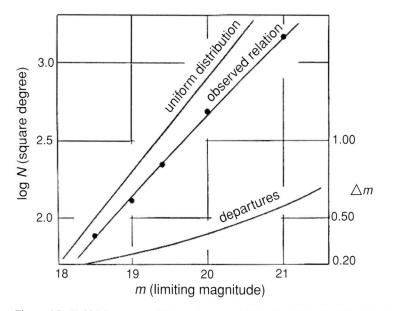

Figure 6.2: Hubble's counts of faint galaxies published in 1936 (Hubble, 1936). The line labelled 'Uniform Distribution' corresponds to the relation $\log N(\leq m) = 0.6m +$ constant. The points represent the observed counts with a best-fitting line shown. At the bottom of the diagram is the difference in magnitude between the uniform counts and the observations, which Hubble interpreted as the effect of redshift.

In Hubble's monograph, he described the number counts of faint galaxies made with the 100-inch telescope, which, by 1935, extended to an apparent magnitude limit of 21, the faintest counts feasible with the 100-inch telescope and the available photographic plates. The counts followed the expected relation $N(\leq m) = 0.6m +$ constant for a uniform distribution of galaxies down to 18th magnitude (see Section A5.1), but at fainter magnitudes fewer galaxies were observed than were expected for a uniform distribution (Figure 6.2). Hubble correctly concluded that the counts extended to such faint magnitudes, and consequently large distances, that the effects of redshift upon the number counts had to be taken into account. He also correctly concluded that the counts were evidence for the overall homogeneity of the Universe as far the 100-inch telescope could observe distant galaxies. In the last pages of the book, he speculated that the convergence of the number counts may be associated with the curvature of space. His conclusion that the Universe must have positive curvature was incorrect, but this can be attributed to the fact that it took some time before the proper relativistic formulation of the relations between observables and the intrinsic properties of galaxies were worked out.[18] His concluding remarks strike a resonance with anyone who has attempted these difficult cosmological observations:

Eventually, we reach the dim boundary – the utmost limits of our telescopes. There, we measure shadows, and we search among the ghostly errors of measurement for landmarks that are scarcely more substantial. The search will continue. Not until the empirical resources are exhausted, need we pass on to the dreamy realms of speculation.

6.5 The Robertson–Walker metric

The discovery of the velocity–distance relation for galaxies acted as a major stimulus to the study of the Friedman models. Of prime importance was the need to place the world models of de Sitter, Friedman and Lemaître on a firm theoretical foundation. There remained confusion about the notions of time and distance to be used in the world models of general relativity because the basis of that theory was that the field equations could be set up in any frame of reference one pleased. The principle of special relativity meant that observers located on galaxies moving relative to one another could not agree on the synchronisation of their clocks. By 1935, the problem was solved independently by Robertson and Arthur Walker (1909–2001) (Robertson, 1935; Walker, 1936).

A key concept was enunciated by Hermann Weyl (1885–1955) in 1923 and is known as *Weyl's postulate* (Weyl, 1923). To eliminate the arbitrariness present in the choice of coordinate frame, Weyl introduced the idea that, to quote Hermann Bondi (1919–2005),

The particles of the substratum (representing the nebulae) lie in space-time on a bundle of geodesics diverging from a point in the (finite or infinite) past.[19]

The most important aspect of this statement is the postulate that the geodesics which represent the world-lines of galaxies do not intersect, except at a singular point in the past. Note that this postulate predates Hubble's discovery of the velocity–distance relation, but follows the pioneering works of Lanczos and Friedman. By the term 'substratum' Bondi meant an imaginary medium which can be thought of as a fluid which defines the kinematics of the system of galaxies. The consequence of Weyl's postulate is that there is only one geodesic passing through each point of space-time, except at the origin. Once this postulate is adopted, it is possible to assign a notional observer to each world-line, known as a *fundamental observer*, and to define a time coordinate, known as cosmic time. Each fundamental observer can be provided with a standard clock which measures proper time along the geodesic, or world-line, of that observer. The clocks can be synchronised at the time when the geodesics were all together at the singular point at the origin, so that *cosmic time* is defined to be the proper time measured by such a fundamental observer.

One further assumption is essential before the framework of the standard models can be derived. This is known as the *cosmological principle* and is the statement that our Galaxy is not located in a privileged or special position in the Universe. In other words, a fundamental observer located on any other galaxy at the same epoch would observe the same large-scale features of the Universe that we observe. The implication is that our Galaxy is located at a typical point in the Universe. We therefore have to decide what large-scale features should be common to all fundamental observers.

One requirement is that all observers should observe the same velocity–distance relation at the same cosmic epoch. A second requirement is that, on large enough scales, the Universe should present the same appearance in all directions and that, on average, matter and radiation should be homogeneously distributed. Hubble's galaxy counts were important empirical evidence that the distribution of galaxies is isotropic and homogeneous, since they followed closely the relation $\log N = 0.6m + (\text{constant})$ expected of a uniform distribution (see Section A5.1). Although there is clustering in the distribution of galaxies on a local scale, it appeared that, on a large enough scale, these irregularities average out and the galaxies are isotropically distributed. It was a simple calculation to show that, in any uniformly expanding homogeneous fluid, an observer located on any particle of the fluid observes the same velocity–distance relation for the relative motion of other particles of the fluid.

Putting these concepts together, Robertson and Walker independently showed that the metric for any isotropically expanding substratum had to have the form

$$ds^2 = dt^2 - \frac{R^2(t)}{c^2} \left[\frac{dr^2}{(1 + \kappa r^2)} + r^2(d\theta^2 + \sin^2 \theta \, d\phi^2) \right] \qquad (6.16)$$

(Robertson, 1935; Walker, 1936). There are a number of important features of this metric, which is appropriately known as the *Robertson–Walker metric*.

- The first is that it can be derived solely from Weyl's postulate, the cosmological principle, the special theory of relativity and the assumptions of isotropy and homogeneity and that the Universe expands isotropically. It contains all the permissible isotropic geometries consistent with the assumptions of isotropy and homogeneity and these are described by the curvature $\kappa = \mathcal{R}^{-2}$, where $\mathcal{R}$ is the radius of curvature of the spatial sections of the isotropic curved space. If κ is positive, the geometry is spherical; if κ is zero, the geometry is flat; if κ is negative the geometry is hyperbolic – no other geometries are allowed. The point of importance is that this metric is of very general form and is correct whatever forces act upon the substratum. In particular, it does not depend upon the dynamics of the Universe being described by general relativity.
- The physics of the expansion has been absorbed into the scale factor, $R(t)$, which, in the form of equation (6.16), takes the value unity at the present epoch. An important result, which can be derived directly from this metric, is the relation between the scale factor, R, and redshift, z: $R = (1 + z)^{-1}$. This formula elucidates the significance of redshift in cosmology – it provides directly the value of the scale factor of the Universe when the radiation was emitted and is independent of the cosmological model.
- Care has to be taken over the definition of the radial coordinate, r. In the above form, r is defined to be the *radial comoving distance coordinate*. Once this metric of isotropic expanding space-time was derived, it was a straightforward task to use it to derive relations between the intrinsic properties of objects and their observed properties.[20]

It was some time before all the subtleties of the analysis were fully appreciated.

6.6   Milne–McCrea and Einstein–de Sitter

The most important solutions for the variation of the scale factor, $R(t)$, with cosmic time are those derived from general relativity, including those in which the cosmological constant is non-zero. The solutions for $R(t)$ can be written down formally as the integral of Friedman's second equation (6.14):

$$t = \int_0^R \frac{\mathrm{d}R}{\frac{8\pi G\rho}{3}R^2 - \frac{c^2}{\mathcal{R}^2} + \frac{1}{3}\lambda R^2}.$$

(6.17)

Note that it is assumed that the pressure, p, can be neglected. There is no simple general closed solution for this equation, and it has been the subject of a great deal of study.

One of the most important contributions to understanding the physical content of the solutions was provided by Milne and McCrea in 1934 (Milne and McCrea, 1934a,b). They showed that, despite the fact that Newtonian mechanics cannot provide a fully self-consistent cosmological model, simple ideas from Newtonian physics can provide insight into the solutions derived from equation (6.14). They realised that the requirements of isotropy and homogeneity are very powerful constraints upon the properties of the models. In the simplest form of their argument, it is supposed that our Galaxy is located at the centre of a uniformly expanding sphere. This is precisely what we, and any other fundamental observer anywhere in the Universe, must observe. Suppose we work out the deceleration of a galaxy of mass m at distance x from our Galaxy. Applying Gauss's theorem for gravity without bothering about the boundary conditions at infinity, we can find the deceleration of the galaxy as follows:

$$m\frac{\mathrm{d}^2x}{\mathrm{d}t^2} = -\frac{4\pi x^3}{3}\frac{Gm\rho}{x^2}.$$

(6.18)

The mass of the galaxy, m, cancels out on either side of the equation – the dynamics refer to the sphere as a whole. Now, we replace the x and ρ by their values at some reference epoch, t_0, by writing $x = R(t)r$, where r is the comoving radial distance coordinate, which is a label attached to a galaxy for all time, and $\rho = \rho_0 R^{-3}(t)$. It is convenient to set $R(t) = 1$ at the present epoch t_0, and then

$$\ddot{R} = -\frac{4\pi GR}{3}\rho = -\frac{4\pi G\rho_0}{3R^2}.$$

(6.19)

The first integral of this equation is given by

$$\dot{R}^2 = \frac{8\pi G\rho}{3}R^2 - C = \frac{8\pi G\rho_0}{3R} - C,$$

(6.20)

where C is a constant of integration. We have derived formulae of exactly the same forms as are found in the full theory. Not unexpectedly, the Newtonian argument cannot cope with the curvature of space, and a relativistic expression for the inertial mass density is needed which includes the pressure of the gas. It turns out that the constant in equation (6.20) involves the curvature of space at the present epoch, $C = c^2\kappa = c^2/\mathcal{R}^2$. It is straightforward to add terms representing the cosmological constant, λ, to equations (6.19) and (6.20) to recover equations (6.13) and (6.14).

The reason this argument works is that, because of the postulates of isotropy and homogeneity, local physics is also cosmic physics and is applicable on the large scale. Every fundamental observer would perform exactly the same calculation and obtain the same answer. The analysis of Milne and McCrea was of considerable importance because it showed that, despite the problems with the boundary conditions at infinity, the Newtonian model can be used successfully on large scales in the Universe, and, in particular, on scales less than the horizon scale, $r \approx ct$, where t is cosmic time, it is perfectly adequate to use Newtonian arguments.

With the discovery of the expansion of the system of galaxies and Hubble's law, Einstein regretted the inclusion of the cosmological constant into the field equations. According to Gamow, Einstein stated that the introduction of the cosmological constant 'was the greatest blunder of my life' (Gamow, 1970). In 1932, Einstein and de Sitter demonstrated that one particularly simple solution of the field equations for an expanding Universe seemed to be in good accord with observations (Einstein and de Sitter, 1932). They noted that, if the cosmological constant is set equal to zero, there is a special solution of the equations in which the spatial curvature is zero, $\kappa = 0$ and $\mathcal{R} \to \infty$, corresponding to Euclidean space sections. This model is often referred to as the *Einstein–de Sitter model*, and it has particularly simple dynamics, $R(t) = (t/t_0)^{2/3}$, where $t_0 = 2/3H_0$ and H_0 is Hubble's constant, the constant of proportionality in the velocity–distance relation. The model has average density at the present epoch, $\rho_0 = 3H_0^2/8\pi G$. This density is often referred to as the *critical density*, and the Einstein–de Sitter model is often referred to as the *critical model*, because it separates the ever-expanding models with open, hyperbolic geometries from the models which eventually collapse to a singularity and which have closed, spherical geometry (Figure 6.3). When Einstein and de Sitter inserted the value of Hubble's constant from Hubble's observations, $H_0 = 500$ km s^{-1} Mpc^{-1}, into the expression for the critical density, they found a mean value for the density of the Universe of 4×10^{-25} kg m^{-3}. Although recognising that this value was somewhat greater than the value derived by Hubble, they argued that the density was of the correct order of magnitude and, in any case, there might well be considerable amounts of what would now be called 'dark matter' present in galaxies and in the Universe at large.

Evidence of dark matter in the Universe was not long in coming. Astronomers of the 1930s had two ways of measuring the masses of individual galaxies. The best procedure was to measure the *rotation curves* of the discs of spiral galaxies, meaning the variation of the rotational speed of the disc as a function of distance from its centre. Since the rotational velocities can be assumed to be in centripetal equilibrium about the centre of the galaxy, the distribution of mass and total mass can be found by equating the centripetal acceleration to the gravitational force needed to maintain that circular velocity. The second, less satisfactory, method was to assume that there is a constant ratio between the masses of galaxies and their luminosities so that, once the mass-to-luminosity ratio of one galaxy, or one class of galaxies, has been determined, the masses of others can be found from their luminosities.

In 1933, Zwicky, working at the Mount Wilson Observatory, made the first studies of rich clusters of galaxies, in particular of the Coma cluster, which is one of the largest regular clusters in the northern sky (Zwicky, 1933, 1937). The method Zwicky used to estimate the total mass of the cluster had been derived by Eddington in 1916 to estimate the masses of star clusters. Using methods familiar in the theory of gases, Eddington derived the *virial theorem*,[21] which relates the total internal kinetic energy, T, of the stars or galaxies in a cluster to the total gravitational potential energy, $|U|$, assuming the system is in a state of

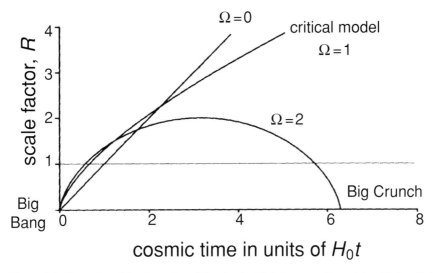

Figure 6.3: Examples of the dynamics of the standard Friedman world models with $\lambda = 0$. The scale factor, $R(t)$, has been normalised to unity at the present epoch. The critical model, with density parameter $\Omega = 1$, separates the re-collapsing models with $\Omega > 1$ from those which expand forever and which have $\Omega \geq 1$. In this presentation, the trajectories of the world models have been scaled to the same value of Hubble's constant at the present epoch, $R = 1$.

statistical equilibrium under gravity (Eddington, 1916a). The kinetic energy can be written $T = \frac{1}{2}M\langle v^2 \rangle$, where $\langle v^2 \rangle$ is the mean-square velocity of the stars, or galaxies, and $|U| = GM^2/2R_{\rm cl}$, where $R_{\rm cl}$ is some suitably defined radius that depends upon the distribution of mass in the cluster. For a cluster of stars or galaxies in statistical equilibrium, Eddington showed that $T = \frac{1}{2}|U|$. Therefore, if the cluster is known to be in statistical equilibrium, the total mass of the cluster can be found from the virial theorem, $M \approx 2R_{\rm cl}\langle v^2 \rangle/G$.

In rich regular clusters of galaxies, such as the Coma cluster, there is convincing evidence from the radial distribution of galaxies within the cluster that they have reached statistical equilibrium and so good estimates can be made of its total mass. Zwicky measured the velocity dispersion of the galaxies in the Coma cluster and found that there was much more mass in the cluster than could be attributed to the visible masses of galaxies. In solar units of $M_\odot/L_\odot$, the ratio of mass-to-optical luminosity of a galaxy such as our own is about 3, whereas for the Coma cluster the ratio was found to be about 500. In other words, there must be about 100 times more dark, or hidden, matter as compared with visible matter in the cluster.

It was some time before Zwicky's results were accepted by the astronomical community,[22] but they have been confirmed by all subsequent studies of rich clusters of galaxies. The nature of the dark matter remains an open and crucial question for physics and cosmology. It is now generally agreed that much of the mass in the Universe is in some form of dark matter, but its nature is unknown.

6.7   Eddington–Lemaître

Despite Einstein's renunciation of the cosmological constant, this was very far from the end of the story, because there remained one grave problem for world models in which the cosmological constant is set equal to zero. It is a simple calculation to show that, if $\lambda = 0$, the age of the Universe must be less than H_0^{-1}. Using Hubble's estimate of $H_0 = 500\,\mathrm{km\,s^{-1}\,Mpc^{-1}}$, the age of the Universe had to be less than 2×10^9 years old, a figure in conflict with the age of the Earth derived from the ratios of abundances of long-lived radioactive species. The present best estimate for the age of the Earth is about 4.6×10^9 years.

Eddington and Lemaître (Eddington, 1930; Lemaître, 1931b) immediately recognised that this problem could be eliminated if the cosmological constant is positive. The effect of a positive cosmological constant is to counteract the attractive force of gravity when the Universe grows to a large enough size. There are special solutions of the integral in equation (6.17) which correspond to the Einstein stationary Universe, but not necessarily at the present epoch. It was possible to find models which had remained in the static Einstein state for an arbitrarily long period in the past and which then began to expand away from that state under the influence of the cosmological term. In this type of *Eddington–Lemaître* model, the age of the Universe could be arbitrarily long. As Eddington expressed it, the Universe would have a 'logarithmic eternity' to fall back on,[23] and so resolve the conflict between estimates of Hubble's constant and the age of the Earth. The dynamics of the Eddington–Lemaître model, and a closely related Lemaître model, are illustrated in Figure 6.4. In the Lemaître model, the Universe does not quite attain a static state, but undergoes a long 'coasting phase' when its velocity of expansion is small and the total age of the model can be much greater than H_0^{-1}.

6.8   The cosmological problem in 1939

Thus, by the end of the 1930s, the basic problems of what I call *classical cosmology* had been clearly identified. The solution of the cosmological problem lay in the determination of the parameters which defined the Friedman world models. This was the goal of the great programmes of observation to be carried out by the 200-inch telescope and the subsequent generation of 4-metre class telescopes. The challenge was to measure the following parameters which characterise the Universe:

- *Hubble's constant,* $H_0 = \dot{R}/R$ at the present epoch, is a measure of the present rate of expansion of the Universe;
- *the deceleration parameter,* $q_0 = -\ddot{R}/\dot{R}^2$ at the present epoch, describes the present deceleration of the Universe, noting that if q_0 is negative, the Universe is accelerating;
- *the curvature of space* $\kappa = \mathcal{R}_0^{-2}$;
- *the mean density of matter in the Universe,* ρ, in particular the question of whether or not it attains the critical density, ρ_{crit};
- *the age of the Universe,* T_0, as given by the integral in equation (6.17);
- *the cosmological constant,* λ.

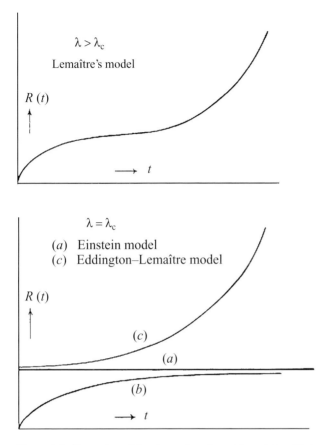

Figure 6.4: Examples of the dynamics of world models with $\lambda \neq 0$. (After H. Bondi, *Cosmology* (Cambridge: Cambridge University Press, 1960), p. 84.) The Einstein static model is illustrated by the model for which $R(t)$ is constant for all time. The Eddington–Lemaître model, in which the Universe expanded from the Einstein static universe in the infinite past, is illustrated, as is a Lemaître model, in which the value of λ is slightly different from that of the static model. In model (b) the Universe approaches the stationary Einstein model from a singular origin at $t = 0$.

These are not independent if the Friedman models are a correct description of the large-scale dynamics of the Universe. For example, for the models which include the cosmological constant,

$$\kappa = \mathcal{R}^{-2} = \frac{(\Omega - 1) + \frac{1}{3}(\lambda/H_0^2)}{(c/H_0)^2}, \quad q_0 = \frac{\Omega}{2} - \frac{1}{3}\frac{\lambda}{H_0^2}, \tag{6.21}$$

where $\Omega = \rho/\rho_{\mathrm{crit}}$ is known as the *density parameter*. Note that, if $\lambda = 0$, there is a simple one-to-one relation between the geometry of the world models, their densities and dynamics, $q_0 = \Omega/2$ and $\kappa = \mathcal{R}^{-2} = (\Omega - 1)/(c/H_0)^2$.

The determination of these parameters has turned out to be among the most difficult observational challenges in the whole of astronomy, and progress was much slower than the optimists of the 1930s must have hoped. In compensation, completely new vistas were to open up after the Second World War as the whole of the electromagnetic spectrum became available for astronomical observation. The precise determination of these parameters only became possible in the very last years of the twentieth century by quite different approaches from those envisaged by the pioneers of the 1930s.

Notes to Chapter 6

1 The history of cosmology and the development of relativistic cosmologies is described in the monographs by J. D. North, *The Measure of the Universe* (Oxford: Clarendon Press, 1965), H. Bondi, *Cosmology*, 2nd edn (Cambridge: Cambridge University Press, 1960) and E. Harrison, *Cosmology: The Science of the Universe* (Cambridge: Cambridge University Press, 2001).

2 A delightful discussion of this correspondence is given by E. R. Harrison, *Darkness at Night: A Riddle of the Universe* (Cambridge: Cambridge University Press, 1987), Chapter 6.

3 This remark appears in the second edition of Newton's *Principia Mathematica* as a result of the exchange of letters with Richard Bentley.

4 An excellent history of the discovery of non-Euclidean geometry is provided by R. Bonola in his book *Non-Euclidean Geometry, and the Theory of Parallels by Nikolas Lobachevsky, with a Supplement Containing the Science of Absolute Space by John Bolyai*, trans. H. S. Carslaw (New York: Dover, 1955).

5 See North, *The Measure of the Universe*, pp. 74–75.

6 It is often stated that Gauss carried out the measurements himself, but this is probably a myth. In his biography of Gauss, Bühler writes as follows (W. K. Bühler, *Gauss: A Biographical Study* (Berlin: Springer-Verlag, 1981), p. 100): 'The often-told story according to which Gauss wanted to decide the question [of whether space is non-Euclidean] by measuring a particularly large triangle is, as far as we know, a myth. The great triangle Hohenhagen–Inselberg–Brocken was a useful control for the smaller triangles which it contains. Gauss was certainly aware of the fact that the error of measurement was well within the possible deviation from 180 degrees from which, under strict conditions, one could have derived the non-Eucludean nature of space.'

7 I have given a simple introduction to isotropic curved spaces in Chapter 17 of Malcolm Longair, *Theoretical Concepts in Physics* (Cambridge: Cambridge University Press, 2003).

8 A. Pais, *Subtle is the Lord . . . The Science and Life of Albert Einstein* (Oxford: Clarendon Press, 1982).

9 J. Stachel, History of relativity, in *Twentieth Century Physics*, vol. 1, eds L. M. Brown, A. Pais and A. B. Pippard (Bristol: Institute of Physics Publishing and New York: American Institute of Physics Press, 1995), pp. 249–356.

10 In fact, Henry Cavendish (1731–1810) appears to have come to essentially the same result as von Soldner in an unpublished manuscript of about 1784, inspired by John Michell's paper of the previous year on the escape of light from a massive body (Michell, 1784). Clifford Will provides an intriguing comparison of how Cavendish may have derived his result with that of von Soldner in C. Will, Henry Cavendish, Johann von Soldner, and the deflection of light, *American Journal of Physics*, **56**, 1988, 413–415.

11 An account of the trials and tribulations associated with these famous expeditions and the analysis of the results is presented by P. Coles, Einstein, Eddington and the 1919 eclipse, in *Historical Development of Modern Cosmology*, ASP Conference Series 252, eds V. J. Martínez, V. Trimble and M. J. Pons-Bordeía (San Francisco: ASP, 2001), pp. 21–41.

12 A fourth test of general relativity was proposed by Irwin Shapiro (b. 1929) in 1964 (Shapiro, 1964). There is a small time delay for electromagnetic waves which pass by the limb of the Sun at grazing incidence; this is because, according to an observer at infinity, the speed of light is variable in a changing gravitational field. The time delay amounts to only 200 μs for the round trip

to Mars and back, but very precise time delays within the gravitational field of the Sun have now been measured and are in agreement with the predictions of standard general relativity within the accuracy of the measurements, which is within 0.2% of the expected delay.

13 See the discussion in North, *The Measure of the Universe*, p. 80.

14 In Einstein's original paper of 1917, the cosmological constant is written as a lower-case λ. It is common practice to write the term as upper-case Greek Λ nowadays.

15 See E. A. Tropp, V. Ya. Frenkel and A. D. Chernin, *Alexander A. Friedman: The Man Who Made the Universe Expand* (Cambridge: Cambridge University Press, 1993). Note that I prefer to spell Friedman's name with one 'n' since he was Russian, not German.

16 I have given a detailed discussion of these equations and their solutions in Malcolm Longair, *Galaxy Formation* (Berlin: Springer Verlag, 1998).

17 An unfortunate convention adopted by the early cosmologists was to convert the redshift, which is a splendid dimensionless number, into a velocity by multiplying by the speed of light. As discussed in Section 6.5, in all isotropic, homogeneous world models, the cosmological redshift is directly related to the scale factor, R, at which the radiation was emitted through the relation $R = 1/(1 + z)$.

18 The correct results for any cosmological model are given in Longair, *Galaxy Formation*, Section 17.2.

19 See H. Bondi, *Cosmology*, p. 100.

20 See, for example, Longair, *Galaxy Formation*, Section 5.5 or Longair, *Theoretical Concepts in Physics*, Section 19.4.

21 I have given a simple proof of this result for clusters of stars and galaxies in Section 3.4 of Longair, *Galaxy Formation*.

22 Zwicky was a remarkable and original astronomer who ploughed his own furrow and, in the process, rubbed up many distinguished astronomers the wrong way. A classic example of his vitriolic style can be found in the Preface to F. Zwicky, *Catalogue of Selected Compact Galaxies and Post-eruptive Galaxies* (Guemlingen, Switzerland: F. Zwicky, 1968).

23 See North, *The Measure of the Universe*, p. 125.

Part III

The opening up of the electromagnetic spectrum

7 The opening up of the electromagnetic spectrum and the new astronomies

7.1 Introduction

Until 1945, astronomy meant optical astronomy. The commissioning of the Palomar 200-inch telescope in 1949 highlighted the dominance of the USA in observational astrophysics in the period immediately after the Second World War. The need for greater light-gathering power to detect faint galaxies for cosmological studies led to George Ellery Hale's concept of the 200-inch telescope (Hale, 1928). Hale symbolised the entrepreneurial approach of US astronomers to the sponsorship of private US observatories, such as the Lick, Harvard, Yerkes and Mount Wilson Observatories, which began in the late nineteenth century. James Lick (1796–1876), for example, was a successful maker and seller of pianos and an enthusiast for astronomy who, on his death in 1876, left a bequest of $700 000 to build 'a powerful telescope, superior to and more powerful than any telescope ever yet made ... and also a suitable observatory connected therewith'. The observatory was constructed on Mount Hamilton and officially opened in 1888 with the completion of the 36-inch telescope, under which James Lick was buried, according to the terms of his bequest.

Hale's record of observatory and telescope construction is remarkable by any measure.[1] He persuaded Charles T. Yerkes (1837–1905), the entrepreneur who built and electrified the Chicago street-train system and who was regularly on the verge of legal embarrassment, to provide the funds to build and equip the Yerkes Observatory as part of the University of Chicago. At Mount Wilson, Hale first constructed the 60-inch telescope and then the Hooker 100-inch telescope, completed in 1917, with funds provided by John D. Hooker and the Carnegie Institution of Washington. At the time, each of these was the largest telescope of its type in the world. Hale's greatest achievement was, however, the construction of the 200-inch telescope on a good, dark site on Mount Palomar in Southern California. Technologically, the 200-inch telescope was a masterpiece of engineering which stretched mirror and telescope technology to the limit.[2]

The programme of construction was delayed by the Second World War, but the 200-inch telescope was finally completed in 1948, ten years after Hale's death. Like most of the other major observatories in the USA, the 200-inch telescope was a private telescope which was used more or less exclusively by the astronomers employed at the host institutions, the Astrophysics Department of the California Insitute of Technology, the Mount Wilson

Observatory and the Carnegie Institute of Washington. Thus, in the early 1950s, the most important telescope in the world for all types of astrophysical and cosmological research was in the hands of a relatively small group of privileged astronomers. There is no question that the 200-inch telescope dominated observational astrophysics and cosmology from the time it was commissioned until the 1980s, when a new generation of 4-metre class telescopes on better sites provided astronomers with superior observing facilities.

The astronomical scene was, however, about to change with the development of new ways of tackling astrophysical and cosmological problems. There were several reasons for this major change in outlook of observational and theoretical astrophysicists since 1945.

(i) The expansion of the wavebands for astronomical observation

The most important reason has been the expansion of the wavebands that have become available for astronomical observation. A plot of the temperature of a black body against the frequency (or wavelength) at which most of the radiation is emitted is shown in Figure 7.1(a). In the lower panel, Figure 7.1(b), the transparency of the atmosphere to radiation as a function of frequency is presented, showing how high a telescope must be placed above the surface of the Earth for the atmosphere to become transparent to radiation of different wavelengths. With the development of radio astronomy and the capability of placing telescopes for different wavebands in space, the expansion of the accessible electromagnetic spectrum has led to a vast increase in the range of *temperatures* which are accessible for astronomical study. In turn, this had led to a much more complete description of our physical Universe and to the discovery of new physical phenomena which are important for fundamental physics as well as astronomy.

Figure 7.1(a) shows that observations in the optical waveband correspond to studying the Universe in a rather narrow wavelength interval, 300–800 nm, and hence to black-body temperatures in the range 3000–10 000 K. Of course, a somewhat wider range of temperatures can be studied since bodies at temperatures outside this range emit some radiation in the optical waveband, but this is a fair representation of the temperatures of most of the objects observed at optical wavelengths, for example stars, hot gas clouds and their associations into galaxies, clusters and so on. The capability of making observations from above the Earth's atmosphere opened up the far-infrared, ultraviolet, X-ray and γ-ray wavebands, so that very much hotter and cooler objects could be studied. It comes as no surprise that, as observational capabilities in these new wavebands developed, new and unexpected phenomena were discovered which added important new dimensions to astrophysical and cosmological research.

(ii) Non-electromagnetic astronomy

Equally important has been the development of non-electromagnetic means of tackling astrophysical and cosmological problems. The oldest of these is the study of *cosmic rays*, the high-energy electrons, protons and nuclei accelerated in a variety of astrophysical environments, including the Sun, supernovae and active galaxies. In addition, different approaches to observational astronomy have been developed. *Neutrino astronomy* has already made

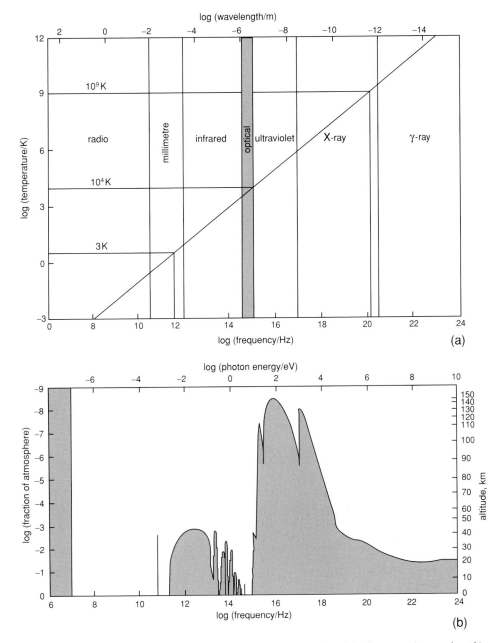

Figure 7.1: (a) The relation between the temperature of a black body and the frequency (or wavelength) at which most of the energy is emitted. The frequency (or wavelength) plotted is that corresponding to the maximum of a black body at temperature T. Convenient expressions for this relation are $\nu_{max} = 10^{11}\,(T/K)\,Hz$ or $\lambda_{max}\,T = 3 \times 10^6\,nm\,K$. The ranges of wavelength corresponding to the different wavebands – radio, millimetre, infrared, optical, ultraviolet, X-rays and γ-rays – are shown. (b) The transparency of the atmosphere for radiation of different wavelengths. The solid line shows the height above sea-level at which the atmosphere becomes transparent for different wavelengths. After R. Giacconi, H. Gursky and L. P. van Speybroeck, *Annual Review of Astronomy and Astrophysics*, **6**, 1968, 373. Both diagrams are from M. S. Longair, The new astrophysics, in *The New Physics*, ed. P. C. W. Davies (Cambridge: Cambridge University Press, 1988), p. 94.

spectacular contributions to astrophysics and fundamental physics. In addition, there are a number of emerging astronomies which are bound to have fundamental significance for astrophysics. *Gravitational-wave astronomy* is fully expected to become a major tool for the high-energy astrophysicist, while *astroparticle physics*, the search for stable massive particles predicted by theories of elementary particles, has a key role to play in understanding the dark matter problem as well as in fundamental physics.

(iii) Technological advance and computation

None of these developments would have been possible without remarkable technological developments in the design and construction of telescopes, instruments and detectors for all wavebands. In optical astronomy, the photographic plate has been largely replaced by highly efficient digital detectors. In many of the new astronomies, technologies were imported from non-astronomical disciplines and modified for the special needs of astronomical observation.

As in all the sciences, the semiconductor and computer revolutions have been crucial to the advance of observation, data collection and analysis, interpretation and theory. Astronomers were among the very first to capitalise upon the possibilities opened up by fast digital computers. For example, the rapid growth of the radio technique of aperture synthesis was wholly dependent upon the availability of the first computers to be made generally available to the scientific community. The role of computation in astrophysics and cosmology has completely changed many aspects of research in these disciplines. An important consequence has been that theory and observation can now be compared with a precision which would have been quite inconceivable to the pioneers of the pre-War years.

(iv) The growth of the astronomical community

There has been a huge increase in the volume of activity in the astronomical sciences.[3] At least part of the growth has been associated with an influx of physicists whose research interests and expertise led them to consider astrophysical problems. By the same process of symbiosis between the astrophysical and laboratory sciences, which is a recurring theme throughout this history, astronomy has assimilated new tools from physics and theoretical physics, most obviously in general relativity and particle physics, but also from fields such as chemistry, solid state physics, plasma physics, superconductivity and biophysics.

Some measure of this increase in activity is provided by the membership of the International Astronomical Union, which is open to all professional astronomers and which was founded in 1919. At the first General Assembly held in Rome in 1922, there were just over 200 members from 19 adhering countries. By 1938, the numbers had risen to 550 from 26 countries. The number was roughly the same immediately after the Second World War. By the time of the 2003 General Assembly held in Sydney, Australia, the membership had risen to 9100 from 67 adhering countries.

(v) Astronomy as 'big science'

Astronomy, astrophysics and cosmology have become one of the 'big sciences'. The case has already been made that research in astrophysics has always been 'big science'. At the beginning of the modern era, huge resources were needed by Tycho Brahe to advance the study of the motions of the Sun, Moon and planets and, in the nineteenth century, Pickering's vast projects, albeit with small telescopes, needed very considerable resources to place the science of astrophysics on a secure foundation. The 200-inch telescope broke all records for the cost of an individual instrument in the 1930s and 1940s.

After the Second World War, there was a spectacular increase in investment in basic research in the USA, largely stimulated by the huge contributions which the very best research scientists had made during the period of hostilities and the realisation of the enormous potential for economic growth, as well as strategic defence requirements, which the fruits of basic research could bring. In Europe, it took somewhat longer to recover from the ravages of the War, but, in due course, these countries began to invest heavily in pure and applied research. The attitude of many of the best research workers had been changed by their wartime experiences. To quote Bernard Lovell (b. 1913), they adopted an approach to research that was (Lovell, 1987)

... utterly different from that deriving from the pre-war environment. The involvement with massive operations had conditioned them to think and behave in ways which would have shocked the pre-war university administrators. All these facts were critical in the large-scale development of astronomy.

In due course, the astronomers rode the post-war wave of investment in the fundamental sciences, but these initiatives had to be seen in a national or international context rather than as sponsorship by private institutions, as had occurred in the USA. Whilst the increase in the numbers of astronomers alone made the construction of more large telescopes a priority, the major discoveries of radio, X-ray and γ-ray astronomy, as well as the rise of high-energy astrophysics in the 1960s and 1970s, had a considerable impact upon the case for increasing investment in large astronomical facilities. Examples of the culmination of this historical progression were the NASA–ESA Hubble Space Telescope, which in the end cost in excess of $2 billion, and the European Southern Observatory's Very Large Telescope, consisting of four 8-metre optical–infrared telescopes on Cerro Paranel in Chile, which cost roughly $500 million. Thus, the telescopes needed to carry out frontier research have become very complex and costly, international collaboration often being essential to construct and operate them.

In Part III, we trace the development of astronomy from the point of view of the opening up of new regions of the electromagnetic spectrum and the changing ways in which astronomical research is carried out. At the start of this period, astronomical research was primarily carried out by small groups of astronomers working with their own dedicated telescopes. By the end of the period, most of the large telescopes for all wavebands were national or international facilities, operated by specialist teams for the benefit of the communities of astronomers. Most of the world-leading telescopes are now very high-technology instruments of great sophistication, and astronomy is more than ever one of the 'big sciences'. The science, however, is still small science in the sense that the big facilities provide the data needed by a very wide range of different astronomical disciplines and by a large population of astronomers. Typically, each large facility will provide data for hundreds of different astronomical projects, ranging from our own Solar System to the earliest phases of the Big Bang.

7.2   The discovery of subatomic particles and cosmic rays

The first hints that there was more to the Universe than just gas, dust and stars came from the discovery of *cosmic rays*, the history of which is intimately related to the understanding of subatomic particles. Let us first review some of the key discoveries which were to lead to the new astronomies in the post Second World War period.

7.2.1   *The discovery of α-, β- and γ-rays and the neutron*

In 1895, Wilhelm Röntgen (1845–1923) discovered, by accident, that wrapped unexposed photographic plates left close to Crookes discharge tubes were darkened. In addition, fluorescent materials left close to them glowed in the dark. Röntgen came to the correct conclusion that both phenomena were associated with some new form of radiation emitted by the Crookes tube, and he named these X-rays (Röntgen, 1895).[4] His discovery caused an immediate sensation when the first X-ray photographs showing the bones of the body were published. Overnight, X-rays became a matter of the greatest public interest and were very rapidly incorporated into the armoury of the doctor's surgery. The X-rays were more penetrating than cathode rays, since they could blacken photographic plates at a considerable distance from the hot spot on the Crookes tube, which was known to be their source. Their identification with 'ultra-ultraviolet' radiation was only convincingly demonstrated in 1906, when Charles Barkla (1877–1944) found that the X-radiation was polarised (Barkla, 1906) and, even more convincingly, when Max von Laue (1879–1960) had the inspiration of looking for their diffraction by crystals in 1912 (Friedrich, Knipping and Laue, 1912; Laue, 1912), in the process opening up the new field of X-ray crystallography.

The association of X-rays with fluorescent materials led to the search for other sources of X-radiation. In 1896, Henri Becquerel (1852–1908) tested several known fluorescent substances before investigating some samples of potassium uranyl disulphate. The photographic plates were wrapped in several sheets of black paper, the phosphorescent material was exposed to sunlight and then the plate was developed to find if it had been darkened by X-rays. Becquerel's remarkable discovery was that the plates became darkened even when the phosphorescent material was not exposed to light. This was the discovery of *natural radioactivity* (Becquerel, 1896). In further experiments carried out in the same year, Becquerel showed that the amount of radioactivity was proportional to the amount of uranium in the substance and that the radioactive flux of radiation was constant in time. Another important discovery was that the radiation from the uranium compounds discharged electroscopes.

Other radioactive substances were soon identified. Thorium was discovered in 1898 (Schmidt, 1898) and then followed the isolation of polonium and radium, both much stronger sources of radioactivity than uranium, by Pierre Curie (1859–1906) and Marie Sklodowska-Curie (1867–1934) (Curie and Sklodowska-Curie, 1898; Curie, Sklodowska-Curie and Bémont, 1898). In his first publication on radioactivity, Rutherford established that there are at least two separate types of radiation emitted by radioactive substances (Rutherford, 1899). He called the component which is most easily absorbed α-radiation (or α-rays) and the much more penetrating component β-radiation (or β-rays). It took another ten years before Rutherford conclusively demonstrated that the α-radiation consisted of what we now know as the nuclei of helium atoms (Rutherford and Royds, 1909). In contrast, β-rays were convincingly shown by Walter Kaufmann (1871–1947) to have the same mass-to-charge ratio as the recently discovered electron (Kaufmann, 1902). Subsequently, γ-radiation was discovered in 1900 by Paul Villard (1860–1934) as an extremely penetrating form of radiation emitted in radioactive decays (Villard, 1900a,b). The γ-rays were conclusively identified as electromagnetic waves 14 years later when Rutherford and Edward Andrade (1887–1971) observed the reflection of γ-rays from crystal surfaces (Rutherford and Andrade, 1913).

The α-, β- and γ-rays were the only known radiations which could cause the ionisation of air. The characteristic properties which distinguished them were their penetrating power. In quantitative terms, they had the following properties.

- The α-*particles* ejected in radioactive decays produce a dense stream of ions and are stopped in air within about 0.05 m. This is called the *range* of the particles.
- The β-*particles* have greater ranges, but there is not a well-defined value for any particular radioactive decay. We now understand that the spread in range is due to the fact that the electrons are emitted as part of a three-body process involving the emission of a neutrino as well as an electron.
- The γ-*rays* were found to have by far the longest ranges, a few centimetres of lead being necessary to reduce their intensity by a factor of 10.

The unravelling of the nature of the atomic nucleus continued throughout the period 1911–1930. It was soon established that typical nuclei have mass about two or more times that which can be attributed to the protons alone. The commonly held explanation for this difference was that the nucleus was composed of electrons and protons, the 'inner' electrons neutralising the extra protons. The fact that certain nuclei ejected electrons in radioactive β-decays supported this point of view. Rutherford had speculated in 1920 that the neutral mass in the nucleus might be in the form of some new type of particle, similar to the proton but with no electric charge (Rutherford, 1920). During the 1920s Rutherford and his colleagues, particularly James Chadwick, made a number of unsuccessful attempts to find evidence for these particles, which became known as *neutrons*.

In 1930, Walther Bothe (1891–1957) and Herbert Becker (b. 1906) in Germany and, in 1932, Irene Joliot-Curie (1897–1956) and her husband Frederic Joliot (1900–1958) in France discovered that neutral penetrating radiation was emitted when light elements were bombarded by α-particles. Both groups believed that the radiation was some form of γ-radiation. Chadwick guessed that the penetrating radiation was a flux of the elusive neutrons. He rapidly performed a classic series of experiments in which the neutral radiation collided with different substances, including hydrogen and nitrogen, and then, from the recoil effects of the collisions between the unseen particles and the ambient gas, he could estimate the mass of the particles. This measurement of the mass of the neutral radiation showed conclusively that it could not be γ-radiation but rather neutral particles ejected from the nucleus with mass roughly the same as that of the proton (Chadwick, 1932). This was the discovery of the neutron.

7.2.2   The discovery of cosmic rays

The cosmic-ray story begins in about 1900, when it was discovered that electroscopes discharged even if they were kept in the dark well away from sources of natural radioactivity.[5] The electroscope was a key instrument in many of the early experiments in radioactivity because the rate at which the leaves of the electroscope came together provided a measure of the amount of ionisation. The origin of this behaviour was a major puzzle, and various ingenious experiments were carried out to discover the origin of the ionising radiation. A good example is this quotation from C. T. R. (Charles) Wilson (1869–1959) (Wilson, 1901):

> The experiments with this apparatus were carried out at Peebles. The mean rate of leak when the apparatus was in an ordinary room amounted to 6.6 divisions of the micrometer scale per hour. An experiment made in the Caledonian Railway tunnel near Peebles (at night after the traffic had ceased) gave a leakage of 7.0 divisions per hour ... There is thus no evidence of any falling off of the rate of production of ions in the vessel, although there were many feet of solid rock overhead.

Later, Rutherford showed that most of the ionisation was due to natural radioactivity, either in rocks or from radioactive contamination of the equipment. The big breakthrough came in 1912 and 1913 when first Victor Hess (1883–1964) and then Werner Kolhörster (1887–1946) made manned balloon ascents in which they measured the ionisation of the atmosphere with increasing altitude (Hess, 1912; Kolhörster, 1913). By late 1912, Hess had flown to 5 km and then, by 1913, Kolhörster had made ascents to 9 km, all these dangerous experiments being carried out in open balloons. It was Hess who discovered the first definite evidence that the source of the ionising radiation was extraterrestrial.

Hess and Kolhörster found the startling result that the average ionisation increased with respect to the ionisation at sea-level above about 1.5 km (see Table 7.1). This was clear evidence that the source of the ionising radiation must be located above the Earth's atmosphere. From the data in Table 7.1, the attenuation constant, α, defined by $n(l) = n_0 \exp(-\alpha l)$, was found to have values of 10^{-3} m^{-1} or less. The ionising radiation was much more more penetrating than the γ-rays found in radioactive decays. Hess made the immediate inference:

> The results of the present observations seem to be most readily explained by the assumption that a radiation of very high penetrating power enters our atmosphere from above, and still produces in the lower layers a part of the ionisation observed in closed vessels.

Even at sea-level, there is residual ionisation due to the extraterrestrial ionising radiation, amounting to about 1.4×10^6 ion pairs m^{-3}.

It was not too much of an extrapolation to assume that the *cosmic radiation*, or *cosmic rays* as they were named in 1925 by Robert Millikan (1868–1953), were γ-rays with greater penetrating power than those observed in natural radioactivity. In 1929, Dmitri Skobeltsyn

Table 7.1. *The variation of ionisation with altitude from the observations of Kolhörster (1913)*

Altitude (km)	Difference between observed ionisation and that at sea-level ($\times 10^6$ ions m^{-3})
0	0
1	−1.5
2	+1.2
3	+4.2
4	+8.8
5	+16.9
6	+28.7
7	+44.2
8	+61.3
9	+80.4

(1892–1992), working in his father's laboratory in Leningrad, constructed a cloud chamber to study the properties of the β-rays emitted in radioactive decays. The experiment involved placing the chamber within the jaws of a strong magnet so that the curvature of their tracks could be measured. Among the tracks, he noted some which were hardly deflected at all and which looked like electrons with energies greater than 15 MeV (Figure 7.2). He identified them with secondary electrons produced by the 'Hess ultra γ-radiation'. These were the first pictures of the tracks of cosmic rays (Skobeltsyn, 1929).

The year 1928 saw the invention of the *Geiger–Müller detector* by Hans Geiger (1882–1945) and Walther Müller (1905–1979) which enabled individual cosmic rays to be detected and their arrival times determined very precisely (Geiger and Müller, 1928, 1929). In 1929, Bothe and Kolhörster carried out one of the key experiments in cosmic ray physics and introduced the important concept of *coincidence counting* to eliminate spurious background events (Bothe and Kohlhörster, 1929). This coincidence technique is now standard practice in many different types of cosmic ray, X-ray and γ-ray experiment. By using two counters, one placed above the other, they found that simultaneous discharges of the two detectors occurred very frequently, even when a strong absorber was placed between the detectors, indicating that charged particles of sufficient penetrating power to pass through both of them were very common. In the crucial experiment, they placed slabs of lead and then gold up to 4 cm thick between the counters and measured the decrease in the number of coincidences when the absorber was introduced. The mass absorption coefficient agreed very closely with that of the atmospheric attenuation of the cosmic radiation. The experiment strongly suggested that the cosmic radiation consists of charged particles. As they wrote in their classic paper (Bothe and Kohlhörster, 1929):

One can perhaps summarise the whole discussion in a single argument: the mean free path of a γ-ray between two electron ejecting processes would be $1/\mu = 10$ m in water $1/\mu = 0.9$ m in lead and $1/\mu = 0.52$ m in gold for the high latitude radiation. Hence one can see that a quite exceptional accident must be supposed to happen if two electrons produced by the same γ-ray should display the necessary penetrating power and the correct direction to strike both counters directly.

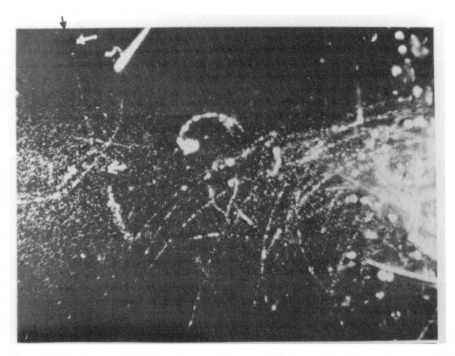

Figure 7.2: An image of the first photographic record of the arrival of a cosmic-ray particle by Skobeltsyn in 1929. The track of the particle in the cloud chamber is indicated by the two white arrows and one black arrow. From Y. Sekido and H. Elliot, eds, *Early History of Cosmic Ray Studies* (Dordrecht: D. Reidel Publishing Company, 1985), p. 47.

They also showed that the flux of these particles could account for the observed intensity of cosmic rays at sea-level and, because of their long ranges in matter, the energies of the particles had to be about $10^9 - 10^{10}$ eV.

The experiments carried out using cloud chambers showed that showers of cosmic ray particles are often observed. Most of the cosmic-ray particles observed at the surface of the Earth are, in fact, secondary, tertiary or higher products of very-high-energy cosmic rays entering the top of the atmosphere. The full extent of some of these *extensive air showers* was established by Pierre Auger (1899–1993) and his colleagues from observations with a number of separated detectors (Auger *et al.*, 1939). To their surprise, they found that the air showers could extend over areas greater than 100 metres on the ground and consist of the arrival of millions of ionising particles. The particles responsible for initiating the showers must have had energies exceeding 10^{15} eV at the top of the atmosphere. This was direct evidence for the acceleration of charged particles to extremely high energies in extraterrestrial sources.

7.2.3   Cosmic rays and the discovery of elementary particles

From the 1930s to the early 1950s, the cosmic radiation provided a natural source of very-high-energy particles which were energetic enough to penetrate into the nucleus. This procedure was the principal technique by which new particles were discovered until the early 1950s. In 1930, Millikan and Anderson used an electromagnet ten times stronger than that used by Skobeltsyn to study the tracks of particles passing through the cloud chamber. Anderson observed curved tracks identical to those of electrons but corresponding to particles with positive electric charge (Anderson, 1932). This discovery was confirmed by Patrick Blackett and Giuseppe Occhialini (1907–1993) in 1933 using an improved technique in which the cloud chamber was only triggered after it was certain that a cosmic ray had passed through (Blackett and Occhialini, 1933). They obtained many excellent photographs of the positive electrons, on many occasions showers containing equal numbers of positive and negative electrons created by cosmic-ray interactions with the body of the apparatus being observed.

The discovery of the *positive electron*, or *positron*, coincided almost exactly with Paul Dirac's theory of the electron (Dirac, 1928a,b). In one of the great theoretical extensions of quantum mechanics, Dirac (1902–1984) succeeded in deriving the relativistic wave equation for the electron which not only predicted its spin and magnetic moment, but also the existence of what we would now call the *antiparticle* to the electron, the positron.[6]

There were more surprises in store, however. Anderson noted that there were often much more penetrating positive and negative particle tracks in the cloud chamber pictures. These particles displayed little evidence of interaction with the gas in the chamber. By 1936, Anderson and Seth Neddermeyer (1907–1988) were sufficiently confident of their results to announce the discovery of particles with mass intermediate between that of the electron and the proton (Anderson and Neddermeyer, 1936). These *mesotrons* had mass between about 50 and 400 times the mass of the electron. This discovery coincided rather nicely with a theoretical prediction by Hideki Yukawa (1907–1981) concerning the strong force which binds neutrons and protons together in the nucleus. According to Yukawa's theory, the strong short-range force could be understood in terms of the exchange of particles about 250 times as massive as the electron (Yukawa, 1935). In fact, the particles discovered by Anderson and Neddermeyer, nowadays known as *muons*, are not the particles which bind nuclei together. The identification was somewhat unsatisfactory because the mesotrons showed so little interaction with the nuclei in the chamber, whereas the exchange particle is expected to show a strong interaction with nuclei.

The same procedures were used immediately after the Second World War by George Rochester (1908–2001) and Clifford Butler (1922–1999), who constructed a new cloud chamber to use with a large electromagnet obtained by Blackett before the War. In 1947 they reported the discovery of two cases of particle tracks in the form of 'V's with apparently no incoming particle (Rochester and Butler, 1947). They correctly suggested that the Vs resulted from the spontaneous decay of an unknown particle, the mass of which could be estimated from the decay products. Both had mass about half that of the proton. To obtain higher fluxes of cosmic radiation, the experiments were repeated at much higher altitudes. Two years later, the experiments were carried out by Blackett's group working at the Pic du Midi Observatory in the Pyrenees and by Anderson and his colleagues on White Mountain in California. Many more examples of Vs were found, and this class of particle became known as *strange particles*. Both neutral and charged strange particles were discovered. Most of them had mass about half that of the proton and are what are now referred to as *charged* and *neutral kaons* (K^+, K^-, K^0). There were a few examples, however, of neutral particles with mass greater than the mass of the protons – these are now known as *lambda particles* (Λ). What puzzled physicists was their long lifetimes, 10^{-8} and 10^{-10} s, many orders of magnitude greater than the timescale associated with the strong interactions.

Meanwhile, another powerful tool for the study of particle collisions and interactions had been developed by Cecil Powell (1903–1969) at Bristol University. Photographic plates had played a key role in the discovery of X-rays and radioactivity in the 1890s. Powell, in collaboration with the Ilford photographic company, developed special 'nuclear' emulsions which were sufficiently sensitive to register the tracks of protons, electrons and all the other types of charged particle which had been discovered. Powell and his colleagues mastered the techniques of producing thick layers of emulsion by stacking layer upon layer of emulsion, resulting in a three-dimensional picture of the interactions taking place in the emulsion. Among the first discoveries using this high-precision technique was that of the *pion* (π) in 1947, which was the particle predicted by Yukawa in 1936 (Lattes, Occhialini and Powell, 1947).

By 1953, accelerator technology had developed to the point where energies comparable with those available in cosmic rays could be produced in the laboratory with known energies and directed precisely onto the chosen target. After about 1953, the future of high-energy physics lay in the accelerator laboratory rather than in the use of cosmic rays. The interest in cosmic rays shifted to the problems of their origin and their propagation in astrophysical environments from their sources to the Earth.

7.3 Radio astronomy

The expansion of the observable electromagnetic spectrum began with Karl Jansky's announcement of the discovery of the radio emission from the Galaxy in May 1933.[7] Working at the Bell Telephone Laboratories at Holmdel, New Jersey, Jansky (1905–1950) was assigned the task of identifying naturally occurring sources of radio noise which would interfere with radio transmissions. In what turned out to be a classic series of observations made at the long wavelength of 14.6 metres (20.5 MHz), he discovered the radio emission from the Galaxy (Figure 7.3) (Jansky, 1933).

This discovery was confirmed by Grote Reber (1911–2002), a radio engineer and enthusiastic amateur astronomer. With his home-built radio antenna and receiving system operating at a wavelength of 1.87 metres (160 MHz), he made a radio scan along the plane of the Galaxy which was published in the *Astrophysical Journal* in 1940 (Reber, 1940). Comparison of Jansky's and Reber's observations showed that the emission could not be black-body radiation, and Reber proposed that it was bremsstrahlung, or free–free emission. In the immediately following paper in the *Astrophysical Journal*, Louis Henyey (1910–1970) and Philip Keenan (1908–2000) showed that, whilst the radiation at 1.87 m might be the bremsstrahlung of gas at 10 000 K, the intensity observed by Jansky at the longer wavelength was far too great for this to be the emission process (Henyey and Keenan, 1940). Other than this negative conclusion, these observations attracted little attention from professional astronomers. The culmination of Reber's work was the publication of the first map of the radio emission from the Galaxy (Figure 7.4) in the *Astrophysical Journal* in 1944 (Reber, 1944).

Figure 7.3: Karl Jansky's radio antenna with which he discovered the radio emission of the Galaxy in 1933. (Courtesy of the US National Radio Astronomy Observatory.)

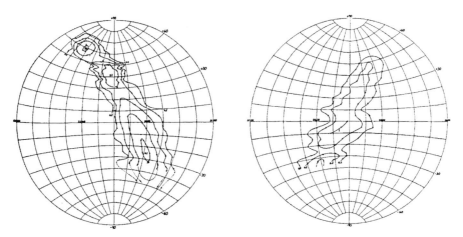

Figure 7.4: Reber's map of the radio emission of the Galaxy, made at a radio frequency of 160 MHz (1.87 m). The contours of radio emission are plotted in celestial coordinates and are more or less coincident with the Milky Way (Reber, 1944).

7.3.1 The first discrete radio sources

The development of radar during the Second World War had two immediate consequences for radio astronomy. Firstly, sources of radio interference which might confuse radar location had to be identified. In 1942, James Hey (1909–2000) and his colleagues at the Army Operational Research Group in the UK discovered intense radio emission from the Sun which coincided with a period of unusually high sun-spot activity (Hey, 1946).[8] Towards the end of the War, Hey and his colleagues continued to improve the sensitivities of the receivers in order to detect incoming V2 rockets. To their consternation, they discovered that the noise performance of the telescope system did not improve. They soon realised that the background radio emission from the Galaxy itself was the factor which limited the sensitivity of the telescope system, and not the receivers. At the end of hostilities, Hey and his colleagues began mapping the sky at 5 m wavelength, and in 1946 they discovered the first discrete source of radio emission, which lay in the constellation of Cygnus – the source became known as Cygnus A (Hey, Parsons and Phillips, 1946). The second consequence was that the extraordinary research efforts to design powerful radio transmitters and sensitive receivers for radar resulted in new technologies which were to be exploited by the pioneers of the new science of radio astronomy, all of whom came from a background in radar.

Immediately after the War, a number of the radar scientists began the systematic study of the astronomical phenomena discovered, more or less by chance, as a result of the War effort. The three main groups were headed by Martin Ryle (1918–1984) at Cambridge University, by Bernard Lovell at Manchester University in the UK and by Joseph Pawsey (1908–1962) at Sydney. Further discrete sources of radio emission were discovered, and radio interferometry provided the best means of measuring their positions with improved precision. In 1948, Ryle and Francis Graham Smith (b. 1923) discovered the most powerful source in the northern hemisphere, Cassiopeia A (Ryle and Graham Smith, 1948), and in 1949 the Australian radio astronomers John Bolton (1922–1993), Gordon Stanley (1921–2001) and Bruce Slee (b. 1924) succeeded in associating three of the discrete radio sources with remarkable nearby astronomical objects. One was associated with the supernova remnant known as the Crab Nebula and the others, Centaurus A and Virgo A, were associated with the strange galaxies NGC 5128 and M87, respectively (Bolton, Stanley and Slee, 1949). In addition to the diffuse radio emission of our own Galaxy, these early surveys established the existence of a population of discrete radio sources, some concentrated towards the plane of the Galaxy, but many lying outside it. There was some uncertainty as to whether the isotropic component of the source population was primarily associated with nearby radio stars in our own Galaxy or with distant extragalactic objects.[9]

The radio astronomers could not answer this question from the radio data alone, since the radio spectra were found to be continuous, without any spectral features from which a redshift could be estimated. Distances could only be determined by finding an associated optical object and measuring its distance. In 1951, Graham Smith measured interferometrically the positions of the two brightest sources in the northern sky, Cygnus A and Cassiopeia A, with an accuracy of about 1 arcmin (Graham Smith, 1951). This led to their optical identification by Walter Baade and Rudolph Minkowski (1895–1976) from observations with the Palomar 200-inch telescope (Baade and Minkowski, 1954). Cassiopeia A was associated with a young supernova remnant in our own Galaxy, while Cygnus A was associated with a faint and distant galaxy. The latter observation immediately showed that the radio sources could be used for cosmological studies. By 1960, another of the brightest radio sources in the sky, 3C 295, had been associated with the brightest galaxy in a cluster of galaxies at the largest redshift, $z = \Delta\lambda/\lambda = 0.461$, measured for any galaxy at that time (Minkowski, 1960b). This remained the largest redshift for any galaxy until the mid 1970s. The cosmological importance of radio astronomical observations of discrete sources was thus apparent by the mid 1950s – fainter radio sources would lie at greater cosmological distances and hence probe the Universe at epochs much earlier than the present.

7.3.2   Synchrotron radiation

The nature of the Galactic radio emission and, by analogy, of the discrete radio sources, was solved in the late 1940s. During the 1930s and 1940s, particle accelerators of increasing size, such as cyclotrons and betatrons, were constructed in which protons or electrons moved in circular paths in a uniform magnetic field. It was realised that radiation losses associated with the centripetal acceleration of electrons in their circular orbits would become important as their energies increased. Dmitri Ivanenko (1904–1994) and Isaak Pomeranchuk (1913–1966) published their calculations in 1944 showing that these losses would limit the maximum energy of a betatron to about 500 MeV (Ivanenko and Pomeranchuk, 1944). The energy loss rate of an accelerated relativistic electron had been worked out by George Schott (1868–1937) in 1912 (Schott, 1912), a modern form of the loss rate formula being

$$-\left(\frac{\mathrm{d}E}{\mathrm{d}t}\right) = \frac{e^4\gamma^2 B^2 v^2 \sin^2\theta}{6\pi\epsilon_0 c^3 m_{\mathrm{e}}^2}, \tag{7.1}$$

where $\gamma = E/mc^2 = (1 - v^2/c^2)^{-1/2}$ is the Lorentz factor of the electron, B is the magnetic flux density and θ is the pitch angle of the electron, the angle between its direction of motion and the magnetic field direction. This energy loss was first observed in the 100 MeV betatron at the General Electric Research Laboratory in Schenectady, New York, by John Blewett (1910–2000) in 1946, but the radiation of the electrons themselves was not observed (Blewett, 1946). It was thought at that time that most of the radiation would be emitted at low harmonics of the electrons' orbital frequency and so radio receivers sensitive in the 50 to 1000 MHz waveband were used. In the meantime, Julian Schwinger (1918–1994) had worked out in great detail the expected radiation spectrum of highly relativistic electrons orbiting in a uniform magnetic field and had shown that, because of the extreme effects of aberration, the radiation is most intense at a very much higher frequency, $v \approx \gamma^2 v_{\mathrm{g}}$, where $v_{\mathrm{g}} = eB/2\pi m_{\mathrm{e}}$ is the non-relativistic gyrofrequency of the electron (Schwinger, 1946, 1949). The next accelerator built at the General Electric Laboratory was a 70 MeV synchrotron accelerator with a transparent glass vacuum tube. Intense optical radiation from the synchrotron accelerator was first seen in April 1947 and was named *synchrotron radiation* (Elder *et al.*, 1947). The characteristic properties of the radiation are that its spectrum is a broad-band continuum and that it is highly polarised and directional.

The first application of synchrotron radiation in an astronomical context was proposed by Hannes Alfvén (1908–1995) and Nicolai Herlofson (1916–2004), who, in 1950, suggested that the emission of the 'radio stars', which had just been discovered, might be the synchrotron radiation of high-energy electrons gyrating in magnetic fields with flux density 10^{-10}–10^{-9} T within a 'trapping volume' of about 0.1 light-year radius about the star (Alfvén and Herlofson, 1950). Then, Karl-Otto Kiepenheuer (1910–1975) and Vitali Ginzburg (b. 1916) made the much better suggestion that the Galactic radio emission observed by Jansky and Reber is the synchrotron radiation of ultrarelativistic electrons gyrating in the interstellar magnetic field (Kiepenheuer, 1950; Ginzburg, 1951). By the mid 1950s, the power-law spectrum of the Galactic radio emission and its high degree of polarisation convinced everyone of the correctness of the synchrotron hypothesis, a story which is taken up in Section 11.1. Radio emission is observed throughout the disc of the Galaxy and so provides direct evidence for an interstellar flux of very-high-energy electrons being present throughout the disc of the Galaxy. At that time, it was likely that the cosmic ray protons and nuclei, which had been detected in balloon flights at high altitude, were of interstellar origin, but the flux of high-energy electrons was difficult to distangle from secondary electrons created in the upper atmosphere.

The power-law spectra of the discrete radio sources and the polarisation of their radio emission were naturally interpreted as evidence for synchrotron radiation of ultra-relativistic electrons, but the energy requirements of some of the most luminous radio emitters, such as Cygnus A, were enormous. Furthermore, the radio emission did not originate from the body of the galaxies but from vast radio lobes, very often of much greater dimension than the galaxies themselves. Somehow, the radio galaxies were capable not only of accelerating huge fluxes of electrons to ultra-relativistic energies but also of ejecting them into intergalactic space. These observations and their interpretation in terms of the properties of relativistic plasmas and magnetic fields provided a powerful stimulus for the new discipline of high-energy astrophysics.

7.3.3 Radio telescopes, aperture synthesis and VLBI

From the early 1950s onwards, radio astronomy developed as a discipline in its own right. Large steerable reflectors were constructed, the radio analogues of the large optical reflectors. The culmination of these efforts were the construction of the Jodrell Bank 76-metre telescope in the UK (1957), now named the Lovell Telescope, the Parkes 64-metre telescope in Australia (1961), the NRAO 300-foot telescope at Greenbank (1962)[10] and the Effelsberg 100-metre telescope in Germany (1971). These telescopes had excellent sensitivity, but the angular resolution, θ, was limited by the diameter, D, of the telescope, according to the Rayleigh criterion $\theta \approx \lambda/D$.

Higher angular resolution could be obtained by radio interferometry, which had been developed as part of the radar development programmes during the Second World War and which led to the concept of *aperture synthesis*. As Peter Scheuer has remarked,[11]

By the beginning of 1954 the principles of aperture synthesis were fully understood all over the world, but the world of radio astronomy was then very small, and the world I mean, in which radio astronomy was controlled by radio engineers who were learning astronomy, was smaller still. In the Netherlands and in the United States radio astronomy was in the hands of real astronomers, to whom a telescope meant a paraboloidal mirror and nothing else; their contribution was of a different kind. So the little world that understood aperture synthesis consisted of CSIRO Radiophysics Division in Sydney, the English radio astronomers at Cambridge and Manchester, and the French Group at Nançay.

Although the underlying principles of aperture synthesis at radio wavelengths had been clearly set out by Ronald Bracewell (b. 1921) and James Roberts (b. 1927) in 1954 (Bracewell and Roberts, 1954), there were many technical problems to be overcome before these concepts could be converted into a reality. The important realisation was that, by measuring

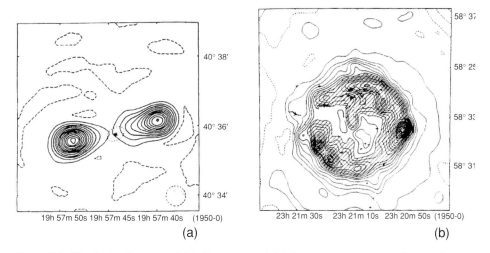

Figure 7.5: The first radio maps of (a) Cygnus A and (b) the supernova remnant Cassiopeia A, as observed by the Cambridge One-Mile Telescope (Ryle *et al.*, 1965). The observations were made at a frequency of 1.4 GHz, at which the angular resolution of the telescope was 23 arcsec.

both the amplitudes and phases of the incoming signals, the distribution of radio brightness across celestial sources could be completely reconstructed. The technical problems were solved by Martin Ryle and his colleagues at Cambridge. There were two major hurdles to be overcome. The first arose from the need to add together coherently the signals from separated telescopes. As the separate telescopes follow the same patch of sky, varying delays had to be switched into the cables from each antenna to compensate for the varying electrical path-length from the radio source to the correlator. The second was the need for high-speed computation. The essence of aperture synthesis is that the correlated signals between different pairs of telescopes sample the Fourier transform of the radio brightness distribution on the sky and so, to reconstruct the image, a two-dimensional inverse Fourier transformation of the correlated signals had to be made. The availability of the first high-speed digital computers by the end of the 1950s made this computational challenge a feasible undertaking. The principles of Earth-rotation aperture synthesis were demonstrated by the remarkable image of the region about the north celestial pole created by Ryle and Ann Neville (Ryle and Neville, 1962). The success of this programme led to the construction of the Cambridge One-Mile Telescope, the first Earth-rotation aperture synthesis telescope system with fully steerable telescopes. The first images of radio sources taken with the telescope were a startling achievement (Ryle, Elsmore and Neville, 1965), the angular resolution of 23 arcsec corresponding to a fully filled aperture of diameter one mile (Figure 7.5).

The success of this programme led to the construction of a number of large-aperture synthesis radio telescopes, including the Westerbork Synthesis Telescope in the Netherlands (1970) and the next generation 5-km telescope, subsequently named the Ryle Telescope, in Cambridge (1971). The culmination of these efforts was the construction of the Very Large Array in New Mexico in the USA (1981), the Australia Telescope (1988) and the Giant Metrewave Radio Telescope in India (1999). During the 1960s, interferometric techniques were extended to intercontinental baselines, the technique known as *very long baseline interferometry (VLBI)*, resulting in angular resolutions of the order of a milliarcsec. In this technique, the receivers at the different observing stations were equipped with very precise stable clocks, which enabled the radio signals to be recorded separately on magnetic tape and correlated at a later time. The first successful VLBI observations were carried out in 1967 (Broten *et al.*, 1967; Moran *et al.*, 1967).

The new discipline of radio astronomy resulted in many key discoveries for contemporary astrophysics, those of quasars, the cosmic microwave background radiation, neutron stars as the parent bodies of radio pulsars, interstellar molecules through their millimetre line emission and superluminal motions being of special importance. The history of these developments is described in Parts IV and V.

7.4 X-ray astronomy

Immediately after the Second World War, those physicists and astronomers interested in ultraviolet and X-ray astronomy made the first observations from above the Earth's atmosphere.[12] The atmosphere is opaque to all radiation with wavelengths shorter than about 300–310 nm and so ultraviolet, X-ray and γ-ray astronomy have to be conducted from above the Earth's atmosphere. Above about 150 km, absorption by the Earth's atmosphere is no longer important (see Figure 7.1). The German V2 rocket programme had made enormous strides in rocket technology during the War, and the German scientists who had built them, led by Werner von Braun (1912–1977), as well as 300 box cars full of V2 parts, were taken to the USA where they formed the core of the US Army's rocket programme. The US Army announced that these rockets would be available for scientific research.[13]

One of the prime targets of the early rocket experiments was the ultraviolet and X-ray emission of the Sun.[14] It was known that the Sun possessed a very hot corona, and it was surmised that the Earth's ionosphere might be ionised by its ultraviolet and X-ray emission. The first successful rocket ultraviolet observations of the Sun were made in October 1946 by the group led by Richard Tousey (1908–1997) at the Naval Research Laboratory (Baum *et al.*, 1946). Then, in September 1949, Herbert Friedman (1916–2000) and his colleagues made the first successful X-ray observations of the Sun, confirming the expectation that the Sun's corona is very hot (Friedman, Lichtman and Byram, 1951). These rocket experiments continued throughout the 1950s and elucidated many of the X-ray properties of the Sun.

The flights of Sputniks 1 and 2 in late 1957 and the orbital flight by Yuri Gagarin (1934–1968) in 1961 came as a profound shock to the US administration, which realised that the USA had fallen behind the USSR in space technology and therefore was strategically vulnerable. The US response was to set up the National Aeronautics and Space Administration (NASA) in July 1958 as a civilian organisation to begin the process of catching up with the USSR. As part of that endeavour, the American Science and Engineering group (AS&E) was set up in association with the Massachusetts Institute of Technology to work on military and civilian contracts.

The AS&E group, led by Riccardo Giacconi (b. 1931), developed plans for making astronomical observations in the X-ray waveband, but their theoretical calculations did

Figure 7.6: The payload of the rocket containing the X-ray detectors which made the first observation of the discrete X-ray source Sco X-1 and the X-ray background radiation in June 1962. The payload was constructed by the AS&E group (American Science and Engineering). From W. Tucker and R. Giacconi, *The X-ray Universe* (Cambridge, Massachusetts: Harvard University Press, 1985).

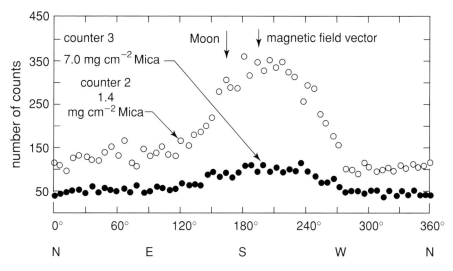

Figure 7.7: The discovery record of the X-ray source Sco X-1 and the X-ray background emission by Giacconi and his colleagues in a rocket flight of June 1962. The prominent source was observed by both detectors, as was the diffuse background emission (Giacconi *et al.*, 1962).

not promise much success with the sensitivities available at that time. The best target seemed to be the search for fluorescent X-rays from the Moon, which would result from the impact of streams of energetic solar particles hitting its surface.[15] The first successful flight took place in June 1962 (Figure 7.6). In the five minutes of observing time during which the rocket payload was above the Earth's atmosphere, Giacconi and his colleagues failed to detect any X-rays from the Moon, but discovered an intense discrete source of emission in the constellation of Scorpius, which became known as Sco X-1 (Figure 7.7) (Giacconi *et al.*, 1962). In addition, an intense background of X-rays was observed which was remarkably uniformly distributed over the sky (Gursky *et al.*, 1963). These observations were soon confirmed by other rocket flights by the AS&E group as well as by Friedman's group at NRL, which also discovered X-rays from the supernova remnant, the Crab Nebula (Bowyer *et al.*, 1964). As in the case of radio astronomy, these were entirely unexpected discoveries.

The next decade saw a flurry of activity in which a dozen or more X-ray astronomy groups made numerous rocket flights of increasing sophistication to understand the nature of the X-ray sky. These experiments were of considerable ingenuity and involved making optimum use of the five minutes observing which was possible during each rocket flight. The Lockheed group introduced the technique of spin-stabilisation, which enabled small regions of the sky to be scanned slowly with much greater sensitivity. By 1967, more than 30 X-ray sources were known, including the detection of X-rays from a number of supernovae, the quasar 3C 273 and the radio galaxy M87. The angular resolution of the X-ray telescopes was improved by the development of sophisticated X-ray collimators which were placed in front of the X-ray detectors, and this enabled the optical counterpart of the bright X-ray source Sco X-1 to be detected – it was a faint blue, variable nova-like star (Sandage *et al.*, 1966). These pioneering observations provided tantalising glimpses of the richness of the X-ray sky, but the picture was confused. Some of the sources were highly variable, since they would be present in one rocket flight and then disappear on the next.

These problems were resolved with the launch, in December 1970, of the UHURU X-ray observatory, which was the first satellite dedicated to X-ray astronomy and which initiated the successful series of Explorer satellites sponsored by NASA. The satellite was built by Giacconi and his colleagues at AS&E and was designed as a simple, robust survey telescope with two gas-filled proportional counter detectors with angular resolutions of $0.5° \times 5°$ and $5° \times 5°$. The UHURU observatory conducted the first survey of the X-ray sky and revealed the true nature of the X-ray population (Giacconi *et al.*, 1971b). The X-ray sources turned out to include a wide variety of very hot objects – X-ray binaries with neutron stars and black holes as the 'invisible' companion, supernova remnants, young radio pulsars, active galactic nuclei and the intergalactic gas in clusters of galaxies. Some impression of the variety of sources present in the X-ray sky can be gained from the plot of the sources listed in the fourth UHURU catalogue (Figure 7.8). The history of the astrophysics of these sources is described in Part IV.

Over the next seven years following the launch of UHURU seven satellites with X-ray detectors were flown, including the Netherlands ANS satellite and the UK Ariel V satellite. The latter satellite included an X-ray spectrometer, which made the first detection of the X-ray emission line of 26 times ionised iron, Fe^{+26}, from the hot gas in the Perseus cluster of galaxies (Mitchell *et al.*, 1976). The next major survey instrument, launched in August 1977, was the NASA High Energy Astrophysical Observatory-A, HEAO-A, which can be considered to be a super-UHURU with about seven times greater sensitivity.

The next step was the development of X-ray telescopes with imaging capabilities, and this was achieved by the NASA HEAO-B satellite, which was named the *Einstein X-ray Observatory*. The problem to be overcome was the fact that ordinary mirrors do not reflect X-rays but absorb them. The only means of focussing X-rays with energies of about 1 keV is to make use of the phenomenon of grazing incidence reflection, in which the X-rays are deflected through angles of less than about 5°. As a result, imaging X-ray telescopes

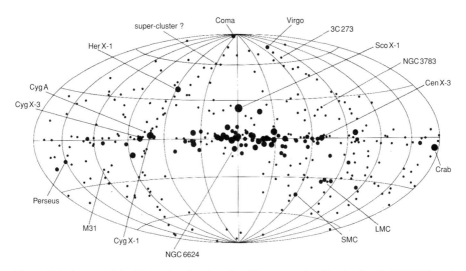

Figure 7.8: A map of the X-ray sky showing the objects contained in the fourth UHURU catalogue (Forman *et al.*, 1978). The names of X-ray sources associated with well known astronomical objects are indicated.

are very long, often consisting of a paraboloid–hyperboloid configuration to focus the X-rays onto the distant detector. These concepts were tested in rocket flights in the mid 1960s and were used for solar X-ray studies in the *Skylab* mission in the early 1970s. In addition, microchannel plate array detectors were developed to register the X-ray images. The Einstein X-ray Observatory was a complete X-ray observatory consisting of a suite of telescopes and instruments for high- and low-angular-resolution imaging as well as X-ray spectroscopy. The telescope was launched in November 1978 and opened up the detailed study of the astrophysics of the classes of source so far detected. Among the most intriguing finding was that essentially all classes of star can be X-ray emitters. The imaging quality of a few arcsec was achieved by the high-resolution camera, among the most impressive images being those of well known supernova remnants (Figure 7.9).

There was a natural progression from facilities such as the Einstein Observatory to large dedicated observatories for use by the astronomical community at large. In the NASA astronomy programme, the concept was developed of a series of 'Great Observatories' which were designed to be long-lived observatories in space. The four observatories involved in this programme were the Hubble Space Telescope (HST), the Gamma-Ray Observatory (GRO), the Advanced X-ray Astronomy Facility (AXAF) and the Space Infrared Telescope Facility (SIRTF). They were all planned to take advantage of the launch and service capabilities provided by the NASA Space Transportation System, more commonly known as the Space Shuttle. These very large and expensive missions were very major undertakings and were all subject to long development and construction phases.

Returning to the development of X-ray capabilities in space, the AXAF observatory was launched by the *Columbia* Space Shuttle in July 1999 and was then boosted into an elliptical high-Earth orbit, allowing long-duration uninterrupted exposures of X-ray sources. The observatory was named the *Chandra X-ray Observatory*. In parallel, the

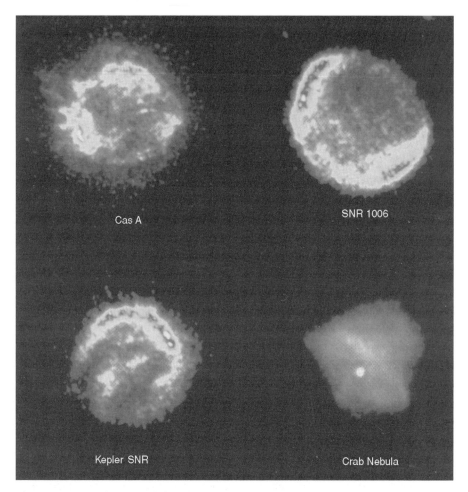

Figure 7.9: X-ray images of four well known supernova remnants observed by the Einstein X-ray Observatory. Notice the intense central source in the Crab Nebula which is associated with a rapidly rotating neutron star. From W. Tucker and R. Giacconi, *The X-ray Universe* (Cambridge, Massachusetts: Harvard University Press, 1985).

European Space Agency developed the concept of the XMM mission, the acronym meaning X-ray Multi-Mirror Telescope. The primary objective of this mission was high sensitivity and high spectral resolution, and this was achieved by a telescope system involving 58 nested paraboloid–hyperboloid mirrors and high-sensitivity CCD X-ray detectors. In addition, the telescope package included an optical monitor, allowing simultaneous optical–X-ray observations. These capabilities complemented those of the Chandra Observatory. The XMM mission was launched in December 1999 and renamed the *XMM–Newton X-ray Observatory*. Both missions have been very successful and have provided astronomers with data of extraordinarily high quality for the study of X-ray sources of all types.

At the same time, it was understood that, while these powerful facilities for X-ray astronomy enabled individual objects to be studied in exquisite detail, there remained the need to carry out surveys of the whole sky to understand in detail the nature of the population of X-ray sources. This objective was achieved by the German-led *ROSAT Observatory*, ROSAT standing for *Roentgen Satellit*, which was launched in June 1990. After nine years in space, an all-sky catalogue of 150 000 objects was compiled, many of these sources providing important targets for the Chandra and XMM–Newton Observatories.

7.5 Gamma-ray astronomy

In the early 1960s there was already military interest in placing γ-ray detectors in space in order to monitor the atmospheric nuclear test-ban treaties concluded between the USA and the USSR.[16] The Vela series of satellites was launched for this purpose in the 1960s, but there was no intention that they should have any astronomical role. Cosmic γ-rays were first detected in observations made by the Explorer II satellite in 1965 (Kraushaar *et al.*, 1965), but this experiment did little more than show that there existed γ-rays which originated from beyond the Earth's atmosphere. The first important astronomical observations were made by the third Orbiting Solar Observatory (OSO-III) launched in March 1967. The prime discovery of this mission was the detection of γ-rays with energies $E_\gamma > 100\,\mathrm{MeV}$ from the general direction of the Galactic centre (Clark, Garmire and Kraushaar, 1968). This γ-ray flux was convincingly interpreted as the γ-ray emission associated with the decay of neutral pions created in collisions between relativistic protons and the cold plasma of the interstellar gas.

These pioneering observations were followed up by balloon observations, but these suffered from severe contamination problems because of the production of secondary γ-rays as a result of interactions of primary cosmic rays with the nuclei of atoms in the atmosphere. These experiments provided important experience with compact spark chambers, which had originally been developed as detectors in high-energy physics experiments. The Small Astronomical Satellite, SAS-2, was launched in November 1972 and included an array of spark chambers to detect the electron–positron pairs created when an incoming γ-ray is converted into a pair within the instrument. Although it operated for only eight months and detected about 8000 γ-rays of cosmic origin, these were sufficient to make a number of key astronomical discoveries (Fichtel, Simpson and Thompson, 1978). Firstly, it was confirmed that there is a general concentration of γ-rays towards the plane of the Galaxy. Secondly, discrete sources of γ-rays were present, in particular two of the sources were associated with the pulsars in the Crab and Vela supernova remnants. Thirdly, evidence was found for diffuse extragalactic γ-ray background radiation.

The SAS-2 mission was followed in 1975 by the equally successful COS-B satellite lauched by a European consortium. It also consisted of an array of spark chambers sensitive to γ-rays with energies greater than about 70 MeV. It continued to take data continuously for six and a half years and resulted in a detailed map of the Galactic plane as well as evidence for 24 discrete γ-ray sources (Mayer-Hasselwander *et al.*, 1982).

The first evidence of γ-ray line emission came from balloon-borne telescopes in the early 1970s by the Rice University Group (Johnson and Haymes, 1973). In 1977, balloon observations confirmed that the line was the electron–positron annihilation line at 511 keV originating in the direction of the Galactic centre (Leventhal, MacCallum and Stang, 1978). Then, in 1984, definitive observations were made of the 1.809 MeV line of radioactive ^{26}Al by the HEAO-C satellite (Mahoney et al., 1984), this line also being detected from the direction of the Galactic centre.

To everyone's surprise, the Vela satellites, which were designed as γ-ray monitors for atmospheric nuclear tests, led to a key astronomical discovery in its own right. During the course of the monitoring, γ-ray bursts of astronomical origin were discovered, each γ-ray burst lasting typically less than one minute (Klebesadel, Strong and Olson, 1973). During that time, each burst was the brightest γ-ray source in the sky. The first of these was detected by the Vela satellite in 1967, but the bursts were not reported in the scientific literature until 1973. Their nature remained a mystery since the angular resolution of the γ-ray telescopes was very low and the short duration of the bursts made follow-up observations a very serious challenge.

The second of NASA's Great Observatories, the Gamma-Ray Observatory, was successfully launched by the Space Shuttle in April 1991 and was renamed the *Compton Gamma-ray Observatory* (CGRO), providing γ-ray astronomers with four different types of detector to explore different aspects of the γ-ray sky. These included a scintillation spectrometer for the detection of γ-ray emission lines (OSSE), a Compton telescope for exploring the difficult energy range 1–30 MeV (COMPTEL), an energetic γ-ray experiment consisting of the spark chamber detector (EGRET) sensitive to γ-rays with energies $E_\gamma \geq 100$ MeV and eight detectors placed at the corners of the satellite designed to monitor γ-ray transients and γ-ray bursts (BATSE). For all types of γ-ray study, these instruments were about an order of magnitude more sensitive than the previous generation of γ-ray telescopes. The result was a definitive map of the whole sky in γ-rays (Figure 7.10), showing clearly the plane of our Galaxy and a variety of discrete Galactic and extragalactic sources – in the third EGRET catalogue over 250 sources with photon energies greater than 100 MeV were listed. It was discovered that the distribution of the γ-ray bursts was isotropic over the sky, an important clue to their nature. In addition, the plane of the Galaxy was mapped in the radioactive decay line of ^{26}Al, and the most extreme active galactic nuclei were established as the sources of the intense, variable extragalactic γ-ray sources. The 13-ton space observatory remained nine years in orbit before being destroyed in a controlled atmospheric burn-up in June 2000.

As discussed in Section 7.2.2, extensive air-showers are initiated by very-high-energy cosmic rays entering the atmosphere, and among the products of the collisions between these particles and the nuclei of nitrogen and oxygen atoms are neutral pions, which subsequently decay into electron–positron pairs. In turn, the electrons and positrons produce high-energy γ-rays by bremssstrahlung, which are then converted into electron–positron pairs in interactions with nuclei, and so on. In 1948, Blackett realised that the speeds of the ultrarelativistic electrons and positrons were so close to the speed of light that they exceeded the local speed of light in the atmosphere, $v = c/n$, where n is the refractive index of the atmosphere. As a result, the ultrarelativistic electrons and positrons should emit optical Cherenkov radiation (Blackett, 1948). This prediction was confirmed several years later

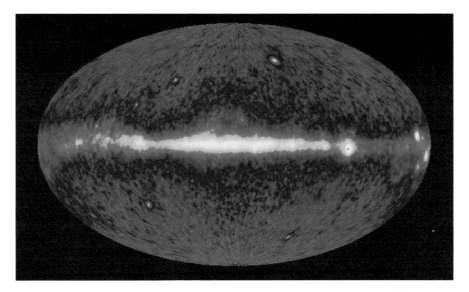

Figure 7.10: This map of the γ-ray sky at energies $E \geq 100\,\mathrm{MeV}$ was made by the EGRET telescopes of the Compton Gamma-ray Observatory. The emission from the Galactic plane consists mostly of γ-rays produced in the decay of neutral pions, π^0, generated in collisions between cosmic ray protons and nuclei and the interstellar gas. In addition, various Galactic and extragalactic discrete sources of γ-rays have been detected, including the pulsars in the Crab and Vela supernova remnants, the strange object Geminga and the luminous active galactic nuclei such as the quasar 3C 273. (Courtesy of NASA and the CGRO Science Team.)

by William Galbraith (b.1925) and John Jelley (1918–1997) (Galbraith and Jelley, 1953). Exactly the same technique can be used to detect ultra-high-energy γ-rays incident upon the top of the atmosphere.[17]

This technique can be used to detect γ-rays with energies typically in the range 300 GeV to 30 TeV, but it is a particularly demanding discipline since the fluxes of γ-rays are very low and have to be distinguished from very much more common optical pulses associated with extensive air-showers excited by cosmic rays. The Cherenkov light is distributed over an area similar to that of an extensive air-shower, and so a large light collector or array of smaller light collectors is used to detect the weak signal. For many years instruments such as the 10-metre Fred Lawrence Whipple Telescope at Mount Hopkins in the USA, the 18-mirror array at the Tata Institute of Fundamental Research at Ootacamund in India and the four-element γ-ray telescope of the University of Durham located at Dugway, Utah, produced results at the few sigma significance level, but it was not wholly convincing that significant detections had been made.

The breakthrough came with the development of imaging cameras for the light collectors, which enabled weak γ-ray sources to be discriminated with high efficiency against the background of cosmic-ray-induced extensive air-showers. In 1989, the group at the Fred Lawrence Whipple Observatory used a 37-pixel camera to detect γ-rays from the Crab Nebula at the 9σ level (Weekes *et al.*, 1989). The higher the angular resolution of the camera, the better the discrimination against background events. Two years later, with an

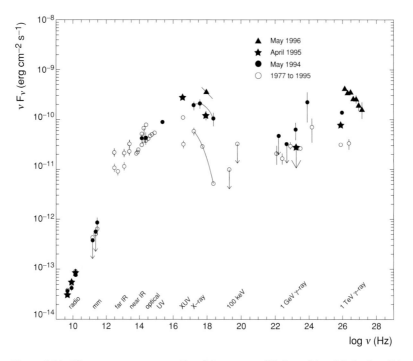

Figure 7.11: The energy spectrum, νF_ν, of the extreme BL Lac object Markarian 421 observed from the radio to the ultra-high-energy γ-ray waveband. The γ-ray source is highly variable. The lower energy γ-ray observations were made by the Compton Gamma-ray Observatory and the high-energy γ-ray points from observations by Atmospheric Cherenkov Imaging Telescopes. From M. Catanese and T. C. Weekes, Very high energy gamma ray astronomy, *Publications of the Astronomical Society of the Pacific*, **111**, 1999, 1193–1222.

upgraded camera with 109 pixels, the signals were detected at the 20σ level (Vacanti *et al.*, 1991). The detected photon rate from the Crab Nebula amounted to about two photons per minute. These detectors, known as Atmospheric Cherenkov Imaging Telescopes, were successfully developed by a number of groups and all detected the signal from the Crab Nebula. The importance of the Crab Nebula is that the flux of ultra-high-energy γ-rays is constant, being associated with the synchro-Compton radiation of high-energy electrons within the diffuse nebula as a whole, and so the source can be used as a calibrator. In subsequent years, ultra-high-energy γ-rays have been detected from the extreme BL Lac objects Markarian 421 and 501. Unlike the Crab Nebula, these sources are highly variable, but a large fraction of their emitted energy takes place in the $E \geq 300\,\text{GeV}$ energy band (Figure 7.11). The success of these observations has led to the development of a number of next-generation Atmospheric Cherenkov Imaging Telescopes.[18]

7.6   Ultraviolet astronomy and the Hubble Space Telescope

Among the earliest beneficiaries of the opening up of space for astronomy were the ultraviolet astronomers. The central figure in this story is Lyman Spitzer (1914–1997), who, in 1946,

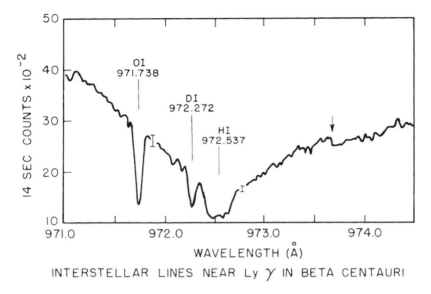

INTERSTELLAR LINES NEAR Ly γ IN BETA CENTAURI

Figure 7.12: The discovery of the Lyman-γ absorption line of deuterium in the interstellar medium in the far-ultraviolet spectrum of β Cen observed by the *Copernicus* satellite (Rogerson and York, 1973). The broad absorption feature is the interstellar Lyman-γ line of hydrogen centred at 97.2537 nm. The deuterium line is the weaker feature at wavelength 97.2272 nm.

was invited by the RAND project of the US Air Force to write a report on the utilisation of space for astronomical purposes.[19] Although it was many years before any of these ideas could become a reality, the seeds were sown early. The development of space ultraviolet astronomy followed the same pattern as that of X-ray astronomy. Firstly, there were rocket experiments, which gave a flavour of the scientific potential of the new waveband. Then, in 1957, as soon as the USSR space achievements galvanised the US space programme into action, Spitzer and his colleagues planned a series of three space observatories, to be known as the *Orbiting Astrophysical Observatories* (OAO), which were to be dedicated to spectroscopy in the ultraviolet waveband between 90 and 330 nm.

Unlike the other new astronomical wavebands, the astrophysical objectives of ultraviolet astronomy were well defined. The resonance transitions of essentially all the common elements lie in the ultraviolet rather than the optical waveband, and so studies of the different phases and chemical composition of interstellar matter are very effectively carried out in this waveband. OAO-3 was named the *Copernicus* satellite and was the great success of the series. The spectrographs were of very high spectral resolution and had the capability of exploring the wavebands to the short-wavelength side of the Lyman-α line at 121.6 nm (Rogerson *et al.*, 1973a). This capability was of special significance because, among the many resonance lines, those of deuterium are of great cosmological importance – one of the key discoveries of the mission was the detection of interstellar Lyman-β to Lyman-ε absorption lines of deuterium in the spectra of luminous blue stars such as β Cen (Figure 7.12) (Rogerson and York, 1973). In addition, observations of these deuterium transitions towards different stars showed that its interstellar abundance is remarkably constant wherever one looks in the local interstellar medium, corresponding to an abundance by mass relative to hydrogen of about 1.5×10^{-5}, a key result for astrophysical cosmology.

In addition, abundances of the common elements were measured in the interstellar medium for the first time. The mission also found evidence for a hot component of the interstellar gas through observations of the absorption lines of highly ionised oxygen, O^{5+}. The OAO observatories led in turn to the launch of the International Ultraviolet Explorer (IUE) in 1978, a joint UK–European Space Agency–NASA project. This was a spectacularly successful space astronomy mission and has had an impact upon essentially all branches of astronomy.[20]

In many ways the IUE was the precursor of the Hubble Space Telescope (HST).[21] Optical astronomers had long been aware of the fact that large ground-based telescopes never achieve their theoretical angular resolution because of refractive index fluctuations in the atmosphere which blur the images of stars to typically about 1 arcsec, the phenomenon known as astronomical 'seeing'. This figure should be compared with the theoretical angular resolution of a 4-metre telescope of about 0.03 arcsec. In addition to providing very much sharper astronomical images, the increase in angular resolution brings with it an increased sensitivity to point sources since, even on the darkest sites, a diffuse sky background is present due to atmospheric light scattering and emission. These problems are eliminated by placing the telescope above the Earth's atmosphere.

In the 1960s, plans were formulated for the construction of a Large Space Telescope, which would be an all-purpose optical–ultraviolet space observatory and which would build on the achievements of space ultraviolet astronomy. The plans were initially for a 3-metre space telescope, but this was reduced to 2.4 metres when the decision was taken to launch the telescope using the Space Shuttle. Not only would the telescope be launched by the Shuttle, but it would also be regularly serviced by it, enabling malfunctioning components to be replaced and new scientific instruments to replace the old. Since the telescope would be placed in low Earth orbit, it had to be reboosted into a higher orbit every few years during the servicing missions.

The approval process for the Hubble Space Telescope was not straightforward. In many ways, the leap from the small OAO class specialist missions of the 1960s to a fully equipped multi-purpose 2.4-metre telescope, which would operate at the diffraction limit, was a huge one and brought with it many technical problems. The biggest problem, however, was the fact that the project was very expensive, the cost estimates being greater than for any pure science programme ever undertaken. In the end, international collaboration was secured with the European Space Agency, which negotiated a 15% involvement in the telescope for European astronomers. The Ford administration approved the project in 1977, but the budget was very tight indeed.

The programme encountered major technical and financial difficulties within a couple of years of approval. The initial launch date was scheduled for the last quarter of 1983, but this soon proved to be over-optimistic. The crisis came in 1981 when the programme almost ran out of money. Managerial changes were made and a new, more realistic, budget was set for the programme. The programme was further delayed by the tragic loss of the *Challenger* Space Shuttle in 1986, which resulted in a cessation of Shuttle launches for a few years. Eventually, the telescope was launched in April 1990. Within weeks, it was found that the primary mirror had been figured to the wrong shape – the telescope system had an unacceptable amount of spherical aberration that blurred the images so that, without computer enhancement of the data, they were little better than what could be obtained from the ground. By the process of deconvolution, it proved possible to recover the full angular resolution capabilities of the telescope for imaging, but with very much reduced sensitivity.

NASA and the Space Telescope Science Institute immediately set about seeking solutions to the spherical aberration problem and came up with the concept of introducing correction optics into the optical train for the scientific instruments, which would restore the full capability of the HST for astronomical imaging and spectroscopy. On the first servicing mission of 1993, the correction optics were successfully installed, along with a new Wide Field Camera with correction optics built into the optical design. The full capability of the HST was restored and the results have been quite spectacular. The astronomical results have far exceeded the most optimistic expectations of the astronomers, and there is no field of astronomy which has not been impacted by its discoveries. NASA made the wise decision to make a significant investment in the public dissemination of the science and images obtained by the HST, with the result that the international public was immediately exposed to some of the most important and spectacular images ever taken by an astronomical telescope. There is no more remarkable picture than that of the Hubble Ultra-Deep Field, the results of a three-month exposure taken with the most recent camera, the Advanced Camera for Surveys (ACS), and the Near Infrared Camera and Multi-object Spectrometer (NICMOS) (Figure 7.13).

The remarkable success of the Hubble Space Telescope has encouraged NASA to plan the next leap forward in astronomical imaging and spectroscopy from space with the development of a 6–8-metre space telescope, optimised for observations in the near-infrared waveband, 1–5 μm, with capabilities to wavelengths as short as 0.5 nm. It is planned that this telescope, named the *James Webb Space Telescope*, will be launched in 2011 and will be placed at the second Lagrange point (L2), approximately 1.5 million kilometres from Earth, outside the orbit of the Moon. The region about L2 is a gravitational saddle point, where the telescope will remain at a roughly constant distance from the Earth throughout the year through a series of small spacecraft manoeuvres.

7.7 Infrared astronomy

The infrared radiation of the Sun was first detected in 1800 by William Herschel in his famous experiments in which he placed mercury-in-glass thermometers with blackened bulbs beyond the red end of an optical spectrum of the Sun. He found a greater temperature increase in what he termed the 'ultra-red' as compared with the red region of the spectrum and noted that these rays were refracted less than optical light (Herschel, 1800a–d). In his words,

there are rays coming from the Sun, which are less refrangible than any of those which affect the sight. They are invested with a high power of heating bodies, but with none of illuminating objects; and this explains why they have hitherto escaped unnoticed.

Figure 7.13: This image of the Hubble Ultra-Deep Field (HUDF) was obtained in a three-month set of observations using the Advanced Camera for Surveys (ACS) and the Near Infrared Camera and Multi-object Spectrometer (NICMOS) of the Hubble Space Telescope. Within a square area of angular size 3 arcmin, about 10 000 very distant galaxies have been detected. (Courtesy of NASA, ESA, Dr Steve Beckwith, the HUDF Team and the Space Telescope Science Institute.)

Note that this work was carried out two years before Thomas Youngs' pioneering papers on the 'Theory of light'. Because of the 'invisibility' of infrared radiation, this type of astronomy depended wholly upon the development of either 'thermal' detectors, in which the incident radiation gave rise to a temperature rise in the detector, or of 'non-thermal' devices, in which chemical or electronic transitions were excited by the infrared radiation.[22]

Highlights of nineteenth-century developments in infrared astronomy include measurements by Claude-Servais Pouillet (1790–1868) of the total heat flux of the Sun using his pyrheliometer, a small waterbath in a blackened enclosure with a thermometer to measure the temperature rise (Pouillet, 1838). Including a correction for absorption in the Earth's atmosphere, he found a value for the solar constant, the incident solar energy flux at the top of the atmosphere, of $1.44 \, \text{kW m}^{-2}$, in remarkable agreement with the present value of $1.37 \, \text{kW m}^{-2}$. Soon after, John Herschel discovered the broad absorption bands in the spectrum of the Sun, due to molecular absorption in the Earth's atmosphere, now called the *telluric* bands (Herschel, 1840). Herschel's experiments were carried out using a sheet of paper blackened by soot, which had been soaked in alcohol. The dispersed spectrum of the Sun dried out the alcohol in regions where the solar spectrum was strong and left damp the regions where four deep absorption troughs were found. In 1847, this remarkable observation was confirmed by Fizeau (1819–1896) and Foucault, who used sensitive alcohol thermometers with miniature bulbs (Fizeau and Foucault, 1847). In this same paper, they also showed that the infrared rays displayed the same properties as light, namely interference, polarisation and diffraction.

Thermoelectricity was discovered by Thomas Seebeck (1770–1831) at Jena in 1822 and resulted in the invention of the *thermocouple*, which consisted of pairs of dissimilar metallic strips, such as bismuth and copper, and which provided a better means of measuring tiny temperature differences than Herschel's thermometers. The thermocouples could be built into arrays which were known as *thermopiles*. The first observations of stars in the infrared waveband were made by William Huggins in 1868–1869, who used a thermopile with a small number of elements (Huggins, 1869). Huggins used the technique which would now be called 'nodding' to make observations of the star. This involved observing a blank field close to the star, allowing the galvanometer needle to settle, then moving the telescope onto the star, making the observation and then reobserving the same blank field. In his words,

The needle was then watched during five minutes or longer; almost always the needle began to move as soon as the image of the star fell upon it. The telescope was then moved, so as to direct it again to the sky near the star. Generally, in one or two minutes, the needle began to return to its original position. In a similar manner twelve to twenty observations of the same star were made.

By this means, Huggins made successful observations of Regulus, Arcturus, Sirius and Pollux, later confirmed by Edward Stone (1831–1897).

Perhaps the most important nineteenth-century figure in developing the techniques of infrared astronomy was Samuel Pierpoint Langley (1834–1906), who perfected the use of the bolometer for astronomical spectroscopy in the infrared region of the spectrum. Langley's bolometer used the fact that the resistance of platinum is highly temperature dependent. Platinum strips a few microns wide could be used and the tiny resistance changes were measured by a precision Wheatstone bridge so that it was possible to measure temperature changes as small as $10^{-4} \, \text{K}$. Langley devoted most of his efforts to bolometric observations of the Sun and, from observations made from the summit of Mount Whitney, extended measurements of the solar spectrum to $5.3 \, \mu\text{m}$ (Langley, 1886). In subsequent observations, he mapped the absorption lines in the solar spectrum out to $5.3 \, \mu\text{m}$, observing about 700 lines and measuring accurate wavelengths for 222 of them (Langley, 1900).

Towards the end of the nineteenth century, yet another approach to infrared photometry was developed based upon the radiometer invented by William Crookes (1832–1919). The device consisted of vanes which were blackened on one side and mounted on a support, such that when radiation was incident upon the vanes, they rotated. In the torsion radiometer, the vanes were blackened on the same side and, if there was a difference in the intensity of radiation on one compared with the other, there would be a net deflection of the vanes.

This device was used by Charles Abbot (1872–1973) to carry out the first stellar spectrora-diometry of stars. Using the device at the Coudé focus of the 100-inch Hooker telescope at Mount Wilson, Abbot measured the spectra of 9 stars in 15 wavebands in the wavelength region 0.437 to 2.224 μm, including four wavebands longer than 1 μm. This was the first time that spectrophotometric energy distributions were available for stars over a wide wave-length range, and they enabled black-body curves to be fitted to their spectra (Abbot, 1924). He found that the temperatures ranged from about 2500 K for α Herculis to 16 000 K for Rigel. Assuming that the stars behave like black bodies, their angular sizes could be worked out using the Stefan–Boltzmann law, and then, knowing the distances of the stars, their physical sizes could be estimated. Whereas stars such as Procyon had diameters similar to the Sun, Rigel and α Herculis had radii several hundred times that of the Sun. The results were generally in good agreement with the sizes found by Michelson and Pease's stellar interferometric observations (Michelson and Pease, 1921).

The final part of the pre Second World War story developed from the invention, by Peyotr Lebedev (1866–1912), of the vacuum thermocouple, which had markedly superior performance to the standard thermocouple because of the large decrease in thermal losses by conduction and convection. These detectors were optimised for use in astronomy by Herman Pfund (1878–1949) at the Allegheny Observatory and by William Coblentz (1873–1962) at the US Bureau of Standards. The most important of these observations were carried out by Edison Pettit (1889–1962) and Seth Nicholson (1891–1963), who overcame the inherent low efficiency of these detectors by using the Hooker 100-inch telescope at Mount Wilson (Pettit and Nicholson, 1928). During the 1920s, they carried out a programme of observations of 124 bright stars and came to similar conclusions to those of Abbot, but with much better statistics. The diameters of stars determined by the photometric technique were compared with those derived by Pease by optical interferometry, and reasonable agreement was found, the discrepancies being attributed to the fact that the photometric technique assumed that the spectra were black bodies.

Like many astronomical disciplines, infrared astronomy benefited from technological developments stimulated by the needs of the military, in particular the need to develop heat-seeking missiles. The first of a new generation of detectors for infrared astronomy was the lead sulphide cell (PbS), which was pioneered by Charles Oxley and Robert Cashman (1906–1988) and which could operate to wavelengths as long as 3.6 μm. The great advantage of this new class of semiconductor detector was that they were about 1000 times more sensitive than a thermopile. The first infrared spectroscopy of bright stars and planets was carried out by Gerard Kuiper (1905–1973) and his colleagues in 1947 (Kuiper, Wilson and Cashman, 1947), who incorporated the technique of rapid chopping of the beam to overcome the problem of detecting faint sources against the bright infrared background, which was also incident on the detector.

One of the major astronomical advantages of observing in the infrared waveband is that interstellar dust, which obscures many of the most interesting regions in gas clouds and galaxies, becomes transparent. The dependence of interstellar extinction upon wavelength had been determined in the optical waveband by Trumpler in the early 1930s (Trumpler, 1930) and, in more detail, by Stebbins, Hufford and Whitford using photoelectric techniques in 1940 (see Section 5.6). Interstellar extinction can be described by an extinction coefficient,

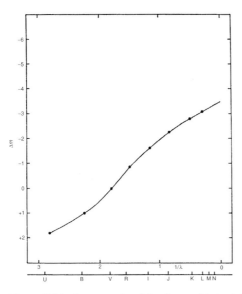

Figure 7.14: An example of the form of interstellar extinction curve derived by Johnson in 1965 (Johnson, 1965). Frequency is plotted on the abscissa in units of λ^{-1}, where the wavelength, λ, is measured in micrometres. The extinction is measured in magnitudes relative to magnitude 0 in the V waveband. At infinite wavelength, the extinction is zero, corresponding to $\Delta m = -3.5$.

α, such that $I = I_0\, e^{-\alpha r}$, where r is the distance. Joel Stebbins and his colleagues found that α was inversely proportional to wavelength in the optical waveband (Stebbins, Hufford and Whitford, 1940). This work was extended by Whitford, who observed distant supergiant stars and concluded that this power-law extinction law was a reasonable approximation out to 2 µm (Whitford, 1948) (see Figure 7.14).

The background radiation in the detector and its enclosure could be significantly reduced by cooling, and Harold Johnson (1921–1980) carried out the first systematic surveys of the stars of a wide range of spectral types, first using cooled PbS detectors and then, after 1961, a cooled indium antimonide detector, which had the advantage of operating at wavelengths as long as 5 µm. In 1962, he defined what have become the standard broad-band J (1.2 µm), K (2.2 µm), L (3.6 µm) and M (5.0 µm) infrared wavebands by means of interference filters (Johnson, 1962).

Also in 1961, Frank Low (b. 1933) pioneered observations at yet longer wavelengths through his development of the gallium-doped germanium bolometer, which operated at liquid helium temperatures (Low, 1961). These developments led to the definitions of the N (10 µm) and Q (22.2 µm) wavebands, at which the atmosphere has sufficient transparency to allow observations to be made from good high-altitude, ground-based sites (Low and Johnson, 1964; Low, 1966). In the late 1950s and early 1960s, Johnson and his colleagues measured the magnitudes of several thousand stars in a programme of UBVRIJKLMN photometry (Johnson et al., 1966). These careful studies confirmed the strong dependence of interstellar extinction upon wavelengths out to 2.2 µm (Figure 7.14). In addition, they enabled effective temperatures and bolometric corrections to be determined for stars of a wide range of spectral types.

At about the same time, Gerry Neugebauer (b. 1932) and Robert Leighton (1919–1997) of the California Institute of Technology began a major survey of the sky north of declination $\delta = -33°$ at 2.2 µm with a 62-inch telescope, or rather light-collector, which they built themselves (Neugebauer and Leighton, 1969). The images were about 4 arcmin in size but, using an array of eight lead sulphide detectors matched to the beam of the telescope, they found 5612 infrared sources brighter than K = 3. The *Two-micron Sky Survey* was of particular importance for astronomy because of its reliability and completeness. In many cases, the infrared emission represented the extension of the optical spectra of stars into the infrared waveband but, in addition, many more strong infrared emitters were discovered than were expected. About 50 stars were found to have (I–K) colour temperatures of only about 1000 K, while others turned out to be very intense emitters at far-infrared wavelengths, including objects such as the late M-supergiant NML Cygni and the heavily reddened carbon star IRC+10216.

At longer wavelengths, 4, 10 and 20 µm, the infrared sky was surveyed using helium-cooled germanium bolometers in a series of rocket flights by Russell Walker (b. 1931) and Stephan Price (b. 1941) of the Air Force Cambridge Research Laboratories (AFCRL) (Walker and Price, 1975). An important result of the AFCRL survey was that many of the brightest 20 µm sources were extended regions of ionised hydrogen, such as the Orion Nebula, with temperatures of only about 250 K, associated with hot dust grains within the ionised clouds.

In 1966, a key discovery was made by Eric Becklin (b. 1940) and Neugebauer, who made the first painstaking maps of the Orion Nebula at 1.65, 2.2, 3.5 and 10 µm with the Palomar 200-inch telescope (Becklin and Neugebauer, 1967). To their surprise, a very intense infrared 'star' was detected, not from the prominent optical nebula but from an obscured area to the north of the four trapezium stars which are responsible for illuminating the visible nebula (Figure 7.15(a)). The source was as luminous as the prominent trapezium stars, but there was no detectable optical emission from it. Their preferred interpretation was that the object was a massive protostar which was still enshrouded in a dusty envelope that absorbed the energy emitted by the protostar and reradiated it at far-infrared wavelengths. In another heroic paper of 1968, Becklin and Neugebauer made the first infrared maps of the Galactic centre in the H, K and L wavebands (Becklin and Neugebauer, 1968). The optical extinction to the Galactic centre was found to be about 25 magnitudes but, because of the rapid decrease in extinction with increasing wavelength, the Galactic centre region itself was observable for the first time (Figure 7.15(b)). They found evidence for an increase in the stellar density towards the centre as well as a compact region, coincident with the compact radio source Sagittarius A, which is associated with the dynamical centre of our Galaxy.

The great potential of the infrared waveband for these types of study led to the construction of telescopes optimised for observations in the infrared region of the spectrum. The UK Infrared Telescope (UKIRT) and the NASA Infrared Telescope Facility (IRTF), both located on the summit of Mauna Kea, began operation in the late 1970s and played a major role in making infrared observations an integral part of observational astrophysics. In the mid 1970s, lead sulphide detectors were replaced by the more sensitive indium antimonide detectors, and, by the late 1970s, detector technology and observing techniques had advanced to such an extent that galaxies as faint as K = 18 could be detected.

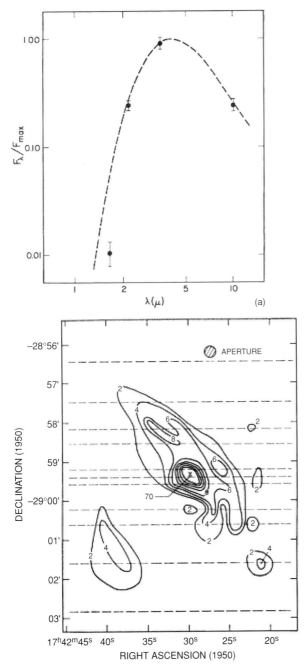

Figure 7.15: (a) The spectrum of the bright infrared star in the Orion Nebula, which became known as the Becklin–Neugebauer object. The dashed line shows the spectrum of a black body at a temperature of 700 K (Becklin and Neugebauer, 1967). (b) The first infrared map of the Galactic centre region mapped with an angular resolution of 0.25 arcmin at a wavelength of 2.2 μm. The central source coincides with the bright radio source Sagittarius A, which is located at the dynamical centre of the Galaxy (Becklin and Neugebauer, 1968). The dashed lines show the directions of the scans made to construct the infrared map.

In the mid 1980s, infrared array technology was declassified by the US military agencies, who had invested heavily in this technology for use as guidance devices for cruise missiles. These arrays had to be specially modified for astronomical use, particularly with regard to their uniformity and the reduction of the dark current in the detectors to very low values. With these developments, it at last became possible to build infrared array cameras and spectrographs with which to take images and spectra in the infrared waveband.[23]

The far-infrared regions of the spectrum cannot be observed from the surface of the Earth because of atmospheric absorption (see Figure 7.1). Pioneering experiments were carried out from high-flying aircraft and from balloon-borne platforms in the 1970s. Low carried out a number of exploratory programmes from a modified executive Learjet (Low and Aumann, 1970; Low, Aumann and Gillespie, 1970), which led in due course to the development by NASA of the Kuiper Airborne Observatory. This facility consisted of a Lockheed C-141 transport aircraft with a hole cut in the side to enable observations to be made with a 91-cm diameter telescope. Typically, the aircraft flew at an altitude of about 13 km, and observations could be carried out for about eight hours at high altitude.

The next natural step was to construct a dedicated satellite to undertake a systematic survey of the far-infrared sky. The Infrared Astronomy Satellite (IRAS), an international venture involving the Netherlands, the USA and the UK, was launched in January 1983, and the mission lasted ten months in space until the cryogens were exhausted. The whole sky was mapped in those infrared wavebands which are inaccessible from the ground, namely bands centred on 12, 25, 60 and 100 μm. About 250 000 infrared sources were discovered, and broad-band colours were measured for many of these. These observations have had a major impact upon essentially all branches of astronomy, but the outstanding contributions were made in the study of regions of star formation and the realisation that many galaxies emit as much radiation in the far-infrared waveband as they do at optical wavelengths.[24]

The success of the IRAS mission led to the development of the Infrared Space Observatory (ISO) by the European Space Agency. This satellite observatory operated in space from November 1995 to May 1998 and was a cryogenically cooled telescope which could operate in conditions of very low thermal background into the far-infrared region of the spectrum. The duration of the mission was determined by the lifetime of the cryogens which maintained the mirror and structure of the telescope at 4 K. The observatory was 1000 times more sensitive than IRAS with 100 times better angular resolution at 12 μm. In following up the IRAS survey, this mission began the exploitation of the science of the infrared wavebands which are inaccessible from the ground with array detectors and spectrographs.[25] This mission was followed by the NASA Space Infrared Telescope Facility (SIRTF), subsequently named the *Spitzer Space Telescope*, which was successfully launched in August 2003. The European Space Agency will launch the Far-Infrared Space Telescope (FIRST), named the *Herschel Space Observatory*, in 2008.

While these developments were taking place in the near-, mid- and far-infrared wavebands, millimetre astronomers were pushing their techniques to higher and higher frequencies. The story of the discovery of molecular lines and their role in understanding the physics of the interstellar medium is taken up in Chapter 9. Radio telescopes operating at centimetre and millimetre wavelengths were constructed, for example the Kitt Peak 12-metre telescope, but to make observations in the submillimetre waveband considerably

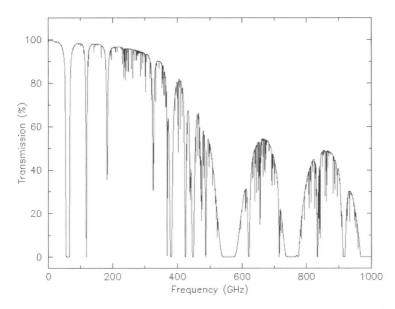

Figure 7.16: Illustrating the transparency of the atmosphere in the millimetre and submillimetre wavebands for 0.5 mm precipitable water vapour, a very good figure for a high, dry observing site such as the summit of Mauna Kea or the Cajnantor plateau in Chile. At frequencies greater than 300 GHz (1 mm wavelength), the main broad-band windows for which the transmission is adequate for astronomical observations are centred at about 350 GHz (850 μm), 450 GHz (650 μm), 650 GHz (450 μm) and 850 GHz (345 μm). For optimum detection of the astronomical signals, the submillimetre filters are matched to the transmission function of the atmosphere. (Data taken from the Caltech Submillimetre Observatory's atmospheric transmission estimator.)

higher surface accuracy was required. As a result, many early observations in submillimetre astronomy were carried out on optical or infrared telescopes. New types of heterodyne receivers had to be developed to enable spectral line observations at wavelengths less than 1 mm to be carried out. Schottky barrier diodes were superseded by SIS detectors, which made enormous strides in sensitivity throughout the period 1980 to 2000. These detectors enabled spectral line observations to be made in all the submillimetre 'windows' which are accessible to ground-based telescopes from high, dry observing sites (Figure 7.16). The big advantage of observing in these wavebands is that the spectral lines are relatively stronger than at millimetre wavelengths and probe deeper into regions which are optically thick at the longer wavelengths. In addition, the higher rotational transitions of molecules can be observed, for example the $j = 3 \rightarrow 2$ rotational transition of CO at 346 GHz and the $j = 6 \rightarrow 5$ transition at 691 GHz. There is a myriad of molecular lines in the submillimetre waveband, and, in high-resolution, high-sensitivity spectral scans, the noise signal is associated with the multitude of weak spectral lines in this region of the spectrum.

The opening up of the submillimetre waveband for continuum astronomical observations required the development of sensitive bolometric detectors which operated at liquid helium temperatures (4 K) or less in order to minimise the effects of thermal noise in the detectors. Germanium bolometer detectors were successfully developed by Low in the early 1960s and were used by Low and Hartmut Aumann (b. 1940) to make observations in the

50–300 μm waveband from the Learjet observatory (Low and Aumann, 1970). The important breakthrough came with the construction of *composite bolometers*, in which a tiny piece of germanium crystal acts as a very sensitive thermometer to detect the small temperature rise of an absorbing film consisting of a conducting metal film deposited on a dielectric substrate. These single-element detectors were developed during the 1970s and were used successfully on the UKIRT and IRTF telescopes in Hawaii, both of which had the great advantage of having chopping secondary mirrors. The first common-user bolometer for the submillimetre wavebands, UKT14, was built for UKIRT in the early 1980s. Heroic observations with this instrument established the fact that dust emission from star-forming regions and star-forming galaxies can be successfully observed in the submillimetre waveband. The potential of these line and continuum observations led to the construction of dedicated submillimetre telescopes, such as the 15-metre James Clerk Maxwell Telescope and the 10.4-metre CalTech Submillimeter Observatory (CSO) on Mauna Kea in Hawaii and the 15-metre Swedish–European Southern Observatory Telescope (SEST) in Chile.

The discipline of submillimetre astronomy really came of age in the mid 1990s with the development of the first common-user detector arrays, which were built into the massive SCUBA submillimetre camera on the James Clerk Maxwell Telescope (JCMT). The SCUBA array consisted of two back-to-back arrays of composite bolometers, principally for use in the 850 and 450 μm wavebands, and enabled observations in these wavebands to be made 1000 times more rapidly than had been previously possible. Among the most important discoveries was a population of very distant star-forming galaxies, which provide important information about the star-formation history of the Universe (see Chapter 14). The success and importance of these observations led to the Atacama Large Millimetre Array (ALMA) project, which involves constructing an array of 64 12-metre submillimetre telescopes operating as a synthesis array on the Cajnantor plateau at 5000 m altitude in the Atacama Desert in Chile. This international venture involving the USA, the European Southern Observatory and the Japanese National Astronomy Observatory is planned to begin observations in about 2011 and is expected to do for submillimetre astronomy what the HST has done for optical astronomy.

7.8   Optical astronomy in the age of the new astronomies

In parallel with the technical advances which expanded enormously the wavebands available for astronomical observation, optical astronomy developed out of all recognition over the same period. In the 1960s, the construction of a number of 4-metre class optical telescopes was begun with the intention of providing improved access for many more astronomers to world-class observing facilities.

In the USA, the need for national telescopes in addition to the private observatories was recognised with the founding in May 1960 of the Association of Universities for Research in Astronomy (AURA). This federally funded organisation had the responsibility of building and operating telescopes for the US astronomical community, the principal facilities being the 4-metre Mayall Telescope at the Kitt Peak National Observatory (KPNO) in Arizona, the Sacramento Peak Solar Observatory in New Mexico and the 4-metre Blanco telescope at the Cerro Tololo Inter-American Observatory (CTIO) at La Serena in Chile, building upon the American astronomers' long association with observatories in Chile.[26]

Within Europe, it was realised that, while the USA had capitalised upon its advantage in the northern hemisphere, the many riches of the southern hemisphere, including the Galactic centre and the Magellanic Clouds, had yet to be explored with large telescopes. The European Southern Observatory (ESO) was set up in 1962 'to establish and operate an astronomical observatory in the southern hemisphere, equipped with powerful instruments with the aim of furthering and organising collaboration in astronomy'. Operation of the 3.6-metre telescope at La Silla, a 2400 m mountain bordering the southern extremity of the Atacama desert in Chile, began in 1977.

The UK had long-standing interests in the southern skies, dating from the time of the Herschels and the foundation of the Cape and Pretoria Observatories in South Africa. The UK did not become a founding member of ESO but joined with Australia to construct the 3.9-metre Anglo-Australian Telescope on Siding Spring Mountain in New South Wales, which was completed in 1975.[27] France, Canada and Hawaii agreed to collaborate in the construction of the 3.6-metre telescope on the summit of Mauna Kea in Hawaii, one of the best and highest all-round sites for astronomy in the world, and the telescope began operations in 1979. The UK, Spain and the Netherlands constructed an observatory for the northern skies on the island of La Palma in the Canary Islands, including the 4.2-metre William Herschel Telescope, which was completed in 1987.

7.8.1 *Electronic detectors for astronomical telescopes*

Just as important as the availability of large telescopes for the community as a whole was the development of electronic detectors, which, by the end of the twentieth century, had largely replaced the photographic plate as the preferred means of recording astronomical images and spectra.[28] The photoelectric effect was discovered by Heinrich Hertz (1857–1894) while he was carrying out his brilliant experiments which demonstrated that electromagnetic waves have all the properties of optical light in 1885–1887. It was not until the 1920s, however, that photoelectric photometry began to make an impact upon astronomy with the development of electronic vacuum tubes. These devices had the advantage of having a linear response over a wide dynamic range and so enabled the calibration of the magnitudes of stars and galaxies to be carried out much more effectively. The first photomultiplier tubes which had a major impact upon astronomy were constructed by Vladimir Zworykin (1889–1953) at the RCA laboratories. The principle of these devices is that the incoming photon causes a secondary electron cascade to take place though a series of dynodes so that each detected photon results in a very short burst of electrons at the anode. The efficiency of detection of the photons is only limited by the quantum efficiency of the first stage of photon detection. In fact, the primary use of these photomultiplier tubes was for the soundtracks of movies in the motion picture industry. These devices were first used by Albert Whitford and Gerald Kron (b. 1913) as an autoguider for the 60-inch telescope at Mount Wilson, but became the preferred means of calibrating magnitude scales after the Second World War.

The next step in the application of advanced electronics to optical astronomy came with the development of image intensifiers. These were off-shoots of the television industry

Figure 7.17: Willard S. Boyle (left) and George E. Smith, the inventors of the charge-coupled device (CCD), demonstrating the imaging capabilities of their patented CCD camera in 1974. (Courtesy of the AT&T Laboratories.)

and, in particular, their development for low-light level applications for military purposes during the 1960s and 1970s. The principle of these devices is that each photon detected by the photocathode results in an electron cascade, as in the photomultiplier, but now the electron beam is focussed onto a light-emitting screen which is scanned by a television camera. The arrival of each detected photon is registered and the image reconstructed by photon counting. These types of systems, including the Vidicon System developed at the Westinghouse Corporation and the Image Photon Counting Systems developed by Alexander Boksenberg (b. 1936), completely transformed the spectroscopy of faint objects during the 1970s. They are ideal for faint objects since the counting rate is limited to about one photon per pixel during the time it takes the television system to register the arrival of the photon. The Faint Object Camera of the Hubble Space Telescope used this technology for the imaging of faint objects in the ultraviolet region of the spectrum.

In 1969, the charge-coupled device (CCD) was invented by Willard Boyle (b. 1924) and George Smith (b. 1936), who were working at the Bell Telephone Laboratories at Murray Hill, New Jersey (Boyle and Smith, 1970). Their objective was to develop the technology for a 'Picturephone', which would enable telephone callers to see each other (Figure 7.17). The semiconducting materials which detect the photons can have very high quantum efficiencies and then the ejected electrons are stored in potential wells within the semiconductor material. The problem is how to extract the signals without undue losses. This is where the process of charge-coupling plays a key role. Once the signal is accumulated on the chip, the electrons are shuffled along the rows of the detector array and are read out by a single amplifier at the end of the row. The first 100×100 arrays were introduced in 1973 and the patent for the device was received in 1974. The astronomers realised the potential of these devices for astronomy and they were developed under contract from the Jet Propulsion Laboratory by Texas Instruments, who built the first devices specifically for astronomy in 1976. The development of these devices for astronomy received an enormous boost by their selection as the preferred detectors for the Wide Field Camera of the Hubble Space Telescope in 1977. Since then, CCDs have dominated optical astronomy in providing directly digital images with very high quantum efficiency. Rather than the 1% achievable by photographic plates, the CCDs can have quantum efficiencies of up to about 80%, the equivalent of increasing the collecting area of the telescope by a factor of 80. In parallel with their development for astronomy, CCD detectors have come to dominate the domestic camera market, and digital cameras now contain CCD chips with many millions of pixels. In the same way, the sizes of astronomical quality CCD chips have become very large, so that, except for very-wide-field imaging, the days of the photographic plate have passed into history.

7.8.2 New technologies for large telescopes

The Palomar 200-inch telescope was the ultimate in traditional telescope building. As one writer put it, 'the Hale telescope stands among all telescopes as the climax of dreadnought design' and represented the pre-Second World War 'brute force' approach to telescope construction. There have been many changes in the concepts for large-telescope design since then. During the 1970s, for example, telescopes were built short and stubby, which had the advantage of making the structure more rigid for a given amount of steel and also meant that the building which enclosed the telescope could be much smaller. These changes both brought with them large cost advantages.

For many years, 4 metres was regarded as the largest feasible aperture for an optical telescope because, if the traditional approach is taken to maintaining the figure of the mirror as it points at different elevations and to preventing the telescope structure bending under gravity, the cost of the telescope increases as a high power of its diameter, D, roughly proportional to D^4. The most important revolution in telescope design, which began in the 1980s, was the realisation that the cost–diameter relation can be profoundly changed if the telescope and mirror are allowed to deform under gravity, but computer-controlled actuators change the figure of the mirror and the pointing of the telescope so that the telescope always remains in correct focus as it points to different parts of the sky. To express this in another way, the cost of high-speed computers decreased so dramatically that it proved much more cost effective to put the money into compensating for the floppiness of the mirror and telescope by computer control, rather than by making them rigid structures. These concepts were developed by ESO in the design and construction of the 4-metre New Technology Telescope (NTT). The first 8–10-metre-class telescope to exploit similar concepts was the Keck 10-metre telescope, which has a segmented mirror consisting of 36 hexagonal off-axis mirrors which are computer-controlled and which provide excellent subarcsecond imaging at the Mauna Kea site in Hawaii.

In 1987, the European Southern Observatory obtained approval for the programme to construct the Very Large Telescope (VLT), to consist of four 8.2-metre telescopes located at Cerro Paranal in the Atacama Desert in northern Chile. Their combined collecting aperture is equivalent to a 16-metre optical–infrared telescope. The thickness of each 8-metre mirror was less than 20 cm and their figures were maintained by computer-controlled actuators in the mirror cell. A similar approach was adopted by the Japanese 8-metre Subaru Telescope located on Mauna Kea and the two 8-metre telescopes of the International Gemini Observatory, one located on Mauna Kea and the other at Cerro Pachon close to Cerro Tololo in Chile. The latter project involves the USA, the UK, Canada, Brazil, Australia, Argentina and Chile. Several other 8-metre telescope projects are currently approaching completion. These telescopes are very complex systems involving a great deal of computer control. Likewise, the astronomical instruments are very large and complex so that it is most effective if the observations are made by the experts at the telescopes rather than in the traditional mode in which the astronomer travels to the telescope to make the observations. The high cost of observing time makes it much more cost effective to carry out the observing programme in 'queue mode', meaning that observations are planned to maximise the time the telescope is taking astronomical data for a large number of approved programmes, rather than each programme being allocated a specific observing period.

Another great advance has been a much deeper understanding of the phenomenon of astronomical seeing and ways in which it can be minimised for observations with ground-based telescopes. In the design of the new generation of 8–10-metre telescopes, many precautions are taken to eliminate the effects of local seeing caused by the fact that the telescope is located within a telescope dome. For example, the temperature of the mirror is carefully controlled so that it is not a source of thermal convection cells and the telescope domes have huge thermal vents so that, when observing, the telescope is essentially in the open air (Figure 7.18). The net result of all these precautions is that the intrinsic seeing of the new generation of large telescopes is about 0.4 arcsec, this residual figure being caused by refractive index fluctuations in the upper layers of the atmosphere.

The next challenge facing the astronomical technologists has been to eliminate the effects of astronomical seeing. Once local effects associated with the enclosure have been eliminated, the residual seeing at optical–infrared wavelengths is due to refractive index fluctuations in the upper atmosphere which cause distortions of the wavefronts of the incoming signals. The scale of these distortions is measured by Fried's parameter, r_0, which is the diameter over which the root-mean-square wavefront fluctuation is 1 radian (Fried, 1965). For a good astronomical site, $r_0 \sim 15$–20 cm at 500 nm, giving a seeing-limited resolution of $\theta \approx \lambda/r_0 \sim 0.6$ arcsec. The Fried parameter increases with wavelength as $r_0 \propto \lambda^{6/5}$ so that the intrinsic seeing improves at longer wavelengths as $\lambda^{-1/5}$. The principle of *adaptive optics* is to compensate for these wavefront distortions by measuring them in real time and introducing compensating wavefront corrections in order to 'flatten' the wavefront, thus obtaining diffraction-limited images. Although these principles have been understood for a number of years, their effective implementation for astronomy only became a priority with the construction of the new generation of 8-metre telescopes and the need to understand how to construct the next generation of 20–50-metre telescopes.

Another important development in astronomical technology has been the development of aperture synthesis techniques for optical and infrared wavelengths. The principles of optical interferometry had been laid down by Michelson and implemented in his pioneering measurements of the angular diameters of red giant stars (Michelson and Pease, 1921). In

Figure 7.18: A view of the interior of the Gemini North telescope enclosure at sunset showing the fully open thermal vents and the fully open observing slit. (Courtesy of the International Gemini Observatory.)

these observations, the visibility of the target star at different baselines was used to estimate the angular size of the star. Michelson built a larger interferometer with a baseline of 50 feet (15.2 m), but this was at the limit of mechanical stability (Pease, 1931).

The techniques of aperture synthesis at radio wavelengths were well understood by the time Antoine Labeyrie (b. 1943) published his seminal paper in which he observed interference fringes from the bright star Vega using a pair of small telescopes separated by 12 metres (Labeyrie, 1975). These observations indicated how optical interferometry could produce images of very high angular resolution, but there were many technical problems to be overcome. The sky fluctuations, which cause the stars to twinkle, needed to be recorded at kilohertz frequencies and required the development of sensitive photon-counting detectors. In order to combine coherently the light from separated telescopes, micrometre-level metrology of variable optical delay lines had to be constructed, which only became possible when stabilised lasers became available. Finally, in order to produce images, the phases as well as the amplitudes of the correlated signals had to be measured. This problem had been solved by the radio astronomers involved in VLBI observations using the technique of *closure phases*, which enables phases to be determined when many separate baselines are available (Rogers *et al.*, 1974). This procedure is now standard in radio interferometry and is referred to as *self-calibration* (Pearson and Readhead, 1984). These technical challenges required advanced control systems engineering, state-of-the-art detectors and high-speed computation.

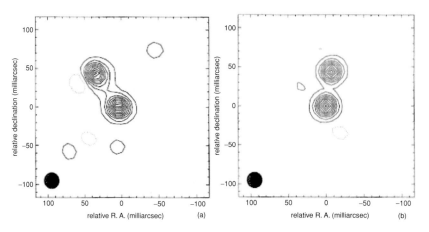

Figure 7.19: The first images from an optical aperture synthesis array made with three telescopes of the COAST interferometer at a wavelength of 830 nm (Baldwin *et al.*, 1996). These images of the binary star system Capella were taken on (a) 13 September 1995 and (b) 28 September 1995. The 20 arcsec restoring beam is shown in the bottom left of each image.

From the mid 1980s onwards, John Baldwin (b. 1931) and his colleagues at Cambridge used their experience of aperture synthesis at radio wavelengths to construct an optical aperture synthesis array, COAST, which enabled these technical problems to be overcome. Their first maps at an angular resolution of about 20 milliarcsec were published in 1996, showing milliarcsecond motion of the stars of the binary system Capella over an interval of 15 days (Figure 7.19) (Baldwin *et al.*, 1996). Since that time, a number of imaging optical–infrared aperture synthesis arrays have been developed and have produced important scientific results. These include precise measurements of stellar diameters, orbital studies of close binary stars enabling precise masses to be determined, brightness profiles across stellar discs, providing tests of the theory of stellar atmospheres, and angular diameter changes in pulsating stars, testing the theory of stellar pulsation. The next generation of optical interferometers will be observatory-class facilities with the sensitivity to measure features in active galactic nuclei on the scale of a milliarcsecond.[29]

7.8.3 Survey astronomy

Astronomical surveys of the whole sky are at the heart of many important studies because they provide statistical information about the relative importance of different classes of stars and galaxies and are also the means for discovering rare classes of astronomical objects such as quasars. In all the new wavebands described in the previous sections, sky surveys were among the most important priorities in opening them up for scientific exploration.

In the optical waveband, large-scale sky surveys were first carried out using very-wide-field telescopes, which enable large regions of sky to be observed in a single exposure. The widest field telescopes for all sky surveys were the Schmidt telescopes, which used an innovative optical design invented by Bernhard Schmidt (1879–1935) in 1929 (Schmidt, 1931). Observations with this type of telescope were pioneered by Zwicky in the 1930s (see Section 8.10). Immediately after the Second World War, a large Schmidt telescope of effective aperture 1.2 metres (48 inches) was constructed to support observations made with the 200-inch telescope at Mount Palomar. The size of each plate was 14 inches, corresponding to about 6° on the sky and, over a period of eight years, this telescope completed a photographic survey of the whole sky north of declination −20° in blue and red wavebands. Photographic copies of the survey plates were made available to the worldwide community of astronomers, and the resulting *Palomar Sky Atlas* proved to be an important research tool for astronomers from all wavebands.

The northern hemisphere had a monopoly of large telescopes, and it was only in the 1960s that this imbalance began to be rectified. The European Southern Observatory and the UK constructed Schmidt telescopes similar to the Palomar telescope in the southern hemisphere to carry out the same type of survey as had been completed in the northern sky. With the use of new emulsions, particularly the IIIaJ emulsions, these Schmidt telescopes were able to reach significantly fainter magnitudes than the northern surveys. The surveys also took about seven years to complete and have provided crucial databases for the whole of astronomy, including an astrometric database for observations to be made with the Hubble Space Telescope. These large sky surveys contain an enormous amount of statistical data of importance for astronomy, but quantitative data could only be extracted if suitable high-speed measuring machines were built for this purpose. UK astronomers took the lead in these developments with the construction of the COSMOS High-Speed Measuring Machine at the Royal Observatory, Edinburgh, and the Automatic Plate Measuring Machine (APM) at Cambridge. These studies have provided many of the most important targets to be observed by the 4-metre-class telescopes, for example in the discovery of large complete samples of radio-quiet quasars.

While these surveys provided targets for large telescopes, they yielded little spectral information. In particular, the redshifts of the galaxies and quasars found in these surveys had to be determined individually. To overcome this problem, multi-object spectrographs were developed to enable large numbers of spectra of faint objects to be obtained in a single exposure. An excellent example of this approach was the 2° field (2dF) multi-object spectrograph designed for the Anglo-Australian Telescope. The top end of the telescope was redesigned to provide a 2° field of view and within that area 400 spectra of faint galaxies and quasars could be measured simultaneously. The resulting 2dF Galaxy Redshift Survey (2dFGRS) and the 2dF quasar survey were major spectroscopic surveys which took full advantage of these unique capabilities. The 2dF survey obtained spectra for almost 250 000 objects, mainly galaxies, including about 25 000 quasars. These data have enabled a very wide range of astronomical and cosmological questions to be addressed.

Even more ambitious is the Sloan Digital Sky Survey, which has been carried out by a dedicated 2.5-metre telescope with a 3° field of view at the Apache Point Observatory, in Sunspot, New Mexico. Unlike previous sky surveys in the optical waveband, the survey used entirely digital CCD detectors to map about one-quarter of the whole sky. The survey uses the technique of shifting the image on the detector electronically at the sidereal rate as the sky moves over the telescope, a technique known as the Time-Delay and Integrate (TDI) mode. This technique was first used to undertake quasar searches using the 200-inch telescope in transit mode by Donald Schneider (b. 1955), Maarten Schmidt and James Gunn (Schmidt,

Schneider and Gunn, 1986; Schneider, Schmidt and Gunn, 1994). The Sloan Survey detector array consists of 30 2048 × 2048 CCD detectors which provide simultaneous observations in five different wavebands. The telescope can also be operated in spectroscopic mode, enabling 600 spectra to be taken in a single exposure. It is planned that the positions and magnitudes of more than 100 million celestial objects will be determined. The telescope will also measure the redshifts of up to a million galaxies, providing a three-dimensional map of the universe, as well as measuring the redshifts of about 100 000 quasars. The results of the Sky Survey are made available to the scientific community electronically, both as images and in the form of catalogues.

7.9   Other types of astronomy

In addition to exploiting the unique properties of electromagnetic radiation for astrophysical studies, other disciplines have already contributed to astrophysics and cosmology in important ways. The development of cosmic ray physics was discussed in Section 7.2, and these observations provide direct evidence for high-energy particles accelerated in cosmic environments. In addition, *neutrino studies* of the Sun have provided information of importance for particle physics and astrophysics and have become an important growth area. The significance of neutrino astrophysics was reinforced by the remarkable discovery of neutrinos from the supernova 1987A. *Astroparticle physics* has become a major growth area through the attempt by laboratory physicists to detect the particles which may constitute the dark matter in our Galaxy. *Gravitational waves* have been inferred to be emitted by the binary pulsar PSR 1913+16, but gravitation waves themselves have not yet been detected directly. The new generation of large gravitational-wave detectors is now attaining the sensitivities at which a positive result can be reasonably expected in the near future.

The history of these developments will be told in their astrophysical context in Parts IV and V.

Notes to Chapter 7

1 For more details of Hale's remarkable contributions, see H. Wright, J. N. Warnow and C. Weiner, eds, *The Legacy of George Ellery Hale* (Cambridge, Massachusetts: MIT Press, 1972).
2 From the historical perspective, the 200-inch telescope included a number of major advances which were built into succeeding generations of large telescopes.
   - The primary mirror was constructed of Pyrex$^{®}$ by the Corning Glass Works company because of its low coefficient of expansion.
   - The mirror was mass-reduced by creating a hexagonal cellular structure within the Pyrex$^{®}$.
   - The f-ratio of the telescope was reduced to $f/3.3$ to reduce the length of the telescope tube and so reduce the size of the enclosure.
   - The weight of the telescope was supported on oil-pads rather than floated in mercury.
   - Serrurier trusses were used to maintain the separation of the primary and secondary mirrors. The result was that, even when the telescope tube bent under gravity, the primary and secondary mirrors remained parallel and aligned.
   - The mirrors were coated with aluminium rather than silver, with the result that the mirror needed to be recoated less frequently.

3 Information on the development of astronomy internationally can be obtained from the *Proceedings of the General Assemblies of the International Astronomical Union*, which have been held at regular three-yearly intervals since 1922, except during the years of the Second World War.

4 Röntgen's paper was published in December 1895. It was also published, in English, in 1896 in *Nature*, **53**, 274–276.

5 An excellent documentary history of studies of cosmic rays is provided by A. Michael Hillas, *Cosmic Rays* (Oxford: Pergamon Press, 1972).

6 In Dirac's two great papers of 1928, he used the relativistic wave equation, known in his honour as the Dirac equation, to show that the magnitude of the intrinsic angular momentum, or spin, of the electron is given by $\hbar s(s+1)$, where $s = 1/2$, that its magnetic moment is $e\hbar/m$ and that there exist negative-energy solutions as well. There was considerable debate before the physical nature of the negative solutions was understood. At first, they were thought to correspond to protons, but they had to have the same mass as the electron. Only in 1931 did Dirac come down decisively in favour of the interpretation that the negative solutions correspond to positively charged electrons, the positrons (Dirac, 1931). Anderson discovered these particles in the following year.

7 The early history of radio astronomy has been surveyed in W. T. Sullivan III, ed., *The Early Years of Radio Astronomy* (Cambridge: Cambridge University Press, 1984). Sullivan has also edited a compilation of important early papers on radio astronomy: W. T. Sullivan III, *Classics in Radio Astronomy* (Dordrecht: D. Reidel Publishing Company, 1982). An amusing account of some of the personalities involved in the history of radio astronomy is contained in the conference volume: K. Kellermann and B. Sheets, eds, *Serendipitous Discoveries in Radio Astronomy* (Green Bank, Virginia: National Radio Astronomy Observatory Publications, 1983).

8 This paper reported the observation of intense radio emission associated with a large solar flare which occurred on 27 and 28 February 1942. The information was declassified after the War. Hey wrote his own account of the early history of radio astronomy in James Hey, *The Evolution of Radio Astronomy* (New York: Science History Publications, 1973). He also wrote a touching autobiographical account of these discoveries in his short book, James Hey, *The Secret Man* (Eastbourne: Care Press, 1992).

9 See the discussion 'The origin of cosmic radio noise' at the Conference on *Dynamics of Ionised Media* held in 1951 at University College, London.

10 The original 300-foot telescope, which operated in transit mode, collapsed in 1988 and has been replaced by a new, fully steerable, 100-metre telescope which can operate at frequencies as high as 100 GHz (3 mm wavelength).

11 The history of Martin Ryle's discovery of Earth-rotation aperture synthesis is delightfully told by Peter Scheuer in his article 'The development of aperture synthesis at Cambridge' in W. T. Sullivan III, ed., *The Early Years of Radio Astronomy* (Cambridge: Cambridge University Press, 1984), pp. 249–265.

12 The history of X-ray astronomy is told by W. Tucker and G. Giacconi, *The X-ray Universe* (Cambridge, Massachusetts: Harvard University Press, 1985).

13 The history of space exploration is described in M. Rycroft, ed., *The Cambridge Encyclopaedia of Space* (Cambridge: Cambridge University Press, 1990).

14 For the early history of ultraviolet observations of the Sun, see H. Friedman, *Sun and Earth* (New York: Scientific American Library, 1986).

15 In fact, this fluorescent emission was observed by the ROSAT soon after its launch in 1991 (see Figure 14.5(a)).

16 The development of γ-ray astronomy is described in P. Ramana Murthy and A. A. Wolfendale, *Gamma-ray Astronomy* (Cambridge: Cambridge University Press, 1993).

17 Reviews of the recent history of ultra-high-energy γ-ray astronomy are given by F. A. Aharonian and C. W. Akerlof, Gamma-ray astronomy with imaging atmospheric Cherenkov telescopes, *Annual Reviews of Nuclear Science*, **47**, 1997, 273–314, and M. Catanese and T. C. Weekes,

Very high energy gamma-ray astronomy, *Publications of the Astronomical Society of the Pacific*, **111**, 1999, 1193–1222.

18 A survey of planned Atmospheric Cherenkov Imaging Telescopes, many of which have now come into operation, is given by B. L. Dingus, *26th International Cosmic-ray Conference*, Conference Proceedings 516 (Melville, New York: American Institute of Physics, 2000), pp. 351–364.

19 See R. W. Smith, *The Space Telescope: A Study of NASA, Science, Technology and Politics* (Cambridge: Cambridge University Press, 1989), p. 30.

20 A comprehensive survey of the many areas of astronomy in which the IUE made major contributions is contained in Y. Kondo, ed., *Exploring the Universe With the IUE Satellite* (Dordrecht: D. Reidel Publishing Company, 1987).

21 The history of the Hubble Space Telescope project is splendidly told by Smith, *The Space Telescope*. The second edition (1994) includes a discussion of the problem of the spherical aberration of the primary mirror and its solution using correction optics.

22 A splendid history of the trials and tribulations of the pioneers of infrared astronomy is given in Chapter 6 of John Hearnshaw, *The Measurement of Starlight: Two Centuries of Astronomical Photometry* (Cambridge: Cambridge University Press, 1996). An introduction to the post-War history of infrared astronomy is given by D. A. Allen, *Infrared: The New Astronomy* (Shaldon, Devon: Keith Reid Ltd, 1975).

23 Some of the very first astronomical images taken with infrared array cameras in the infrared waveband are included in C. G. Wynn-Williams and E. E. Becklin, eds, *Infrared Astronomy with Arrays* (Honolulu: Institute for Astronomy, University of Hawaii, 1987). The papers in this volume indicate the technological challenges that had to be overcome to enable these array technologies to be adapted for astronomical purposes.

24 The proceedings of the first symposium dedicated to the results of the IRAS mission were reported in F. P. Israel, ed., *Light on Dark Matter* (Dordrecht: D. Reidel Publishing Company, 1985).

25 Good examples of the quality of the science achieved by the ISO mission are given in the review by R. Genzel and C. Cesarsky, Extragalactic results from the Infrared Space Observatory, *Annual Reviews of Astronomy and Astrophysics*, **38**, 2000, 761–814.

26 The history of AURA's involvement in these telescope projects is recounted by F. K. Edmondson in *AURA and its US National Observatories* (Cambridge: Cambridge University Press, 1997).

27 The history of the Anglo-Australian Telescope is described in detail in the excellent book by S. C. B. Gascoigne, K. M. Proust and M. O. Robins, *The Creation of the Anglo-Australian Observatory* (Cambridge: Cambridge University Press, 1990).

28 An excellent introduction to all types of electronic imaging detectors is given by I. S. McLean in *Electronic Imaging in Astronomy: Detectors and Instrumentation* (Chichester: Wiley-Praxis Series in Astronomy and Astrophysics, 1997).

29 An excellent review of the principles and status of optical aperture synthesis is given by J. E. Baldwin and C. A. Haniff, 'The application of interferometry to optical astronomical imaging', *Philosophical Transactions of the Royal Society of London*, **A360**, 2002, 969–986.

Part IV

The astrophysics of stars and galaxies since 1945

Many astrophysicists and cosmologists refer to the years since 1945 as the 'golden age' of astrophysics and cosmology. The areas pioneered before the Second World War began to flourish vigorously, and completely new vistas were opened up as a result of the expansion of the wavebands accessible for astronomical observation and of discoveries in fundamental physics. The background to these developments was outlined in Part III, and, in Part IV, the astrophysics of stars, the interstellar gas, galaxies, clusters of galaxies and high-energy astrophysical phenomena are discussed. Part V is devoted to the achievements of astrophysical cosmology.

8 Stars and stellar evolution

8.1 Introduction

By 1945, many of the physical processes involved in the evolution of stars on the main sequence were beginning to be understood, but there remained an enormous amount of detailed work to be undertaken before a precise comparison between theory and observation could be made. To build detailed models of the stars, three types of data are required. The first is the equation of state of the material of the star; the second are accurate nuclear reaction rates; and the third is the opacity of stellar material for the transfer of radiation. These quantities need to be known for the wide ranges of temperature and density encountered inside the stars. Then, the problems of radiation transfer through the body of the star and its surface layers have to be solved so that meaningful comparisons can be made between the theory and observations. As a result, the astrophysicists had to have access to a very wide range of data from nuclear, atomic and molecular physics, which began to become available with the great expansion in the funding for the physical sciences after the Second World War.

Then, there was the need to develop models for the evolution of stars from one region of the Hertzsprung–Russell diagram to another. It was a daunting task, but there was light at the end of the tunnel with the development of high-speed digital computers in the 1950s and 1960s, which was to convert the study of the structure and evolution of the stars into a precise astrophysical science. The new wavebands brought important new insights into many of the key phases of stellar evolution using techniques which could not have been imagined by the pioneers of the first half of the twentieth century.

8.2 Nucleosynthesis and the origin of the chemical elements

Two of the problems in the development of stellar astrophysics discussed above were closely related. The first concerned the processes responsible for the synthesis of the chemical elements, and the second related to the nuclear processes responsible for energy generation once stars had moved off the main sequence. Although the CNO cycle could convincingly account for the synthesis of helium in stars with masses greater than about $1 M_\odot$, how were the carbon, nitrogen and oxygen created in the first place?

The big problem was that there are no stable isotopes with mass numbers 5 and 8. As a result, there is no straightforward way in which protons, neutrons and α-particles can be added successively to helium nuclei as the first of a sequence of reactions which lead to the formation of carbon. The solution was first proposed by Öpik in 1951 and was worked out independently in more detail by Salpeter in 1952. They pointed out that, when the central temperature of the star reaches about 4×10^8 K, the triple-α reaction, in which three α-particles come together to form carbon, can take place (Öpik, 1951; Salpeter, 1952). The process may be thought of as consisting of the formation of ^{8}Be, which, being highly unstable, exists for a very short time before disintegrating into two α-particles; during that time, however, there is a small probability that a third α-particle reacts with the ^{8}Be to form ^{12}C. There remained the problem that the cross-section for the reaction ^{8}Be $+\alpha \rightarrow ^{12}$C was too small to create a significant abundance of carbon.

The problem was solved by Fred Hoyle in 1953, who realised that the cross-section for the interaction would be increased if there is a resonance associated with formation of ^{12}C in an excited state (Hoyle, 1954). Hoyle estimated that the excited state of ^{12}C should occur at about 7.7 MeV. This was a remarkable prediction in that, at that time, models of nuclei were not sufficiently well developed that any resonance state of any nucleus could be predicted. Ward Whaling (b. 1923) and his colleagues were persuaded to search for the resonance and found it at exactly the energy predicted by Hoyle (Dunbar et al., 1953). Subsequent experiments by William Fowler (1911–1995) and his colleagues in 1957 established the details of the sequence of reactions. The energetics of the formation of ^{12}C are as follows:

$$2\alpha + 94 \text{ keV} \rightarrow {}^8\text{Be}, \tag{8.1}$$

$$^8\text{Be} + \alpha + 278 \text{ keV} \rightarrow {}^{12}\text{C}^*, \tag{8.2}$$

$$^{12}\text{C}^* \rightarrow {}^{12}\text{C} + 2\gamma + 7.654 \text{ MeV}. \tag{8.3}$$

The inclusion of the carbon resonance increased the cross-section for the formation of carbon by the triple-α process by a factor of 10^7. Hoyle went on to show that helium burning can take place at a temperature of 10^8 K, the temperature deduced by Allan Sandage (b. 1926) and Martin Schwarzschild (1912–1997) for the cores of red giant stars at the tip of the giant branch (Sandage and Schwarzschild, 1952).

Öpik and Salpeter realised that once the carbon had been created, heavier elements, such as oxygen and neon, could be created by the successive addition of α-particles. In his paper of 1954, Hoyle went on to argue that, once the star had exhausted the helium in its core, massive enough stars would continue to contract, increasing the central temperature in the star so that the nuclear burning of ^{12}C into ^{24}Mg would take place and, at a slightly higher temperature, ^{16}O would be converted into ^{32}S. The process of nuclear burning would continue in massive enough stars until all the nuclear energy resources were used up, that is when the core of the star consists of ^{56}Fe, the element with the greatest nuclear binding energy of the chemical elements.

In 1956, Hans Suess (1909–1993) and Harold Urey published their detailed analysis of the cosmic abundances of the elements (Suess and Urey, 1956). The primary sources for their abundance determinations were the chondritic meteorites, which have abundances similar to those of the photosphere of the Sun – it is commonly assumed that these meteorites have preserved the primordial chemical composition out of which the Sun and the Solar System were formed. These abundances were in reasonable agreement with the solar abundances

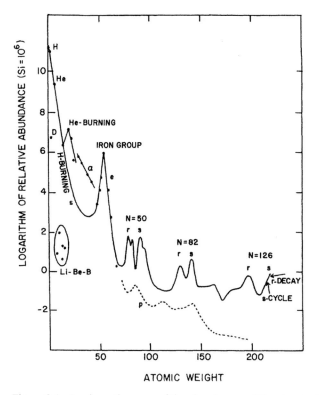

Figure 8.1: A schematic curve of the abundances of the chemical elements as a function of atomic weight based on the data of Suess and Urey (1956), who used relative isotopic abundances to determine the slope and general trend of the curve. Burbidge and her colleagues drew special attention to the overabundances associated with α-particle nuclei with atomic weights $A = 16$, 20 and 40, the peak at the iron group, and the twin peaks at $A = 80$ and 90, at 130 and 138 and at 194 and 208 (Burbidge *et al.*, 1957).

and those of Population I stars. Suess and Urey not only worked out elemental abundances, but also the isotopic abundances, which proved to be important discriminators of different processes of nucleosynthesis. The abundances of the chemical elements fall off rapidly with increasing atomic weight, but there are important features of the abundance curves which provide clues to the processes of nucleosynthesis (Figure 8.1).

The nuclear processes involved in the synthesis of the elements were described in two famous papers published in 1957, one by Margaret Burbidge (b. 1919), Geoffrey Burbidge (b. 1925), Fowler and Hoyle, commonly known by the acronym B^2FH, and the other by Alastair Cameron (1925–2005) (Burbidge *et al.*, 1957; Cameron, 1957). The B^2FH paper drew attention to the overabundance of the 'α-particle' nuclei such as those with 16, 20 and 32 nucleons, as well as to the iron-group elements and the peaks of stability at $N = 50$, 82 and 126. These peaks of stability corresponded to the 'magic numbers' of nuclear physics, which had already been noted by Walter Elsasser (1904–1991) in 1933 (Elsasser, 1933). In the B^2FH paper, eight nuclear processes by which the elements could be synthesised were described. In addition to hydrogen burning, helium burning and the α-process, they drew

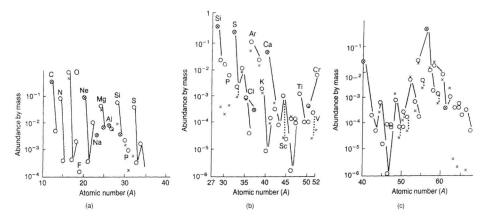

Figure 8.2: Examples of the products of explosive nucleosynthesis from calculations by Arnett and Clayton (1970). In these computer simulations, shells of carbon, oxygen and silicon were heated rapidly to a very high temperature, as in a supernova explosion, and the nucleosynthesis takes place in an expanding cooling shell. The peak temperatures reached were (a) 2×10^9 K in the case of carbon burning, (b) 3.6×10^9 K in the case of oxygen burning and (c) $4.7–5.5 \times 10^9$ K in the case of silicon burning. The circles represent the observed solar abundances and the crosses show the products of explosive nucleosynthesis.

special attention to processes involving the addition of neutrons to pre-existing nuclei, the slow (s) and rapid (r) processes. These reactions provide the means by which nuclei with mass numbers greater than the iron group can be synthesised. If iron nuclei are found in neutron-rich environments, they can absorb neutrons, and the process is termed 'rapid' or 'slow' depending upon whether or not the product nucleus decays before another neutron is added to the nucleus. In the r-process, several neutrons are added before decay occurs. At high enough temperatures and densities, the inverse β-decay process results in the formation of large numbers of neutrons. Supernova explosions were identified as the sites in which such reactions could take place. The s-process was believed to occur at an earlier stage in the evolution of stars on the giant branch.

The general picture that emerged from these nucleosynthesis studies was that, the more massive the star, the further it would proceed through the sequence of nuclear burning before it collapsed to some form of dead star, such as a white dwarf or neutron star. Thus, many of the abundant elements, such as carbon, oxygen and silicon, are synthesised through steady nuclear burning, which occurs at an advanced stage of evolution on the giant branch. Cameron drew particular attention to the importance of nucleosynthesis in supernova explosions, a process that is now referred to as *explosive nucleosynthesis*. He realised that different chemical abundances are created if the process of nucleosynthesis takes place in a non-stationary manner, as in the case of supernova explosions. With the development of high-speed computers, it became possible to quantify these predictions. Although it was not possible to simulate the explosion of complete stars, it was possible to carry out explosive nucleosynthesis calculations for shells of particular elements. In 1970, David Arnett (b. 1940) and Donald Clayton (b. 1935) showed how many of the element abundances could be naturally attributed to explosive nucleosynthesis for shells of carbon, oxygen and silicon[1] (Figure 8.2) (Arnett and Clayton, 1970).

8.3   Solar neutrinos

While the astrophysicists were refining their models of the Sun on the basis of the study of its surface properties, Raymond Davis (b. 1914) suggested in 1955 that it might be possible to search for the electron neutrinos liberated in the nuclear reactions which take place in the CNO cycle (see Section 3.5) (Davis, 1955). This proposal was made before the neutrino, predicted as long ago as 1934 by Fermi, was first measured experimentally in the laboratory by Clyde Cowan (1919–1974) and Frederick Reines (1918–1998) in 1956 (Cowan *et al.*, 1956; Reines and Cowan, 1956). Because of their very small cross-sections for interaction with matter, neutrinos escape essentially unimpeded from their point of origin within the central 10% of the Sun by radius, and thus the detection of the flux of solar neutrinos provides a direct test of the processes of nucleosynthesis. Davis proposed detecting the solar neutrinos by the nuclear transformations which they would produce in a fluid which contained a large number of chlorine atoms. Specifically, the nuclear reaction

$$^{37}\text{Cl} + \nu_e \rightarrow \ ^{37}\text{Ar} + e^- \tag{8.4}$$

has a threshold energy of 0.814 MeV. The argon created in this reaction is radioactive, and the amount produced can be measured from the number of radioactive decays of the ^{37}Ar nuclei. Unfortunately, as pointed out by Isadore Epstein (1919–1996) and Beverley Oke (1928–2004), the p–p chain rather than the CNO cycle is the principal source of energy in the Sun (Epstein, 1950; Oke, 1950). Neutrinos are, however, emitted in the p–p chain:

$$\text{p} + \text{p} \rightarrow \ ^{2}\text{H} + e^+ + \nu_e \quad : \quad ^{2}\text{H} + \text{p} \rightarrow \ ^{3}\text{He} + \gamma. \tag{8.5}$$

The first reaction, in which deuterium is formed, is the principal source of neutrinos from the Sun; these neutrinos are of low energy, however, the maximum energy being 0.420 MeV, and so could not be detected by a chlorine detector. In 1958, Cameron and Fowler independently pointed out that more energetic neutrinos are emitted in a side-chain of the main p–p chain (Cameron, 1958; Fowler, 1958). There are three routes for the formation of helium, the most straightforward and likely being the pp1 branch:

$$\text{pp1} : \ ^{3}\text{He} + \ ^{3}\text{He} \rightarrow \ ^{4}\text{He} + 2\text{p}. \tag{8.6}$$

The other routes involve the formation of ^{7}Be as a first step:

$$^{3}\text{He} + \ ^{4}\text{He} \rightarrow \ ^{7}\text{Be} + \gamma. \tag{8.7}$$

Then, ^{7}Be can either interact with an electron (the pp2 branch) or, very rarely, a proton (the pp3 branch) to form two ^{4}He nuclei:

$$\text{pp2} : \ ^{7}\text{Be} + e^- \rightarrow \ ^{7}\text{Li} + \nu \ : \ ^{7}\text{Li} + \text{p} \rightarrow \ ^{4}\text{He} + \ ^{4}\text{He}, \tag{8.8}$$

$$\text{pp3} : \ ^{7}\text{Be} + \text{p} \rightarrow \ ^{8}\text{B} + \gamma \ : \ ^{8}\text{B} \rightarrow \ ^{8}\text{Be}^* + e^- + \nu_e, \tag{8.9}$$

$$^{8}\text{Be}^* \rightarrow 2\ ^{4}\text{He}. \tag{8.10}$$

The electron neutrinos emitted in the decay of ^{8}B nuclei have maximum energy 14.06 MeV and so could be detected in the type of experiment proposed by Davis.

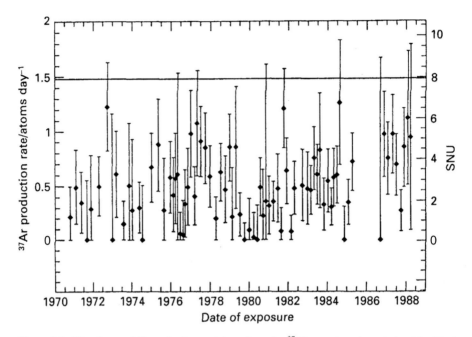

Figure 8.3: The observed flux of solar neutrinos from the ^{37}Cl experiment carried out by Davis and his colleagues for the period 1970–1988. The solid line at 8 SNU is the expectation of the standard solar model of Bahcall and Ulrich (Bahcall, 1989).

The first detailed predictions of the solar neutrino flux were made by John Bahcall (1934–2005) in 1964 (Bahcall, 1964), and, at about the same time, the famous *solar neutrino experiment* was begun by Davis and his colleagues using a 100 000 gallon tank of perchloroethylene C_2Cl_4 located at the bottom of the Homestake gold-mine in South Dakota. Davis found the neutrino flux to be significantly less than the value predicted by the solar models. Indeed, by the time of their 1976 review, Bahcall and Davis found a positive signal at only the one-sigma level (Bahcall and Davis, 1976). The predictions of the neutrino flux improved over the years as the solar models were refined and the nuclear cross-sections determined with greater accuracy.[2] As the statistics improved over succeeding years, a significant flux of neutrinos was detected but, over the 18 years illustrated in Figure 8.3, it corresponded to only about one-quarter of the flux predicted by the standard solar models. This discrepancy is the famous *solar neutrino problem*. The results quoted by Bahcall in 1989 were as follows:

$$\text{observed flux of neutrinos}: \qquad 2.1 \pm 0.9 \text{ SNU}, \qquad (8.11)$$
$$\text{predicted flux of neutrinos}: \qquad 7.9 \pm 2.6 \text{ SNU}, \qquad (8.12)$$

where 1 SNU = 1 Solar Neutrino Unit = 10^{-36} absorptions per second per ^{37}Cl nucleus (Bahcall, 1989). The errors quoted are formal 3σ errors for both the observations and the predictions.

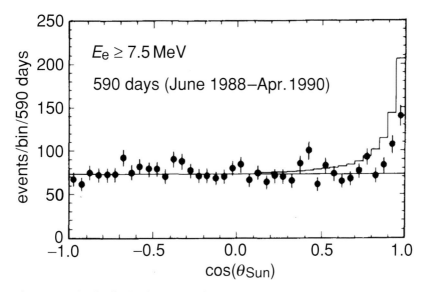

Figure 8.4: The distribution in $\cos(\theta_{Sun})$ for the 590-day sample for $E_e \geq 7.5$ MeV, where θ_{Sun} is the angle between the momentum vector of an electron observed at a given time and the direction of the Sun. The isotropic background, which is roughly 0.1 events day^{-1} bin^{-1}, is due to spallation products induced by cosmic ray muons, γ-rays from outside the detector and radioactivity in the detector water. The angular resolution of the detector system has been taken into account in calculating the expected distribution of arrival directions of the neutrinos from the Sun, which is indicated by the histogram (Hirata et al., 1990).

The origin of this discrepancy became one of the most controversial topics in astrophysics after Davis's results were first reported in 1968. They suggested that there must be something wrong either with the nuclear physics, or with the astrophysics of the Sun, or with both.[3]

Confirmation that the flux of high-energy neutrinos indeed originated within the Sun was provided in 1990 by the Japanese Kamiokande II experiment, which had the great advantage that the arrival directions of the incoming neutrinos are measured (Hirata et al., 1990). The high-energy neutrinos scatter electrons which recoil with relativistic velocities. The Cherenkov detectors which line the walls of the Kamiokande II experiment measure the direction of travel of the scattered electrons, and thus the arrival directions of the neutrinos can be found. The results of 590 days of observation are shown in Figure 8.4. There was a small, but significant, excess flux of neutrinos coming from the direction of the Sun, but it was less than that expected from the standard solar model of Bahcall and Roger Ulrich (b. 1942) (Bahcall and Ulrich, 1988). As the Kamiokande II team stated:

These provide unequivocal evidence for the production of ^{8}B by fusion in the Sun.

The final results quoted by the Kamiokande II team from 1036 days of observations from January 1987 to February 1995 were:

$$\text{measured flux of neutrinos} = 2.56 \pm 0.16\,\text{(stat)} \pm 0.16\,\text{(syst)}, \qquad (8.13)$$

where (stat) refers to the statistical errors and (syst) refers to the systematic errors (Fukuda et al., 1996).

The Kamiokande II experiment was upgraded with an active volume of 32 000 tons of pure water and renamed super-Kamiokande. The rate of detection of high-energy neutrinos was greatly enhanced and, from 1258 days of observation, their flux was found to be:

$$\text{measured flux of neutrinos} = 2.32 \pm 0.03 \, (\text{stat})^{+0.008}_{-0.007} \, (\text{syst}), \qquad (8.14)$$

again in agreement with the earlier results and those of Davis (Fukuda *et al.*, 2001).

A key test of the solar models is the detection of the low-energy neutrinos from the first interaction, (8.5), of the p–p chain, since this is the essential first step in the synthesis of helium and is much more directly related to the luminosity of the Sun than the high-energy neutrinos. The best approach for measuring the much more plentiful low-energy neutrinos is to use gallium as the detector material and to measure the neutrino flux from the number of radioactive germanium nuclei created by the neutrino interaction:

$$\nu_e + {}^{71}\text{Ga} \rightarrow e^- + {}^{71}\text{Ge}. \qquad (8.15)$$

During the early 1990s, two international collaborations, GALLEX and SAGE, reported the results of these demanding experiments, which typically require the use of 30 tons of gallium to produce a significant result. Over the period 1992 to 1997, the GALLEX collaboration provided successively improved estimates for the flux of low-energy neutrinos from the detector located in the Laboratori Nazionali del Gran Sasso in the Abruzzi region of central Italy. The final result of the experiment was:

$$\text{measured flux of neutrinos} = 77.5 \pm 6.2 \, \text{SNU} \qquad (8.16)$$

(Hampel *et al.*, 1999), significantly less than the flux of $129 \, {}^{+8}_{-6}$ SNU expected from the improved standard solar models of Bahcall and his colleagues (Bahcall *et al.*, 1997). The result reported by the SAGE experiment, located at the Baksan Neutrino Observatory in the northern Caucasus mountains of Russia, was similar:

$$\text{measured flux of neutrinos} = 70.9^{+5.3}_{-5.2} (\text{stat})^{+3.7}_{-3.2} (\text{syst}) \qquad (8.17)$$

(Abdurashitov *et al.*, 2002, 2003).

There had been a great deal of speculation about the solution of the solar neutrino problem, but, by the early 1990s, helioseismological experiments were beginning to demonstrate that the standard models of the Sun were a remarkably precise description of its internal structure right into the central nuclear-burning regions (see Section 8.4), and so the focus of theorists began to centre upon the physics of the neutrino. In 1990, Bahcall and Bethe proposed that the phenomenon of *neutrino oscillations* could account for the observed discrepancy (Bahcall and Bethe, 1990). The phenomenon known as the MSW effect had been discussed by Lincoln Wolfenstein (b. 1923) and by Stanislav Mikheyev (b. 1940) and Alexei Smirnov (b. 1951) and involves physics beyond the standard model of particle physics, in which neutrinos can change their type in the presence of matter (Wolfenstein, 1978; Mikheyev and Smirnov, 1985). Thus, although the neutrinos created in the nuclear reactions in the Sun are electron neutrinos, and this is the number predicted by the standard solar models, as they propagate through the matter of the Sun, they can change into muon and tau neutrinos. A consequence of this process is that the neutrinos have finite rest mass.

The test of this picture is to determine what fraction of the detected high-energy neutrinos are electron neutrinos; this experiment has been carried out by combining the results of the super-Kamiokande experiment with those of the Sudbury Neutrino Observatory (SNO) located at Sudbury in Ontario, Canada. The SNO detector consists of 1000 tons of heavy water, D_2O, and can be operated in a mode which is sensitive to the electron neutrinos alone, whereas super-Kamiokande is sensitive to all types of neutrino. The SNO consortium found that the electron neutrino flux was indeed significantly less than the flux quoted in equation (8.14), that is

$$\text{measured flux of } \nu_e = 1.75 \pm 0.07 \,(\text{stat})\,^{+0.12}_{-0.11}\,(\text{syst}) \pm 0.05 \,(\text{theory}) \qquad (8.18)$$

(Ahmad *et al.*, 2001). Using the ratio of these fluxes, they estimated that the total flux of electron neutrinos emitted by the Sun is 5.44 ± 0.99 SNU, in good agreement with the most recent estimates of the predicted neutrino flux by Bahcall and his colleagues. This is undoubtedly one of the most remarkable discoveries of modern astrophysics and demonstrates again the role of astrophysics in making discoveries which strike right to the heart of fundamental physics. More recently, the same phenomenon of neutrino oscillations has been measured in laboratory experiments (Eguchi *et al.*, 2003).

8.4 Helioseismology

The study of the internal structure of the Sun remained the province of theoretical astrophysics with little prospect of testing the theory directly until the discovery of *solar oscillations* in the 1960s. These oscillations were first observed by Robert Leighton and his colleagues (Leighton, 1960; Leighton, Noyes and Simon, 1962) who discovered 'five-minute' oscillations in their studies of the velocity field of the solar atmosphere. The nature of these oscillations was an unsolved problem throughout the 1960s. In a prescient paper of 1968, Edward Frazier (b. 1939) suggested that the oscillations were trapped acoustic waves in the outer layers of the Sun (Frazier, 1968). The first detailed analyses of the normal modes of oscillation of the Sun were carried out by Roger Ulrich in 1970 and by John Leibacher (b. 1941) and Robert Stein (b. 1935) in 1971 (Ulrich, 1970; Leibacher and Stein, 1971). The 'five-minute' oscillations were identified with standing acoustic waves confined to the outer layers of the Sun. Frazier had made the first plot of the modes of oscillation of the Sun on a frequency–wave number plot, but it showed no structure. The first analysis to show clearly the 'ridges' in the dispersion relations for the different modes of oscillation of the Sun, the (k, ω) diagram, was carried out by Franz-Ludwig Deubner (b. 1934) in 1975 (Deubner, 1975). He compared these results with predictions of improved models of the spectrum of solar oscillations by Hiroyasu Ando (b. 1946) and Yoji Osaki (b. 1938) (Ando and Osaki, 1975) and found a disagreement between the observations and the theory. The discrepancy was resolved by Douglas Gough (b. 1941) in 1978, who pointed out that the depth of the convection zone in the standard models of the Sun had been underestimated by about 50% (Gough, 1977). These studies revealed the great potential of this approach for investigating the internal structure of the Sun.

In 1976, Andrei Severny (1913–1987) at the Crimea Astrophysical Observatory and the Birmingham group led by George Isaak (1933–2005) reported the discovery of oscillations with much longer periods of 20–60 minutes and 160 minutes which had quite different properties from the 'five-minute' oscillations (Brooks, Isaak and van der Raay, 1976; Severny, Kotov and Tsap, 1976). Oscillations of the diameter of the Sun were also reported by Henry Hill (b. 1933) in 1976 as a by-product of his high-precision experiments to measure the oblateness of the Sun (Brown, Stebbins and Hill, 1976).

It is no exaggeration to say that studies of the astrophysics of the Sun were revolutionised by these discoveries. In terrestrial seismology, the resonance modes of the Earth can be found by tracing the paths of sound waves inside the Earth. Exactly the same procedure can be employed to study the resonant modes of the Sun, and this new astrophysical discipline was named *helioseismology*. The theory of the modes of oscillation of the Sun is a beautiful example of the power of classical mathematical physics applied to an astrophysical problem. It is remarkable that many of the basic techniques of analysis, involving the description of the normal modes of oscillation of the Sun in terms of associated Legendre polynomials, is contained in the sixth edition of 1932 of the classic text *Hydrodynamics* by Horace Lamb (1849–1934) (Lamb, 1932). The modes of oscillation of the Sun can be thought of as standing waves resulting from the interference of oppositely directed propagating waves. The modes of oscillation are of two types: *acoustic* or *p modes*, in which the restoring force is provided by pressure gradients; and *gravity* or *g modes*, for which the restoring force is buoyancy. The modes of greatest interest for the study of the internal structure of the Sun are the acoustic modes of small degree, l, since these probe into the central nuclear burning regions. The existence of these low-degree modes was deduced by Jörgen Christensen-Dalsgaard (b. 1950) and Gough from the data published by the Birmingham group in 1979 (Christensen-Dalsgaard and Gough, 1980).

During the 1980s and 1990s, there were major observational campaigns to determine the dispersion relations for the modes of oscillation of the Sun. Examples of the quality of the data available at that time are shown in Figures 8.5(a) and (b). The first diagram shows the (k, ω) diagram determined from four months of observation in 1986 and 1988 with the helioseismology telescope at the Big Bear Observatory, showing the 'ridges' in the (k, ω) relation associated with different modes of oscillation of the Sun (Woodard and Libbrecht, 1988). The error bars shown are 1000σ errors, giving some impression of the extraordinary power of these techniques. The second is the power spectrum of the total luminosity of the Sun obtained from the IPHIR experiment on board the PHOBOS space probe while in transit between the Earth and Mars in 1988 and 1989 (Toutain and Frölich, 1992). The power spectrum displays both large and small splittings of the spectral lines.

Ground-based campaigns continued through the 1980s and 1990s with projects such as the GONG and Birmingham global networks of solar telescopes providing 24-hour coverage of the solar oscillations. These studies were greatly advanced by the launch of the European Space Agency's Solar and Heliospheric Observatory (SOHO) in 1995, which provided an ideal observatory for continuous observation of the Sun in space. The instrumental payload included a Michelson Doppler Imager (MDI), which enabled the two-dimensional velocity structure of the solar oscillations on the surface of the Sun to be imaged. These observations enabled the three-dimensional internal structure of the Sun to be determined with remarkable precision.

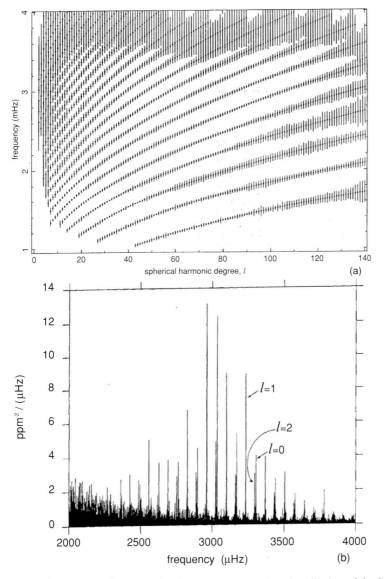

Figure 8.5: (a) The (k, ω) relation for the normal modes of oscillation of the Sun determined from four months of observations in 1986 and 1988 with the helioseismology telescope at the Big Bear Solar Observatory. The modes are p modes, and each 'ridge' contains modes with a fixed number of radial nodes, n, the lowest frequency ridge having $n = 1$. The error bars shown are 1000σ (Woodard and Libbrecht, 1988). (b) An example of the frequency spectrum of solar oscillations showing some of the normal modes of oscillation of the Sun. These data were derived from 160 days of observation by the IPHIR experiment on board the PHOBOS spacecraft (Toutain and Frölich, 1992). This power spectrum of the low-degree p modes shows an alternating pattern of double and single peaks; the double peaks are the $l = 0, 2$ modes and the single peaks the $l = 1$ modes.

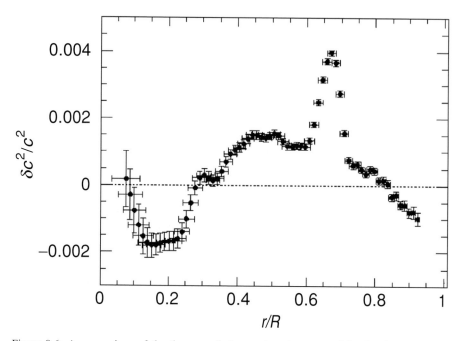

Figure 8.6: A comparison of the theory and observations in terms of the fractional deviations of the sound speed inferred from observation relative to the best-fitting standard model of the interior structure of the Sun. The agreement is better than 0.2% throughout most of the Sun, except at the boundary between the radiation and convective zones at $0.7R_\odot$. (Courtesy of ESA and the SOHO Science Team.)

A measure of the quality of data obtained from the SOHO measurements is provided by the precision with which the speed of sound can be determined as a function of radius within the Sun. As in the case of terrestrial seismology, the frequencies of the standing waves depend upon the variation of the speed of sound as a function of depth within the Sun. Modes of different order, l, sample different volumes of the solar interior, and so the solar oscillation data can be inverted to determine how the speed of sound varies with radius in the Sun. In Figure 8.6, the difference between the square of the speed of sound and the predictions of a standard solar model are shown as a function of radius within the Sun. The observations are in agreement with the predictions of the standard solar models, within a few tenths of one percent.

One important feature is the prominent deviation from the predicted relation at about 70% of the solar radius. This region corresponds to the boundary between the inner regions of the Sun, in which the transfer of energy is by radiation, and the outer regions, in which energy is transferred by convection. This sharp feature has been associated with a turbulent layer that results from the transition from radiative to convective energy transport. The remarkable success of the standard solar models has enabled non-standard models, for example those involving large amounts of core mixing, rapid rotation of the core of the Sun, or the presence of weakly interacting massive particles in the core of the Sun, to be ruled out.

Figure 8.7: A cross-section through the Sun showing the internal relative rotation speeds of different regions. Dark grey regions are slightly slower and the light grey regions are slightly faster relative to the mean velocity distribution given by equation (8.19). (Courtesy of ESA and the SOHO Project.)

These results were of crucial importance in understanding the origin of the *solar neutrino problem*, discussed in Section 8.3. The helioseismological data are now of such precision that the astrophysics of the solar interior is very well defined and the origin of the deficit of solar neutrinos must lie in some new type of neutrino physics, rather than in uncertainties about the astrophysics of the temperature distribution inside the Sun, as discussed in Section 8.3.[4]

The SOHO observatory also mapped the two-dimensional structure of oscillations on the surface of the Sun, and this has enabled remarkable three-dimensional mapping of the kinematics of its interior. An example of the results of this mapping is shown in Figure 8.7. The dark grey and light grey regions indicate deviations from a mean angular velocity distribution described by the expression

$$\Omega = \Omega_0 (1 - \alpha_2 \cos^2 \theta - \alpha_4 \cos^4 \theta), \tag{8.19}$$

where θ is the angle measured from the pole of the Sun; the constants α_2 and α_4 are both about 0.14. It can be seen that, both on its surface and in its interior, the speed of rotation is less at the poles than it is at the equator, relative to the above expression. In addition, the pattern of rotation is different within the interior radiation-dominated zone and the exterior convection-dominated zone. Within the convective zone, there is a broad rapidly rotating band, as well as polar convection currents, which are indicated by the black flow-lines. These phenomena provide clues to the nature of the dynamo processes responsible for maintaining the Sun's magnetic field, one of the most challenging areas of contemporary cosmic magnetohydrodynamics.

One of the most exciting prospects for the future study of the stars is the extension of these techniques to the internal structures of nearby stars, the discipline known as *astero-seismology*. Our understanding of other types of star will only become secure when similar types of observation and analysis can be carried out on nearby stars, and this is already within the grasp of present state-of-the-art technology.

8.5   Evolving the stars

Eddington's pioneering studies of the internal structure of the stars (Eddington, 1926b) and Hoyle and Lyttleton's construction of homologous stellar models (Hoyle and Lyttleton, 1942) put the structure of main-sequence stars on a secure physical basis, but these were no more than the first steps in what was to grow into a major industry in the second half of the twentieth century. To do better than scaling relations, it was necessary to integrate the equations of stellar structure through the star using the best available data on the opacities, equations of state and nuclear energy generation rates for stellar material. In addition, although the homologous models had finite radii, the boundary conditions at the surface of the star needed careful attention. On top of these challenges, Öpik had come to the correct conclusion that the solution of the red giant problem was to consider inhomogeneous stars, which he argued must form when the nuclear energy sources were exhausted in the central regions of the star (Öpik, 1938). The inevitability of this process was demonstrated by the calculations of Schönberg and Chandrasekhar, who constructed models in which energy was generated in a thin shell about a central inert core and found that there were no solutions when the core constituted more than 10% of the mass of the star (Schönberg and Chandrasekhar, 1942).

The integration of the equations of stellar structure by numerical calculation was a laborious procedure before the development of digital computers. As Martin Schwarzschild remarked in 1958 (Schwarzschild, 1958),

A person can perform more than twenty integration steps per day ... so that for a typical single integration of, say, forty steps, less than two days is needed.

In 1952, Allan Sandage and Schwarzschild developed theoretical evolutionary tracks for stars in globular clusters on the Hertzsprung–Russell diagram, stimulated by Sandage's important observations of this diagram for old globular clusters in which a continuous distribution of stars from the main sequence onto the giant branch was observed (Arp, Baum and Sandage, 1952). These computations demonstrated the rapid evolution of the star from the main sequence to the giant branch, and they used the main-sequence termination point to estimate the ages of the clusters, which turned out to be about 3×10^9 years (Sandage and Schwarzschild, 1952). This has remained one of the most powerful tools for estimating the ages of star clusters. The theoretical astrophysics of these studies was taken very much further by Hoyle and Schwarzschild, who followed the detailed physics of the evolution of stars from the main sequence to the tip of the giant branch, showing that, at this point, the temperature of the core would be sufficiently high to initiate helium burning (Hoyle and Schwarzschild, 1955).

The necessary atomic, molecular and nuclear data that were required to perform more exact computations gradually became available as a result of the work of the national laboratories and because of the military interest in understanding the physics of nuclear explosions. For example, Arthur Cox (b. 1927) and his collaborators working at Los Alamos National Laboratory carried out a major programme to determine the opacities of stellar material using all the best available data, and these became the standard opacities used by stellar modellers for many years (Cox, 1965).

By the late 1950s and early 1960s, electronic computers had become available to theoretical astrophysicists, and numerical procedures were developed to integrate the equations of stellar structure accurately and efficiently. The pioneers of these numerical procedures included Louis Henyey, Rudolph Kippenhahn (b. 1926), Icko Iben (b. 1931) and Robert Christy (b. 1916). Thanks to their efforts, the study of the stars became one of the most precise of the astrophysical sciences. The issues which had to be addressed are clearly described by Kippenhahn and Weigert in their authoritative book *Stellar Structure and Evolution*.[5]

The full complexity of stellar evolution once hydrogen burning in the centre of the star is completed became apparent. All numerical studies showed the formation of red giant stars, but the exact physical reason for this behaviour has been controversial since a number of different processes take place almost simultaneously. The complexity of post main-sequence evolution is most simply illustrated by the diagrams showing the evolution of a $5M_\odot$ star by Kippenhahn and Alfred Weigert (b. 1927) (Figure 8.8). The key episodes in the star's history are as follows.

- From A to C, the energy source is hydrogen burning in the core, all the hydrogen in this region being exhausted by the point C.
- Immediately following C, the core begins to collapse and hydrogen burning takes place in a shell about the collapsing core. This phase is very rapid and the core collapse is accompanied by a huge expansion in the radius of the envelope, a factor of 25 in this simulation, resulting in the formation of a red giant star. The rapid evolution of the star across this region of the Hertzsprung–Russell diagram accounts for the fact that very few stars are found in this region, a phenomenon known as the *Hertzsprung gap*. A deep outer convective zone is formed, corresponding to the star approaching the Hayashi limit, which is the locus of fully convective stars (see Section 9.3). This process enables mixing of the products of nucleosynthesis to take place in the outer envelope of the star.
- In this massive star, the core is non-degenerate and so heats up until, at $T \approx 10^8$ K, the onset of helium core-burning begins at E. The process of nuclear helium burning, creating carbon as well as oxygen and neon, continues around the loop E → F → G.
- Helium-core burning ends at G and then helium-shell burning takes over, with the result that energy generation takes place in two nuclear burning shells.

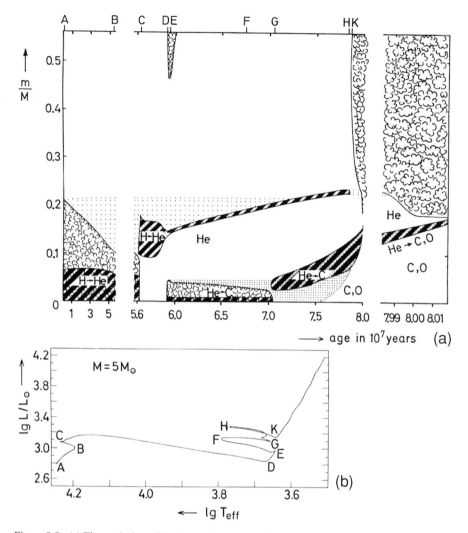

Figure 8.8: (a) The evolution of the internal structure of a $5M_\odot$ star, illustrating the nuclear processes taking place in its interior. The abscissa shows the age of the model star after the ignition of hydrogen in its core in units of 10^7 years. The ordinate shows the radial coordinate in terms of the mass, m, within a given radius relative to the total mass, M, of the star. The cloudy regions indicate convective zones. Heavily hatched areas indicate high rates of nuclear energy generation. (b) The corresponding positions of the star on a Hertzsprung–Russell diagram at each stage in its evolution. From R. Kippenhahn and A. Weigert, *Stellar Structure and Evolution* (Berlin: Springer-Verlag, 1990).

- Expansion of the outer envelope cools the hydrogen burning shell, which ceases nuclear energy generation and a deep outer convective zone forms, enabling further mixing of the processed material to take place in the envelope of the star. This process is referred to as the *second dredge up*.

In the case of low-mass stars, the evolution is quite different. The core is in radiative, rather than convective, equilibrium and so nuclear burning results in the growth of the helium core. Once hydrogen core burning ceases, the core contracts to form an isothermal degenerate core and nuclear energy generation takes place in a hydrogen-burning shell about this inert core. The slow growth of the core and the gradual increase in the luminosity of the star does not result in the rapid movement across the Hertzsprung–Russell diagram seen in Figure 8.8(b). Furthermore, the low-mass stars are closer to the Hayashi limit as they gradually move away from the main sequence. Ultimately, as the luminosity of the hydrogen-burning shell increases, the temperature of the degenerate core reaches $\approx 10^8$ K, at which temperature helium burning takes place in a thermal runaway, known as the *helium flash*, following which the degeneracy is relieved and a deep outer convection zone is formed in the star's envelope.

These features of the evolution of stars of different masses could only be understood in detail once the numerical calculations had pointed the way to the physical understanding of post-main-sequence evolution.[6] One of the important developments was the understanding of pulsating stars, such are the RR-Lyrae and Cepheid variables. Both classes of pulsating star lie within a narrow band in the Hertzsprung–Russell diagram, which is known as the *instability strip*, when stars of different masses evolve onto the horizontal branch. The first serious analysis of the stability of stars and the problem of maintaining their pulsations was carried out by Eddington in *The Internal Constitution of the Stars* (Eddington, 1926b). He demonstrated that the pulsations would be damped out with a characteristic timescale of 8000 years, a figure which was subsequently shown to be a very significant overestimate.[7] He proposed two mechanisms which would provide a driving mechanism for the pulsations, the more important being a 'valve' mechanism associated with pulsations in the envelope of the star. In its simplest form, the mechanism involves reducing the heat leakage during the compression phase of the pulsation and increasing it during the expansion phase. It was first proposed by Sergei Zhevakin (1916–2001) that the appropriate opacity changes could be associated with the ionisation zones of ionised helium in the outer envelope of the star in which $He^+ \rightleftharpoons He^{++}$ (Zhevakin, 1953). Many studies, including those by John Cox (1926–1984), Norman Baker (b. 1931) and Kippenhahn, demonstrated the correctness of this picture, which can account for the pulsational properties of most types of pulsating star, including the RR-Lyrae and Cepheid variable stars. The appropriate ionisation conditions are found in the envelopes of stars of different masses as they cross the instability strip when they have joined the horizontal branch. The process of pulsation is inherently non-linear, and powerful computer codes were needed to determine the light-curves of the stellar pulsations. Robert Christy successfully developed numerical codes to account for the observed properties of RR-Lyrae variable stars (Christy, 1964), and he and Robert Stobie (1941–2002) developed the corresponding models for Cepheid variable stars (Christy, 1968; Stobie, 1969).

These very considerable achievements made stellar evolution computations a key tool for the understanding of the evolution of the stellar populations of galaxies. The ages of star clusters, for example, could be estimated from the forms of their Hertzsprung–Russell diagrams. There were, however, inconsistencies which a number of authors suspected might be associated with uncertainties in the stellar opacities and the equations of state. In the 1980s, two large programmes, the OPAL project and the Opacity Project (OP), were undertaken to improve knowledge of the opacities used in stellar structure and evolution calculations. The

OPAL code was developed at the Lawrence Livermore National Laboratory to compute opacities of elements with low to mid atomic numbers and included a very large number of improvements as compared with the earlier opacities (Rogers and Iglesias, 1994).[8] The approach adopted by the Opacity Project, an international consortium led by Michael Seaton (b. 1923), was to calculate opacities based on a new formalism for the equation of state and on the computation of accurate atomic properties such as energy levels, f-values and photoionisation cross-sections (Hummer and Mihalas, 1988; Seaton *et al.*, 1994).

These massive programmes resulted in much improved opacities for all types of astrophysical application. Interestingly, the two projects took quite different approaches to the determination of the opacities, but they ended up being in very good agreement. As Forrest Rogers (b. 1938) and Carlos Iglesias (b. 1951) described in their review paper of 1994, the new opacities were generally somewhat larger than the previous values, much of the increase being associated with transitions of iron, and this had the effect of resolving a number of astrophysical puzzles.

- Some Cepheid variables pulsate in different modes at different frequencies, and the ratios of these frequencies are sensitive to the mass of the star. Using the old Los Alamos opacities, the deduced masses were up to about 50% smaller than expected. Using the new opacities, this discrepancy has been removed and the masses agree with those expected from stellar evolution from the giant branch without mass loss. Similar results are found for other classes of Cepheid variable.
- A similar discrepancy was found for the RR-Lyrae stars. It was thought that the new opacities would have only a very small effect because these stars are metal-poor. Despite this, the new opacities result in masses which are now consistent with the masses expected from the theory of stellar evolution in globular clusters.
- The new opacities provide better agreement with the p-mode frequencies observed in helioseismic experiments. The resulting changes in the sound speed deduced from the solar models now place the inner boundary of the outer convective zone at 0.713 ± 0.003 of the solar radius.
- The new opacities have also contributed to understanding the lithium depletion problem observed in the Sun and hotter stars. The new opacities increase the depth of the outer convective zone so that the ^{7}Li is exposed to temperatures $T \approx 2.4 \times 10^6$ K at which the reaction ^{7}Li $+$ p $\rightarrow$ 2 ^{4}He can take place.

8.6   The discovery of neutron stars

The existence of neutron stars had been predicted as long ago as 1934 by Baade and Zwicky (1934a), soon after the discovery of the neutron, but models of neutron stars suggested that the only detectable emission would be the thermal X-ray radiation from their surfaces. Neutron stars were expected to be very hot on formation but, after an initial period of rapid cooling by neutrino emission, they remain very hot and cool slowly by thermal emission from their surfaces. Because they are very compact stars, with radii only about 10–20 km, it was believed that the only chance of observing them would be as weak X-ray emitters. In a prescient paper of 1967, Franco Pacini (b. 1939) predicted that they might be observable at long radio wavelengths if they were magnetised and were oblique rotators (Pacini, 1967).

The neutron stars were discovered, more or less by chance, as the parent bodies of radio pulsars by Antony Hewish (b. 1924) and Jocelyn Bell (b. 1943) in 1967 at Cambridge, UK (Hewish et al., 1968).[9] Hewish had pioneered the technique of observing the scintillation or 'twinkling' of radio sources at long radio wavelengths due to irregularities in the plasma density along the lines of sight to the radio sources. His work in this area had been fundamental in understanding the influence of irregularities in the electron density in the ionosphere upon radio astronomical observations. By the early 1960s, it became apparent that the same technique could be used to study the properties of the interplanetary plasma, the phenomenon of *interplanetary scintillation*. It turned out that this technique could be used, not only as a means of studying density fluctuations in the interplanetary plasma and their motions, but also as a means of discovering radio quasars, many of which are compact radio sources and hence expected to display large radio scintillations at low frequencies.

Hewish designed a large array to undertake these studies and was awarded a grant of £17 286 by the Department of Scientific and Industrial Research to construct it, as well as outstations for measuring the velocity of the solar wind. To obtain adequate sensitivity at the low observing frequency of 81.5 MHz (3.7 m wavelength), the array had to be very large, 4.5 acres (1.8 hectares) in area, in order to record the rapidly fluctuating intensities of bright radio sources on a timescale of one-tenth of a second. This was the key technological development which led to the discovery of radio pulsars since normally radio astronomical observations require long integrations to detect faint sources.[10]

The first sky surveys began in July 1967, and Jocelyn Bell, Hewish's research student, discovered a strange source which seemed to consist entirely of scintillating radio signals (Figure 8.9(a)). The source was not always present, and its nature remained a mystery. In November 1967, the source reappeared and was observed using a receiver with a much shorter time-constant. It was found to consist entirely of a series of pulses with a pulse period of about 1.33 s (Figure 8.9(b)). This pulsating radio source, PSR 1919+21, was the first to be identified, and over the next few months three further sources were discovered with pulse periods ranging from 0.25 to almost 3 seconds. The name *pulsar* was coined soon after the announcement of the discovery.[11]

Within a year, more than 20 more pulsars were discovered and a flood of theoretical papers appeared. The favoured picture for the nature of the pulsar phenomenon was described by Thomas Gold (1920–2004) in 1968, similar in many respects to Pacini's proposal of 1967, and consisted of an isolated, rotating, magnetised neutron star in which the magnetic axis of the star and its rotation axis are misaligned (Pacini, 1967; Gold, 1968). The radio pulses were assumed to originate from beams of radio emission emitted along the magnetic axis. The key observations that supported this picture were the very short stable periods of the pulses and the observation of polarised radio emission within the pulses. Two of the pulsars discovered in 1968 were of special importance. A pulsar of period 0.089 s was discovered in the young supernova remnant in the constellation of Vela (Large, Vaughan and Mills, 1968), and soon afterwards a pulsar with the very short period of 0.033 s was discovered in the centre of the Crab Nebula, the supernova which exploded in 1054 and which was extensively studied by the Chinese astronomers at that time (Staelin and Reifenstein, 1968). Optical

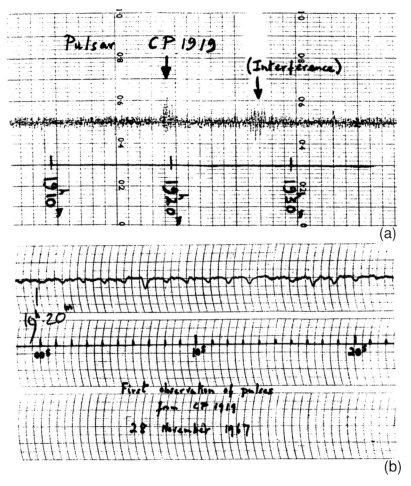

Figure 8.9: The discovery records of the first pulsar to be discovered, PSR 1919+21 (Hewish *et al.*, 1968; see also note 10). (a) The first record of the strange scintillating source labelled CP 1919. Note the subtle differences between the signal from the source and the neighbouring signal due to terrestrial interference. (b) The signals from PSR 1919+21 (top trace) observed with a shorter time-constant than the discovery record, showing that the signal consists entirely of regularly spaced pulses with period 1.33 s. The lower trace shows one-second time markers.

pulses from the Crab Nebula pulsar with precisely the same pulse period were discovered in 1969 (Cocke, Disney and Taylor, 1969), and within three months rocket flights by the teams from the Naval Research Laboratory (Fritz *et al.*, 1969) and the Massachesetts Institute of Technology (Rossi, 1970) had discovered X-ray pulses as well.[12]

The short periods of these pulsars proved beyond any shadow of doubt that the parent bodies of the pulsars had to be neutron stars. By pushing all the parameters to their limits, it had just been possible to find models of white dwarfs which could rotate with periods of about 1 s before they would break up, but periods as short as 0.1 s were excluded. Furthermore, the formation of neutron stars in supernova explosions was conclusively demonstrated by the coincidence of these short-period pulsars with young supernova remnants.

One further prediction made by Gold was soon confirmed observationally. He had noted that a spinning magnetic dipole emits magnetic dipole radiation and that this energy is extracted from the rotational energy of the neutron star. Consequently, the periods of pulsars should increase as they radiate away magnetic dipole radiation. In 1969, the period of the pulsar in the Crab Nebula was found to be increasing steadily with time (Richards and Comella, 1969).

This result solved another long-standing problem in understanding the physics of the Crab Nebula. The polarisation and non-thermal spectrum of the optical radiation from the Crab Nebula had been interpreted as the synchrotron radiation of high-energy electrons (Shklovsky, 1953). The problem with this interpretation was that the high-energy electrons would lose all their energy in a time short compared with the age of the nebula. The problem was exacerbated by the fact that the X-ray emission from the nebula was attributed to the same mechanism and the lifetimes of these electrons were even shorter. In consequence, some means had to be found of providing a continuous supply of energy to the nebula. In 1969, Gold pointed out that energy was being supplied to the nebula by the slowing down of the pulsar. The rate of loss of rotational energy by the pulsar turned out to match almost precisely the rate at which energy had to be supplied to the nebula to account for the non-thermal radio, optical and X-ray emission (Gold, 1969).

In the same year, the first papers exploring the electrodynamics of pulsars were published by Peter Goldreich (b. 1939) and William Julian (b. 1939) and by James Gunn (b. 1938) and Jeremiah Ostriker. In 1968, Pacini had shown that the magnetic field strengths at the surfaces of neutron stars had to be enormous, $B \sim 10^6 - 10^8$ T (Pacini, 1968). These magnetic fields were so strong that the Lorentz ($\mathbf{v} \times \mathbf{B}$) force extracted electrons from the surface layers of the neutron star so that electric currents must be present in its magnetosphere (Goldreich and Julian, 1969). Ostriker and Gunn showed how electrons could be accelerated to very high energies in the strong electromagnetic waves emitted by the rotating magnetised neutron star (Ostriker and Gunn, 1969).

Studies of the internal structure of neutron stars were pursued with renewed vigour as theorists realised that they were no longer theoretical ornaments but an integral part of observational astrophysics. The equation of state of neutron matter had been the subject of numerous studies prior to the discovery of pulsars. An equation of state had been given by Kent Harrison (b. 1934) and John Wheeler (b. 1911) in 1958 (Harrison, Wakano and Wheeler, 1958) and in convenient analytic forms by Harrison and his colleagues in 1965 (Harrison et al., 1965). Much improved calculations were carried out by Gordon Baym (b. 1935), Hans Bethe and David Pines (b. 1924) in 1971, and these were used to construct the standard models for neutron stars (Baym, Bethe and Pethick, 1971a; Baym, Pethick and Sutherland, 1971b). The equation of state was well understood up to about nuclear densities, $\rho \sim 3 \times 10^{17} \, \text{kg m}^{-3}$, but there remained uncertainties at higher densities which may be found in the centres of the most massive neutron stars.

In addition, the interior is threaded by an intense magnetic field. The magnetic field does not have a strong influence upon the structure of the neutron star, but it does influence its internal dynamical properties. Long before the pulsars were discovered, Arkadii Migdal (1911–1993) in 1959, as well as Vitali Ginzburg and David Kirzhnits (1926–1998) in 1964 and Vittorio Canuto (b. 1937) and Györgi Marx[13] in 1965, proposed that the interiors of neutron stars should be superfluid (Migdal, 1959; Ginzburg and Kirzhnits, 1964). Studies of the properties of the neutron–proton–electron fluid in different regimes inside the neutron stars showed that the inner crust and the neutron liquid phases are superfluid and that the protons are superconducting. In 1969, Baym, Christopher Pethick (b. 1942) and Pines showed that the magnetic field is therefore quantised into vortices which are pinned to the crust of the neutron star (Baym, Pethick and Pines, 1969a).

The link between the interior structure of the star and observable phenomena was provided by the phenomenon of *glitches*, which were discovered in the Vela pulsar in 1969 (Radhakrishnan and Manchester, 1969; Reichley and Downs, 1969). All radio pulsars were known to slow down, but, occasionally, discontinuous changes in the slow-down rate were observed in which the period decreased abruptly. A natural interpretation of this phenomenon was to associate it with internal changes in the structure of the neutron star. One possibility proposed by Malvin Ruderman (b. 1927) and by Baym, Pethick and Pines in 1969 was that, as the neutron star slows up, its crust takes up a new shape in a 'star-quake' in which the structure changes discontinuously to take up a new equilibrium configuration (Baym *et al.*, 1969b; Ruderman, 1969). This process cannot, however, be the whole story since, in pulsars such as the Vela pulsar, the glitches occur too frequently. The preferred interpretation is that the glitches may be associated with the unpinning of the superconducting vortices with the crust of the neutron star. One important aspect of the glitches is that the recovery of the rotation speed to a steady value is remarkably slow. This is direct observational evidence that the bulk of the moment of inertia of the neutron star must be in some superfluid form, which is only weakly coupled to the crust. Thus, the significance of these observations goes far beyond astronomy since this is the only environment we know of in which matter in bulk can be studied at nuclear densities.

Intriguingly, one of the most difficult problems of pulsar studies was the understanding of the mechanism of radio emission. The brightness temperature of the radio pulses often exceeded 10^{28} K, indicating that the emission must involve some form of coherent radiation mechanism. The most promising model was proposed by Venkataraman Radhakrishnan (b. 1929) and his colleagues in 1969 and involved the radiation of high-energy electrons as they streamed from the poles of the magnetised neutron star along the curved field lines (Radhakrishnan and Cooke, 1969; Radhakrishnan *et al.*, 1969). The radiation process, known as *curvature radiation*, is similar in many ways to synchrotron radiation, but the radius of curvature of the particles' trajectories is defined by the curvature of the field lines emerging from the poles of the neutron star and must involve coherent bunches of electrons to produce the extraordinarily high brightness temperatures. This model can naturally account for the observed variation of the direction of the linear polarisation of the radiation as the beamed radiation sweeps past the observer.

At the 1970 IAU General Assembly, the distinguished Soviet astrophysicist Iosef Shklovsky asked me to introduce him to Jocelyn Bell. He told her, 'Miss Bell, you have made the greatest astronomical discovery of the twentieth century.' In many ways, it is difficult to disagree with his assessment. The existence of these stars shows that neutron stars form as a result of stellar collapse. Furthermore, these are the last stable stars. The radius of a solar mass neutron star is only about three times the radius of a black hole of the same mass, and so the neutron stars may be thought of as objects which have just failed to become black holes. Even in neutron stars, general relativity is no longer a small correction factor but is crucial in determining their stability.

8.7 X-ray binaries and the search for black holes

The next event, which was to have a profound influence upon thinking in high-energy astrophysics, was the discovery of binary X-ray sources by the UHURU satellite in 1971.[14] From previous sounding rocket experiments, it had been suspected that some of the X-ray sources were variable in intensity because sometimes they were present and at other times they seemed to have faded or disappeared. To resolve these problems, it was necessary to carry out long-term systematic studies of these sources. The UHURU X-ray observatory was the first satellite dedicated exclusively to X-ray astronomy, and it carried out the first complete survey of the whole sky. The nature of the variability of the sources was one of the key objectives of the mission. In late 1970, the variable source Cygnus X-1 (Cyg X-1) was observed and, although for some time it was thought that it had an X-ray periodicity of 0.073 s, this result was spurious and the source displayed somewhat random variability on timescales as short as 100 ms, indicating that the source region must be very compact (Oda *et al.*, 1971; Rappaport, Doxsey and Zaumen, 1971).

Much more remarkable were the observations of the source Centaurus X-3 (Cen X-3) which were first made in January 1971. Unlike the case of Cyg X-1, Cen X-3 showed a clear periodicity with a pulse period of about 5 s, longer than that of any known radio pulsar. Furthermore, the pulsation period was not stable but seemed to vary with time (Giacconi *et al.*, 1971a). The source was reobserved in May 1971, and it was found that the period of the X-ray pulsations varied sinusoidally with a period of 2.1 days. This suggested that the X-ray source was a member of a binary system, the change in period of the pulses being due to the Doppler shift of the X-ray pulses in the binary orbit. Then, on 6 May, the source disappeared, only to reappear half a day later. This pattern repeated roughly every two days. Clearly, the X-ray source was being occulted by the primary star in the binary system (Schreier *et al.*, 1972). With these clues, Wojciech Krzeminski (b. 1933) was able to identify the primary star, which turned out to be a massive blue star with the same binary period of 2.1 days as the X-ray source (Krzeminski, 1973, 1974). Soon after this discovery, another similar source was discovered (Tananbaum *et al.*, 1972), the source Hercules X-1 (Her X-1), which had a pulse period of 1.24 s and an orbital period of 1.7 days (Figure 8.10).

The short period of the X-ray source in Her X-1 was strong evidence that the parent body must be a neutron star, similar to those of the radio pulsars. Furthermore, the source of energy for the system was immediately identified as accretion. The idea of accretion as a source of energy for the X-ray sources had already been suggested by Satio Hayakawa (1923–1992) and Masaru Matsuoka (b. 1939) in 1964 (Hayakawa and Matsuoka, 1964), who considered normal close binary systems, and by Shklovsky in 1966 (Shklovsky, 1967), who proposed accretion from a binary companion onto a neutron star as the energy source for the brightest X-ray source in the sky, Sco X-1. In 1968, Kevin Prendergast (b. 1929) and Geoffrey Burbidge made the important point that, in the accretion of matter from the primary star onto a compact secondary in a binary system, the accreted matter would

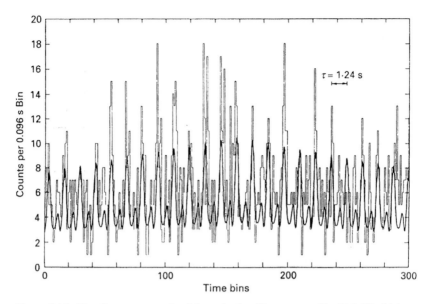

Figure 8.10: The discovery records of the pulsating X-ray source Her X-1. The histogram shows the number of counts observed in successive 0.096 s bins. The continuous line shows the best-fitting harmonic curve to the observations, taking account of the varying sensitivity of the telescope as it swept over the source (Tananbaum *et al.*, 1972).

necessarily have a considerable amount of angular momentum and so an *accretion disc* would form about the compact star (Prendergast and Burbidge, 1968). Accretion of matter from the primary star onto a compact neutron star is a very powerful energy source. A simple Newtonian calculation shows that the luminosity due to accretion onto an object of mass M and radius r is roughly $0.5\,\dot{m}c^2(r_g/r)$, where $r_g = 2GM/c^2$ is the Schwarzschild radius of an object of mass M and $\dot{m}$ is the mass accretion rate (see Section A11.1.2). According to this simple estimate, the accretion of matter onto a $1\,M_\odot$ neutron star with radius 10 km can liberate about 5–10% of the rest mass energy of the infalling matter. When the effects of general relativity are taken into account, the upper limit to the energy release is 5.72% for accretion onto a solar mass black hole, roughly an order of magnitude greater than can be liberated by nuclear fusion reactions.[15]

The study of pulsating X-ray binaries was of particular importance astrophysically because the masses of the neutron stars could be estimated using the standard procedures of celestial mechanics. The gratifying result was found that the masses of the seven binary X-ray sources for which this analysis was possible lay in the range 1.2 to 1.4 $M_\odot$, entirely consistent with the upper limit to the masses of neutron stars, which is similar to the Chandrasekhar limit for white dwarfs (Rappaport and Joss, 1983).

Distances could be estimated for many of the X-ray sources and so their X-ray luminosities could be found. In 1973, Bruce Margon (b. 1948) and Ostriker showed that the luminosities of the binary sources extended up to about $L = 10^{31}$ W, which is very close to the Eddington limiting luminosity for spherical accretion onto objects with mass $1\,M_\odot$, the precise limit being $L \leq 1.3 \times 10^{31}(M/M_\odot)$ W (Margon and Ostriker, 1973). This was a key result because it demonstrated that sources existed which could radiate X-rays at luminosities close to the maximum permissible luminosity (see Section A11.1.3). Furthermore, it was natural that these sources should emit most of their radiation in the X-ray waveband. Assuming that the X-ray emission originates close to the surface of the neutron star, application of the Stefan–Boltzmann law showed that the temperature of the emitting region had to be greater than about 10^7 K to produce such large luminosities, and so it was natural that the energy of accretion should be emitted in the X-ray waveband.

From these considerations, a standard picture of the nature of the pulsating binary X-ray sources developed.[16] In the case of a massive companion, such as a blue supergiant star, the neutron star is located in a strong stellar wind, and the matter within a certain radius of the neutron star is accreted onto it. In the low-mass binary systems, mass transfer takes place through the process of *Roche lobe overflow*, in which the primary star fills its Roche lobe, the equipotential surface joining the two stars, and so matter attains a lower gravitational potential by collapsing to form an accretion disc about the neutron star.[17] The X-ray pulsations are attributed to accretion onto the poles of the rotating neutron star, the strong non-aligned magnetic field channelling the matter into the polar regions. Evidence for the presence of strong magnetic fields in the source Her X-1 was found by Joachim Trümper (b. 1933) and his colleagues in 1978; they identifed a cyclotron radiation feature in its X-ray spectrum at about 58 keV (Trümper *et al.*, 1978). This feature was subsequently observed by the Japanese GINGA satellite and was interpreted as an absorption feature at 34 keV, corresponding to a magnetic field strength of 3×10^8 T (Mihara *et al.*, 1990).

The next obvious step was to ask whether or not there was any evidence for black holes among the binary X-ray sources. Isolated black holes are very difficult to detect, and it is only when they are close to sources of fuel that their presence can be readily detected. In 1965, Yakov Zeldovich and Oktay Guseynov (b. 1938) had proposed that the observation of X-rays or γ-rays from single-line spectroscopic binaries might be the signature of either a neutron star or a black hole (Zeldovich and Guseynov, 1966). In these binary systems, only the bright component is observable, and the unseen secondary must be some form of dark star, either a very-low-mass star, a neutron star or a black hole. In 1969, Virginia Trimble (b. 1943) and Kip Thorne (b. 1940) investigated whether or not any of the dark companions of known single-line spectroscopic binaries could be massive enough to be black holes, but no likely candidate was found and none coincided with known X-ray sources (Trimble and Thorne, 1969).

The first strong candidate for a black hole companion was found in the bright X-ray source Cyg X-1. The positions of sources determined by the UHURU observatory were not normally accurate enough to make an identification of the optical counterpart unless there was some identifiable feature, such as a binary orbital period. In 1971, with an improved position provided by an MIT rocket flight, radio astronomers at the National Radio Astronomy Observatory in the USA and at the Westerbork Observatory in the Netherlands searched the field for a variable radio source which might be associated with Cyg X-1. These searches were successful, and resulted in an accurate radio position which coincided with the 9th-magnitude blue supergiant star (Braes and Miley, 1971; Hjellming and Wade, 1971). In the next year, Louise Webster (1941–1990) and Paul Murdin (b. 1942) and, independently, Thomas Bolton (b. 1943) showed that the star was the primary star of a binary system with

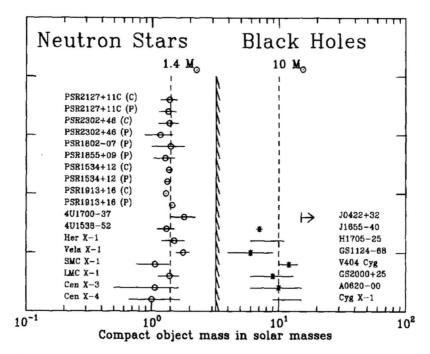

Figure 8.11: The mass distribution of neutron stars and black holes. Those systems which are known to possess neutron stars all have masses about $1.4M_\odot$. The masses of the black holes in X-ray binary systems for which good mass estimates are available all exceed the upper limit for neutron star masses of $M \approx 3M_\odot$ (Charles, 1998).

period 5.6 days (Bolton, 1972; Webster and Murdin, 1972). These observations enabled the ratio of the masses of the two components to be estimated. Assuming the mass of the supergiant B star was greater than $10M_\odot$, the mass of the invisible companion had to be greater than $3M_\odot$, the most likely masses being $20M_\odot$ for the blue supergiant and $10M_\odot$ for the unseen companion. The latter mass exceeded the upper limit for stability as a neutron star, and it was concluded that it must therefore be a black hole.

Over the succeeding years, three other good examples of X-ray binaries with massive invisible companions were discovered: the X-ray binary sources LMC X-1, LMC X-3 and A0620-00 (Cowley, 1992; McClintock, 1992). In each of these, the X-ray intensity exhibits short-period variability but no signature of pulsed X-ray emission. In a recent compilation of neutron star and black hole mass estimates, the numbers of black hole candidates has continued to increase, with the black hole candidates having quite different X-ray spectral and variability characteristics as compared with those cases in which pulsating X-rays are found and which are confidently associated with accreting neutron stars (Figure 8.11) (Charles, 1998). It is now generally accepted that the simplest interpretation of the properties of these systems is that they contain black holes. Their great advantage is that they are relatively nearby systems and so the behaviour of matter in the vicinity of the strong gravitational fields associated with black holes can be studied in some detail. There will be much more to say about the astrophysics of black holes in Chapter 11 in the context of the physics of active galactic nuclei.

8.8 Radio pulsars and tests of general relativity

By a strange twist of astronomical fortune, it turned out that the radio pulsars provide some of the very best tests of general relativity. Observations by Joseph Taylor (b. 1941) and his colleagues made with the Arecibo radio telescope demonstrated that the arrival times of the radio pulses from pulsars are among the most stable clocks available to us (Taylor, 1992). The most important systems from the point of view of testing general relativity are the binary pulsars. As the techniques of pulsar discovery and precise timing became more and more refined, many binary pulsars were discovered, the *Australia Telescope National Facility Pulsar Catalogue* of 2003 showing that, of almost 1500 pulsars listed, 90 are members of binary systems.

The most important binary pulsars are those in which both stars are neutron stars in a close binary orbit. The first of these, the pulsar PSR 1913+16, was discovered by Russell Hulse (b. 1950) and Taylor in 1974 (Hulse and Taylor, 1975),[18] and it has been observed with very precise timing since then. The system has a binary period of only 7.75 hours and the orbital eccentricity is large, $e = 0.617$. This system is pure gold for the relativist – to test general relativity a perfect clock in a rotating frame of reference is needed, and systems such as PSR 1913+16 are ideal for this purpose. The neutron stars are so inert and compact that the binary system is very 'clean', in the sense that the neutron stars behave like point masses in their mutual gravitational fields. Various parameters of the binary orbit can be measured very precisely, and these provide estimates of different quantities which involve the masses of the two neutron stars, M_1 and M_2, in different ways. The fact that the loci of these quantities intersect precisely at one point in the (M_1, M_2) plane indicates that there is no discrepancy with the expectations of general relativity (Figure 8.12) (Taylor, 1992; Will, 2001). Assuming general relativity is the correct theory of gravity, the mass estimates of the two neutron stars are the most accurate values for any stars, besides the Sun: $M_1 = 1.4411 \pm 0.0007 M_\odot$ and $M_2 = 1.3873 \pm 0.0007 M_\odot$.

A second remarkable measurement has been the rate of loss of rotational energy of the binary system because of the emission of gravitational radiation. The rate of loss of rotational energy can be precisely predicted once the masses of the two neutron stars are known and the binary orbit determined. The rate of change of the angular frequency, Ω, of the orbit due to gravitational radiation energy loss is precisely known, $-d\Omega/dt \propto \Omega^5$ (see Section A8). The change of orbital phase of the system PSR 1913+16 has been observed since its discovery in 1974, and the observed changes agree precisely with the predictions of general relativity (Figure 8.13) (Taylor, 1992; Will, 2001). The data up to 1992 were presented by Taylor in his 1993 Nobel prize lecture. The gap in the data between 1992 and 1998 was caused by the refurbishment of the Arecibo radio telescope with which these observations were made. From the data presented by Clifford Will (b. 1946) in his review of the confrontation between general relativity and experiment, the change of phase has continued to follow the expected relation after 1998, as can be seen in Figure 8.13 (Will, 2001). Thus, although the gravitational waves themselves have not been detected, exactly the correct energy loss rate has been observed. This is an important result since it enables a wide range of alternative theories of gravity to be excluded. For example, the gravitational waves derived from standard general relativity are quadrupolar in nature, and

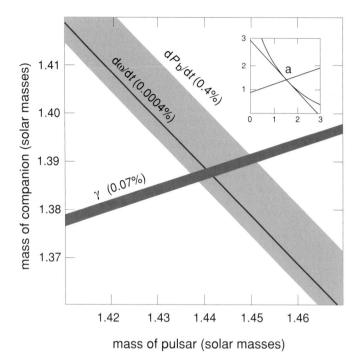

Figure 8.12: Constraints on masses of the pulsar PSR 1913+16 and its companion from precise timing data, assuming general relativity to be valid. The width of each strip in the (M_1, M_2) plane reflects the observational uncertainty, shown as a percentage. The inset shows the three constraints on the complete (M_1, M_2) plane; the intersection region (a) has been magnified 400 times in the main figure (Will, 2001).

any theory which, say, predicted the emission of dipole or scalar gravitational radiation can be excluded.

The same techniques of accurate pulsar timing can also be used to determine whether or not there is any evidence for the gravitational constant varying with time (Taylor, 1992). These tests are slightly dependent upon the equation of state used to describe the interior of the neutron star, but, for the range of plausible equations of state, the limits to $\dot{G}/G$ are less than about 10^{-11} year^{-1}. Thus, there can have been little change in the value of the gravitational constant over cosmological timescales. Einstein's standard theory of general relativity has therefore passed the most precise tests devised so far and can be used with some confidence in the study of black holes and cosmology.

8.9   The search for gravitational waves

It is generally assumed that Figure 8.13 is convincing evidence for the existence of gravitational waves, and this has acted as a spur to their direct detection by the gravitational wave observatories currently coming into operation. The detection of gravitational waves is one of the most demanding challenges facing astronomical technologists.

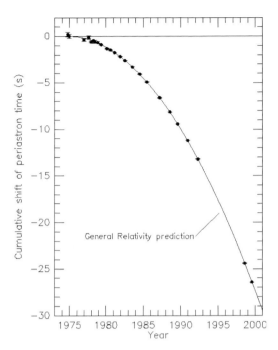

Figure 8.13: The change of orbital phase as a function of time for the binary neutron star system PSR 1913+16 compared with the expected change due to gravitational radiation energy loss by the binary system (solid line) (Taylor, 1992; Will, 2001).

The search for these waves was begun by Joseph Weber (1919–2000) in a pioneering set of experiments carried out in the 1960s. Weber was an electrical engineer by training and, having become fascinated by general relativity in the late 1950s, he was inspired to devise the means of relating the theory to experiments which could be carried out in the laboratory. He published these ideas in his book *General Relativity and Gravitational Radiation* (Weber, 1961) and set about constructing an aluminium bar detector to measure gravitational radiation from cosmic sources (Weber, 1966). He estimated that with this detector he could measure strains as small as 10^{-16}, this figure referring to the fractional change in the dimensions of the detector caused by the passage of gravitational waves. His first published results caused a sensation when he claimed to have found a positive detection of gravitational waves by correlating the signals from two gravitational wave detectors separated by a distance of 1000 km at the University of Maryland and the Argonne National Laboratory (Weber, 1969). In a subsequent paper, he reported that the signal originated from the general direction of the Galactic centre (Weber, 1970). These results were received with considerable scepticism by the astronomical community since the reported fluxes far exceeded what even the most optimistic relativists would have predicted for the flux of gravitational waves originating anywhere in the Galaxy. The positive effect of Weber's experiments was that a major effort was made by experimentalists to disprove his results, and, in the end, his results could not be reproduced.

The challenge to the experimental community was important in stimulating interest in how the extremely tiny strains expected from strong sources of gravitational waves could be detected. The outcome was the approval of a number of major national and international projects in order to detect the elusive gravitational waves. Many new technologies had to be developed, and the programmes initiated in the 1980s can best be regarded as development programmes which were to be fully approved in the 1990s. For example, the LIGO project, an acronym for Laser Interferometer Gravitational-Wave Observatory, was initiated by Kip Thorne and Rainer Weiss (b. 1932) in 1984, but final approval by the US National Science Foundation was only given in 1994 at a revised cost for development, construction and early operations of $365 million.[19] The project consists of two essentially identical interferometers, each with 4 km baselines located at Livingston, Louisiana and Hanford near Richland, Washington. Similarly, the VIRGO project is a French–Italian collaboration to construct an interferometer with a 3 km baseline at a site near Pisa, Italy. The GEO600 experiment is a German–British interferometer project with a 600 m baseline, while the Japanese TAMA project is a 300 m baseline interferometer located at Mitaka, near Tokyo. For all these projects, there was a long development programme to reach the sensitivities at which there is a good chance of detecting gravitational waves from celestial sources. In the best commissioning runs from experiments carried out in 2003, strain sensitivities in the range 10^{-20} to 10^{-21} have been achieved. The importance of having a number of gravitational wave detectors operating simultaneously is that they form a global network and any significant events should be detectable by all of them. By precise timing, the direction of arrival of the gravitational waves can be estimated.

At the time of writing, all the gravitational wave observatories are entering their operational phases with more or less their design sensitivities. None of them has yet detected gravitational waves, but it will be no surprise if they are discovered in the next few years. The potential sources of detectable radiation include the collapse of stellar cores in supernova explosions, collisions and coalescences of neutron stars or black holes, rotations of neutron stars with deformed crusts, the continuous emission of very close binary neutron stars and black holes (see Section A8) and primoridal gravitational radiation created during the very earliest phases of our Universe. In addition, like all other new astronomies, the observational study of gravitational waves is likely to produce some real surprises.

8.10   Supernovae

When Walter Baade and Fritz Zwicky wrote their famous papers of 1934 about the existence of 'super-novae' (Baade and Zwicky, 1934a,b), their conclusions were based upon the properties of 12 supernovae which had occurred between 1900 and 1930, as well as a few historical supernovae, including Tycho Brahe's 'new star' of 1572.[20] Zwicky began the systematic search for supernovae in 1934, first with a $3\frac{1}{4}$-inch Wollensack lens camera mounted on the roof of the Robinson Astrophysics Laboratory of the California Institute of Technology. Then, in 1936, he supervised the construction of a wide-angle 18-inch Schmidt telescope, which was built in the workshops of the Laboratory and sited at the new observatory at Palomar. The first supernova was discovered in NGC 4157 in March 1937 and the second in the dwarf spiral galaxy IC 4182 on August 26 1937. As chance would have it, this second supernova reached apparent magnitude 8.4, six magnitudes brighter than the luminosity of the dwarf galaxy, and it remained the brightest supernova discovered in the twentieth century until the appearance in 1987 of the supernova SN 1987A. Zwicky continued to discover about four supernovae per year in relatively nearby galaxies with the 18-inch Schmidt telescope. With the construction of the 48-inch Schmidt telescope at Palomar in 1949, searches could be made for fainter supernovae and typically about 20 were discovered each year.[21]

8.10.1 Type I and Type II supernovae

In 1938, Baade and Zwicky gave the first description of the typical light-curves of supernovae consisting of an initial outburst lasting for a few weeks, following which the brightness decreases exponentially with a half-life of about 60 days (Baade and Zwicky, 1938). The light-curves of the first dozen supernovae discovered by Zwicky were remarkably similar, but in 1941 Minkowski discovered that there are two quite distinct types of supernovae (Minkowski, 1941). The primary distinction between the two types concerns differences in their spectral evolution.

- The spectra of *Type I supernovae* consist of broad emission bands, the nature of which were not understood until almost 30 years after Minkowski's paper when Kirshner and Oke showed that they could be interpreted as the superposition of hundreds of lines of Fe^+ and Fe^{++} (Kirshner and Oke, 1975). A key feature of their spectra is the absence of hydrogen lines. The spectral and luminosity evolution of the Type I supernovae are essentially identical, so that it is possible to tell how old the supernova is from its luminosity and spectrum. They are found in all types of galaxy.
- In contrast, the *Type II supernovae* show the Balmer series of hydrogen soon after maximum light. They display a much wider range of properties than the Type II supernovae and are only found in spiral galaxies, generally within the spiral arms.

In 1960, Minkowski showed that there are also differences in the masses ejected by the two types of supernovae from fragmentary evidence on historical supernova remnants. Typically, the expanding shells of the Type I supernovae have masses about $0.1 M_\odot$ and expansion velocities of about 1000 km s^{-1}, resulting in a kinetic energy of about 10^{41} J, whereas for the Type II supernovae the ejected masses amount to several solar masses and typical velocities are about 6000 km s^{-1}, corresponding to energies of the order of 10^{44} J (Minkowski, 1960a). Although some of his classifications would probably be different nowadays, Minkowski had come to the correct conclusion.

In both cases, the energetics of the explosions were so great that they must involve the collapse of the star to some form of compact remnant, either a neutron star or a black hole. The Type II supernovae were identified with the collapse of the central regions of very massive stars, $M \geq 8 M_\odot$, which have relatively short lifetimes and cannot have travelled far from the spiral arm regions within which they formed. To account for the presence of the hydrogen absorption lines, it was assumed that the explosion of the core takes place within a red giant, and models by Sydney Falk (b. 1947) and David Arnett showed that their light curves could be accounted for by the outward passage of a strong shock wave through a red giant atmosphere (Falk and Arnett, 1973).

In contrast, for the Type I supernovae, the most attractive picture was that they are formed by the accretion of mass onto a white dwarf in a binary system. The process by which the explosion takes place is not established, but it is thought to be associated with the heating of the surface of the white dwarf by the infalling matter. If nothing else happened, the process of accretion would eventually take the total mass of the star over the Chandrasekhar limit, and collapse to a neutron star would then ensue. It is believed, however, that before this can happen the temperature of the surface layers resulting from the process of accretion becomes high enough for nuclear burning to begin in the star's surface layers. There then ensues a nuclear deflagration which propagates through the star, causing a violent explosion in which the whole star is disrupted. It is plausible that this critical point occurs when the total mass of the white dwarf has reached a certain critical mass and this can account for the uniformity of their properties. This scenario can also account for the facts that no hydrogen absorption lines are observed, that their spectra are iron rich and that they are observed in all types of galaxy. The formation of neutron stars in binary systems can also account for the high space velocities of radio pulsars if the binary is disrupted in the explosion.

Just how uniform the population of Type IA supernovae is was demonstrated by Mark Phillips (b. 1951), who analysed the light-curves of nine examples and showed that there is indeed a small dispersion in their absolute luminosities at maximum, amounting to 0.6 and 0.5 magnitudes in the V and I wavebands, respectively. In addition, he found that the peak luminosity is correlated with the rate at which the supernova subsequently decayed (Phillips, 1993), a relation first suggested by Yuri Pskovskii (1926–2004) (Pskovskii, 1977, 1984). Another version of this relation using the light-curve shapes of the supernova outbursts was developed by Adam Riess (b. 1969), William Press and Robert Kirshner (b. 1949) (Riess, Press and Kirshner, 1995). This procedure took account of the correlation between the luminosity of the supernova at maximum and the increasing timescale of the outburst, and resulted in a 'corrected' dispersion in the maximum luminosity of only 0.21 magnitudes in the V waveband, indicating how uniform the population of these supernovae must be. These supernovae have turned out to be important 'standard candles' for measuring distances out to large redshifts, and an enormous effort has been devoted to discovering many more of them, particularly at large redshifts, and in understanding the astrophysics of these explosions in detail (see Chapter 13).

The long exponential decay of the luminosities of supernovae following their initial outbursts was a puzzle since the characteristic decay time was the same for essentially all supernovae. In 1962, Titus Pankey suggested in his Ph.D. dissertation that the decay might be associated with the decay of radioactive nuclides created in the explosion (Pankey, 1962). This proposal was put on a firm astrophysical basis by Stirling Colgate (b. 1925) and Chester McKee (b. 1942) in 1969 (Colgate and McKee, 1969). The basic idea was that, in the process of collapse to form a neutron star, explosive nucleosynthesis takes place, and among the products is the radioactive isotope of nickel, ^{56}Ni. This isotope then decays as follows:

$$^{56}\text{Ni} \xrightarrow{\beta^+} {}^{56}\text{Co} \xrightarrow{\beta^+} {}^{56}\text{Fe}. \tag{8.20}$$

The first β-decay of ^{56}Ni has a half-life of only 6.1 days, while the second β-decay, which has a half-life of 77.1 days, is presumed to be the source of energy for the exponential decay of the luminosity of the supernova, 3.5 MeV being liberated in the form of γ-rays in each decay of a ^{56}Co nucleus. Thus, the exponential decay of the luminosity of the supernova is attributed to the creation of radioactive nickel in the explosion, which is ejected into the expanding envelope of the supernova and so contributes to the enrichment of the abundance of heavy elements in the interstellar medium.

8.10.2 Supernova 1987A

Unquestionably, the most important event in supernova studies in the twentieth century was the explosion of a supernova in one of the dwarf companion galaxies of our own Galaxy, the Large Magellanic Cloud. This supernova, known as SN 1987A, was first observed optically on 24 February 1987 and reached about third visual magnitude by mid May 1987.[22] It was the brightest supernova since Kepler's supernova of 1604 and the first bright supernova to be studied with all the power of modern instrumentation. Ironically, it appears to have been a peculiar Type II supernova because the light-curve showed a much more gradual increase to maximum light than is typical of Type II supernovae. It took 80 days to reach maximum light, and its bolometric luminosity then remained roughly constant for about two months after that time, despite the fact that there was a rapid decline in its surface temperature. It was also subluminous as compared with the typical Type II supernova.

The supernova coincided precisely with the position of the bright blue supergiant star Sanduleak − 69 202, which disappeared following the supernova explosion. This observation indicated that the progenitor of the supernova was a massive early-type B3 star. The fact that the progenitor was a highly luminous blue star was a surprise because it might have been expected to be a red supergiant. The early phases of development of the light-curve suggested that the progenitor star must have been massive, $M \approx 20 M_\odot$, consistent with the mass of the B star Sanduleak −69 202. Stellar evolution models, which begin with this mass, were developed in which the progenitor first becomes a red giant and then, because of strong mass loss, moves to the blue region of the Hertzsprung–Russell diagram for 10^4 years before exploding as a supernova. According to this picture, the star must have had a smaller envelope than is usual for the B star and a lower abundance of heavy elements than the standard cosmic abundances, roughly one-third the solar value. This abundance is consistent with the general trend of heavy element abundances in the Large Magellanic Cloud.

One piece of great good fortune was that, at the time of the explosion, neutrino detectors were operational at the Kamiokande experiment in Japan and at the Irvine–Michigan–Brookhaven (IMB) experiment located in an Ohio salt-mine in the USA. Both experiments were designed for an entirely different purpose, which was to search for evidence of proton decay, but the signature of the arrival of a burst of neutrinos was convincingly demonstrated in both experiments.[23] Only 20 neutrinos with energies in the range 6 to 39 MeV were detected, 12 at Kamiokande and 8 at IMB, but they arrived almost simultaneously at the two detectors, the duration of this pulse being about 12 seconds. The timescale of 12 seconds is consistent with what would be expected when allowance is made for neutrino trapping in the core of the collapsing star. What made this identification of the neutrino pulse with the supernova convincing was that the supernova was only observed optically some hours after the neutrino pulse. The neutrinos escape more or less directly from the centre of the collapse

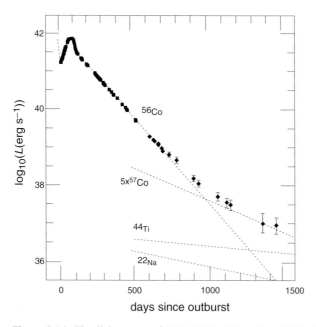

Figure 8.14: The light-curve of SN 1987A (Chevalier, 1992). The bolometric luminosity of the supernova in the ultraviolet, optical and infrared wavebands during the first five years. The energy deposited by radioactive nuclides (dashed lines) is based upon the following initial masses: $0.075 M_\odot$ of ^{56}Ni (and subsequently ^{56}Co), $10^{-4} M_\odot$ of ^{44}Ti, $2 \times 10^{-6} M_\odot$ of ^{22}Na and $0.009 M_\odot$ of ^{57}Co, the last being five times the value expected from the solar ratio of ^{56}Fe/^{57}Fe.

of the progenitor star, whereas the optical light has to diffuse out through the supernova envelope. This observation, coupled with the measured energies of the neutrinos, enabled limits to be set to the rest mass of the neutrino. If the electron neutrino had a finite rest mass, the more energetic neutrinos would be expected to arrive before the less energetic ones since they would have velocities closer to that of light. The limits derived from these considerations corresponded to $m_{\nu_e} \leq 20$ eV.

The observation of the neutrino flux from the supernova is uniquely important for the theory of stellar evolution. Adopting standard cross-sections for the neutrino interactions and the size of the detectors, it turns out that the neutrino luminosity of the supernova was of the same order as that expected from the formation of a neutron star ($E \approx 10^{46}$ J). Note that these observations provide strong support for the essential correctness of our understanding of the late stages of stellar evolution.

Figure 8.14 shows how the bolometric light curve of SN 1987A evolved over the five years since the initial explosion. After the initial outburst, the luminosity decayed exponentially with characteristic half-life of 77 days until roughly 800 days after the explosion when the rate of decline decreased (Chevalier, 1992). One of the most intriguing observations has been the search for the products of the radioactive decay chains in the γ-ray and optical–infrared spectra of the supernova. To account for the luminosity of the supernova, about $0.07 M_\odot$ of ^{56}Ni must have been deposited in the supernova envelope, and this figure agrees very well with the theoretical expectations of explosive nucleosynthesis. It is therefore expected that the ratio of abundances of ^{56}Ni and ^{56}Co to iron should decrease as these radioactive elements decay. Evidence was found for the γ-ray line of ^{56}Co at the 1238 keV line within six months of the explosion from observations with the γ-ray spectrometer on board the Solar Maximum Mission (Matz *et al.*, 1988). In addition, the fine-structure lines of singly ionised cobalt at 10.52 μm and singly ionised nickel at 10.62 μm appeared in the infrared spectrum of the supernova and increased in strength once the exponential decrease in luminosity began (Aitken *et al.*, 1988).

There are other important features of the light-curve shown in Figure 8.14. The optical light-curve showed a break at about 500–600 days but, at that same time, the far-infrared flux increased so that the total luminosity continued to decrease exponentially. In addition, observations of the near-infrared lines of iron showed that less than $0.075 M_\odot$ of iron was present. At about the same time, the emission lines showed absorption of the redshifted gas, indicating that absorption was taking place within the supernova. All these observations are consistent with the formation of dust within the supernova ejecta after about 500 days.

After about 900 days, the rate of decline of the total luminosity of the supernova decreased. The natural interpretation of this phenomenon is that another, longer lived, radioactive nuclide had taken over from ^{56}Co as the energy source for the remnant, and the expected candidate is ^{57}Co. In Figure 8.14, the expected light-curve is shown, assuming that ^{57}Co is five times more abundant than its local abundance; also shown are the expected contributions of the longer lived radionuclides ^{44}Ti and ^{22}Na. The totality of these observations provides direct confirmation for the radioactive origin of supernova light-curves and the formation of iron peak elements in supernova explosions.

Observations of SN 1987A have also provided a great deal of information about the pre-history of the explosion and the surrounding interstellar medium. The burst of radiation associated with the explosion illuminated the material ejected in previous mass loss events, particularly from the period of strong mass loss during the red giant phase. Evidence has been found for bipolar emission structures, as well as for 'light echoes' from clouds or dust 'sheets' along the line of sight to the supernova.

Perhaps the most remarkable feature has been the discovery by the Hubble Space Telescope of a ring of emission in the line of doubly ionised oxygen [OIII] about the supernova (Figure 8.15). This ring was excited by the initial outburst of ultraviolet radiation from the supernova and is in the form of a perfect ellipse. The ultraviolet spectrum of the supernova had been monitored regularly by the International Ultraviolet Explorer, and, at a certain time after the explosion, forbidden ultraviolet emission lines appeared which increased in strength over the subsequent weeks. It was natural to associate this event with the illumination of the ring. These observations enabled a rather precise estimate of the distance of the supernova to be made, a value which was entirely consistent with independent estimates of the distance of the Large Magellanic Cloud (Panagia *et al.*, 1991).

It was predicted by Richard McCray (b. 1937) that eventually the ring would be hit by the blast-wave from the supernova, which travels at a speed much less than the speed of light, and then it would be expected that the ring would be illuminated at optical and X-ray wavelengths (McCray, 1993). As he remarked in that review,

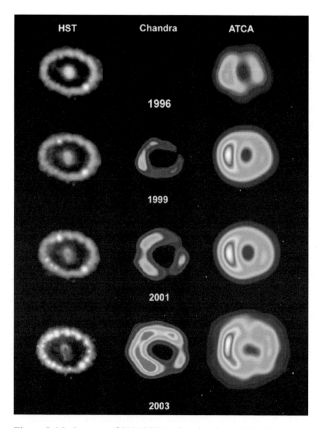

Figure 8.15: Images of SN 1987A taken by the Hubble Space Telescope, the Chandra X-ray Observatory and the Australia Telescope National Facility, showing the evolution of the structure of the ring over the period 1996 to 2003 at optical, X-ray and radio wavelengths. (Courtesy of NASA, the Space Telescope Science Institute, the Chandra Science Team and the ATNF.)

... the impact will occur $c.2004 \pm 3$ AD, allowing for the uncertainty in the density, $n_b = 8\,\text{cm}^{-3}$, of the shocked wind from the blue giant progenitor of SN 1987A.

Observations by the Hubble Space Telescope, the Chandra X-ray Observatory and the Australia Telescope in its compact configuration have observed this remarkable event (Figure 8.15). In all three wavebands, enhanced emission has been observed, the optical image showing a 'ring of pearls' where the shock-wave has encountered higher density regions within the ring. The collision of the shock-wave with the ring has heated the stationary material to high temperatures and accounts for the increase in X-ray intensity observed from the ring. Equally impressive is the fact that the radio emission, which is associated with the synchrotron radiation of high-energy electrons accelerated in the shock-wave, shows a clear increase in intensity, providing a real-time example of the acceleration of high-energy electrons in a supernova shock-wave (see Section 11.4 for a discussion of the acceleration process).

Notes to Chapter 8

1 This work was carried out during a remarkable period in the late 1960s when Fred Hoyle founded the Institute of Theoretical Astronomy in Cambridge, UK. During the summer months, Fowler, Arnett, Clayton, Wagoner, the Burbidges and their students were regular visitors. Many of the key problems of nuclear astrophysics could be addressed by the new generation of high-speed digital computers.

2 A historical review of the solar models and how the predicted solar neutrino flux changed from the 1960s to 2003 is given by J. N. Bahcall, Solar models: an historical review, *Nuclear Physics B (Proceedings Supplement)*, **118**, 2003, 77–86.

3 An excellent history of the solar neutrino problem up to 1989 is given by Bahcall in his book *Neutrino Astrophysics* (Bahcall, 1989).

4 Bahcall has provided an interesting commentary on the subject of why it was so long before the particle physicists took the solar neutrino problem seriously. He argues that the particle physicists did not realise just how robust the helioseismological determinations of the internal structure of the Sun were; see Bahcall, Solar models.

5 An excellent survey of the role of computation in developing models of stellar structure is given by R. Kippenhahn and A. Weigert, *Stellar Structure and Evolution* (Berlin: Springer-Verlag, 1990).

6 The physics of the evolution of stars from the main sequence to the red giant branch has remained one of the classic 'unsolved' problems of stellar astrophysics. John Faulkner has provided deep insights into the physical cause of the phenomenon in his paper 'Low-mass red giants as binary stars without angular momentum' (Faulkner, 2001). He identifies the key feature as the ratio of the densities in the core and in the hydrogen-burning shell and the need to be able to store a considerable amount of mass in the red giant envelope. The consequence of these is an *anti-homology theorem*, according to which, under these conditions, the density of the envelope must decrease as the core density increases.

7 An excellent survey of the history of the understanding of the theory of stellar pulsation is included in John P. Cox, *The Theory of Stellar Pulsation* (Princeton: Princeton University Press, 1980).

8 On the OPAL web-site, the following message indicates what was included in these massive computations. 'Briefly, the calculations are based on a physical picture approach that carries out a many-body expansion of the grand canonical partition function. The method includes electron degeneracy and the leading quantum diffraction term as well as systematic corrections necessary for strongly-coupled plasma regimes. The atomic data are obtained from a parametric potential method that is fast enough for in-line calculations while achieving an accuracy comparable to single configuration Dirac–Fock results. The calculations use detailed term accounting; for example, the bound–bound transitions are treated in full intermediate or pure LS coupling depending on the element. Degeneracy and plasma collective effects are included in inverse bremsstrahlung and Thomson scattering. Most line broadening is treated with a Voigt profile that accounts for Doppler, natural width, electron impacts, and for neutral and singly ionized metals broadening by H and He atoms. The exceptions are one-, two-, and three-electron systems where linear Stark broadening by the ions is included.'

9 Hewish was awarded the Nobel prize for physics for the discovery of neutron stars. The prize was awarded jointed with Martin Ryle, the citation describing his work in developing the principles of aperture synthesis.

10 The history of the discovery of pulsars is described by A. Hewish, The pulsar era, *Quarterly Journal of the Royal Astronomical Society*, **27**, 1986, 548–558. A delightful description of the discovery of pulsars is given by Jocelyn Bell-Burnell, The discovery of pulsars, in *Serendipitous Discoveries in Radio Astronomy*, eds. K. Kellermann and S. Sheets) (Green Bank, West Virginia: National Radio Astronomy Observatory Publications), pp. 160–170.

11 According to William McCrea, the term 'pulsar' was invented by Anthony Michaelis, then science correspondent for the *Daily Telegraph*.

12 A survey of observations and the physics of pulsars, including references to many original papers, is given by A. G. Lyne and F. Graham Smith, *Pulsar Astronomy* (Cambridge: Cambridge University Press, 1998).

13 The preprint on this subject is referred to by A. G. W. Cameron in his review Neutron stars *Annual Review of Astronomy and Astrophysics*, **8**, 1970, 179–208.

14 The history of the UHURU satellite and its discoveries is described by W. Tucker and R. Giacconi, *The X-ray Universe* (Cambridge, Massachusetts: Harvard University Press, 1985).

15 I have given a proof of this result in Section 17.10 of Malcolm Longair, *Theoretical Concepts in Physics* (Cambridge: Cambridge University Press, 2003).

16 For details of the physics of accreting binary systems, see S. I. Shapiro and S. A. Teukolsky, *Black Holes, White Dwarfs and Neutron Stars: The Physics of Compact Objects* (New York: Wiley Interscience, 1983).

17 Many more details of these processes are contained in J. Frank, A. King and D. Raine, *Accretion Power in Astrophysics*, 3rd edn (Cambridge: Cambridge University Press, 2002).

18 Hulse and Taylor were awarded the Nobel prize for physics in 1993 for this discovery.

19 A very readable account of black holes, gravitational waves and their significance for physics and astronomy is contained in Kip Thorne's book *Black Holes and Time Warps: Einstein's Outrageous Legacy* (New York: W. W. Norton and Company, 1994).

20 For a comprehenive review of the properties of supernovae, see V. Trimble, Supernovae. Part 1: the events, *Reviews of Modern Physics*, **54**, 1982, 1183–1224 and V. Trimble, Supernovae. Part II: the aftermath, *Reviews of Modern Physics*, **55**, 1983, 511–563.

21 For an entertaining description of Zwicky's researches on supernovae, see F. Zwicky, Review of the research on supernovae, in *Supernovae and Supernova Remnants*, ed. C. Battali Cosmovici (Dordrecht: D. Reidel Publishing Company, 1974), pp. 1–16.

22 There is a vast literature on SN 1987A. Symposia devoted to the supernova include: *Supernova 1987A in the Large Magellanic Cloud*, eds M. Kafatos and A. G. Michalitsianos (Cambridge: Cambridge University Press, 1988) and *Supernova 1987A and Other Supernovae*, eds I. J. Danziger and K. Kjär (Garching bei München: European Southern Observatory, 1991). Excellent summaries of the observations up to 1992 are given by R. Chevalier, *Nature*, **355**, 1992, 691–696, and by R. McCray, *Annual Reviews of Astronomy and Astrophysics*, **31**, 1993, 175–216.

23 An excellent discussion of these observations is given by Bahcall (1989).

A8   Explanatory supplement to Chapter 8

A8.1   *Gravitational radiation loss rate by the binary pulsar*

The observation that the rate of change of the angular velocity, Ω, of the binary pulsar PSR 1913+16 is proportional to Ω^5 is considered to be very strong evidence that the system is losing energy by the radiation of gravitational waves. The following elementary arguments give an impression of how the result comes about. The excellent essay by Bernard Schutz (b. 1946) can be recommended as a gentle introduction to the physics of gravitational radiation.[1] The theory of gravitational radiation is highly non-trivial and has not been without controversy.[2] These notes are intended to provide some insight into how this dependence comes about and the energies involved.

We use the analogy between the inverse square laws of electrostatics and gravity to work out the rate of radiation of quadrupole radiation from a system of charges or masses. In 1906, J. J. Thomson (1856–1940) presented a simple argument to derive the formula for the instantaneous rate of loss of energy by radiation of an accelerated charged particle.[3] I have used this approach in my book *Theoretical Concepts in Physics*, the result being, in SI notation,

$$-\left(\frac{dE}{dt}\right) = \frac{|\ddot{p}|^2}{6\pi\epsilon_0 c^3},\tag{A8.1}$$

where p is the dipole moment of the accelerated charge relative to some origin. In the case of an oscillating dipole, we can write $p = p_0 e^{-i\Omega t} = qr_0 e^{-i\Omega t}$, where Ω is the angular frequency of oscillation and $p = qr$. Therefore, the expression for the time-averaged rate of radiation loss is given by

$$-\left(\frac{dE}{dt}\right) = \frac{q^2\Omega^4|r_0|^2}{12\pi\epsilon_0 c^3} = \frac{\Omega^4|p_0|^2}{12\pi\epsilon_0 c^3}.\tag{A8.2}$$

The equivalent expression for gravity results in no energy loss because a gravitational dipole has zero dipole moment, both 'gravitational charges' having the same sign. Thus, there is no dipole emission of gravitational waves.

A system of gravitating masses can, however, have a finite quadrupole moment, and, if this is time varying, there is an energy loss by gravitational radiation. The corresponding equation to equation (A8.2) for the time-averaged rate of loss of energy by electric quadrupole radiation is given by

$$-\left(\frac{dE}{dt}\right) = \frac{\Omega^6}{1440\pi\epsilon_0 c^5}\sum_{j,k}|Q_{jk}|^2,\tag{A8.3}$$

where the elements of the electric quadrupole moment tensor, Q_{jk}, are given by

$$Q_{jk} = \int (3x_j x_k - r^2\delta_{jk})\rho(x)\,d^3x,\tag{A8.4}$$

and $\rho(x)$ is the electric charge density.[4]

We obtain an approximate expression for quadrupole gravitational radiation in this 'Coulomb approximation' if we use the equivalence $e^2/4\pi\epsilon_0 \to Gm^2$; then,

$$-\left(\frac{dE}{dt}\right)_{gr} = \frac{G\Omega^6}{360c^5}\sum_{j,k}|Q_{jk}|^2,\tag{A8.5}$$

where the Q_{jk} are now the components of the gravitational quadrupole moment tensor. The exact expression quoted by Schutz is as follows:

$$-\left(\frac{dE}{dt}\right)_{gr} = \frac{G}{5c^5}\left(\sum_{j,k}\dddot{Q}_{jk}\dddot{Q}_{jk} - \tfrac{1}{3}\dddot{Q}^2\right),\tag{A8.6}$$

where the Q_{jk} are the components of spatial quadrupole tensor, or matrix, the second moment of the mass, or charge, distribution,[5]

$$Q_{jk} = \int \rho x_j x_k\,d^3x.\tag{A8.7}$$

Following Schutz's pleasant order-of-magnitude estimates, let us estimate the rate of radiation of a binary star system, each star having mass M moving in a circular orbit of radius r at speed v about their common centre of mass. To order of magnitude, the third derivative of the quadrupole moment is then given by

$$\frac{\mathrm{d}^3 Q}{\mathrm{d}t^3} \sim \frac{Q}{t^3} \sim \frac{Mr^2}{t^3} \sim \frac{Mv^3}{r}, \tag{A8.8}$$

where r and t are the typical spatial and time scales over which the quadrupole moment varies. Inserting this value into equation (A8.6) for the term in large brackets on the right-hand side, we find the gravitational radiation loss rate:

$$-\left(\frac{\mathrm{d}E}{\mathrm{d}t}\right)_{\mathrm{gr}} \sim \frac{G}{5c^5}\left(\frac{Mv^3}{r}\right)^2. \tag{A8.9}$$

Since $v = \Omega r$, we obtain the result

$$-\left(\frac{\mathrm{d}E}{\mathrm{d}t}\right)_{\mathrm{gr}} \sim \frac{G}{5c^5} M^2 r^4 \Omega^6. \tag{A8.10}$$

Note that equations (A8.5) and (A8.10) have exactly the same dependence upon the quantities which appear in the loss rate for gravitational radiation since $Q_{jk} \sim Mr^2$. The angular velocity of the binary star is given by $\Omega^2 = GM/4r^3$, and so equation (A8.10) can be rewritten as follows:

$$-\left(\frac{\mathrm{d}E}{\mathrm{d}t}\right)_{\mathrm{gr}} \sim \frac{c^5}{80G}\left(\frac{GM}{rc^2}\right)^5, \tag{A8.11}$$

where, following Schutz, it is assumed that four comparable terms contribute to the total gravitational loss rate. As Schutz points out, the term c^5/G corresponds to a luminosity of 3.6×10^{52} W, an enormous value, which is modified by the 'relativistic factor', $(GM/rc^2)^5$. Thus, really close binary neutron stars are expected to be extraordinarily powerful sources of gravitational waves. In particular, the very last phases of inspiralling of a pair of neutron stars in a very close binary system, before they coalesce into a black hole, is one of the most important targets for observation by the present generation of gravitational wave detectors.

The key relation from our present perspective is that equations (A8.5) and (A8.10) show that the rate of energy loss by gravitational waves is proportional to Ω^6. This energy is extracted from the binding energy of the binary system, $E = -\frac{1}{2}I\Omega^2$, and so we can write

$$-\left(\frac{\mathrm{d}E}{\mathrm{d}t}\right)_{\mathrm{gr}} = \left[\frac{\mathrm{d}(\frac{1}{2}I\Omega^2)}{\mathrm{d}t}\right] \propto \Omega^6; \tag{A8.12}$$

$$\dot{\Omega} \propto \Omega^5. \tag{A8.13}$$

This relation was found by Hulse and Taylor for the binary pulsar PSR 1913+16 (see Figure 8.13). They did much more than this because the orbits of the two neutron stars and their masses were very well determined by their precise timing observations, and so the exact radiation loss rate for their highly elliptical orbits, $e = 0.615$, could be estimated and was found to be in excellent agreement with the observed spin-up rate of the binary system.

Notes to Section A8

1 B. Schutz, Gravitational radiation, *Encyclopaedia of Astronomy and Astrophysics*, vol. 2, ed.
   P. Murdin (Bristol and Philadelphia: Institute of Physics Publishing and London: Nature Publishing
   Group, 2001), pp. 1030–1042.

2 The care with which the theory has to be formulated is carefully described by Ray D'Inverno in
   Introducing Einstein's Gravity (Oxford: Clarendon Press, 1995). A brief history of some of the
   problems involved in understanding the existence of gravitational waves in general relativity is
   given by John Stachel in his review History of relativity, in *Twentieth Century Physics*, vol. 1, eds
   L. M. Brown, A. Pais and A. B. Pippard (Bristol and Philadelphia: Institute of Physics Publishing
   and New York: American Institute of Physics Press, 1995), pp. 249–356.

3 The argument was first presented by Thomson in his book *Electricity and Matter* (London:
   Archibald Constable and Company, 1906), the published version of his Silliman Memorial Lectures delivered at Harvard in 1903. This same formula was subsequently used by Thomson in his
   book *Conduction of Electricity through Gases* (Cambridge: Cambridge University Press, 1907)
   to derive the expression for the cross-section for the scattering of X-rays by free electrons, the
   Thomson cross-section.

4 Equation (A8.3) is derived by J. D. Jackson in his excellent book *Classical Electrodynamics*, 3rd
   edn (New York: John Wiley and Sons, Inc., 1999), Section 9.3.

5 Note the difference of a factor of 3 between the definitions of the quadrupole moment tensor for
   electromagnetism, equation (A8.4), and gravity, equation (A8.6). There is a compensatory factor of
   3 in the term in brackets on the right-hand side of equation (8.6). The difference in the gravitational
   case is because of technicalities about the choice of gauge to be used in the full development of
   the theory of gravitational radiation.

9 The physics of the interstellar medium

9.1 The photoionisation of the interstellar gas

By 1939, the existence of various forms of interstellar matter had been established. From the study of interstellar absorption lines and the variation of interstellar extinction with distance, it was known that diffuse gas and dust are present in the interstellar medium (Plaskett and Pearce, 1933; Joy, 1939). Gaseous nebulae had been known to be constituents of the Galaxy since the time of Huggins' pioneering observations in the 1860s. During the first two decades of the twentieth century, Edward Barnard (1857–1923) made extensive studies of the forms of the dark clouds apparent in photographs of the Milky Way (Barnard, 1919). The nature of these clouds was studied by Max Wolf (1863–1932), who determined the amount of extinction they cause by making star counts in their vicinity (Wolf, 1923). He correctly attributed the extinction to dust grains rather than gas because in the latter case the strong dependence of Rayleigh scattering upon wavelength would have resulted in much greater reddening of background stars than was observed.

On the theoretical side, it was recognised in the early 1920s that both the central stars of planetary nebulae and the O stars are very hot and so radiate a great deal of energy in the ultraviolet waveband. Russell suggested that the excitation of the emission lines seen in gaseous nebulae and planetary nebulae were due to photoexcitation (Russell, 1921), and Eddington showed that, as a result, the gas would attain a temperature of about 10 000 K (Eddington, 1926b). In 1926, Donald Menzel applied this model to planetary nebulae and suggested that the Balmer emission lines were associated with the photoionisation of hydrogen by the ultraviolet radiation of the central star followed by recombination of the protons and electrons (Menzel, 1926). He believed, incorrectly, that the temperatures of the central stars would have to be unreasonably high for this process to be effective. In 1927, Herman Zanstra (1894–1972), however, had no hesitation in postulating that the central stars of planetary nebulae had temperatures of about 30 000 K (Zanstra, 1926). In a subsequent paper, he described the processes of photoionisation and recombination of hydrogen, in which each photon with energy greater than 13.6 eV ionises one hydrogen atom (Zanstra, 1927). In recombining, the electron does not necessarily return to the ground state but cascades through a number of energy levels, emitting the Balmer series of hydrogen in the process.

In 1928, Ira Bowen (1898–1973) at last identified the 'nebulium' lines which had baffled spectroscopists since their discovery in 1868 (Bowen, 1927, 1928). In a brilliantly argued paper, he showed that the nebulium lines can all be associated with what are now termed *forbidden transitions* between low-lying metastable states of the ions of common elements and the ground state. For example, the pair of lines at 500.7 and 495.9 nm are associated with forbidden transitions of doubly ionised oxygen [OIII]. These lines had not been observed in laboratory experiments because the densities are so high that the metastable states are de-excited by collisions. In constrast, in the diffuse nebulae, the rate of collisions can be very low and so, although the transition probabilities are low, radiative transitions are the means by which the ions are de-excited and reach the ground state.

These discoveries set the scene for the development of the theory of photoionisation and of recombination and forbidden-line spectra under interstellar conditions during the 1930s and 1940s, studies particularly associated with the names of Menzel and his colleagues Lawrence Aller (1913–2003) and James Baker (b. 1914).[1] The result of these studies was to establish powerful means of estimating the densities and temperatures of regions of ionised gas. The importance of the process of photoionisation for clouds of ionised gas in the interstellar medium was described by Bengt Strömgren (1908–1987) in 1939 (Strömgren, 1939). He solved the problem of the dependence of the ionisation of the interstellar gas upon its density and the temperature of the exciting star. The radius of the ionisation zone about a star is a very strong function of its surface temperature because it depends upon the flux of ionising radiation with wavelengths shorter than the Lyman limit at $\lambda \leq 91.6$ nm. For example, if the interstellar density were 10^6 m^{-3} and the radius of the star the same as that of the Sun, an O5 star would ionise a region of radius 54 pc, whereas, for a B5 star, the radius would be only 1.6 pc. Strömgren also showed that the regions of ionised gas, often referred to as *Strömgren spheres*, have very sharp edges. As he expressed it (Strömgren, 1939):

Once the proportion of neutral atoms begins to increase, the absorption of the ionising radiation increases, leading to an accelerated increase of neutral atoms.

These conclusions were in very good agreement with the observed properties of the regions of ionised hydrogen known as HII regions. The hottest O stars are found embedded in regions of ionised hydrogen, the most extensive regions being excited by a number of O stars.

This picture of the interstellar gas, as it was understood in 1939, was entirely derived from optical observations. Little did these pioneers realise that they had detected only a tiny fraction of the total amount of material present in the interstellar medium. After the Second World War, observations became possible in the radio, infrared, ultraviolet and X-ray wavebands and revealed the full complexity of the interstellar medium.

9.2 Neutral hydrogen and molecular line astronomy

The prediction of the 21 cm line of neutral hydrogen is one of the more remarkable stories attending the birth of radio astronomy. Jan Oort was the director of the Leiden Observatory throughout the period of the German occupation of Holland during the Second World War, and he kept astronomical activity alive by encouraging theoretical studies among those staff and students who had escaped detention. Conditions were very difficult, and Oort was personally in considerable danger because of his support for Jewish professors in the University. Astronomical seminars were held in secret in the basement of the Leiden Observatory. Copies of the *Astrophysical Journal* somehow continued to reach Leiden, and among the papers was Reber's continuum map of the Galaxy. Oort asked Hendrik van de Hulst (1918–2000) the following question:[2]

Is there a spectral line at radio frequencies we should in principle be able to detect? If so, because at radio wavelengths absorption should be negligible, we should be able to derive the structure of the Galaxy. We might even be able to detect spiral arms, if they exist.

Van de Hulst took up the challenge and studied the many ways in which atoms, ions and molecules could radiate line emission in the radio waveband (Van de Hulst, 1945). His results were presented at the last of the underground seminars held in April 1944. The most significant prediction was that neutral hydrogen should emit line radiation at a wavelength of 21.106 cm because of the minute change of energy when the relative spins of the proton and electron in a hydrogen atom change, that is, when transitions take place between its hyperfine states. Although this is a highly forbidden transition with a spontaneous transition probability of only once every 12 million years, so much neutral hydrogen was expected to be present in the Galaxy that there was a good chance that the line would be detectable. Oort and Lex Muller (1923–2004) would probably have been the first to detect the 21 cm line, but the first receiver was destroyed in a fire. The first detection of the line was made by Harold Ewen (b. 1922) and Edward Purcell (1912–1997) at Harvard University in 1951 (Ewen and Purcell, 1951); six weeks later, Oort and Muller measured the same line (Muller and Oort, 1951). The neutral hydrogen line proved to be one of the most important tools for diagnosing velocity fields in our own and other galaxies. The first maps of the distribution of neutral hydrogen in the Galaxy and the determination of the Galactic rotation curve appeared in 1952. By 1953, neutral hydrogen had been detected in the Magellanic Clouds, and in 1954 the high-velocity features in the Galactic centre and 21 cm absorption spectra were first measured.[3]

In 1958, Oort, Frank Kerr (1918–2000) and Gart Westerhout (b. 1927) published their famous map of the distribution of neutral hydrogen in the plane of the Galaxy (Figure 9.1) (Oort, Kerr and Westerhout, 1958). This map was made by combining observations made in the Netherlands with those made at Sydney, Australia. The derivation of the density distribution along the line of sight through the Galaxy was not a trivial exercise because the neutral hydrogen observations do not provide any distance measure. Nonetheless, subsequent analyses have generally agreed with the large-scale features seen in Figure 9.1. Neutral hydrogen is omnipresent throughout the plane of the Galaxy and extends well beyond the Sun's radius. Various 'spiral features' can be distinguished on the map, although there is only a general correlation with the arms outlined by the O and B stars and other spiral arm indicators. Observations of the 21 cm line have also enabled the Galactic rotation curve to be determined. Of particular interest is the rotation curve beyond the Sun's orbit about the Galactic centre because it is flat, $v(r) = $ constant, suggesting the presence of dark matter in the halo of the Galaxy (Fich and Tremaine, 1991).

Long before the advent of radio astronomy, it was known that there exist significant abundances of molecules in interstellar space. The molecules CH, CH^+ and CN possess

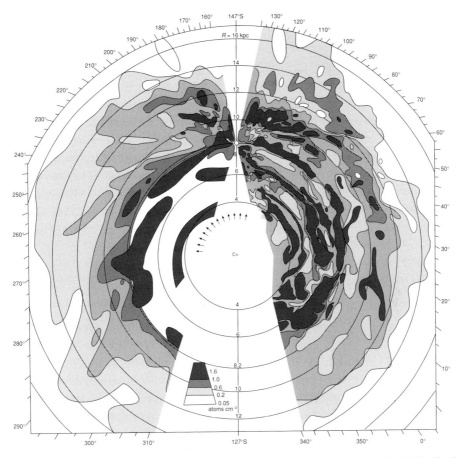

Figure 9.1: The distribution of atomic hydrogen in the Galactic plane (Oort *et al.*, 1958). The Sun–Galactic centre distance is assumed to be 8.2 kpc. The numbers around the outside of the figure denote Galactic longitudes. In deriving this map, it is assumed that the neutral hydrogen rotates in circular orbits about the Galactic centre, with velocities given by a standard rotation curve.

electronic transitions in the optical waveband, and absorption features associated with these were well known in the spectra of bright stars. As radio astronomy technology advanced, it became possible to observe at higher and higher frequencies. The first interstellar molecule to be detected at radio wavelengths was the hydroxyl radical OH, which was observed in absorption against the bright radio source Cassiopeia A by Sander Weinreb (b. 1936), Alan Barrett (1927–1991) and their colleagues at a wavelength of 18 cm in 1963 (Weinreb *et al.*, 1963). Soon afterwards, in 1965, the hydroxyl lines were observed in emission by Harold Weaver (b. 1917) and his group at Berkeley (Weaver *et al.*, 1965; Weinreb *et al.*, 1965). The surprise was that the sources were very compact and variable in intensity. The corresponding brightness temperatures were very great indeed, $T_b \geq 10^9$ K, implying that the emission process must involve some form of maser action. This was the beginning of the intensive search for other interstellar molecules. In 1968, ammonia, NH_3, was detected (Cheung *et al.*, 1968), and in the following year water vapour, H_2O, and formaldehyde, H_2CO, were discovered (Cheung *et al.*, 1969; Snyder *et al.*, 1969). The processes of line formation in these molecules involve doubling processes, and the observed intensities involved various forms of maser action.

A key discovery was the great intensity of the carbon monoxide molecule, CO, which was first observed by Robert Wilson (b. 1936), Keith Jefferts (b. 1931) and Arno Penzias (b. 1933) in 1970 (Wilson, Jefferts and Penzias, 1970). CO is one of the simplest molecules which emits line radiation by electric dipole transitions between neighbouring rotational states. Rotational transitions are only observed from molecules which have a finite dipole moment, and so the molecule which is expected to be by far the most abundant in the interstellar gas, molecular hydrogen, H_2, does not emit rotational molecular lines. Therefore, CO is expected to be the most abundant species which can be detected by its millimetre and submillimetre line emission and it acts as a tracer for the distribution of molecular hydrogen. Molecular abundances and the temperatures of the clouds can be determined since the molecules can be assumed to be in collisional equilibrium. Since that time, the number of detected molecular species has multiplied rapidly. Among these has been the detection of ethyl alcohol in the Galactic centre (Zuckerman *et al.*, 1975), the discovery paper remarking that this

truly astronomical source of ethyl alcohol ... would yield approximately 10^{28} fifths at 200 proof.

By August 2004, 125 different molecules had been found in the interstellar medium, a number of the species being unstable in the laboratory but able to survive in the low-density conditions of interstellar space. The discovery of interstellar molecules came as a surprise because it had been assumed that the ultraviolet radiation in the diffuse interstellar medium would dissociate any but the most tightly bound molecules. This argument had neglected the shielding role of dust, which protects the molecules from the ultraviolet radiation. Molecular gas is present throughout the plane of the Galaxy, and, as soon as the first maps of its distribution were made, it was found that a large fraction of the molecules belong to *giant molecular clouds*, which typically have masses about $10^6 M_\odot$. By the late 1970s, it was apparent that these clouds contained a great deal of fine-scale structure and that these are the sites of star formation. The giant molecular clouds are transparent to millimetre and submillimetre radiation and so the narrow molecular lines are excellent probes of the internal dynamics of the clouds. The presence of so many different molecular species and the large densities of interstellar dust gave rise to the new discipline of *interstellar chemistry*.[4] To understand the existence and abundances of the many molecules now observed in molecular clouds requires an understanding, not only of gas-phase reactions, but also of molecular processes occurring on the surfaces of dust grains.

Among the more remarkable aspects of these discoveries was the interaction between chemists and astronomers. Harold Kroto (b. 1939) and his colleagues at the University of Sussex in the UK had successfully synthesised the linear acetylenic chain molecule cyanoacetylene, HC_5N, in the laboratory and measured its rotational spectrum. Despite expectations that the molecule would only be observed in very low abundances in the interstellar medium, Kroto and his Canadian colleagues from the National Research Council discovered the molecule in the giant molecular cloud Sagittarius B2 towards the Galactic centre (Avery *et al.*, 1976). Next, they performed the same exercise for the molecule HC_7N, expected to be even rarer, but again they found the molecule with very much greater abundance than expected from simple thermodynamic arguments (Kroto *et al.*, 1978). Similar success was achieved in the same year in detecting HC_9N, the longest chain molecule yet detected (Broten *et al.*, 1978). The importance of these observations was that they demonstrated conclusively that the long-chain molecules could not be synthesised by equilibrium gas-phase reactions. The clue was provided by sources such as IRC+10216, a cool red giant carbon star, which was a plentiful source of the acetylenic chain molecules.

The sequel to this story is just as remarkable. In 1985, Kroto carried out experiments to find out if the chain molecules could be created in the laboratory using a cluster beam machine constructed by Richard Smalley (1943–2005) and his colleagues at Rice University, Houston, modified to use a carbon target. This was successfully achieved in 1985 (Kroto *et al.*, 1987), confirming the idea that these molecules could be created in cool red giant envelopes. In the same experiments, which involved creating a hot nucleating carbon plasma, they also found evidence for a very large abundance of molecules consisting of 60 carbon atoms, the discovery of the C_{60} molecule, buckminsterfullerene (Kroto *et al.*, 1985).[5]

While much of the study of interstellar dust grains had focussed upon the properties of particles roughly 1 μm in size, evidence was found by Kristen Sellgren (b. 1955) for a population of very much smaller grains from studies of the infrared continuum spectra of reflection nebulae (Sellgren, 1984). She found that the colour temperature of the emission was high, about 1000 K, and could not be attributed to the standard continuum emission processes. Her proposal was that the emission was associated with transient heating of very small dust grains. In the case of grains with dimension 1 μm, the energy of the absorbed photons is thermalised and reradiated at the temperature to which the grains are heated. In the case of grains only about 1 nm in size, this is no longer the case. An incident ultraviolet photon can raise the temperature of the grain to a temperature of about 1000 K and then cool very rapidly, resulting in a quite different continuum spectrum. Sellgren showed that the necessary number of very small dust grains could be explained as an extrapolation of the grain size distribution from larger sizes. These tiny grains can be thought of as large molecules, rather than as solid materials.

This idea was taken further by Alain Leger (b. 1943) and Jean-Loup Puget (b. 1947), who sought to explain the strong unidentified emission features observed in the infrared region of the spectrum (Leger and Puget, 1984). These prominent lines are observed at wavelengths of 3.28, 6.2, 7.7, 8.6 and 11.3 μm in the spectra of a wide variety of Galactic and extragalactic sources (Figure 9.2). Leger and Puget proposed that these lines are associated with various bending and stretching modes of the small aromatic molecules known as *polycyclic aromatic hydrocarbons*, or PAHs. These molecules typically consist of about 50 carbon atoms in the form of planes of benzene rings. Taking as an example the PAH coronene, they computed that, at a temperature of 600 K, spectral features should be observed at 3.3, 6.2, 7.6, 8.8 and 11.9 μm. These features were identified as follows: the feature at 3.3 μm with the $C-H$ stretching mode, those at 6.2 and 7.7 μm with the $C-C$ stretching modes, that at 8.6 μm with the in-plane bending mode and that at 11.3 μm with the $C-H$ out-of-plane bending mode. In the last case, other features are expected depending upon the number of nearby hydrogen atoms. The excitation of these modes is again associated with the absorption of a single UV photon, which transiently raises the temperature of the molecule to about 1000 K.

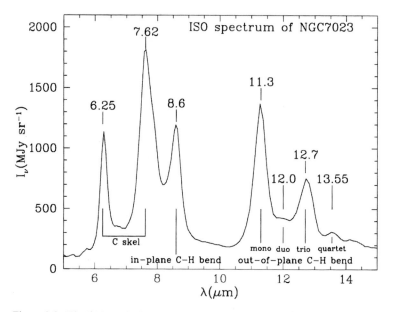

Figure 9.2: The PAH emission features in the 5–15 µm spectrum of the reflection nebula NGC 7023 obtained with the ISO Observatory by Diego Cesarsky and his colleagues (Draine, 2003).

Other features in the absorption spectrum of dust, such as the diffuse interstellar bands seen in the optical region of the spectrum and the prominent feature at 217.5 nm, have so far eluded agreed identification. The net result of these studies is that interstellar dust must be composed of a number of different components. An excellent discussion of the range of different types of dust particles necessary to account for the observations is given by Bruce Draine (b. 1947) (Draine, 2003).

9.3   The multi-phase interstellar medium

In the years immediately after the Second World War, significant progress was made in understanding the properties of the diffuse interstellar medium. The realisation that interstellar extinction was due to dust particles had led Carl Schalén (1902–1993) to apply the theory of Mie scattering to the problem of interstellar extinction in the 1930s, and he found that metallic particles with size roughly 10^{-6} m could reproduce the observed dependence of extinction upon wavelength (Schalén, 1936). The problem was addressed in much more detail during and after the War by Hendrik van de Hulst, who considered a very wide range of possible properties for interstellar dust, ranging from perfectly scattering dielectric spheres to perfectly absorbing metallic spheres (Van de Hulst, 1949b). He showed how the observed extinction as a function of wavelength provides information about both the size distribution of the particles and their chemical composition. It is certain that a range of particle sizes must be present in the interstellar medium, but the exact chemical composition of the grains has remained the subject of controversy.

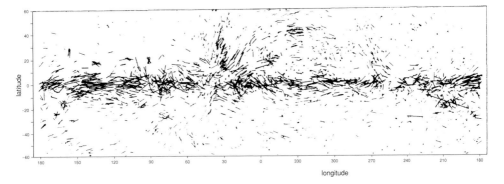

Figure 9.3: The polarisation of stars as a function of galactic coordinates. The magnitudes of the vectors are proportional to the percentage polarisations, and the directions of the vectors indicate the plane of polarisation of the light (Matthewson and Ford, 1970).

These studies were relevant to the discovery of interstellar polarisation by William Hiltner (1914–1991) and John Hall (1908–1991) in 1949 (Hall, 1949; Hiltner, 1949). Their objective was to observe the polarisation of light emitted from the limbs of stars due to electron scattering in their atmospheres, an observation suggested by Chandrasekhar in 1946 (Chandrasekhar, 1946). The idea was to search for polarisation effects in binary star systems as one of the stars is occulted by the other. Hiltner indeed discovered large polarisations, but they were independent of the phase of the binary star, and so he attributed the result to interstellar polarisation. Hall, who was collaborating with Hiltner in this project, showed that the percentage polarisation was correlated with the extinction by dust.

In the same year, Van de Hulst proposed that the polarisation could be attributed to dust grains which are elongated and aligned so that there is preferential absorption of one of the states of linear polarisation of the starlight (Van de Hulst, 1949a). This led to a search for the alignment mechanism for the grains, and in 1951 Leverett Davis Jr (1914–2003) and Jesse Greenstein (1909–2002) proposed that the alignment is produced by the presence of a large-scale interstellar magnetic field (Davis and Greenstein, 1951). The orientation of the dust grains depends upon the dielectric properties of the grains but, in their preferred model, the grains are aligned with their long axes perpendicular to the magnetic field and so the magnetic field would be expected to lie parallel to the polarisation vectors of the starlight. Davis and Greenstein also estimated the strength of the magnetic field necessary to align the grains and found values of about 0.2–0.3 nT. Magnetic field strengths of the same order were found by interpreting the diffuse non-thermal radio emission from the Galaxy as synchrotron radiation, which is also strongly polarised. The most convincing evidence for the presence of a large-scale magnetic field in the vicinity of the Sun came from polarisation observations of about 7000 stars by Donald Matthewson (b. 1929) and Vincent Ford (b. 1943) in 1970 (Matthewson and Ford, 1970) (Figure 9.3). This picture is broadly consistent with the polarisation properties of the Galactic radio synchrotron emission.

The theory of the diffuse interstellar medium was the subject of a series of papers by Lyman Spitzer beginning in 1948. In 1950, Spitzer and Malcolm Savedoff (b. 1928) studied in detail the processes of heating and cooling of the diffuse interstellar gas (Spitzer and Savedoff, 1950). While they confirmed the results of Eddington and Strömgren that, in the vicinity of O and B stars, the temperature of the ionised gas is about 10 000 K, they also showed that the temperature of the diffuse medium between these phases was determined by the balance between heating by the ionisation losses of cosmic rays and by energy losses associated with the low-lying excited states of CI, CII and SiII. The typical temperature of the diffuse interstellar medium was found to be about 60 K. These low temperatures were confirmed some years later by observations of the 21 cm line of neutral hydrogen.

There remained the problem of confining the cool neutral hydrogen clouds, and this was elegantly solved by George Field (b. 1929), Donald Goldsmith (b. 1943) and Harm Habing (b. 1937) in 1969, who considered the thermal stability of a medium heated by interstellar high-energy particles (Field, Goldsmith and Habing, 1969). In 1965, Field had made a detailed study of thermal instabilities in different astrophysical contexts (Field, 1965) and he had shown that, in the intermediate range of temperatures between about 100 and 3000 K, cosmic ray heating in the presence of cooling by electron excitation of the low-lying states of neutral and singly ionised species results in an unstable situation, the stable phases consisting of a low-density, high-temperature phase at $T \approx 8000$ K in pressure balance with a low-temperature phase at $T \approx 20$ K. This became known as the *two-phase model* for the interstellar medium. The temperature of the diffuse neutral hydrogen clouds was too low, and one solution was that the interstellar carbon, which is one of the principal coolants of the gas, was depleted relative to its cosmic abundance because of condensation into interstellar grains.

This picture of the diffuse interstellar medium was soon confronted by observations from the Copernicus ultraviolet spectroscopic satellite, which was launched in 1972. Spitzer and his colleagues had designed the Copernicus satellite specifically to make very high spectral resolution observations of the interstellar medium and, in particular, to study the resonance absorption lines of common elements which lie in the wavelength range $91.2 < \lambda < 120$ nm. Many important discoveries resulted from these studies. The depletion of the chemical abundances of the heavy elements in the interstellar medium as compared with their cosmic abundances was confirmed (Field, 1974). A very hot component of the medium was detected through observations of the lines of O^{5+} in absorption in the direction of hot stars in the Magellanic Clouds (Rogerson et al., 1973b). There was evidence that this hot gas formed a hot flattened halo about the Galaxy. Molecular hydrogen was also detected in the direction of reddened stars. It was found that the greater the extinction, the greater the column density of molecular hydrogen, suggesting that its formation is catalysed by dust grains (Rogerson and York, 1973).

The discovery of the soft X-ray background from the interstellar medium further complicated the picture. The first observations of the soft X-ray background were made by Stuart Bowyer (b. 1934), Field and John Mack (b. 1942) in 1968 (Bowyer, Field and Mack, 1968), and soft X-ray maps of the sky were made by William Kraushaar (b. 1920) and his colleagues at Wisconsin from a series of rocket flights in 1972 and 1973 (Sanders et al., 1977). These observations showed a strong anti-correlation between the intensity of the X-ray emission and the column density of neutral hydrogen, the inference being that the soft X-rays suffer photoelectric absorption by interstellar neutral hydrogen. Despite the absorption, this was clear evidence that there exists diffuse soft X-ray background emission from the interstellar medium.

In 1974, Donald Cox (b. 1943) and Barham Smith (b. 1947) provided a convincing explanation for this component (Cox and Smith, 1974). The UHURU satellite had shown that young supernova remnants are strong X-ray sources in the 1–10 keV waveband, and this was naturally interpreted as the bremsstrahlung of gas heated to a very high temperature by the shock-wave resulting from the supernova explosion. Cox and Smith worked out the subsequent evolution of these supernova remnants and found that supernovae occur sufficiently often in our Galaxy that old, hot supernova remnants overlap and, by percolation, result in a series of tunnels of hot gas through the interstellar medium. This picture can naturally account for the soft X-ray emission of the Galaxy. This model can also account for the presence of a halo of hot gas about the Galaxy, since bubbles of the hot gas rise up the potential gradient out of the Galactic plane. Detailed studies of the local distribution of neutral and hot gas have suggested that the Sun is located in a local hole of hot gas of radius about 50 pc which was evacuated by a supernova explosion more than 10^6 years ago.[6]

These ideas were synthesised into a picture of the *violent interstellar medium*, an ambitious attempt being described by Christopher McKee (b. 1942) and Ostriker in 1977 (McKee and Ostriker, 1977). The very hot component at 10^6 K, the hot neutral gas at about 10^4 K and the cool diffuse medium at about 100 K are roughly in pressure balance but, in addition, there are the giant molecular clouds. The medium is constantly being buffeted by supernova explosions, which may give rise to the formation and cooling of the giant molecular clouds. The gas is also strongly perturbed by the gravitational influence of spiral arms, which results in large density enhancements on the trailing edges of the arms.

9.4   The formation of stars

While impressive progress was being made in understanding the evolution of stars once nuclear burning in their cores begins, the process of star formation proved a tougher proposition. In 1945, Alfred Joy had drawn attention to the T-Tauri variable stars, which are embedded in nebulous clouds of dust and gas (Joy, 1945). Although there are prominent emission lines, these are superimposed upon an absorption-line spectrum consistent with those of stars with mass roughly equal to the mass of the Sun. In 1947, Viktor Ambartsumian (1908–1996) argued that the T-Tauri stars were low-mass main-sequence stars in the process of formation (Ambartsumian, 1947), and in 1952 George Herbig (b. 1920) suggested that they are low-mass stars which lay above the main sequence (Herbig, 1952). This idea was proved correct by Merle Walker (b. 1926), who studied the extremely young star cluster NGC 2264. In addition to luminous O and B stars, Walker studied the fainter T-Tauri stars in the cluster and showed that they lay above the standard main sequence (Walker, 1956). Walker interpreted these results as indicating that the T-Tauri stars were in the process of contracting towards the main sequence.

The models which Walker had used to study the evolution of pre-main-sequence stars were due to Salpeter, Henyey and their colleagues, and they assumed that, as the star contracted quasi-statically towards the main sequence, energy transport through the star was by radiation. This conclusion was shown to be incorrect by Chushiro Hayashi (b. 1920) in 1961. Hayashi's analysis concerned the stability of quasi-static models of stars and, in particular, those in which energy transport is entirely by convection rather than by radiation (Hayashi, 1961). He had already studied the stability of stars on the giant branch and had shown that there is a limiting locus for stellar models when the energy transport by convection extends throughout the whole of the star. This *Hayashi limit* is what eventually stops the expansion of the envelopes of red giant stars. The corresponding Hayashi tracks for stars on the Hertzsprung–Russell diagram are almost vertical and occur well to the right of the diagram, in the region of low surface temperatures, $T \sim 3000-5000\,\mathrm{K}$. As Kippenhahn and Weigert remark,[7]

One may even say that the importance of the Hayashi track is only surpassed by that of the main sequence.

Once a star evolves off the main sequence and moves across the Hertzsprung–Russell diagram, its passage to the right is halted at the point at which it reaches the Hayashi limit, when it is observed as a red giant star. The star then reorganises its internal structure and moves up the Hayashi track (see Section A9.1).

The opposite process occurs during star formation. There are no quasi-static solutions for stars to the right of the Hayashi track (see Section A9). Rather, the stars take up internal structures in which the outward energy transport is by convection and they evolve down the Hayashi track until energy transport by radiation becomes more important than convection as the central temperature rises. Examples of the evolutionary tracks for stars of different masses worked out by Hayashi in his famous paper are shown in Figure 9.4. Hayashi showed that the T-Tauri stars lie precisely in the regions of the Hertzsprung–Russell diagram spanned by his evolutionary tracks.

It was already apparent that star formation was associated with gas clouds in which there are large amounts of dust, but the central role of dust in the star-formation process was only appreciated when Eric Becklin and Gerry Neugebauer discovered what they identified as an infrared protostar in the Orion Nebula (Becklin and Neugebauer, 1967). Neugebauer was one of the pioneers of infrared astronomy, and, during the winter of 1965, a map was made of the Orion Nebula by scanning across it with a single-element detector operating at 2.2 μm with angular resolution 13 arcsec. Seven point sources were found which could be identified with known stars, but one was not associated with any optical object. During the winter of 1966, photometric observations were made of this source at infrared wavelengths of 1.65, 2.2, 3.4 and 10 μm, and its temperature turned out to be 700 K, well below the surface temperature of the coolest stars (Figure 7.15(a)). Observations were attempted of the same object by Douglas Kleinmann (b. 1942) and Frank Low at a wavelength of 20 μm in the same year, but they failed to find it. Instead, just to the south of the Becklin–Neugebauer object, they discovered a source with an enormous far-infrared luminosity at the low temperature of only 70 K (Kleinmann and Low, 1967). The story developed very rapidly from this point onwards. Although the mapping of regions of star formation with single-element detectors was time consuming, intense compact far-infrared sources were found in the vicinity of regions of star formation. The far-infrared luminosities of some of these sources turned out to be enormous, $10^3-10^5 L_\odot$. Molecular-line surveys had revealed dense condensations

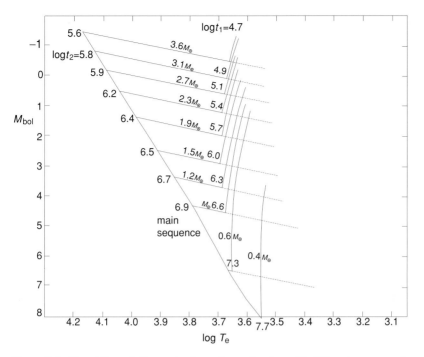

Figure 9.4: Hayashi's evolutionary tracks and ages for stars with different masses under gravitational contraction; the times t_1 and t_2 denote the ages, in years, at the turning point and at the point where the star joins the main sequence (Hayashi, 1961).

within giant molecular clouds and within dark clouds, and extremely luminous infrared sources were often associated with these.

In their discovery paper, Becklin and Neugebauer (1967) remarked

It is well known that the Orion Nebula is a very young association and that the probability of finding a star in the process of forming should be relatively high ... Thus an attractive interpretation of the observation is that the infrared object is a protostar.

Although this picture was attractive, it proved to be very difficult to demonstrate that the Becklin–Neugebauer object is indeed a protostar. Part of the problem arose from the use of words – what exactly was meant by a 'protostar'? An unambiguous definition would be that it is a star in which the energy is derived from the release of gravitational energy of the matter accreted by the star and not from nuclear energy. In cases like the Becklin–Neugebauer source, it was quite possible that an O or B star has already formed but that it is still embedded deep within the dense dusty molecular cloud from which it formed.

The most important contribution to many aspects of these studies came from observations with the Infrared Astronomy Satellite (IRAS), which was launched in 1983. IRAS was designed to carry out the first deep surveys of the whole sky in the far-infrared wavebands at 12, 25, 60 and 100 μm, which are inaccessible from the ground and showed that regions of star formation are among the most intense far-infrared sources. These observations confirmed beyond any doubt the significance of dust in the formation of stars. The dust acts as a transformer, absorbing the optical and ultraviolet radiation of protostars and young stellar objects and reradiating the absorbed energy at the temperature to which the dust is heated. This proves to be an extremely efficient means of getting rid of the gravitational binding energy of matter accreting onto a protostar since the envelope of the star is transparent to radiation in the far-infrared wavebands.

The significance of the IRAS observations was that not only were luminous far-infrared sources found in regions of star formation, but lower luminosity sources were discovered as well with luminosities in the range $1-100L_\odot$ and these are probably the progenitors of solar-mass stars. The infrared spectra of the IRAS sources have provided a number of clues to their evolutionary state. One of the most important early discoveries of the mission was the detection of dust discs about young stars, which are interpreted as the material out of which planets are formed. These discs could be identified by their spectral signatures in T-Tauri stars.

The infrared spectra can be divided into three classes (Figure 9.5), the most extreme class, Type I, consisting of exclusively far-infrared sources in which no optical radiation is observed. These are interpreted as protostars that are still in their accretion phases. The Type II spectra show evidence for dust emission from an accretion disc as well as the photospheric emission from a reddened pre main-sequence star like the T-Tauri stars. The Type III spectra appear to be unshrouded, pre main-sequence stars.

The theory of star formation has been one of the more contentious areas of modern astrophysics because it can no longer be assumed that the star is in a quasi-static state whilst evolving from a density enhancement in a giant molecular cloud to a fully fledged main-sequence star. In 1902, James Jeans had worked out the condition for gravitational collapse, namely that the force of gravity should exceed the force associated with the pressure gradients within the cloud which resist the collapse, the famous *Jeans criterion* for collapse (Jeans, 1902). On large enough scales, gravity will always be the dominant force.

In the modern version of the theory, the collapsing cloud heats up as it collapses but, so long as the cloud is optically thin to radiation, it can cool by radiation. Once the cloud becomes optically thick to radiation, the cloud begins to heat up and the details of the collapse have to be followed by computer simulations. Among the first of these were calculations by Richard Larson (b. 1941) in 1969, who showed that a compact core forms within the collapsing cloud and that the star forms by accretion of matter onto this core (Larson, 1969a,b). He showed that the dust envelope would absorb the optical and ultraviolet radiation, leading to the types of source which had just been discovered by Becklin and Neugebauer. These ideas led in due course to a standard picture of the process of formation of spherically symmetric stars due to Frank Shu (b. 1943) and his colleagues in 1980; this has been widely used to interpret observations of the star-forming regions (Figure 9.6) (Stahler, Shu and Taam, 1980). The core of the protostar comes into hydrostatic equilibrium and matter is accreted onto the core through an accretion shock. The binding energy of the accreted matter is released as radiation, which is absorbed in the dust envelope, which cannot have a temperature greater than about 1000 K or else the dust grains are evaporated. The dust shell reradiates away the absorbed energy in the far-infrared waveband. Shu and his colleagues carried out radiative transfer calculations of the predicted spectra of these objects, and they can account

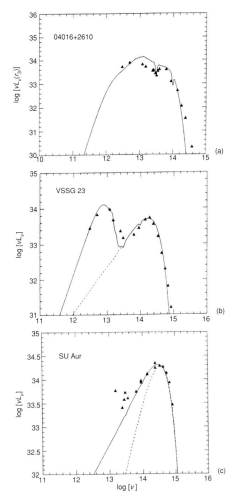

Figure 9.5: The energy spectra of young stellar objects in the Taurus and Ophiuchus molecular clouds. The three sources have mass of the order of $1 M_\odot$. These spectra are interpreted as follows. (a) The protostar during the infall phase, in which the star and protoplanetary disc are embedded in an infalling dust envelope. (b) Intense winds blow away the dust along the rotation axis, revealing the new-born star surrounded by a nebular disc. (c) A T-Tauri star with only a small infrared excess. (Adams, Lada and Shu, 1987.)

naturally for the details of the observed spectra of the far-infrared sources discovered by IRAS.

There must, however, be much more to the story than this. In the late 1970s, maps of young stellar objects were made at millimetre wavelengths, and in 1980 Ronald Snell (b. 1951) and his colleagues discovered that, rather than showing evidence of infall, the molecular gas in the source L1551 seemed to be expelled from the source in the form of what became known as a *bipolar outflow* (Snell, Loren and Plambeck, 1980). These molecular outflows have velocities up to about $150\,\mathrm{km\,s^{-1}}$ and extract large momentum and

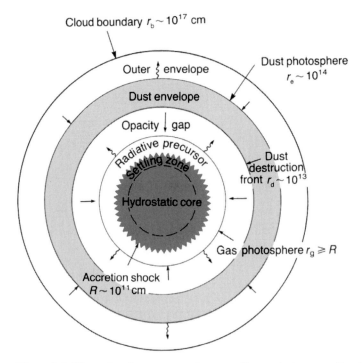

Figure 9.6: Illustrating the structure of an accreting protostar according to Shu and his colleagues (Stahler *et al.*, 1980).

energy fluxes from the star, the total energies corresponding to about 10^{36}–10^{40} J. These outflows are believed to be responsible for a variety of energetic phenomena seen in the vicinity of young stellar objects, including the Herbig–Haro objects, high-velocity water-maser sources, shock-excited molecular hydrogen and optically visible jets. The origin of these outflows is not established, but they seem to be found wherever protostellar or young stellar objects are found. It is suspected that they are associated in some way with two other problems of star formation which have yet to be satisfactorily resolved, namely the way in which protostellar clouds get rid of angular momentum and magnetic fields. In both cases, the process of collapse amplifies the energies associated with both of them. It seems reasonable to suppose that the bipolar outflows are in some way associated with the means by which the protostellar cloud gets rid of its angular momentum and magnetic field, but there is no generally agreed mechanism by which this occurs.

In 1987, Shu and his colleagues put together what has become a standard picture for the formation of stars (Figure 9.7), which seems to account for many aspects of the observations (Shu, Adams and Lizano, 1987).

One of the challenges of the theories of star formation was to account for the *initial mass function*, which describes the rate of formation of stars of different masses, $\xi(M)$. The first analysis of this problem was carried out by Edwin Salpeter in 1955 and involved corrections to the observed luminosity function of main-sequence stars for their different main-sequence lifetimes (Salpeter, 1955). For stars with mass roughly equal to the mass of

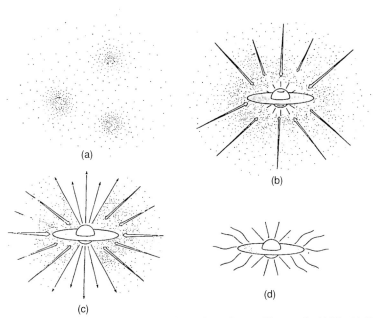

Figure 9.7: A plausible scenario for the formation of stars (Shu *et al.*, 1987). (a) Density inhomogeneities collapse under their own gravity. (b) Main accretion phase, in which an accreting core has formed and infall of matter onto the core takes place. The binding energy of the matter is removed by radiation, which is absorbed by dust and reradiated in the far-infrared waveband. (c) Jets of material burst out of the accreting star along its rotation axis, producing the characteristic bipolar outflows. (d) The accretion of material ceases and the system is left with a young hydrogen-burning star and a rotating dust disc.

the Sun, Salpeter found that the initial mass function could be approximated by a power-law distribution in their masses:

$$\mathrm{d}N = \xi(M)\,\mathrm{d}(\log M) \propto M^{-1.35}\,\mathrm{d}(\log M). \tag{9.1}$$

Salpeter's analysis needed estimates of the main-sequence lifetimes of stars of different masses, and new determinations of the initial mass function were made as better stellar models became available and the stellar statistics improved. In 1979, Glenn Miller (b. 1953) and John Scalo (b. 1948) showed that the initial mass function was better defined by a log-normal distribution over a wider range of masses (Figure 9.8) (Miller and Scalo, 1979). In fact, over the mass range $1 \leq M/M_\odot \leq 10$, the Salpeter initial mass function is a good approximation to the function proposed by Miller and Scalo. One of the goals of the theory of star formation is to account for this initial mass function, and the log-normal distribution has the intriguing property of describing random multiplicative processes. These initial mass functions are useful global averages for the rate of star formation in galaxies, but it is not certain how accurately they describe star formation in individual regions. They are essential for studying the evolution of the ultraviolet, optical and infrared spectra of galaxies and also for studies of the chemical evolution of galaxies (see Section A13.1).

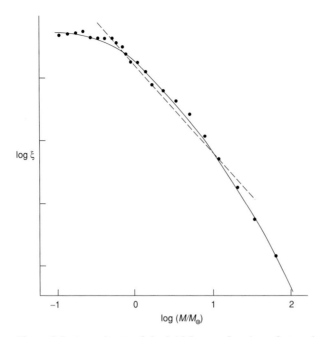

Figure 9.8: An estimate of the initial mass function of stars derived by Miller and Scalo (1979) showing their best-fitting log-normal distribution, $\xi(\log M)\,\mathrm{d}M \propto \exp[-C_1(\log M - C_2)^2]\,\mathrm{d}M$ (solid line). Also shown (by the dashed line) is the initial mass spectrum of power-law form proposed by Salpeter.

9.5   Extrasolar planets and brown dwarfs

The search for planets about nearby stars has been a cherished ambition of astronomers for centuries, but it has proved to be a difficult challenge. In the 1950s, it was claimed that the wobbles reported in the position of the nearby star, Barnard's star, were evidence for a planetary companion, but this turned out to be spurious. The first definite detection of extrasolar planets, or *exoplanets*, came from a quite unexpected direction, the observation of systematic variations in the radial velocity of the pulsar PSR 1257+12 by Alex Wolszczan (b. 1943) and Dale Frail (b. 1961) in 1992 (Wolszczan and Frail, 1992). The reason for their success was the very high precision with which radial velocities of pulsars can be determined by very precise timing of the arrival times of the pulses. In fact, they found evidence for three planets, two with masses roughly that of the Earth and one about 50 times less. This was a wholly unexpected discovery since it was assumed that, when a neutron star forms, any planets orbiting the pre-supernova star would not remain in bound orbits. A favoured view is that these planets formed from an accretion disc about the neutron star after the supernova exploded. In any case, the mere existence of this system shows that planets can be formed in a wide range of different astronomical environments.

The first detection of a Jupiter-mass planet orbiting a normal star was made by Michel Mayor (b. 1959) and Didier Queloz (b. 1966) of the University of Geneva in 1995 (Mayor

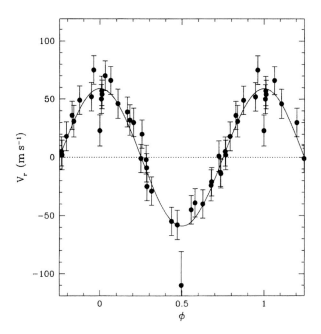

Figure 9.9: The variation of the radial velocity of the star 51 Peg as a function of orbital phase. The period of the planet's orbit about the barycentre of the system is 4.231 days (Mayor and Queloz, 1995).

and Queloz, 1995). Their success can be attributed to the development of very stable spectrographs with very high spectral resolution. For comparison, the Sun orbits the barycentre of the Solar System at a typical speed of about 13 m s^{-1}. Thus, to have a hope of being able to detect the presence of Jupiter-mass planets in systems such as our own, the spectrographs have to be able to resolve radial velocities of a few metres per second. The discovery record of the sinusoidal variation of the radial velocity of the nearby star 51 Peg is shown in Figure 9.9, from which it can be seen that the amplitude of the motion of this solar-type star is very much greater than would be expected of a planetary system such as our own. Furthermore, the period of the planet about the star is only 4.231 days. Analysis of the orbital data has shown that the mass of the planet is at least 0.46 Jupiter masses and its semi-major axis is only 0.052 AU.

This discovery stimulated a huge effort to discover further examples of planets about nearby stars, and it has been extraordinarily successful. By mid 2004, 122 extrasolar planets were known in 107 planetary systems, including 13 multiple-planet systems.[8] If there were any doubts about the correctness of the interpretation of the radial velocity data, they were dispelled by the observation in 1999 of a very small dip in the intensity of HD 209458 due to the transit of the companion across the face of the star, these dips occurring with the same period that was derived from the radial velocity data (Charbonneau *et al.*, 2000). The star HD 209458 is a G0 V dwarf star, similar to the Sun. Assuming the stellar radius and mass are $1.1 R_\odot$ and $1.1 M_\odot$, respectively, the 'eclipse' data shown in Figure 9.10 have been interpreted as being due to the transit of a gaseous giant planet with radius 1.27 times the radius of Jupiter in an orbit with an inclination of $87°$.

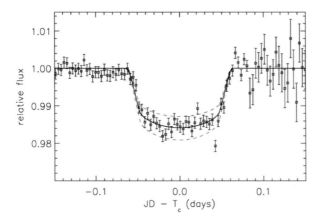

Figure 9.10: The discovery record of the photometric time series for the star HD 209458 for 9 and 16 September 1999 plotted as a function of time. The data have been averaged in five-minute bins. (Charbonneau *et al.*, 2000).

These discoveries have resulted in two major surprises, which have forced the theory of the formation of planetary systems to be considerably revised. The first was the large fraction of Jupiter-like companions at orbital radii about 100 times closer to the parent star than in our Solar System. For example, more than half of these gaseous giant planets orbit within 1 AU of the host star and a significant fraction orbit within 0.1 AU. From the statistics of the detected extrasolar planets, it seems that our Solar System is the odd man out. A favoured solution is that, since such gaseous giants could not have formed so close to the primary star, the Jupiter-sized planets must have been formed much further away and then undergone *orbital migration* under the influence of tidal forces.

The second great surprise was the fact that the orbits of many of the Jupiter-sized planets are highly elliptical. This poses problems for the standard picture of planet formation in which the planets are formed by accretion in a protoplanetary disc. In this picture, dissipative processes rapidly circularise the planetary orbit. Therefore, the elliptical orbits must have come about through some other process. Suggestions have included that they formed directly by gravitational condensation, rather than by accretion within a protoplanetary disc, or their orbits may have been strongly perturbed by a companion star, which may have been the case in a system such as 16 Cyg A and B, or maybe a sling-shot mechanism took place which ejected the gaseous giant into an elliptical orbit through a gravitational encounter with another planet.

The study of extrasolar planets is one of the major growth areas of modern astrophysics and has opened the way to the study of bio-astrophysics and the possibility of determining by observation whether conditions exist in other planetary systems in which life could have formed. The challenge is to measure the spectrum of a very faint companion, about 100 000 times fainter than the star itself and very close to it. Remarkably, evidence has already been found for the presence of an atmosphere in the extrasolar planet associated with HD 209458. Observations with the very stable STIS spectrograph of the Hubble Space Telescope showed that, when the planet transited across the face of the star, the depth of the

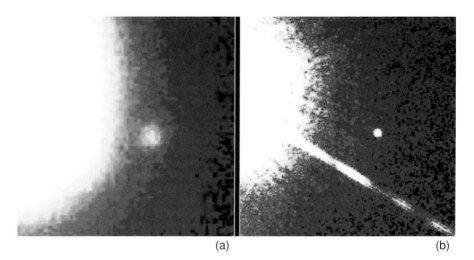

(a)   (b)

Figure 9.11: (a) The discovery image of the faint brown dwarf companion to the solar-type star Gliese 229 obtained at the Palomar Observatory on 27 October 1994 (Nakajima *et al.*, 1995). (b) A confirmatory image taken by the Hubble Space Telescope on 17 November 1995. (Courtesy of T. Nakajima, S. Kulkarni, S. Durrance and D. Golimowski, NASA, ESA and the Space Telescope Science Insitute.)

absorption lines of sodium, the D-lines, increased significantly. This was direct evidence for the presence of sodium in the atmosphere of this extrasolar planet (Charbonneau *et al.*, 2002).

A closely related study has been the search for the class of 'star' known as *brown dwarfs*. These stars have masses less than about $0.08 M_\odot$ and so their central temperatures are too low for the nuclear burning of hydrogen into helium to take place in their cores. There is thus a range of masses between those of Jupiter-like gaseous giant planets and objects about 1000 times more massive in which the central temperatures are not great enough to tap their nuclear energy resource. Conventionally, brown dwarfs are taken to be gravitationally bound objects with masses in the range 0.01 to $0.08 M_\odot$, that is about 10 to 80 Jupiter masses. The lower limit corresponds to the mass at which even deuterium burning ceases to be possible in the core of the 'star'.

Searches for brown dwarfs have been pursued for many years, but the first really convincing case was discovered in 1995 by direct imaging of the companion of the nearby star Gliese 229 at Palomar using the 60-inch and 200-inch telescopes (Figure 9.11) (Nakajima *et al.*, 1995). The spectrum of this faint companion, known as Gliese 229B, showed strong methane and water vapour absorption, similar to the spectrum of Jupiter, the inferred surface temperature being less than 1000 K (Oppenheimer *et al.*, 1995). This surface temperature is too low for nuclear burning to take place in its core and so it must be a *bone fide* brown dwarf. Many further candidates have since be found through systematic sky surveys in the near-infrared wavebands, including the 2 Micron All-Sky Survey (2MASS) and the Sloan Digital Sky Survey. Numerous candidates have also been found in deep-infrared surveys of nearby star-forming regions, such as the Pleiades, Orion and ρ Ophiuchus clusters.

9.6   Cosmic-ray astrophysics and the interstellar medium

The history of cosmic-ray astrophysics and its role in the discovery of elementary particles up to the period immediately after the Second World War was summarised in Section 7.2. As described by Michael Hillas[9] (b. 1932), after the War, cosmic-ray studies were vigorously pursued from high-altitude balloons and by ground-based studies of extensive air-showers. The problem for the cosmic-ray physicists involved in the balloon studies was to discriminate between the primary cosmic rays and the vast numbers of secondary particles created by interactions of very-high-energy cosmic rays with the nuclei of molecules of the atmosphere. Balloons were flown to progressively greater altitudes, until, at about 36 km, the residual atmosphere[10] amounted to only about $50 \, \mathrm{kg \, m^{-2}}$, which was much less than the interaction mean free path for the incoming cosmic rays, which was about $800 \, \mathrm{kg \, m^{-2}}$; for reference, the total depth of the atmosphere is about $10\,000 \, \mathrm{kg \, m^{-2}}$. The detectors used in these balloon experiments consisted of stacks of nuclear emulsions, and these were regularly flown at high altitude. Provided the stacks were safely returned to the ground, the highly developed techniques for the analysis of nuclear emulsion stacks provided information about the energies and charges of the cosmic rays. From the 1960s onwards, cosmic-ray detectors were successfully flown in satellites and space probes, eliminating the problems of contamination by secondary particles.[11] The technologies employed to build the telescopes and detectors were much more akin to those used in high-energy particle physics experiments. The instruments were no longer retrievable and so they had to be miniaturised and space-qualified to survive the harsh environment of space.

The majority of the incoming cosmic rays were protons, but it was immediately apparent that the cosmic radiation also included fluxes of relativistic helium and heavier atomic nuclei (Freier *et al.*, 1948). To complicate the problem, the spectrum of the cosmic rays with kinetic energies less than about 1 GeV per nucleon were strongly influenced by their passage through the interplanetary medium from interstellar space, and these changes varied with the phase of the solar cycle, the phenomenon known as *solar modulation*. The distortions are caused by the scattering of the cosmic rays by irregularities in the interplanetary magnetic field. The variation with the solar cycle is caused by variations in the amplitude of the spectrum of irregularities, more scattering occurring at solar maximum than at solar minimum.

The key results for studies of the astrophysics of our Galaxy and the interstellar medium were the energy spectrum and chemical composition of the primary cosmic radiation. The energy spectra of the cosmic rays at energies greater than 1 GeV per nucleon were found to follow a power-law energy distribution:

$$N(E)\,dE \propto E^{-x}\,dE, \tag{9.2}$$

with $x \sim 2.5$–2.7. This relation was found for protons, helium and heavier nuclei with energies in the range 10^9–10^{14} eV (Figure 9.12) (Simpson, 1983). Such a relativistic gas was inferred to be present throughout the interstellar medium from two quite independent types of observation. Firstly, the Galactic γ-ray emission at energies $E > 100$ MeV can be attributed to the decay of neutral pions, π^0, created in collisions between cosmic-ray protons and nuclei with the nuclei of atoms, ions and molecules in the interstellar gas (see Section 7.5). The agreement between these estimates of the local number density of

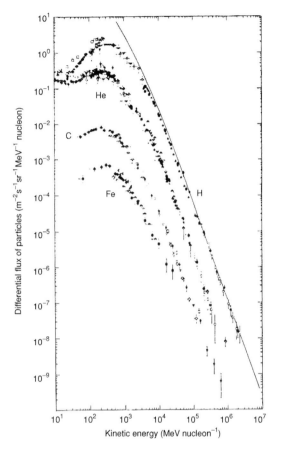

Figure 9.12: The differential energy spectrum of cosmic rays as measured from above the Earth's atmosphere as summarised by John Simpson (1916–2000) (Simpson, 1983). The solid line shows an estimate of the proton spectrum once allowance is made for the effects of solar modulation.

cosmic-ray particles and those of the cosmic-ray particles observed at the top of the atmosphere showed that the latter are part of the population of high-energy particles pervading the whole Galaxy.

Secondly, the synchrotron radiation of the interstellar population of ultra-relativistic electrons gyrating in the Galactic magnetic field is detected in the radio waveband as the Galactic radio emission (see Section 7.3.2). It proved a much more challenging task to measure the primary cosmic-ray electron spectrum since large fluxes of relativistic electrons are produced as secondary particles, even at very high altitudes. For a number of years there was uncertainty about the exact form of the primary electron spectrum, but the observations of William Webber (b. 1929) and his colleagues finally showed that the energy spectrum of very-high-energy electrons with $E \geq 10\,\text{GeV}$ is of power-law form similar to equation (9.1), with a slightly greater spectral index, $x = 3.3$ (Webber, 1983). This spectrum can be smoothly joined onto the spectrum of high-energy electrons inferred to be present in interstellar space for reasonable values of the interstellar magnetic field strength.[12] These

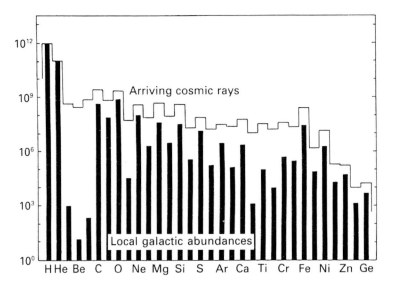

Figure 9.13: The cosmic abundances of the elements in the cosmic rays (solid line) compared with the Solar System abundances (solid histogram). The data have been normalised to a relative abundance of hydrogen of 10^{12} (Lund, 1984).

data are compelling evidence that the primary cosmic rays detected in the vicinity of the Earth are samples of the relativistic gas present in the interstellar medium of our Galaxy.

Following the discovery of heavy nuclei in the cosmic rays, their chemical composition was measured in balloon observations by the Minnesota–Rochester group, and the key results were reported by Helmut Bradt (d. 1950) and Bernard Peters (1910–1993) in 1950 (Bradt and Peters, 1950). The important discovery was the large abundance of the light elements lithium, beryllium and boron relative to their Solar System abundances. The study of the detailed chemical abundances in the cosmic rays continued through the following decades from balloons and satellites, culminating in the measurements by the HEAO-C space observatory, which was launched in 1979. The abundances of the elements in the cosmic rays relative to the Solar System abundances are illustrated in Figure 9.13, which is taken from the summary by Niels Lund (b. 1938) (Lund, 1984). It can be seen that the abundance peaks at the carbon, nitrogen and oxygen group and the iron group are present in both the Solar System and cosmic-ray abundances. The excess of lithium, beryllium and boron in the cosmic rays is very great indeed relative to their cosmic abundances, and there is also an excess of elements with atomic and mass numbers just less than iron. These differences are naturally attributed to the process of *spallation* in the interstellar medium between the sources of the cosmic rays and the arrival of these particles at the top of the Earth's atmosphere. In this process, energetic cosmic-ray nuclei collide with the protons and nuclei in the interstellar medium and nucleons are chipped off, resulting in the formation of lighter nuclei with smaller mass numbers. To carry out a detailed analysis of the products of these nuclear interactions, spallation cross-sections are needed which describe the probability of the formation of different secondary nuclei in each collision. Thus, the primary source of the lithium, beryllium and boron is the spallation of the abundant

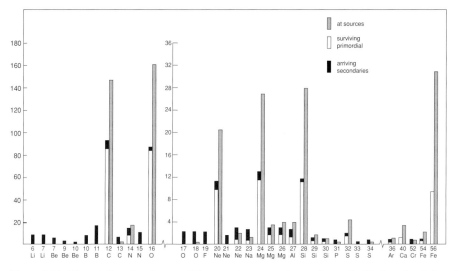

Figure 9.14: The relative abundances of the cosmic rays as observed near the Earth and as inferred to have been present in their sources, once account has been taken of the effects of spallation between their sources and the Solar System. The grey histogram shows the inferred source abundances; the black histogram shows the spallation products; and the white histogram shows the surviving primary elements (Shapiro, 1991).

carbon, nitrogen and oxygen group, while the elements just lighter than iron are naturally formed by the spallation of iron group elements.[13]

In their pioneering paper, Bradt and Peters showed quantitatively that the lithium, beryllium and boron in the cosmic rays could be attributed to the spallation of carbon group elements (Bradt and Peters, 1950). Since that time, these studies have been the subject of increasingly refined calculations which take account of the range of diffusion path-lengths between the sources of the cosmic rays and the Earth, as well as adopting much improved spallation cross-sections. The computations start from the transfer equation for the evolution of the number densities of all types of primary and secondary nuclei as they travel through the interstellar medium. The typical result of this type of calculation is shown in Figure 9.14, which is taken from the survey by Maurice Shapiro (b. 1915) (Shapiro, 1991). On average, the cosmic rays must have traversed about $50 \, \text{kg m}^{-2}$ between their sources and the Solar System, but it is also necessary to assume that there is a distribution of path-lengths in order to obtain the correct observed abundances of the products of spallation of both the carbon and iron group elements. These calculations show how the observed abundances of lithium, beryllium and boron and elements such as ^{15}N, ^{17}O, ^{18}O, ^{19}F and ^{21}Ne can all be accounted for by spallation interactions between the sources of the cosmic rays and the Solar System. It can also be seen that quite large fractions of the common elements, carbon, oxygen, neon, magnesium and iron, survive intact between their sources and the Solar System. These computations show that the cosmic rays were accelerated with chemical abundances similar to the Solar System abundances.

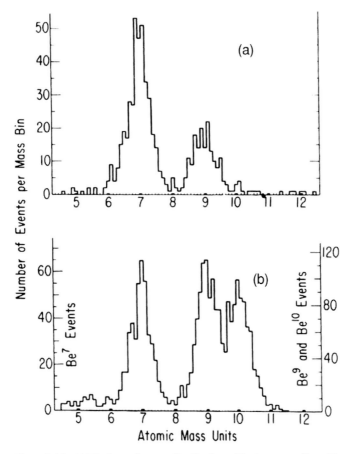

Figure 9.15: (a) The isotopic mass distribution of the isotopes of beryllium as observed by the cosmic-ray telescopes onboard the IMP-7 and IMP-8 space probes. (b) Calibration of the expected distribution of beryllium isotopes in laboratory experiments showing the resolution of the isotopes of ^{9}Be and ^{10}Be. (Garcia-Munoz et al., 1977.)

One of the important issues for the propagation of cosmic rays in the interstellar medium is the typical time they take to diffuse from their sources to the Solar System. Fortunately, some of the species created in the spallation interactions are radioactive and so can act as *cosmic-ray clocks*. The most important of these is the ^{10}Be isotope, which has half-life 3.9×10^6 years. Its abundance can be compared with those of the other stable isotopes of beryllium. This experiment requires excellent resolution of the relative abundances of the isotopes of beryllium, and this was successfully achieved first by Simpson and his colleagues in their experiments onboard the IMP-7 and IMP-8 spacecraft in 1977 (Garcia-Munoz, Mason and Simpson, 1977). Figure 9.15 shows that there is very little ^{10}Be in the cosmic rays and consequently there has been time for most, but not all, of these isotopes to undergo radioactive decay. These data indicate that the average time it takes the cosmic rays to reach the Earth is about 10^7 years. Thus, the cosmic rays cannot have travelled directly from their sources to the Solar System, but rather they must have been scattered many times by irregularities in the interstellar magnetic field. It is simplest to think of the cosmic rays performing a random walk from their sources to the Solar System. Then, if the interstellar density is taken to have a typical value of $3 \times 10^5 \, \mathrm{m}^{-3}$ and the cosmic rays traverse $50 \, \mathrm{kg} \, \mathrm{m}^{-2}$, the distance travelled at the speed of light would be 10^7 light-years, in pleasant agreement with the results of the ^{10}Be experiment.

These observations indicate that relativistic matter and magnetic fields are important components of the interstellar medium. They have roughly equal energy densities, a convenient figure being that both correspond to about $1 \, \mathrm{eV} \, \mathrm{m}^{-3}$.

Notes to Chapter 9

1 These works are summarised in Aller's classic book *Atoms, Stars and Nebulae*, 3rd edn (Cambridge: Cambridge University Press, 1991).

2 Some reminiscences of Oort and the pursuit of Dutch astronomy under the occupation are contained in *Jan Oort Astronomer*, eds J. Katgert-Merkelijn and J. Damen (Leiden: Kleine publicaties van de Leidse Universiteitsbibliotheek no. 35, 2000). The minutes of the underground seminar in which van de Hulst first reported the prediction of the 21 cm line are reproduced in facsimile in D. Hartmann and W. B. Burton, *Atlas of Galactic Neutral Hydrogen* (Cambridge: Cambridge University Press), p. 84. On p. 85, the following translation is given: 'Then Mr Van de Hulst speaks about the spectrum of the Milky Way at wavelengths of several metres . . . Furthermore, one may perhaps expect an observable intensity at several discrete wavelengths as a consequence of transitions between hyperfine structure levels of the ground state of hydrogen . . . After this the chairman [Oort] thanks Mr Van de Hulst for his fine presentation.'

3 Many of these early discoveries using 21 cm hydrogen line observations were summarised at the Paris Symposium on Radio Astronomy (Bracewell, 1959) and in the survey of HI observations by Wim Rougoor (1931–1967) and Oort (Rougoor and Oort, 1960).

4 An introduction is given by W. W. Duley and D. A. Williams, *Interstellar Chemistry* (London: Academic Press, 1984).

5 Kroto gives a splendid blow-by-blow account of the discovery of the C_{60} molecule and the role of interstellar molecules in its discovery in his Nobel prize lecture of December 1996, *Symmetry, Space, Stars and C_{60}*. Kroto, Robert Curl (b. 1933) and Richard Smalley were jointly awarded the Nobel prize for chemistry in 1996 'for their discovery of fullerenes'.

6 See, for example, F. C. Bruhweiler and A. Vidal-Madjar, in *Exploring the Universe with the IUE Satellite*, ed. Y. Kondo (Dordrecht: D. Reidel Publishing Company, 1987), pp. 467–484.

7 See R. Kippenhahn and A. Weigert, *Stellar Structure and Evolution* (Berlin: Springer-Verlag, 1990).

8 An excellent online encyclopaedia of all the known extrasolar planets is maintained by Jean Schneider, entitled *The Extrasolar Planet Encyclopaedia* at http://www.obspm.fr/planets. This web-site also includes many excellent references to different detection methods for discovering extrasolar planets and to theories of the origin of the types of planetary system discovered.

9 Hillas provides an excellent documentary study of the development of cosmic-ray astrophysics from its beginnings up to about 1970. His book describes clearly the many technical difficulties which had to be overcome before reliable estimates could be made of the spectrum, isotropy and chemical composition of the cosmic rays. See A. M. Hillas, *Cosmic Rays* (Oxford: Pergamon Press, 1972).

10 The units used to describe the amount of material traversed by the cosmic ray is the path length, $\int \rho \, \mathrm{d}x$, where ρ is the density of the material and $\mathrm{d}x$ is the increment of distance. The traditional units used are $\mathrm{g} \, \mathrm{cm}^{-2}$, but I have translated these into SI units. For reference, the total depth of the atmosphere is $10\,000 \, \mathrm{kg} \, \mathrm{m}^{-2}$.

11 I have given details of the different types of cosmic ray detectors and telescopes in Malcolm Longair, *High Energy Astrophysics*, vol. 1 (Cambridge: Cambridge University Press, 1994). Several examples are given of the types of data obtained in these experiments.

12 I have discussed the problems of relating the spectrum and emissivity of the Galactic radio emission to the local spectrum of high-energy electrons in Section 18.2 of Malcolm Longair, *High Energy Astrophysics*, Volume 2 (Cambridge: Cambridge University Press, 1994).

13 I have discussed the problems of accounting for the observed abundances of spallation nuclei in Section 20.2 of Malcolm Longair, *High Energy Astrophysics*, Vol. 2.

A9   Explanatory supplement to Chapter 9

A9.1   Notes on the Hayashi track

In Hayashi's pioneering paper, his analysis concerns the stability of fully convective stars. The condition that a region of a star is in convective, rather than radiative, equilibrium is that the temperature gradient exceeds the adiabatic gradient of the stellar material. In this context, the term 'gradient' refers to the derivative of the temperature with respect to pressure, which is a monotonically increasing function of decreasing radius within the star. Conventionally, the temperature gradient is written, in the case of the radiative transport of energy, as

$$\nabla_{\rm rad} = \left(\frac{{\rm d} \ln T}{{\rm d} \ln p} \right)_{\rm rad} \tag{A9.1}$$

and is related to the opacity of the stellar material as discussed in Section A3.2.2. If the stellar material has ratio of specific heat capacities γ, the adiabatic relation is $p \propto T^{\gamma/(\gamma-1)}$ and so $\nabla_{\rm ad} = (\gamma - 1)/\gamma$. If the structure of the star is such that the temperature gradient exceeds this value, the material of the star becomes unstable and convection ensues. The simplest picture of what happens physically is to consider a 'bubble' of material that is slightly compressed and which then rises up the temperature gradient because of the buoyancy of the perturbed region. Convection transports energy more rapidly through the star than does radiation, and the structure of the star reorganises itself under these convective motions until the temperature and pressure stratification satisfy the relation $\nabla_{\rm ad} = (\gamma - 1)/\gamma$. Thus, for stars in which convection is maintained throughout the whole star, the temperature and pressure stratification is given almost exactly by the adiabatic gradient, since even a tiny departure to greater values of $\nabla_{\rm rad}$ results in convective motions.

In Hayashi's brief paper, he considers the stability of the polytropic models of stars in which the adiabatic relation holds throughout the star (Hayashi, 1961). If the perfect gas ratio of specific heats is used, $\gamma = 5/3$, these models have polytropic index $n = (\gamma - 1)^{-1} = 3/2$. He then showed that there is an upper limit to a dimensionless parameter involving the mass, radius, temperature and pressure of the gas, beyond which there exist no quasi-static solutions. This condition translates into the steep regions of the loci shown in Figure 9.4 for stars of different mass. There are no quasi-static solutions to the right of these loci.

A more detailed physical discussion of the structure of fully convective stars is given by Kippenhahn and Weigert,[1] who show that, by adopting the adiabatic relation for the structure of the star, the pressure and temperature structure are decoupled from its luminosity. An

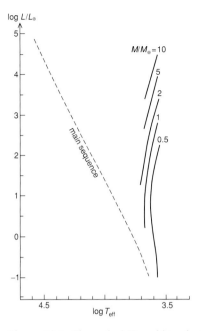

Figure A9.1: Theoretical Hayashi tracks for fully convective stars of different masses presented by Kippenhahn and Weigert, after computations by Ezer and Cameron. From R. Kippenhahn and A. Weigert, *Stellar Structure and Evolution* (Berlin: Springer-Verlag, 1990), Chapter 24.

atmosphere in which energy transport is by radiation is needed to determine the luminosity–temperature relation, and this is found by joining together solutions for the body of the star to its atmosphere. These calculations demonstrate the steepness of the Hayashi track for stars of different mass (Figure A9.1).

Note to Section A9

1 See R. Kippenhahn and A. Weigert, *Stellar Structure and Evolution* (Berlin: Springer-Verlag, 1990).

10 The physics of galaxies and clusters of galaxies

10.1 The galaxies

The Hubble sequence of galaxy types shown in Figure 5.5 gives some impression of the diversity of forms found among the galaxies. Hubble planned to publish an atlas of galaxies illustrating the different galaxy types but, although all the plates for this project were taken with the 60-inch and 100-inch telescopes by 1948, he died in 1953, before what became the *Hubble Atlas of Galaxies* was published. The project was completed by Allan Sandage, who was Hubble's last research assistant, and it was published in 1961 (Sandage, 1961b). The basic Hubble sequence was preserved, including the S0 galaxies, and the irregular galaxies were placed at the end of the sequence.[1]

The morphological classification of large samples of galaxies was pursued by Antoinette (1921–1987) and Gérard de Vaucouleurs (1918–1995), who published a series of *Reference Catalogues of Bright Galaxies*, in which the Hubble classification was refined, the basic linear sequence being preserved (de Vaucouleurs *et al.*, 1991). The distinction between the normal and barred spirals was maintained, but they showed that all intermediate types between pure barred spirals and normal spirals are also observed. What gave this morphological scheme physical significance was the fact that certain physical properties of galaxies are correlated with their position along the sequence. In de Vaucouleurs' survey of 1974, he showed that the mass fraction in the form of gas is a function of position along the sequence, the elliptical galaxies having less than 0.01% of their mass in the form of interstellar gas, whereas in the irregular galaxies as much as 30% of the mass can be in gaseous form (de Vaucouleurs, 1974). The colours of the galaxies also showed a systematic trend along the sequence, the ellipticals being the reddest and the irregulars the bluest galaxies.[2]

The availability of the Palomar 48-inch sky survey plates led Halton Arp (b. 1927) to publish his *Atlas of Peculiar Galaxies* in 1966 (Arp, 1966). As Arp remarked in his introduction to the *Atlas*,

The greatest deviations from the normal are emphasised in this atlas.

The corresponding catalogue for the southern hemisphere was published in 1987 (Arp, Madore and Roberton, 1987). Similar catalogues were prepared by Boris Vorontsov-Velyaminov (1904–1994) entitled *Atlas and Catalogue of Interacting Galaxies*, published in two parts in 1959 and 1977 (Vorontsov-Velyaminov, 1959, 1977). In both cases, the authors drew attention to the fact that there are many pathological types of galaxies which do not fall

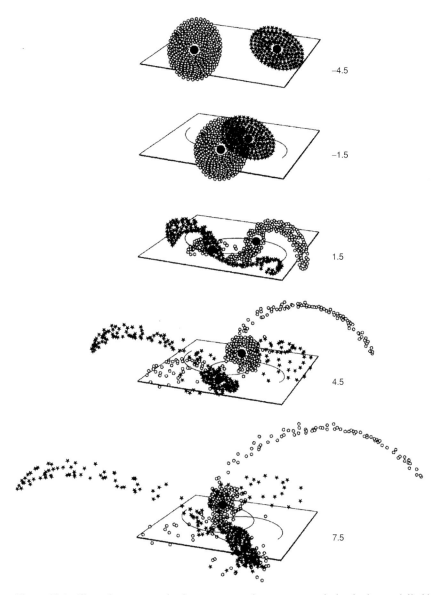

Figure 10.1: Illustating a prograde close encounter between two spiral galaxies modelled by Juri and Alar Toomre (Toomre and Toomre, 1972).

naturally into the Hubble sequence of types. In many cases these strange structures could be interpreted as strong gravitational interactions, or collisions, between galaxies, and in 1972 Juri (b. 1940) and Alar Toomre (b. 1937) showed how even some of the strangest images could be accounted for by such close encounters (Toomre and Toomre, 1972). As an example, Figure 10.1 shows how it is possible to produce long tails and bridges between galaxies as a result of gravitational interactions between them. In this example, the tails are associated with a prograde collision between two spiral galaxies, the outer rings of stars in the spirals being ripped off because, in the prograde collision, the stars in the outer regions feel the same accelerating force for an extended period of time. It is generally the case that most of the objects in the catalogues of peculiar galaxies are the results of collisions or close encounters between normal galaxies and not new types of galaxy. As de Vaucouleurs (1974) remarked,

After a collision, a car is a wreck, not a new type of car.

The study of these galaxies emphasised the importance of interactions between galaxies in their evolution. The interactions were often associated with large amounts of dust and hot gas as the interstellar material in the galaxies came together. A great deal of star formation was to be expected, and this was confirmed by the surveys conducted by the IRAS satellite in the mid 1980s, which showed that many of the most luminous IRAS galaxies in the far-infrared wavebands are those which involve strongly interacting galaxies.[3]

The masses of galaxies range from systems which have mass only about $10^7 M_\odot$ to super-giant elliptical galaxies which, in the most extreme cases, have masses as great as $10^{13} M_\odot$. The dwarf galaxies were recognised by Zwicky in the late 1930s through observations made with the Palomar 18-inch Schmidt telescope (Zwicky, 1942). Although there are some variations, the distribution of luminosities among galaxies of all types per unit volume can be remarkably well described by the form of *luminosity function* introduced by Paul Schechter in 1976 (Schechter, 1976):

$$\Phi(L)\, dL = AL^{-\alpha} \exp(-L/L^*)\, dL, \tag{10.1}$$

where L is the luminosity of the galaxy and L^* is a characteristic 'break' luminosity – for greater values, the numbers of galaxies diminish exponentially.

In 1977, James Felten (b. 1934) surveyed a large number of different determinations of the luminosity function for galaxies and showed that, once account was taken of the different corrections and assumptions made by different authors, they were all consistent with the form of luminosity function proposed by Schechter (Figure. 10.2) (Felten, 1977). The value of L^*, corresponding to the break in the luminosity function, has value $L^* \approx 10^{10} L_\odot$. The index α was found to have a value of about 0.25, indicating that the luminosity function has a long tail which extends to the dwarf galaxies. Our own Galaxy has luminosity about $0.5L^*$, and so it is typical of bright spiral galaxies found in statistical samples, but is by no means among the most luminous galaxies known. Integration over the luminosity function, equation (10.1), shows that most of the background light in the Universe is produced by galaxies with $L \sim L^*$. Both spiral and elliptical galaxies span the whole range of luminosities, but the irregular galaxies are mostly found with $L < L^*$. While the Schechter function is, in general, a good description of the probability distribution of galaxy luminosities, it does not give a good description of the most luminous galaxies in clusters, which are anomalously luminous (see Section 10.5.1).

Extensive studies have been made of correlations between various properties of elliptical galaxies, specifically their luminosities, sizes, central velocity dispersions, their abundances of heavy elements and so on. Of these, two studies are of particular importance. The first is the analysis of Sandra Faber (b. 1944) and Robert Jackson (b. 1949), who in 1976 discovered a strong correlation between luminosity, L, and central velocity dispersion, σ, of the form

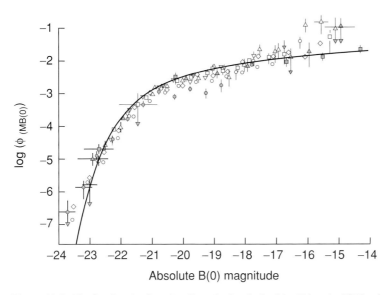

Figure 10.2: The luminosity function for galaxies derived by Felten in 1977 using data from a large number of different surveys of galaxies (Felten, 1977). The solid line shows the best-fitting Schechter function to the data.

$L \propto \sigma^x$, where $x \approx 4$ (Faber and Jackson, 1976). This correlation has been studied by other authors, who have found values of x ranging from about 3 to 5. The significance of this relation is that, if the velocity dispersion, σ, is measured for an elliptical galaxy, its intrinsic luminosity can be inferred from the Faber–Jackson relation and hence its distance found.

This procedure was refined in 1987 by Alan Dressler (b. 1948) and his colleagues and by George Djorgovski (b. 1956) and Marc Davis (b. 1947), who introduced the concept of the *fundamental plane* for elliptical galaxies (Djorgovski and Davis, 1987; Dressler *et al.*, 1987). The fundamental plane lies in a three-dimensional space in which luminosity, L, is plotted against central velocity dispersion, σ, and the surface brightness, Σ_e, within the half-light radius. An even stronger correlation than the Faber–Jackson relation was found when the surface brightness was included:

$$L \propto \sigma^{8/3} \Sigma_e^{-3/5}. \tag{10.2}$$

This empirical formula enables the distances of elliptical galaxies to be determined independent of their redshifts. Dressler and his colleagues estimated that, using these correlations, the distances of individual galaxies can be determined to about 25% and for clusters of galaxies to about 10%.

In 1975, Brent Tully (b. 1943) and Richard Fisher (b. 1943) discovered that, for spiral galaxies, the widths of the profiles of the 21 cm line of neutral hydrogen, once corrected for the effects of inclination, are strongly correlated with their intrinsic luminosities (Tully and Fisher, 1977). In their studies, they correlated the total B luminosities with the corrected velocity width, ΔV, of the 21 cm line and found the relation

$$L_B \propto \Delta V^\alpha, \tag{10.3}$$

where $\alpha = 2.5$. A much larger survey carried out by Marc Aaronson (1950–1987) and Jeremy Mould (b. 1949) in 1983 found a somewhat steeper slope, $\alpha = 3.5$, for luminosities measured in the optical B waveband, and an even steeper slope, $\alpha = 4.3$, in the near-infrared H waveband at 1.65 μm (Aaronson and Mould, 1983). This correlation, the *infrared Tully–Fisher relation*, is very much tighter in the infrared than in the blue waveband because the luminosities of spiral galaxies in the latter waveband are significantly influenced by interstellar extinction within the galaxies themselves – in the infrared waveband the dust becomes transparent. As a result, measurement of the 21 cm velocity width of a spiral galaxy can be used to infer its absolute H magnitude, and hence, by measuring its flux density, its distance can be estimated. This procedure has resulted in some of the best distance estimates for spiral galaxies and has been used in programmes to measure the value of Hubble's constant.

10.2   Dark matter in galaxies

In the simplest picture, the distribution of light in spiral galaxies can be decomposed into two components, the *bulge* and the *disc* (Figure 5.8). The discs are in centrifugal equilibrium, and so measurements of the projected rotational velocity as a function of distance from the centre provides dynamical information about the mass distribution in the galaxy. In the optical waveband, these are very demanding spectroscopic observations because the stellar absorption features in the spectra of galaxies are weak and their surface brightnesses are low. To circumvent this problem, Margaret and Geoffrey Burbidge and Kevin Prendergast used the narrow emission lines of ionised gas clouds in the discs of spiral galaxies as tracers of the velocity distribution. The masses and mass-to-light ratios for the galaxies NGC 5866 and NGC 681 are good examples of these pioneering studies (Burbidge, Burbidge and Prendergast, 1960, 1965). It is conventional to describe the masses of galaxies in terms of the ratio of mass to luminosity, normalised to this ratio for the Sun, $M_\odot / L_\odot$. Typically, mass-to-light ratios in the range 2 to 5 were found, which could be explained by assuming that, within the radius to which the rotation curve had been determined, the light was dominated by old stars somewhat less massive than the Sun.

These observations referred to the central regions of galaxies, but it proved very much more difficult to study the faint outer regions using photographic techniques. From the late 1970s onwards, image tubes and CCD detectors became available which greatly increased the capability for two-dimensional spectroscopy with long spectrographic slits. Vera Rubin (b. 1928) and her colleagues pioneered systematic studies of the rotation curves of galaxies using very long slits, which enabled rotation curves to be determined throughout the bodies of the galaxies, the narrow emission lines again being the preferred tracer of the velocity fields (Rubin, Thonnard and Ford, 1980).[4] This work was complemented by observations of the 21 cm line of neutral hydrogen, which enabled the velocity curves of spiral galaxies to be determined to much greater radial distances than the optical observations (Figure 10.3).

Both types of observation showed that, in the outer regions of galaxies, the velocity curves are generally remarkably flat, $v_{rot} \approx$ constant, as far as the rotation curves could be measured. The significance of this result can be appreciated from a simple Newtonian

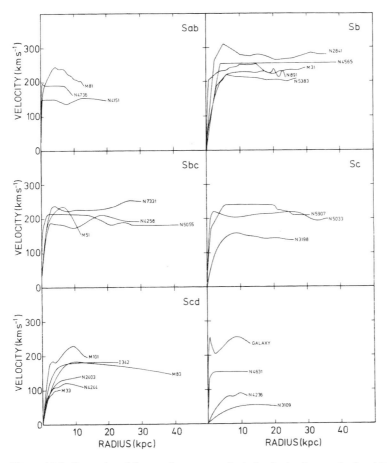

Figure 10.3: Examples of the rotation curves of spiral galaxies from optical and 21 cm neutral hydrogen observations (Bosma, 1981).

calculation. If the galaxy is taken to be spherical and the mass within radius r is $M(<r)$, the circular rotational velocity at distance r is found by equating the inward gravitational acceleration, $GM(<r)/r^2$, to the centripetal acceleration, v_{rot}^2/r, and so

$$v_{rot} = \left[\frac{GM(<r)}{r}\right]^{1/2}. \tag{10.4}$$

Thus, if $v_{rot} = $ constant, it follows that $M(<r) \propto r$ so that the total mass within radius r increases linearly with distance from the centre. This result contrasts strongly with the variation of the surface brightness distributions of spiral galaxies, which decrease much more rapidly with distance from the centre than as r^{-2}. Kenneth Freeman (b. 1940) found that models in which the luminosity per unit surface area decreases exponentially with radius provided an excellent fit to these data (Freeman, 1970). To rephrase this important result, the mass-to-light ratio must increase dramatically in the outer regions of spiral galaxies. The conventional way of expressing this result is to state that there must be a large amount

of *dark matter* in the haloes of galaxies. A typical figure for giant spiral galaxies is that they must contain about ten times as much dark as visible matter.[5]

An astrophysical argument for the existence of dark matter haloes in spiral galaxies was presented in 1973 by Jeremiah Ostriker (b. 1937) and James Peebles (b. 1935), who pointed out that rotating discs are subject to a bar instability, unless a stabilising halo is present which contains a significant fraction of the mass of the system (Ostriker and Peebles, 1973). Their criterion for stability was that the ratio of the ordered kinetic energy of the disc, T_{orb}, to the total potential energy, $|U|$, should be less than 0.14, a result confirmed by subsequent analytic and numerical analyses. A halo of dark matter about spiral galaxies would provide stabilisation of the galactic disc.

The presence of about ten times more dark than visible matter was also found in the nearby giant elliptical galaxy M87 by John Huchra (b. 1948) and Jean Brodie (b. 1953) in 1987 (Huchra and Brodie, 1987) from the velocity dispersions of globular clusters in the halo of the galaxy. A quite different diagnostic tool for studying the mass distribution in galaxies was first used in 1980 by Daniel Fabricant (b. 1952) and his colleagues, who used the X-ray surface brightness distribution of elliptical galaxies as a means of determining the gravitational potential of the system (Fabricant, Lecar and Gorenstein, 1980) (see Section A10.1). Once again, it was inferred that there must be more mass present than would be deduced from the optical surface brightness distribution.[6] We will find the same result for the mass-to-luminosity ratios of rich clusters of galaxies (see Section 10.4).

The nature of the dark matter in galaxies remains unknown. Many ideas have been proposed, including various types of baryonic and non-baryonic dark matter. Examples of low-luminosity discrete objects include planets, brown dwarfs, low-mass stars, neutron stars and small black holes. Non-baryonic examples include the types of particles predicted by particle theorists, for example the lightest supersymmetric particle, massive neutrino-like particles or the gravitino – these types of particles are known collectively as *weakly interacting massive particles*, or *WIMPs*; these have not yet been detected in particle physics experiments. The possibility that some form of WIMP exists gains credence from studies of the origin of the large-scale structure of the Universe (see Chapter 15).

Like the spiral galaxies shown in Figure 10.3, the rotation curve of our own Galaxy is remarkably flat at radii $r > 3$ kpc (Fich and Tremaine, 1991), and so it is expected that there should be a dark matter halo about our Galaxy.[7] One of the most impressive approaches to setting limits to the contribution which discrete low-mass objects, collectively known as *massive compact halo objects*, or *MACHOs*, could make to the dark matter in our own Galaxy has been the search for the gravitational microlensing signatures of such objects as they pass in front of background stars. These are very rare events and so very large numbers of background stars have to be monitored. The beauty of this technique is that it is sensitive to MACHOs with a very wide range of masses, from $10^{-7} M_\odot$ to $100 M_\odot$, and the contributions of a very wide range of candidates for the dark matter can be constrained. Two very-large-scale projects, the MACHO and the EROS projects, have made systematic surveys over a number of years to search for these events. The MACHO project, which ran from 1992 to 1999, used stars in the Magellanic Clouds and in the Galactic bulge as background stars, and millions of stars were monitored regularly (Alcock *et al.*, 1993b). The first example of a microlensing event was discovered in October 1993 (Figure 10.4), the mass

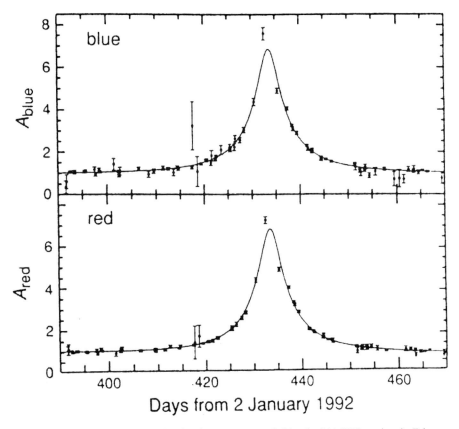

Figure 10.4: The gravitational microlensing event recorded by the MACHO project in February and March 1993. The horizontal axis shows the number of days measured from day zero on 2 January 1992. The vertical axis shows the amplification of the brightness of the lensed star relative to the unlensed intensity in blue and red wavebands. The solid lines show the expected variations of brightness of a lensed star with time. The same characteristic light-curve is observed in both wavebands, as expected for a gravitational microlensing event (Alcock *et al.*, 1993a).

of the invisible lensing object being estimated to lie in the range $0.03M_\odot < M < 0.5M_\odot$ (Alcock *et al.*, 1993a).

By the end of the MACHO project, many lensing events had been observed, including over 100 in the direction towards the Galactic bulge, about three times more than expected. In addition, 13 definite and 4 possible events were observed in the direction of the Large Magellanic Cloud (Alcock *et al.*, 2000). The numbers are significantly greater than the two to four detections expected from known types of star. The technique does not provide distances and masses for individual objects, but, interpreted as a Galactic halo population, the best statistical estimates suggest that the mean mass of these MACHOs is between $0.15M_\odot$ and $0.9M_\odot$. The statistics are consistent with MACHOs making up about 20% of the necessary halo mass, the 95% confidence limits being 8–50%. Somewhat fewer microlensing events were detected in the EROS project, which found that less than 25% of the mass of the standard dark matter halo could consist of dark objects with masses in the range $2 \times 10^{-7} M_{\odot}$ to $1 M_{\odot}$ at the 95% confidence level (Afonso *et al.*, 2003). The most likely candidates for the MACHOs observed by the MACHO project would appear to be white dwarfs, which would have to be produced in large numbers in the early evolution of the Galaxy, but other more exotic possibilities cannot be excluded. The consensus is that MACHOs alone cannot account for all the dark matter in our Galaxy, and so some form of non-baryonic matter must make up the difference.

Useful astrophysical limits can be set to the number densities of different types of neutrino-like particles in the outer regions of giant galaxies and in clusters of galaxies. The WIMPs and massive neutrinos are collisionless fermions, and therefore there are constraints on the phase-space density of these particles, which translate into a lower limit to their masses since, for a given momentum, only a finite number of particles within a given volume is allowed. This calculation was first presented by Scott Tremaine (b. 1950) and James Gunn in 1979, who showed that the masses of neutrino-like particles necessary to bind the halo of our Galaxy would exceed 30 eV, well in excess of the upper limit to the mass of the electron neutrino at that time (Tremaine and Gunn, 1979).

The search for evidence for different types of dark matter particles has developed into one of the major areas of the discipline known as *astroparticle physics*. An important class of experiments involves the search for weakly interacting particles with masses $m \geq 1$ GeV, which could make up the dark halo of our Galaxy. In order to form a bound dark halo about our Galaxy, the particles would have to have velocity dispersion $\langle v^2 \rangle^{1/2} \sim 230$ km s^{-1} and their total mass is known. Therefore, the number of WIMPs passing through a terrestrial laboratory each day is a straightforward calculation. When these massive particles interact with the sensitive volume of the detector, the collision results in the transfer of momentum to the nuclei of the atoms of the material of the detector, and this recoil can be measured in three different ways. (i) There is a small temperature increase, which can be measured in a cryogenically cooled detector, (ii) or the ionisation caused by the recoiling nucleus can be measured in an ionisation chamber, (iii) or the light emitted by the passage of the recoil nucleus through the material can be detected by a scintillation detector. The challenge is to detect the very small number of events expected because of the very small cross-section for the interaction of WIMPs with the nuclei of atoms. A typical estimate is that less than one WIMP per day would be detectable by 1 kg of detector material. These are very demanding experiments, and they have to be located deep underground to avoid contamination by cosmic rays and to be heavily shielded against natural radioactivity in the surrounding rocks. Such experiments have been carried out in deep underground laboratories such as those at Gran Sasso in Italy, the Soudan Underground Laboratory in Minnesota, USA, and the Boulby Underground Laboratory in Yorkshire, UK.

The importance of these measurements for particle physics is that one of the strong candidates for the WIMPs is the lightest particle predicted by supersymmetry theories of elementary particles and which are expected to be stable. Many experiments have been developed to search for these elusive particles, but as yet no convincing positive detection has been made. A good example of the quality of the data now available is provided by the results of the Cryogenic Dark Matter Search at the Soudan Underground Laboratory, which provided 19.4 kg days of observation with a germanium detector cooled to temperatures less than 50 mK. At the most sensitive mass, 60 GeV/c^2, the scalar cross-section for the WIMP–nucleon interaction must be less than $\sigma_w = 4 \times 10^{-47}$ m^2 (Akerib *et al.*, 2004). It is intriguing to compare this value with the weak-interaction cross-section for neutrino–electron scattering, $\sigma = 3 \times 10^{-49}(E/m_ec^2)^2 \; m^2$, where E is the energy of the neutrino. These limits to σ_w are already sufficiently low to rule out some supersymmetry models.

10.3 The dynamics of elliptical galaxies

It might be thought that the internal dynamics of elliptical galaxies would be a relatively straightforward problem. The axial ratios of these galaxies range from 1:1 to about 3:1, and the velocity dispersions of the stars can be measured spectroscopically within the galaxies. These measurements have been compared with the amounts of rotation and internal velocity dispersion which would be expected if the flattening of the elliptical galaxies were wholly attributed to the rotation of an axisymmetric distribution of stars. Francesco Bertola (b. 1937) and Massimo Capaccioli (b. 1944) in 1975 and Garth Illingworth (b. 1947) in 1977 showed, however, that luminous elliptical galaxies rotate only very slowly (Bertola and Capaccioli, 1975; Illingworth, 1977). An analysis of a sample of elliptical galaxies, as well as the bulges of spiral galaxies, by Roger Davies (b. 1954) and his colleagues in 1983 confirmed that, in general, the most luminous elliptical galaxies do not possess enough rotation to account for their ellipticity (Davies *et al.*, 1983) (Figure 10.5). The implication of this result is that the assumptions of an axisymmetric distribution of stars and an isotropic velocity distribution at all points in the galaxies must be wrong. Consequently, these massive elliptical galaxies may well be triaxial systems, that is systems with three unequal axes and with anisotropic velocity distributions. There is no problem in assuming that the velocity distribution is anisotropic because normally the timescale for the exchange of energy between stars through gravitational encounters is greater than the age of the Galaxy. Therefore, if the velocity distribution began by being anisotropic, it would not have been isotropised by now.

Further evidence for the triaxial nature of the stellar distribution in elliptical galaxies has been discovered by Bertola and Giuseppe Galletta (b. 1954), who found that the major axis of elliptical galaxies varies with distance from their centres (Bertola and Galletta, 1979). Furthermore, in some of these galaxies, rotation has been observed along the minor as well as along the major axis (Bertola *et al.*, 1991).

The theoretical position was clarified by Martin Schwarzschild in 1979, who used the procedures of linear programming to determine the orbits of particles in general self-gravitating systems (Schwarzschild, 1979). His analysis showed that there exist stable triaxial systems. In a result similar to that found in classical dynamics, he showed that there exist stable orbits about the largest and smallest axes of a triaxial system, but not about the intermediate axis. Thus, elliptical galaxies can be classified as oblate–axisymmetric, prolate–axisymmetric, oblate–triaxial, prolate–triaxial and so on.

10.4 The large-scale distribution of galaxies

Groups and clusters of galaxies come in a wide variety of different types, ranging from rich regular clusters, which have smooth galaxy density profiles and roughly circular appearance, to irregular systems, which have a ragged appearance without any prominent central

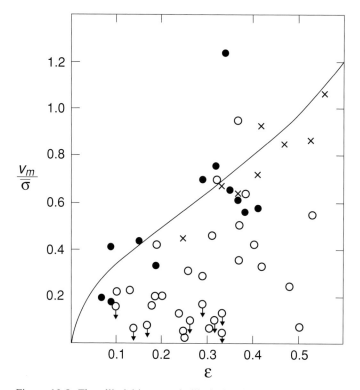

Figure 10.5: The ellipticities, ε, of elliptical galaxies as a function of their rotational velocities, v_m, normalised to the velocity dispersion, $\bar{\sigma}$, of the bulge. The open circles are luminous elliptical galaxies; the filled circles are lower luminosity ellipticals; and the crosses are the bulges of spiral galaxies. If the ellipticity were entirely due to rotation, with an isotropic stellar velocity distribution at each point, the galaxies would be expected to lie along the solid line. The diagram shows that, at least for the massive ellipticals, this simple picture of rotational flattening cannot be correct (Davies *et al.*, 1983).

concentration of galaxies. While the Palomar 48-inch Schmidt Sky Survey was being carried out, the principal observers undertook various research projects on the wide field plates which were taken each night, and, among the most important of these, George Abell (1927–1983) classified and catalogued the rich clusters of galaxies. The catalogue was compiled by visual inspection of the plates according to strict selection criteria such that only the most prominent clusters were included. The apparent magnitudes of the galaxies were estimated by visual inspection. The *Abell Catalogue of Clusters of Galaxies* was published in 1958 and contains about 2400 of the richest clusters of galaxies north of declination −20° away from the Galactic plane (Abell, 1958). The survey was extended to include the southern hemisphere when the UK Schmidt Telescope survey plates were completed, and a catalogue of over 4000 rich clusters over the whole sky was prepared by Abell, Harold Corwin (b. 1943) and Ronald Olowin (b. 1945) in 1989, sadly six years after Abell's untimely death in 1983 (Abell, Corwin and Olowin, 1989).

In Abell's catalogue, there is a bias towards the richest, symmetrical systems, reflecting the strict criteria which he adopted in selecting the clusters. In fact, the clustering of galaxies occurs on a very wide range of physical scales, from small groups containing only a few galaxies to giant clusters and superclusters of galaxies. Following Hubble's pioneering studies of the counts of galaxies in the 1930s, a major effort was made after the Second World War to define the large-scale structure of the Universe of galaxies using the large-scale plates of the Lick Northern Proper Motion Surveys. These large plates had been taken with the 51-cm Carnegie double astrograph for the purpose of measuring the proper motions of stars, but they also included a great deal of information about the number counts of galaxies. Counts of galaxies on the Lick plates were made by Donald Shane (1895–1983), Carl Wirtanen (1910–1990) and their colleagues during the 1950s and were published as counts of galaxies in $1° \times 1°$ boxes for the sky north of $\delta = -23°$ (Shane and Wirtanen, 1957). Jerzy Neyman (1894–1981) and Elizabeth Scott (1917–1988) used correlation functions to analyse the variance of the numbers of galaxies in these cells (Neyman, Scott and Shane, 1954). In turn, these studies led to the use of two-point correlation functions to describe the large-scale clustering properties of galaxies, and these have become the preferred tool for quantifying the large-scale distribution of galaxies statistically. The two-point correlation function, $\xi(r)$, for galaxies can be written as

$$N(r) \, dV = N_0[1 + \xi(r)] \, dV, \tag{10.5}$$

where $N(r)$ is the number density of galaxies at radial distance r from any given galaxy and $\xi(r)$ describes the excess probability of finding a galaxy at distance r over a uniform distribution N_0.

One of the important issues was whether or not there are preferred scales of clustering of galaxies in the Universe. Abell and de Vaucouleurs had shown that there exists non-random clustering of clusters of galaxies, but it was not clear whether or not there is a continuous range of clustering (Abell, 1962; de Vaucouleurs, 1971). Tao Kiang (b. 1928) and William Saslaw (b. 1944) proposed that there were no preferred scales, but that clustering could occur on all scales (Kiang and Saslaw, 1969). In 1969, Hiroo Totsuji (b. 1943) and Taro Kihara (b. 1917–2001) were the first to show that the galaxy correlation function, $\xi(r)$, can be approximated by a power-law over a wide range of scales, and this approach was developed extensively by James Peebles and his colleagues in an important series of papers in the 1970s.[8] The function $\xi(r)$ can be described by a power-law function,

$$\xi(r) = \left(\frac{r}{r_0}\right)^{-\gamma}, \tag{10.6}$$

where $\gamma = 1.77$ and $r_0 = 5h^{-1}$ Mpc, where $h = H_0/100 \, \text{km s}^{-1} \, \text{Mpc}^{-1}$ is Hubble's constant measured in units of 100 km s^{-1} Mpc^{-1}. This function gives a good representation of the clustering of galaxies on scales from about $10h^{-1}$ kpc to $10h^{-1}$ Mpc, but on scales $r \geq 20h^{-1}$ Mpc the function decreases more rapidly with increasing physical size (Figure 10.6)[9] (Maddox et al., 1990; Peebles, 1993).

It should be emphasised that this form of correlation function is spherically symmetric about any point and washes out a great deal of information about the structure of the clustering. Nonetheless, this form of function makes the important point that clustering occurs on a very wide range of physical scales, from small groups of galaxies to systems much greater than even the richest clusters of galaxies. The rich clusters are no more than

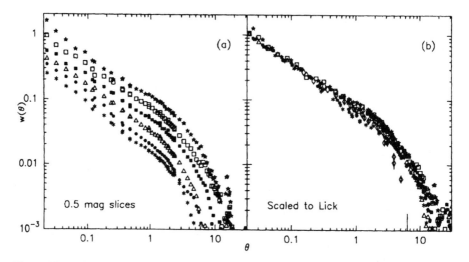

Figure 10.6: The *angular* two-point correlation function for galaxies over a wide range of angular scales. (a) The scaling test for the homogeneity of the distribution of galaxies derived from the APM surveys at increasing limiting apparent magnitudes in the range $17.5 < m < 20.5$. The correlation functions are displayed at intervals of 0.5 magnitudes. (b) The two-point correlation function scaled to the correlation function derived from the Lick counts of galaxies (Maddox *et al.*, 1990).

the most prominent features of a continuous spectrum of clustering. The origin of this form of correlation function is one of the goals of the theory of origin of structure in the Universe.

Figure 10.6 shows that the *angular* two-point correlation function scales with increasing apparent magnitude, as expected, if the distribution of galaxies exhibits the same degree of clustering with increasing distance from our own Galaxy. As Peebles expresses this important result,

the correlation function analyses have yielded a new and positive test of the assumption that the galaxy space distribution is a stationary (statistically homogeneous) random process.

In fact, the distribution of galaxies is much more complicated than this. In the 1970s, Peebles and his colleagues reanalysed the Lick counts of galaxies using the original $10' \times 10'$ cells used by Shane and Wirtanen and demonstrated that clustering exists on a very wide range of scales, and, in particular, that, on scales greater than those of clusters of galaxies, the distribution of galaxies has a stringy, cellular appearance (Seldner *et al.*, 1977) (Figure 10.7).

In parallel with these studies, increasing numbers of redshifts for nearby galaxies were becoming available, and these enabled the three-dimensional distribution of galaxies to be defined directly. Jaan Einasto (b. 1929) and his colleagues were the first to demonstrate the reality of structures on scales very much greater than those clusters of galaxies and to appreciate the significance of these for theories of the origin of structure in the Universe[10] (Jõeveer and Einasto, 1978). The mapping of the local distribution of galaxies proceeded though the 1980s, culminating in the map of the local Universe created by Margaret Geller (b. 1947) and John Huchra, derived from a complete survey of the redshifts of over 14 000

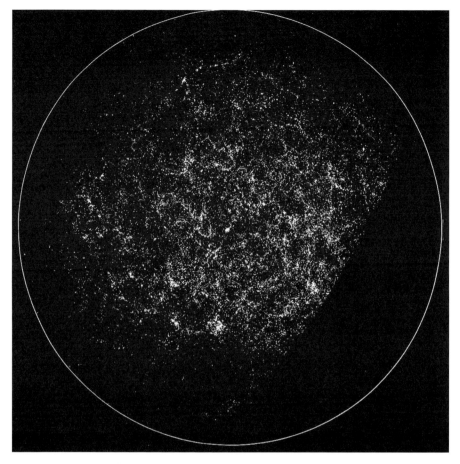

Figure 10.7: A map of the galaxy counts in the northern galactic hemisphere derived by Peebles, Seldner and their colleagues from a reanalysis of the Lick counts of galaxies carried out by Shane and Wirtanen. The northern galactic pole is at the centre; the galactic equator is the white bounding circle. The galactic latitude is a linear function of radius from the pole. Galactic longitude increases clockwise, with $l = 0°$ at the bottom of the map. The prominent 'cluster' in the centre of the image is the Coma cluster. (Seldner *et al.*, 1977.)

bright galaxies (Geller and Huchra, 1989) (Figure 10.8). If the galaxies were uniformly distributed in the local Universe, the points would be uniformly distributed over the diagram. It can be seen that there are large 'holes' in which the local number density of galaxies is significantly lower than the mean and also long 'filaments' or 'walls' of galaxies. The scale of the large holes seen in Figure 10.8 is about 30 to 50 times the scale of a cluster of galaxies.

Richard Gott (b. 1947) and his colleagues (Gott, Melott and Dickinson, 1986) showed that the topology of the distribution of the galaxies on the large scale is 'sponge-like'. The material of the sponge represents the location of the galaxies and the holes represent the large voids. Both the holes and the distribution of galaxies are continuously connected throughout the local Universe. These are the largest known structures in the Universe, and

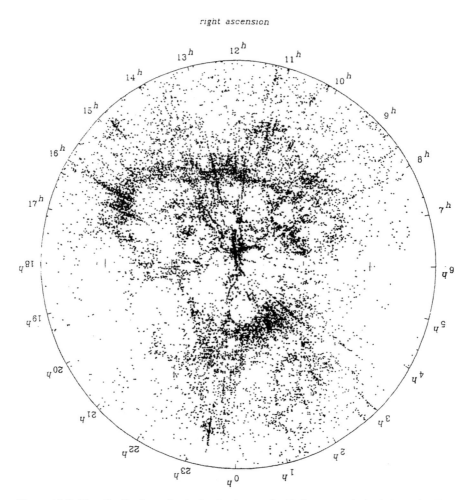

Figure 10.8: The distribution of galaxies in the nearby Universe as derived from the Harvard–Smithsonian Center for Astrophysics survey of galaxies. The map contains over 14 000 galaxies, which form a complete statistical sample around the sky between declinations $\delta = 8.5°$ and 44.5°. Our Galaxy is located at the centre of the map, and the radius of the bounding circle corresponds to a redshift of 0.05, or a distance of $150h^{-1}$ Mpc. The galaxies within this slice around the sky have been projected onto a plane to show the large-scale features in the distribution of galaxies. Rich clusters of galaxies, which are gravitationally bound systems with internal velocity dispersions of about 10^3 km s^{-1}, appear as 'fingers' pointing radially towards our Galaxy at the centre of the diagram. (Courtesy of Margaret Geller and John Huchra.)

one of the great cosmological problems is to reconcile the gross irregularity in the large-scale distribution of galaxies with the remarkable smoothness of the cosmic microwave background radiation.

A key question for cosmology was whether or not this 'cellular' structure persists out to much greater distances, and this was resoundingly answered in the affirmative by a number of large-scale surveys of the distribution of galaxies in various sectors of the sky. Firstly, the Las Campanas Redshift Survey used measurements of the redshifts of 26 418 galaxies

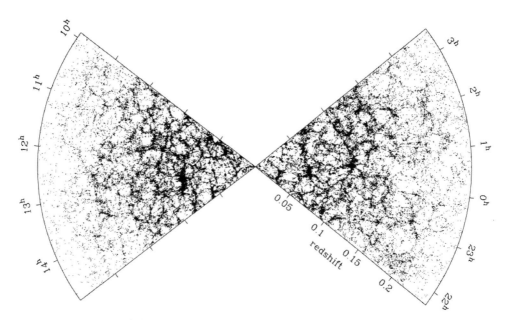

Figure 10.9: A 3° slice through the Anglo-Australian Telescope 2dF galaxy survey showing the cellular structure of the distribution of galaxies extending out to redshift $z \approx 0.25$. This map extends to about five times the distance of the image shown in Figure 10.8. (Courtesy of Dr M. Colless and the 2dF Galaxy Survey Team.)

to extend the depth of the survey to about four times that of the Geller–Huchra survey, and it showed the 'cellular' structure extending out to the limit of their survey (Lin *et al.*, 1996). This was followed by the 2dF survey carried out using the 2° field multi-object spectrograph at the prime-focus of the Anglo-Australian Telescope, which carried out a similar survey of almost 180 000 galaxies, extending the survey to five times that of the Huchra–Geller survey with large statistics (Figure 10.9) (Colless *et al.*, 2001). Most recently, the first redshift maps produced by the Sloan Digital Sky Survey have defined the large-scale distribution of galaxies to a similar depth but with even larger statistics, about 200 000 galaxies being plotted in their first images (Stoughton *et al.*, 2002). Statistical analyses of these maps have shown that the cellular structure is present throughout the distribution of galaxies at the present epoch. Thus, although the distribution of galaxies is highly non-uniform, the same degree of non-uniformity is present throughout the distribution of galaxies, consistent with the overall isotropy and homogeneity of the Universe on large enough scales.

10.5 The physics of clusters of galaxies

The richest clusters, such as the Coma cluster of galaxies, have *crossing times*, $t_c = R/v$, which are much less than the age of the Universe. In this expression, v is the mean speed of a galaxy in the cluster and R is a characteristic size of the cluster. The radial distribution of galaxies is found to be similar to that of an isothermal gas sphere, for which $N(r) \propto r^{-2}$ out to some outer radius. Thus, it is certain that such clusters have come to a state of statistical equilibrium under gravity. It is for this reason that the *virial theorem* can be applied with confidence to these clusters in order to determine their masses (Eddington, 1916a). The masses of the clusters can range up to $3 \times 10^{15} M_\odot$, very much greater than would be inferred from the light of the galaxies. A convenient way of expressing this result is in terms of the mass-to-luminosity ratio of rich clusters of galaxies, which is of the order 200–$300 M_\odot / L_\odot$, very much greater than the values found in the visible regions of galaxies, which are at most about 10–$20 M_\odot / L_\odot$. It is also interesting to compare this figure with the mass-to-luminosity ratio necessary to attain the critical cosmological density, which amounts to $1600 h M_\odot / L_\odot$. These observations provide compelling evidence for the presence of dark matter in the clusters, but not in sufficient quantities for the Universe to attain the critical cosmological density.

10.5.1 Hot gas in clusters of galaxies

One of the most important discoveries of the UHURU X-ray observatory was that some rich clusters of galaxies are intense X-ray sources. In clusters such as the Coma cluster, the X-ray emission is diffuse and fills the core of the cluster (Gursky *et al.*, 1971). This radiation has been convincingly identified with the bremmstrahlung, or free–free emission, of hot intracluster gas, the clinching piece of evidence being the discovery of the emission lines of very highly ionised iron, Fe XXV and Fe XXVI, at 8 keV by the Ariel-V satellite (Mitchell *et al.*, 1976). The hot gas forms an extended atmosphere within the gravitational potential of the cluster, and the same technique already discussed in the context of measuring the masses of galaxies can be used to estimate the mass and mass distribution within the cluster of galaxies (see Sections 10.2 and A10.1). Observations by the ROSAT X-ray observatory enabled maps of the X-ray surface brightness distribution within nearby clusters to be measured and the distributions of the mass in galaxies, in hot gas and in the dark matter, to be determined. Hans Böhringer (b. 1952) applied these procedures to the Perseus cluster of galaxies, in which the X-ray emission could be traced out to a radius $1.5 h^{-1}$ Mpc. From the X-ray observations, it was possible to determine both the total gravitating mass within radius r, $M(<r)$, the mass of gas within radius r, $M_{gas}(<r)$, and then to compare these with the mass in the visible parts of the cluster galaxies (Böhringer, 1994). In Figure 10.10, it can be seen that the mass of hot intracluster gas is about five times greater than the mass in galaxies, but that it is insufficient to account for all the gravitating mass that must be present. Some form of dark matter must be present to bind the cluster gravitationally. Note also that the observation of iron emission from the intracluster gas indicates that the iron created in the stars in galaxies must have been circulated through the intracluster medium.

If the density of the hot intracluster gas is large enough, its cooling rate can be sufficiently large for it to cool over cosmological timescales. At high enough temperatures, the principal energy loss mechanism for the gas is the same thermal bremsstrahlung process that is responsible for the X-ray emission of the cluster. The characteristic cooling time for the gas

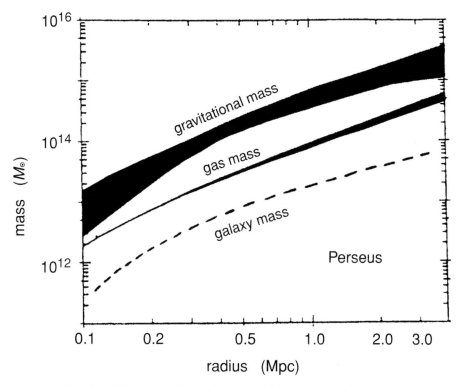

Figure 10.10: Integrated radial profiles for the mass in the visible parts of galaxies, hot gas and total gravitating mass for the Perseus cluster of galaxies, as determined by observations with ROSAT. The upper band indicates the range of possible total masses and the central band shows the range of gaseous masses (Böhringer, 1994).

is given by

$$t_{\text{cool}} = 10^{10} \frac{T^{1/2}}{N} \text{ years,} \qquad (10.7)$$

where the temperature is measured in kelvin and the number density of ions or electrons is measured in particles per cubic metre. Thus, if the typical temperature of the gas is 10^7 to 10^8 K, the cooling time is less than 10^{10} years if the electron density is greater than about 3×10^3 to 10^4 m^{-3}. These conditions are found in many of the clusters of galaxies which are intense X-ray emitters. As a result, the central regions of these hot gas clouds can cool and, to preserve pressure balance, the gas density increases, resulting in the formation of a *cooling flow*. The evidence for these cooling flows was discovered by Andrew Fabian (b. 1948) and his colleagues when it became possible to image clusters of galaxies and determine the variation of the X-ray surface brightness and the temperature of the gas as a function of radius from the centre (Fabian, 1994). A good example of these flows is found in the cluster Abell 478, in which the temperature of the gas in the central regions is less than in the outer regions and a mass inflow rate of about 600–800$M_\odot$ year^{-1} is inferred. Thus, over a period of 10^{10} years, such cooling flows can contribute significantly to the mass of

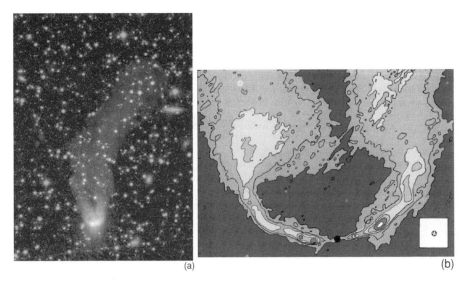

(a)   (b)

Figure 10.11: The radio maps of the radio trail source 3C 83.1B associated with the galaxy NGC 1265 in the Perseus cluster of galaxies. (a) The radio jets are swept back by the ram pressure of the intergalactic gas in the cluster, the galaxy moving at a speed of at least 2200 km s^{-1} relative to the local standard of rest of the cluster. (Courtesy of Dr Alan Bridle and the NRAO/AUI.) (b) A high-resolution map of the core of the radio source showing the radio jets ejected from the core of the galaxy, which is indicated by the black dot.

the central galaxy. According to Fabian, about half of the clusters detected by the Einstein X-ray observatory have high central X-ray surface brightnesses and cooling times less than 10^{10} years. Abell 478 is a particularly massive flow. Typically, the inferred mass flow rates are about 100 to $300 M_\odot$ year^{-1}.

The presence of hot gas in the cluster has a number of important consequences for the physics of the galaxies in clusters. The galaxies are in motion with respect to the intergalactic gas and so the *ram pressure* of the intergalactic gas can sweep the interstellar gas out of spiral galaxies. This process of ram pressure *stripping* of galaxies is a plausible means of forming S0 galaxies, which are found in much greater numbers in rich clusters of galaxies and regions of high galaxy density than in the general field (Gunn, 1978; Dressler, 1980).

Another consequence of the presence of hot gas in the cluster is that the structures of double radio sources, which are powered by beams of relativistic particles, are distorted by the motion of radio galaxies as they pass through the intergalactic gas. In powerful extragalactic double radio sources, the jets penetrate the intergalactic gas to large distances from the nucleus of the active galaxy and create characteristic double structures (see Chapter 11 and Figure 11.1(b)). The jets in the double radio sources associated with nearby cluster galaxies are much less luminous than the most powerful double radio sources, and the relative motion of the radio galaxy through the intergalactic gas causes the beams of energetic electrons to be swept back by the intergalactic gas, resulting in the formation of what are known as *radio trail sources* (Miley et al., 1972) (Figure 10.11).[11]

A further effect of the X-ray-emitting gas in the cluster is that the hot electrons can scatter photons of the cosmic microwave background radiation which pass through the cluster. This process was first described by Rashid Sunyaev (b. 1943) and Yakov Zeldovich in 1970; in this process the photons of the background radiation are scattered to higher energies by Compton scattering (Sunyaev and Zeldovich, 1970a). The result is that the Planck spectrum of the background radiation is shifted to slightly higher energies, resulting in a decrease in the intensity of the background radiation in the direction of the cluster of galaxies in the Rayleigh–Jeans region of the black-body spectrum and an increase in the Wien region of the spectrum. The *Sunyaev–Zeldovich effect* amounts to only about one part in 3000 of the intensity of the background for a rich cluster of galaxies, and it was only observed with certainty by Mark Birkinshaw (b. 1954) and his colleagues in 1990 after many years of difficult observation (Birkinshaw, 1990). Maps of these holes in the microwave background have since been made in the direction of rich clusters, which were known to be X-ray sources, and the shapes of the holes in the background have been determined directly (Jones *et al.*, 1998). The importance of the detection of the Sunyaev–Zeldovich effect is that it enables the pressure of the gas in the cluster to be determined. Combining this observation with the known temperature and emissivity of the gas enables the size of the gas cloud to be determined independent of its distance. This is one of the most promising methods of determining Hubble's constant.

One intriguing aspect of the imaging of these holes in the cosmic microwave background radiation is that the size of the decrement is independent of the redshifts of the clusters, if their properties are independent of redshift. This phenomenon has been observed by John Carlstrom (b. 1957) and his colleagues, who made maps of the Sunyaev–Zeldovich effect in clusters of galaxies spanning the redshift range 0.14 to 0.89 (Carlstrom *et al.*, 2000) (Figure 10.12). Visual inspection of Figure 10.10 shows that the amplitude of the temperature decrement observed in these clusters is independent of redshift. This is therefore a powerful method for detecting clusters of galaxies at very large redshifts, and a number of projects are currently underway to search for these clusters by detecting the holes in the cosmic microwave background radiation.

Most recently, the excess emission expected at wavelengths shorter than the maximum of the cosmic background radiation has been detected from a number of clusters (Benson *et al.*, 2003). These precise measurements enable an upper limit to be set to the peculiar motions of clusters of galaxies with respect to the local standard of rest through observations of the kinetic Sunyaev–Zeldovich effect.

10.5.2 *The dynamical evolution of clusters of galaxies*

The simplest picture of the dynamical evolution of a cluster of galaxies begins by assuming that the galaxies can be considered to be point masses. When the collapse of a protocluster gets underway, large gravitational potential gradients are set up since the collapse is unlikely to be spherically symmetric and the system of galaxies relaxes under the influence of these large-scale perturbations. This process was first described by Donald Lynden-Bell (b. 1935) in 1967 and is known as *violent relaxation* (Lynden-Bell, 1967). He showed that, under

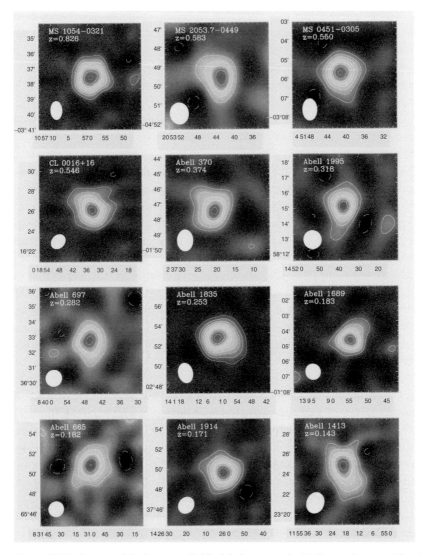

Figure 10.12: Images of the Sunyaev–Zeldovich decrement in 12 distant clusters with redshifts in the range 0.14 to 0.89 (Carlstrom *et al.*, 2000). Each of the images is plotted on the same intensity scale. The data were taken with the OVRO and BIMA millimetre arrays. The filled ellipse at the bottom left of each image shows the full-width half-maximum of the effective resolution used in reconstructing the images.

the influence of these large potential gradients, the galaxies in the cluster rapidly attain an equilibrium configuration in which galaxies of all masses have the same velocity distribution, consistent with observations of the velocities of galaxies of different masses in clusters. In the process of violent relaxation, the system has to get rid of half of its kinetic energy so that the cluster ends up being bound and satisfying the virial theorem.

Just as in the case of a Maxwellian gas of particles, the galaxies can exchange kinetic energy, but now by gravitational encounters. Unlike the case of particles in a gas, the encounters are rather infrequent, but the statistical result is the same in that the galaxies tend towards equipartition of energy so that the more massive galaxies slow up and tend to drift towards the centre of the cluster. This process of deceleration of the galaxies is known as *dynamical friction* and was first discussed by Chandrasekhar in the context of the dynamical evolution of star clusters in 1943 (Chandrasekhar, 1943a–c). This result suggests a reason why, in the regular relaxed clusters, the most massive galaxies are found towards the centre.

The absolute magnitudes of the brightest galaxies in clusters display a remarkably narrow dispersion, resulting in a very well defined velocity distance relation (see Figure 13.1). There was some controversy over the issue of whether this could be explained by randomly sampling the high-luminosity end of the luminosity function, or whether there is some special property of the first-ranked cluster member which is independent of the richness of the cluster. In 1977, Tremaine and Douglas Richstone (b. 1949) compared the dispersion in absolute magnitudes of the first-ranked members with the mean value of the difference in magnitude between the first- and second-ranked members (Tremaine and Richstone, 1977). They showed that, for any statistical luminosity function, there is much less dispersion in the absolute magnitudes of first-ranking cluster galaxies than would be expected if they were simply randomly sampled from the luminosity function, although they recognised that their results were not conclusive. John Kormendy (b. 1948) emphasised that the central cD galaxies in clusters are quite distinct from normal giant elliptical galaxies (Kormendy, 1982).

There are other effects which are important because of the finite sizes of galaxies. Just as in the cases of the peculiar and interacting galaxies, strong tidal effects can cause the disruption of galaxies, and, in particular, large galaxies tend to tear apart and consume smaller galaxies. This process, described by Ostriker and Tremaine in 1975, is often referred to as *galactic cannibalism* and is likely to be particularly important in rich clusters of galaxies (Ostriker and Tremaine, 1975). Computer simulations have shown how the coalescence of galaxies can occur in close encounters between galaxies.

This process, as applied to clusters of galaxies, may explain the effect discovered by Allan Sandage and Eduardo Hardy (b. 1941) in 1973. Laura Bautz (b. 1940) and William Morgan classified rich clusters of galaxies according to the magnitude difference between the brightest member and the next-brightest members (Bautz and Morgan, 1970). Sandage and Hardy found what they termed the Bautz–Morgan effect in which the brighter the brightest galaxy in a cluster, the fainter are the second- and third-brightest members, suggesting that the brightest galaxy has grown at the expense of the next-brightest members of the cluster (Sandage and Hardy, 1973). The process of galactic cannibalism seems to provide an explanation of the origin of the huge giant elliptical galaxies observed at the centre of many of the richest clusters of galaxies, and it has been confirmed by supercomputer simulations of the formation of clusters of galaxies (see Chapter 15). It may also account for the fact that the absolute magnitudes of the brightest galaxies in clusters seem to have remarkably constant absolute luminosity. Marc Hausman and Jeremiah Ostriker showed in 1977 that the galaxies become physically larger and more bloated as they consume galaxies and that therefore they become larger in physical size as well as becoming more luminous (Hausman and Ostriker, 1977). When photometric observations are made of these galaxies within an aperture of fixed physical size at the galaxy, the increase in luminosity due to cannibalism is offset by the fact that the galaxy has grown in size, and these two effects more or less compensate for each other.

Notes to Chapter 10

1 Much more detailed discussions of the properties of galaxies and the physics involved are contained in the books by J. Binney and M. Merrifield, *Galactic Astronomy* (Princeton: Princeton University Press, 1998), L. S. Sparke and J. S. Gallagher, *Galaxies in the Universe* (Cambridge: Cambridge University Press, 2000) and J. Binney and S. Tremaine, *Galactic Dynamics* (Princeton: Princeton University Press, 1987).

2 A review of the global parameters of galaxies as a function of stage along the Hubble sequence was presented by M. S. Roberts and M. P. Haynes, Physical parameters along the Hubble sequence, *Annual Reviews of Astronomy and Astrophysics*, **32**, 1994, 115–152.

3 The proceedings of the first symposium dedicated to the results of the IRAS mission were reported in: *Light on Dark Matter*, ed F.P. Israel (Dordrecht: D. Reidel Publishing Company, 1985).

4 See also V. C. Rubin, Field and cluster galaxies: do they differ dynamically?, in *Large-scale Motions in the Universe*, eds V. C. Rubin and G. V. Coyne (Vatican City: Pontificia Academia Scientiarum, 1988), pp. 541–558.

5 Virginia Trimble provides a comprehensive historical survey of the evidence for dark matter in galaxies up to 1987 in her review Existence and nature of dark matter in the Universe, *Annual Reviews of Astronomy and Astrophysics*, **25**, 1987, 425–472.

6 I have given details of the various approaches to determining the distribution and amount of dark matter in galaxies in Section 4.3 of Malcolm Longair, *Galaxy Formation* (Berlin: Springer-Verlag, 1998).

7 Fich and Tremaine also provide an interesting historical introduction to estimates of the mass of our Galaxy (Fich and Tremaine, 1991).

8 References to the numerous papers of Peebles and his colleagues on correlation functions for galaxies can be found in Peebles' excellent monograph: P. J. E. Peebles, *Principles of Physical Cosmology* (Princeton: Princeton University Press, 1993).

9 The diagrams in Figure 10.6 are shown in terms of the *angular* two-point correlation function on the sky, defined by

$$N(\theta)\, d\Omega = n_g[1 + w(\theta)]\, d\Omega,$$

where $w(\theta)$ describes the excess probability of finding a galaxy at an angular distance, θ, in the solid angle $d\Omega$ from any given galaxy; n_g is a suitable average surface density of galaxies. With a number of reasonable assumptions, $w(\theta)$ can be related to $\xi(\theta)$. For example, if $w(\theta)$ is described by a power law, $w(\theta) \propto \theta^{-x}$, the spatial two-point correlation function is given by $\xi(r) \propto r^{-(x+1)}$. The homogeneity test shown in the diagram follows from the fact that if the same degree of clumpiness is present throughout the region of the Universe surveyed, the angular two-point correlation function scales as

$$w(\theta, D) = \frac{D_0}{D} w_L \left(\theta \frac{D}{D_0} \right),$$

where the function $w_L(\theta)$ has been determined to the distance D_0. This is the scaling tested in Figure 10.6(b).

10 Einasto and his colleagues carried out their studies at the Tartu Observatory in Estonia at a time when relations with the West were strained and it was difficult for scientists from the Soviet Union to communicate their researches and ideas. Einasto (2001) gives some impression of the problems facing astronomers from the former Soviet Union during these years.

11 Figure 10.11(b) was created from the observations made by C. P. O'Dea and F. N. Owen, *Astrophysical Journal*, **301**, 1986, 845 (see M. S. Longair in *The New Physics* ed. P. C. W. Davies (Cambridge: Cambridge University Press 1988), p. 169).

A10 Explanatory supplement to Chapter 10

A10.1 *The mass distribution in galaxies and clusters from X-ray observations*

The X-ray emission of the hot gas in galaxies and clusters of galaxies provides a very powerful probe of their gravitational potentials (Fabricant *et al.*, 1980). It is assumed that the cluster is spherically symmetric so that the total gravitating mass within radius r is $M(<r)$. The gas is assumed to be in hydrostatic equilibrium within the gravitational potential defined by the mass distribution in the cluster, that is by the sum of the visible and dark matter, as well as the gaseous mass. If p is the pressure of the gas and ϱ is its density, both of which vary with position within the cluster, the requirement of hydrostatic equilibrium is given by

$$\frac{dp}{dr} = -\frac{GM(<r)\varrho}{r^2}. \tag{A10.1}$$

The pressure is related to the local gas density, ϱ, and temperature, T, by the perfect gas law:

$$p = \frac{\varrho kT}{\mu m_{\mathrm{H}}}, \tag{A10.2}$$

where m_{H} is the mass of the hydrogen atom and μ is the mean molecular weight of the gas. For a fully ionised gas with the standard cosmic abundance of the elements, a suitable value is $\mu = 0.6$. Differentiating equation (A10.2) with respect to r and substituting into equation (A10.1), we find

$$\frac{\varrho kT}{\mu m_{\mathrm{H}}} \left(\frac{1}{\varrho} \frac{d\varrho}{dr} + \frac{1}{T} \frac{dT}{dr} \right) = -\frac{GM(<r)\varrho}{r^2}. \tag{A10.3}$$

Reorganising equation (A10.3), we obtain

$$M(<r) = -\frac{kTr^2}{G\mu m_{\mathrm{H}}} \left[\frac{d(\log \varrho)}{dr} + \frac{d(\log T)}{dr} \right]. \tag{A10.4}$$

Thus, the mass distribution within the cluster can be determined if the variation of the gas density and temperature with radius are known. Assuming the cluster is spherically symmetric, these can be derived from high-sensitivity X-ray intensity and spectral observations. A suitable form for the bremsstrahlung spectral emissivity of a plasma is given by

$$\kappa_\nu = \frac{1}{3\pi^2} \frac{Z^2 e^6}{\varepsilon_0^3 c^3 m_{\mathrm{e}}^2} \left(\frac{m_{\mathrm{e}}}{kT} \right)^{1/2} g(\nu, T) N N_e \exp\left(-\frac{h\nu}{kT} \right), \tag{A10.5}$$

where N_e and N are the number densities of electrons and nuclei, respectively, Z is the charge of the nuclei and $g(\nu, T)$ is the Gaunt factor, which can be approximated by

$$g(\nu, T) = \frac{\sqrt{3}}{\pi} \ln\left(\frac{kT}{h\nu} \right). \tag{A10.6}$$

The spectrum of thermal bremsstrahlung is roughly flat up to X-ray energies $\varepsilon = h\nu \sim kT$, above which it cuts off exponentially.[1] Thus, by making precise spectral measurements, it is possible to determine the temperature of the gas from the location of the spectral cut-off and the particle density along the line of sight from the emissivity of the gas. In practice, the spectral emissivity has to be integrated along the line of sight through the cluster. Performing this integration and converting it into an intensity, the observed surface brightness at projected radius a from the cluster centre is given by

$$I_\nu(a) = \frac{1}{2\pi} \int_a^\infty \frac{\kappa_\nu(r)r}{(r^2 - a^2)^{1/2}}\, dr. \tag{A10.7}$$

Alfonso Cavaliere[2] (b. 1933) noted that this is an Abel integral which can be inverted to find the emissivity of the gas as a function of radius as follows:

$$\kappa_\nu(r) = \frac{4}{r} \frac{d}{dr} \int_r^\infty \frac{I_\nu(a)a}{(a^2 - r^2)^{1/2}}\, da. \tag{A10.8}$$

Notes to Section A10

1 I have given a derivation of this expression for bremsstrahlung in Malcolm Longair, *High Energy Astrophysics*, vol. 1 (Cambridge: Cambridge University Press, 1992).
2 Cavaliere's paper, Models of X-ray emission from clusters of galaxies, was published in *X-ray Astronomy*, eds R. Giacconi and G. Setti (Dordrecht: D. Reidel Publishing Company, 1980), pp. 217–237.

11 High-energy astrophysics

11.1 Radio astronomy and high-energy astrophysics

The early history of radio astronomy was recounted in Section 7.3; that story ended in the mid 1950s, by which time the Galactic and extragalactic nature of the discrete radio sources was established. From the point of view of astrophysics, the key realisation was that, in most cases, the radio emission was the synchrotron radiation of ultra-high-energy electrons gyrating in magnetic fields within the source regions. The synchrotron radiation process began to be applied to other astronomical objects in which there was evidence for high-energy astrophysical activity.

In 1942, Rudolph Minkowski showed that the emission of the supernova remnant known as the Crab Nebula consists of two components, the filaments, which form a network defining the outer boundary of the remnant, and diffuse continuum emission originating within the nebula, which contributes most of its optical luminosity (Minkowski, 1942). The continuum emission had a featureless spectrum and could not be accounted for by any form of thermal spectrum. In 1949, John Bolton and Gordon Stanley found that the flux density of the Crab Nebula at radio wavelengths was about 1000 times greater than in the optical waveband (Bolton and Stanley, 1949). To account for the continuum emission, Iosif Shklovsky (1916–1985) proposed in 1952 that both the radio and optical continuum was synchrotron radiation, the energies of the electrons radiating in the optical waveband being very much greater than those radiating in the radio waveband (Shklovsky, 1953). One consequence of this hypothesis was that the optical continuum of the nebula should be linearly polarised, and this was discovered by Viktor Dombrovskii (1914–1972) and Mikhail Vashakidze (1909–1956) in 1954 (Dombrovski, 1954; Vashakidze, 1954). This observation was confirmed by Jan Oort and Théodore Walraven (b.1916) in 1956, using superb plates taken with the 200-inch Palomar telescope by Walter Baade in 1955 (Oort and Walraven, 1956). The famous optical jet in the nearby giant elliptical galaxy M87 had been discovered by Heber Curtis in 1918 (Curtis, 1918a), and this strange feature was shown to be linearly polarised by Baade in 1956 (Baade, 1956). The emission of the jet was interpreted as another example of optical synchrotron radiation.

Observations of the diffuse Galactic radio emission were important because they provided a direct quantitative test of the synchrotron hypothesis. The early observations of Jansky and Reber indicated that the radiation was not the thermal emission of hot gas, but the precise determination of its radio spectrum was technically difficult since the emission is diffuse and the observed intensity is sensitive to the exact beam pattern of the radio telescope. The spectrum of the Galactic radio emission was eventually determined using geometrically scaled aerials and was shown to be of power-law form with spectral index $\alpha \approx 0.4$ at low radio frequencies, α being defined by $I(\nu) \propto \nu^{-\alpha}$ (Turtle *et al.*, 1962; Turtle, 1963). According to synchrotron radiation theory, this spectral index is related to the differential energy spectral index of the radiating electrons x by $x = 2\alpha + 1$, where the electron spectrum is described by $N(E)\,dE \propto E^{-x}\,dE$. Thus, the spectral index of the radiating electrons would have to be about 1.8, not too different from that of cosmic-ray protons and nuclei. In fact, the radio spectra of supernovae and extragalactic radio sources were found to be somewhat steeper than this, $\alpha \approx 0.75$, corresponding to $x = 2.5$, precisely the spectral index of the cosmic rays.

A key test of the synchrotron hypothesis was the search for polarised radio emission from the diffuse interstellar medium, and this was observed by Vladimir Razin (b. 1930) in 1958 and by Gart Westerhout and his colleagues at Dwingloo in the Netherlands in 1962 (Razin, 1958; Westerhout *et al.*, 1962). It was to be many years before the spectrum of the cosmic-ray electrons would be measured reliably, but the radio astronomical observations were convincing evidence that very-high-energy electrons and magnetic fields are present throughout the plane of our Galaxy (see Section 9.6). It was inferred that the cosmic rays observed at the top of the atmosphere are a sample of a general Galactic distribution of high-energy particles. Estimates of the strength of the interstellar magnetic field could be obtained from observations of the optical polarisation of starlight and from the Faraday rotation of the polarised emission of discrete radio sources. The inferred energy density of the electrons turned out to be roughly one-hundredth of that in the cosmic-ray protons.

The associations of the radio sources Cassiopeia A and Cygnus A with optical objects by Baade and Minkowski in 1954 were crucial (Baade and Minkowski, 1954). Cassiopeia A was identified with a young supernova remnant which must have exploded about 250 years ago, as deduced from the measured expansion velocities of its optical filaments, though no historical record of this event has been found. Assuming the radio emission of the remnant is synchrotron radiation, the great radio luminosity of Cassiopeia A provided direct evidence for the acceleration of huge fluxes of very-high-energy electrons in supernova remnants, an idea foreshadowed by Baade and Zwicky's remarkable paper of 1934 (Baade and Zwicky, 1934a).

Even more remarkably, the radio source Cygnus A was identified with a galaxy at a redshift of 0.057, implying that its radio luminosity was enormous, more than a million times greater than that of our Galaxy. It must therefore be the source of vast quantities of relativistic material. Just as unexpected was the fact that the radio emission did not originate from the galaxy itself. In 1953, Roger Jennison (b. 1922) and Mrinal Kumar Das Gupta (b. 1923) at Jodrell Bank used interferometric techniques to show that the radio emission originated from two huge lobes (Jennison and Das Gupta, 1953), and it turned out that these were located on either side of the radio galaxy, once the identification was made in the following year (Figure 11.1). Thus, not only must the radio galaxy accelerate an enormous amount of material to relativistic energies, but this material also has to be ejected into intergalactic space in opposite directions.

The intensity of synchrotron radiation depends upon both the magnetic field strength and the flux of high-energy electrons in the source region.[1] In 1959, Geoffrey Burbidge worked out the minimum amount of energy in high-energy particles and magnetic fields

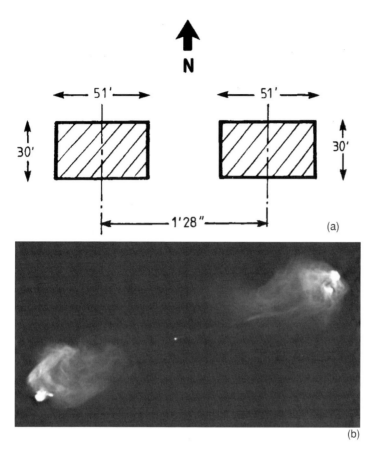

Figure 11.1: (a) A reconstruction of the radio structure of the radio source Cygnus A from radio interferometric observations by Jennison and Das Gupta at a frequency of 125 MHz (Jennison and Das Gupta, 1953). (b) A map of Cygnus A made by the Very Large Array in the USA at a frequency of 5 GHz (Perley, Dreher and Cowan, 1984).

which had to be present in the source regions to account for the radio emission (Burbidge, 1959). The energies proved to be enormous, in some sources corresponding to a rest-mass energy of about $10^6 M_\odot$. In many ways, this result marked the beginning of high-energy astrophysics in its modern guise. Some astrophysical means had to be found for converting a significant fraction of the rest-mass energy of a galaxy into high-energy particles and magnetic fields. The problem was not, however, confined to galaxies – supernovae remnants such as Cassiopeia A were also able to convert a significant fraction of their mass into high-energy particles and magnetic fields.

11.2 The discovery of quasars and their close relatives

The radio astronomical discoveries of the 1950s stimulated a great deal of astrophysical interest and led to major investments being made in the construction of radio telescope

Figure 11.2: Hubble Space Telescope image of the quasar 3C 273, showing the optical jet ejected from the quasar nucleus. The faint smudges to the south of the quasar are now known to be galaxies at the same distance as the quasar. (Courtesy of Dr John Bahcall, NASA and the Space Telescope Science Institute.)

systems. The radio observatories began systematic surveys of the sky in order to understand in more detail both the astrophysics of these intense sources of radio waves and their use as cosmological probes. The discrete sources of intense radio emission were often associated with faint galaxies, and so accurate radio positions were needed to find the associated galaxy among the large numbers of unrelated stars and galaxies. Among the extragalactic radio sources which could be securely identified, most were found to be associated with some of the most massive galaxies known, which are very luminous and so could be observed to large redshifts. By 1960, the largest redshift known for any galaxy was that of the radio galaxy[2] 3C 295 at $z = 0.46$ (Minkowski, 1960b).

By 1962, Thomas Matthews (b. 1927) and Sandage had identified three of the brightest radio sources, 3C 48, 3C 196 and 3C 286, with 'stars' of an unknown type with strange optical spectra (Matthews and Sandage, 1963). The breakthrough came in 1962 when Cyril Hazard (b. 1928) measured very precisely the position of the radio source 3C 273 by the method of lunar occultations using the recently completed Parkes 210-foot (64-metre) radio telescope in New South Wales, Australia (Hazard, Mackey and Shimmins, 1963). These remarkable observations enabled 3C 273 to be identified with what appeared to be a 13th magnitude star (Figure 11.2). The new identification was rapidly relayed to Maarten Schmidt (b. 1929), who used the Palomar 200-inch telescope to obtain its optical spectrum. The spectrum contained prominent emission lines, but at unexpected wavelengths. The clue lay in the familiar pattern of lines, which Schmidt realised was the Balmer series of hydrogen, but shifted to longer wavelengths with redshift $z = \Delta\lambda/\lambda_0 = 0.158$ – this was certainly no ordinary star (Schmidt, 1963); 3C 273 was the first, and brightest, of this class of hyperactive galactic nuclei to be discovered. Its optical luminosity is about 1000 times greater than the luminosity of a galaxy such as our own. To make matters even more intriguing, searches through the Harvard plate archives revealed that the enormous luminosity of 3C 273 varied on a timescale of years (Smith and Hoffleit, 1963). Nothing like this had been observed in astronomy before. This discovery opened the floodgates for the identification of many more examples of radio quasars. The radio source 3C 48 was found to have a redshift of 0.3675 (Greenstein and Matthews, 1963), while 3C 47 and 3C 147 have redshifts $z = 0.425$ and $z = 0.545$, respectively (Schmidt and Matthews, 1964). These sources were termed *quasi-stellar radio sources*, and within a year this term had been contracted to the word *quasar*.

One of the first discussions of the implications of these remarkable discoveries was held at the First Texas Symposium on Relativistic Astrophysics held in Dallas in 1963.[3] For the first time, the optical and radio astronomers got together with the theoretical astrophysicists and, in particular, the general relativists to thrash out what was known about these objects and the role which general relativity might play in these studies. Coincidentally, William Fowler and Fred Hoyle had been investigating the properties of supermassive stars with masses as great as $10^6 M_\odot$, but, although these might radiate enormous luminosities, they proved to be notoriously unstable objects (Hoyle and Fowler, 1963a,b). Perhaps the most important conclusion of the meeting was the realisation that the quasars must involve strong gravitational fields and so general relativity must play a central role in understanding their properties. It was to be a number of years before general relativity became one of the standard tools of the high-energy astrophysicist because many technical questions still had to be addressed, but the writing was already on the wall. At the closing dinner of the 1963 Texas Symposium, Thomas Gold made the following remark:[4]

Everyone is pleased: the relativists who feel they are being appreciated, who are suddenly experts in a field which they hardly knew existed; the astrophysicists for having enlarged their domain, their empire by the annexation of another subject – general relativity.

The remarkable upsurge of interest in general relativity was undoubtedly stimulated by the discovery of radio galaxies and quasars.

For some time, there was concern about the issue of whether the redshifts of the quasars really were of cosmological origin or whether they might be due to some other physical process. The short timescale of variability of their enormous luminosities seemed so extreme that some astronomers, the most prominent of whom were Fred Hoyle, Geoffrey Burbidge and Halton Arp, suggested that the quasars were actually relatively nearby objects and not at the large distances implied by their redshifts. If this were the case, their luminosities would be far less extreme and their variability would be much less of a problem because they would be local objects, and perhaps some form of new variable star. Although widely discussed, this hypothesis never attracted wide support, largely because no other satisfactory explanation for the origin of the redshifts of the quasars was forthcoming.[5] By 1965, the redshifts of the radio quasars had reached $z = 2.012$ for the quasar 3C 9 (Schmidt, 1965).

Gravitational redshifts could not account for the observations, as discussed by Greenstein and Schmidt in 1964 (Greenstein and Schmidt, 1964), and Doppler redshifts were ruled out because of the absence of quasars with large blueshifts. The extreme luminosities and short timescales of variability of quasars pointed inevitably to the presence of very compact luminous sources in the nuclei of galaxies.

In many ways, the quasars were discovered too early. It eventually became apparent that they are among the most extreme examples of what are now termed generically *active galactic nuclei*. Quite by chance, the radio astronomers had stumbled upon a very effective means of discovering the most luminous of them. In fact, the first examples of active galactic nuclei had been discovered in the early 1940s by Carl Seyfert (1911–1960), who studied a number of spiral galaxies, such as NGC 1068 and NGC 4151, which possessed star-like nuclei (Seyfert, 1943). On taking the spectra of these nuclei, he found that they possessed very intense emission lines of the Balmer series and forbidden lines such as [OII], [OIII], [NII], [NeIII], [SII] and [SIII]. Furthermore, the lines were very broad, the Doppler velocities required to broaden the lines being up to $8500 \, \mathrm{km \, s^{-1}}$. The presence and breadth of these lines were quite unlike those observed in regions of ionised hydrogen in galaxies. Furthermore, the spectrum of the continuum radiation was smooth, quite unlike the spectrum of starlight. Seyfert's pioneering work was largely neglected until the 1960s.

Among the galaxies associated with the radio sources was a class of galaxy which William Morgan referred to as N-galaxies, by which he meant elliptical galaxies with star-like nuclei (Matthews, Morgan and Schmidt, 1964). In the N-galaxies, the host galaxy is often barely distinguishable because of the intensity of the star-like nucleus, the optical spectrum of which displays intense broad emission lines, similar to those observed in the Seyfert galaxies.

In 1963, just before the large redshifts of the quasars were discovered, Geoffrey Burbidge, Margaret Burbidge and Allan Sandage surveyed a wide range of evidence concerning activity in the nuclei of galaxies in an influential paper entitled 'Evidence for the occurrence of violent events in the nuclei of galaxies' (Burbidge, Burbidge and Sandage, 1963). Their survey included a very wide range of evidence, including the Seyfert galaxies, the radio galaxies and other galaxies in which explosions seemed to have occurred, such as the irregular galaxy M82 and the giant elliptical galaxy M87, which possesses the prominent optical jet described by Heber Curtis (Curtis, 1918a).

In 1965, radio-quiet counterparts of the radio quasars were discovered by Allan Sandage, and these *radio-quiet quasars* turned out to be about 100 times more common than the radio-loud variety (Sandage, 1965). The similarity between the properties of quasars and the nuclei of Seyfert galaxies was reinforced by the discovery in 1967 that both the continuum emission and the strong emission lines in Seyfert galaxies are variable (Fitch, Pacholczyk and Weymann, 1967). It gradually became apparent that there is a continuous sequence of high-energy activity in the nuclei of galaxies, from weak nuclei, such as the centre of our own Galaxy, to the most extreme examples of quasars in which the starlight of the galaxy is completely overwhelmed by the intense non-thermal radiation from the nucleus.

Among the most extreme examples of active galactic nuclei were the *BL Lacertae* or *BL Lac objects*, which were discovered in 1968 as extremely compact and highly variable radio sources by John McLeod (b. 1937) and Brian Andrew (b. 1939) (McLeod and Andrew,

1968). Their optical spectra are smooth and featureless and display very rapid variability at optical and radio wavelengths from timescales of less than a day to weeks. Indeed, BL Lac itself was so named because it appeared in lists of variable stars. Eventually, when spectrographs with very large dynamic ranges became available, redshifts for the BL Lac objects were measured by detecting faint features in the spectra of the very faint underlying galaxies. These redshifts showed that the BL Lac objects are indeed extragalactic objects, but that their luminosities are not as great as those of the most luminous quasars.

Once the characteristic features of active galactic nuclei were established, many surveys were undertaken to find more examples of them. The surveys for galaxies with ultraviolet excess carried out by Beniamin Markarian (1913–1985) and his colleagues at the Byurakan Observatory in Armenia proved to be a rich source of Seyfert galaxies, Markarian's pioneering surveys beginning in 1967 and culminating in 1981 with paper 15 of the series (Markarian, 1967; Markarian, Lipovetsky and Stepanian, 1981). About 10% of what became known as *Markarian galaxies* are Seyfert galaxies. Special surveys were also undertaken to discover quasars, both the radio-loud and the radio-quiet varieties.

11.3   General relativity and models of active galactic nuclei

In parallel with these great discoveries of observational astronomy, enormous progress was made in understanding the role of general relativity, not only in cosmology, but also in the physics of matter in strong gravitational fields.[6] The exact solution of the Einstein's field equations for a point mass in general relativity was discovered by Karl Schwarzschild[7] in 1916, the year after the final version of Einstein's general theory was published (Schwarzschild, 1916). As Einstein commented on receiving Schwarzschild's paper,

I had not expected that one could formulate the exact solution of the problem in such a simple way.

Schwarzschild had derived the solution in order to provide exact results for tests of general relativity, all of which involve taking the weak field limit at large distances from a point object of mass M. The *Schwarzschild metric* can be written as follows:

$$ds^2 = \left(1 - \frac{2GM}{rc^2}\right) dt^2 - \frac{1}{c^2}\left[\frac{dr^2}{\left(1 - \frac{2GM}{rc^2}\right)} + r^2(d\theta^2 + \sin^2\theta \, d\phi^2)\right], \quad (11.1)$$

where θ and ϕ are polar coordinates and r is a coordinate distance. Schwarzschild did not remark upon the fact that this metric contains two singularities, one at $r = 0$ and the other at coordinate distance $r = 2GM/c^2$. In fact, the second singularity is not a real physical singularity, but is associated with the particular choice of coordinate system in which the Schwarzschild metric was written, as was first demonstrated by Martin Kruskal (b. 1925) in the mid 1950s, but only published in 1960 (Kruskal, 1960). The singularity at $r = 0$ is, however, a real physical singularity in space-time.

If the discovery of quasars in 1963 was symbolic of a turning point in modern astrophysics, the same could be said of general relativity, with the discovery by Roy Kerr (b. 1934) in the same year of what is now known as the *Kerr metric*, one of the most important exact solutions of Einstein's field equations (Kerr, 1963). It turned out that the Kerr solution describes the metric of space-time about a rotating black hole and is a generalisation of the Schwarzschild metric.[8] In 1965, a further generalisation of the Kerr metric was discovered by Ezra (Ted) Newman (b. 1929) and his colleagues for the case of a system with finite electric charge by solving the combined Einstein and Maxwell field equations (Newman *et al.*, 1965). Only later was it realised that the metric describes a rotating black hole with finite electric charge. Whereas the Schwarzschild solution is determined entirely by the mass of the black hole, the Kerr metric depends upon both its mass and its angular momentum. It took some years and a great deal of analysis before it was realised just how powerful these solutions are. In 1971, Brandon Carter (b. 1942) showed that the only possible solutions for uncharged axisymmetric black holes were the Kerr solutions (Carter, 1971), and in 1972 Stephen Hawking (b. 1942) showed that all stationary black holes must be either static or axisymmetric, so that the Kerr solutions indeed included all possible forms of black hole (Hawking, 1972).

These theorems led to important conclusions about the fate of collapsing bodies in general relativity. No matter how complex the object and its properties are before collapse to a black hole, all other properties, except its mass, angular momentum and electric charge, are radiated away during the collapse. Another way of expressing these results is that all multipole moments are radiated away in the process of formation of the black hole, leaving it only with mass, angular momentum and electric charge. This result is often referred to as the *no-hair theorem* for black holes.[9] Thus, as the end points of stellar evolution of massive remnants, black holes are quite remarkably simple objects. Notice that these theorems apply to isolated black holes. Although the black hole cannot possess a magnetic dipole moment, magnetic fields can penetrate into the black hole provided they are firmly attached to the external medium. As shown by Kip Thorne and his colleagues, electrodynamics in the vicinity of black holes can be precisely described by considering the surface of infinite redshift, or *event horizon*, to consist of a membrane with the resistivity of free space, $Z = (\mu_0/\epsilon_0)^{1/2}$ (Thorne, Price and Macdonald, 1986).

One of the key questions addressed by the relativists was whether or not there must be a physical singularity as a result of gravitational collapse. It had been argued by some relativists that the singularity in the Schwarzschild solution was a special case and that, in general, the presence of a singularity might depend upon the initial conditions from which the collapse proceeded. The problem was solved by Roger Penrose (b. 1931) in 1965, who showed quite generally that, once a surface from which light cannot escape outwards has formed, what is known as a *closed trapped surface*, there is inevitably a singularity inside that surface (Penrose, 1965). The nature of the singularity is described by the Kerr metric. These same techniques were applied to the Universe as a whole by Penrose and Hawking, and they were able to show in 1969 that, according to classical general relativity, it is also inevitable that there exists a singularity at the origin of the Hot Big Bang models of the Universe, subject to some rather general physical conditions (Hawking and Penrose, 1969). They later extended their results to a much larger class of theories of gravity, the results of these endeavours being summarised in the book *The Large Scale Structure of Space-Time* by Stephen Hawking and George Ellis (b. 1939) (Hawking and Ellis, 1973). There are numerous versions of the singularity theorems, an example of the Hawking and Penrose theorem stating that:

A space-time which

(i) contains no time-like curves;
(ii) satisfies Einstein's equations (without the cosmological term) and the energy condition ($\varrho + p_i$; $\varrho + \sum p_i \geq 0$);
(iii) is sufficiently general; and
(iv) contains a closed spacelike hypersurface,

cannot be geodesically complete in all timelike and null directions.

In less technical language, applying the singularity theorems to our Universe, particularly to the cosmic microwave background radiation, and subject to some rather general geometrical conditions, there is inevitably a singularity at the origin of the world model, provided gravity is attractive and the equation of state of the matter satisfies the energy condition $\varrho + p_i \geq 0$.

From the point of view of astrophysics, the most important results of the study of black holes, as they were named by John Wheeler in 1968 (Wheeler, 1968), concerned the behaviour of matter in the vicinity of the event horizon and the maximum amount of gravitational binding energy which could be released when matter falls into the black hole from infinity. The Schwarzschild radius, $r_g = 2GM/c^2$, of a spherically symmetric black hole is the surface of infinite redshift, meaning that radiation emitted from this surface is observed with infinite wavelength at infinity. This is the general relativistic version of the insight of John Michell that, for a massive enough star, the escape velocity from the object exceeds the speed of light (Michell, 1767). The interpretation in general relativity is, however, quite different because the geometry changes on crossing the Schwarzschild radius, as can be appreciated by inspection of the metric (11.1). There is also a last stable circular orbit about the black hole at radius $3r_g$ – within this radius there are no stable circular orbits and the matter spirals inevitably through the surface of infinite redshift, adding to the mass of the black hole. The maximum gravitational binding energy which can be released by an element of mass falling into the black hole from infinity is 5.72% of its rest-mass energy.[10]

Corresponding calculations were carried out for Kerr black holes. There is a maximum amount of angular momentum which the hole can possess[11] or else it cannot form, $J_{max} = GM^2/c$. For a maximally rotating black hole, the surface of infinite redshift shrinks to $r_g = GM/c^2$ and, in the case of corotating orbits, up to 42% of the rest-mass energy of the infalling matter can be released. The rotational energy of the black hole can also be tapped, as demonstrated by Roger Penrose in 1969 (Penrose, 1969). The part of the rest-mass energy of the black hole associated with its rotation is, in principle, accessible to external observers and can amount to a maximum of 29% of the rest-mass energy of the black hole. These are very important results astrophysically and showed that the accretion of matter onto black holes is potentially an extremely powerful source of energy – the synthesis of ^{4}He from four protons, for example, can only release 0.7% of the rest-mass energy of the matter.[12]

Immediately following the discovery of quasars, a plethora of models appeared in the literature, all of them attempting to account for their huge variations in luminosity on short timescales. In 1964, Yakov Zeldovich in Moscow and Edwin Salpeter at Cornell independently pointed out that *accretion of matter onto black holes* is potentially a very powerful energy source (Salpeter, 1964; Zeldovich, 1964a) (see Section A11.1.2). An effective way of releasing the binding energy of the infalling material is through the formation of an *accretion disc* about the black hole. These discs form naturally because matter is most unlikely to fall directly into the black hole from infinity since it must acquire some small amount of angular momentum by random gravitational perturbations, which is then greatly amplified by conservation of angular momentum as the material collapses towards the black hole. Therefore, the infalling matter collapses along its rotation axis and forms an accretion disc in centrifugal equilibrium about the black hole. If the disc were frictionless, there would be no energy release. The presence of viscous forces in the disc performs two important functions. The first is that the viscous forces enable angular momentum to be transferred outwards and so the matter of the accretion disc can gradually drift inwards. The second inevitable consequence of this process is that the frictional forces heat up the material disc, and this is the means by which the matter releases its gravitational binding energy.[13]

Surprisingly, it was some time before this model was analysed in detail. In 1969, the first serious analysis of thin accretion discs about black holes was carried out by Donald Lynden-Bell, who showed how, in principle, they could account for the most extreme active galactic nuclei known at that time (Lynden-Bell, 1969). Lynden-Bell had assumed that the black hole was of the static spherically symmetric variety, but James Bardeen (b. 1939) pointed out in 1970 that the black hole is likely to possess angular momentum as the infalling matter brings angular momentum with it (Bardeen, 1970). The energy release could be correspondingly greater, up to 42% of the rest-mass energy of the infalling matter for a maximal corotating black hole, in which the accretion disc could extend in to $r_g = GM/c^2$. The luminosities of the binary X-ray sources, discovered in 1972, could be naturally accounted for by accretion onto neutron stars, and, in a number of cases, these approached the Eddington limiting luminosity, indicating just how effective this energy source could be (Margon and Ostriker, 1973). Detailed models of accretion discs about black holes were published by Nikolai Shakura (b. 1945) and Rashid Sunyaev in 1973 (Shakura and Sunyaev, 1973), and they showed that, although the nature of the viscosity, which is responsible for the outward transport of angular momentum and energy release, was not well understood, many of the properties of thin accretion discs are independent of the specific form of the viscosity. Since that time, accretion models for active galactic nuclei fuelled by accretion discs have been adopted by many authors as their preferred model for active galactic nuclei. Shakura and Sunyaev had pointed out that the viscous stresses might be associated with magnetic stresses within the accretion, and in 1991 Steven Balbus (b. 1953) and John Hawley (b. 1958) revived the instability first discussed by Evgenii Velikhov (b. 1935) and Chandrasekhar and showed that magnetic instabilities in the differentially rotating disc could provide the necessary transfer of angular momentum to provide a physical realisation for the viscous parameter α introduced by Shakura and Sunyaev (Velikhov, 1959; Chandrasekhar, 1981; Balbus and Hawley, 1991).

The thin accretion discs were successfully applied to cataclysmic variable stars and relatively modest active galactic nuclei, but it proved very much more difficult to find self-consistent solutions for the most luminous sources. As the luminosity of the disc increases, most of the thin-disc approximations break down. In particular, the central regions of the disc become very hot and radiation pressure can no longer be neglected. The innermost regions of the disc are inflated and are expected to take up a toroidal configuration rather than a thin disc. An attempt to construct a self-consistent thick-disc model applicable to active galactic nuclei was proposed by Marek Abramowicz (b. 1945) and his colleagues in 1978, in which the torus was inflated to such a degree that there were funnels along the axis of the torus which were thought to be relevant to the formation of the jets observed to be ejected from many active nuclei (Abramowicz, Jaroszyński and Sikora, 1978). Many of these ideas were synthesised in the review of 1982 by Martin Rees and his colleagues, who pointed out the many complications which arose in the attempt to construct self-consistent models of thick discs (Rees *et al.*, 1982). The problem with thick tori, just as in the case of the supermassive stars studied by Hoyle and Fowler in 1964, is that they are grossly unstable. In 1984, John Papaloizou (b. 1947) and James Pringle (b. 1949) showed that the most unstable modes of thick discs are non-axisymmetric and global in character (Papaloizou and Pringle, 1984).

11.4   The spectroscopy of active galactic nuclei

Quasars and the nuclei Seyfert galaxies were ideal objects for spectroscopic study because their spectra are rich in strong emission lines. The Seyfert galaxies were of particular importance because their nuclei are generally much brighter than those of quasars and excellent spectra could be acquired. The objective prism surveys for galaxies with strong blue continua carried out by Markarian and his colleagues at the Byurakan Astrophysical Observatory in Armenia were of special importance because, although about 90% of the objects turned out to be what would now be called star-burst galaxies, the remaining 10% were Seyfert galaxies. In studying these Seyfert galaxies, Eduard Khachikian (b. 1928) and Daniel Weedman (b. 1942) discovered in 1974 that there are basically two types of emission-line spectra (Khachikian and Weedman, 1971, 1974). Those with relatively narrow emission lines, with line widths corresponding to velocities of about 500 km s^{-1}, were classed as Type 2 Seyfert galaxies. In contrast, the Type 1 Seyfert galaxies were found to have very broad permitted emission lines, the line-widths corresponding to about 5000 km s^{-1}. Narrow forbidden lines were also observed in the Type 1 Seyferts, but no forbidden counterparts of the broad permitted lines were detected. These observations suggested that the broad permitted emission lines originate close to the source of excitation in high-density gas clouds in which the forbidden lines are suppressed; the narrow-line regions are of lower density and are located much further away from the nucleus.

The optical continua of quasars were found to be of roughly power-law form and polarised, and so, by the mid 1960s, it was commonly accepted that the smooth continuum radiation observed in the spectra of active galactic nuclei is synchrotron radiation. The intense power-law continuum spectrum suggested that the process of excitation of the emission-line regions was the photoionisation of cool clouds in the vicinity of the nucleus. The power-law nature of the continuum spectrum had the advantage that a wide range of different states of ionisation would be expected to be present in the clouds, in agreement with the observations.[14] This working hypothesis was used throughout the 1970s and was

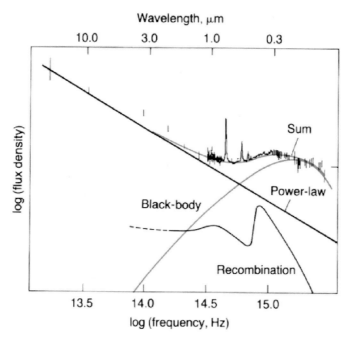

Figure 11.3: The optical–ultraviolet spectrum of the quasar 3C 273. The continuum spectrum has been decomposed into a 'power-law' component, a component associated with recombination radiation, and a 'blue-bump' component, which has been represented by a black-body curve. The prominent Balmer series in the optical waveband which led to the discovery of the large redshifts of quasars can be seen (Malkan and Sargent, 1982).

confirmed by observations of quasars and active galaxies by the International Ultraviolet Observatory (IUE) in the late 1970s and early 1980s. A particularly striking observation made by the IUE in 1982 was of the line and continuum spectrum of 3C 273 which shows a 'blue-bump' continuum component extending into the ultraviolet region of the spectrum (Figure 11.3) (Malkan and Sargent, 1982).

Convincing evidence for this picture was provided by the correlated variability of the continuum ultraviolet radiation and the strength of the broad-line spectrum which was found in a prolonged observing campaign with the IUE (Figure 11.4). More remarkable still was the observation that there is a time delay between an outburst occurring in the nucleus and the surrounding clouds responding to this burst of ionising radiation. In the Seyfert galaxy NGC 4151, the delay is about 10 days, and so the size of the broad emission line regions must be about a few light-days (Ulrich *et al.*, 1984). Combining this dimension with the velocities of the clouds inferred from their Doppler widths, the mass within the central regions was estimated to be about $10^9 M_\odot$.

The natural extension of this technique is to study the time variations of a variety of different emission lines in a single object since these are expected to originate at different distances from the source of ionising radiation.[15] The technique depends upon the fact that, when clouds are excited by incident ultraviolet radiation, they respond essentially instantaneously, and also upon the inference that the line-emitting clouds occupy only a

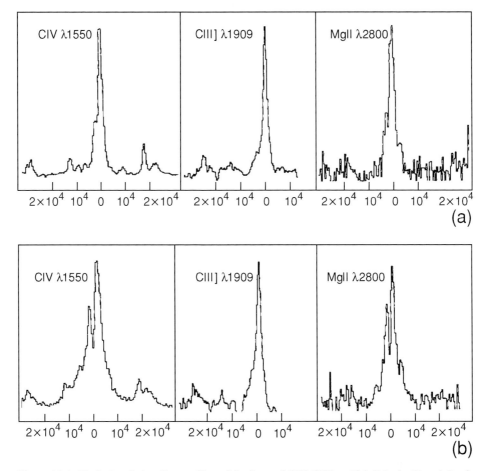

Figure 11.4: Typical emission line profiles of the lines of CIV, CIII] and MgII in the Type 1 Seyfert galaxy NGC 4151. The scales on the abscissa are velocities (km s^{-1}). (a) The spectra in the vicinity of these lines taken on 21 April 1980 when the source was in its low state. The continuum emission from the nucleus is weak and the narrow-line spectra observed. (b) The spectra observed on 19 October 1978. The nucleus was bright, the continuum emission strong and broad wings are observed in the CIV line, as in a Type 1 Seyfert galaxy (Ulrich *et al.*, 1984).

very small fraction of the volume of the nucleus, that is they have a small *filling factor*, so that the ultraviolet ionising photons propagate unhindered from the nucleus to the clouds. Consequently, the time delay between variations in the continuum intensity and those in the line-emitting clouds depend only upon the light-travel time from the nucleus and the geometry of the clouds.

Obtaining high-quality data sets suitable for this type of analysis is a challenge because of the need to coordinate observations from space- and ground-based observatories for observing campaigns extending over periods of months or years. Examples of the application of cross-correlation procedures to a beautiful set of optical and ultraviolet observations of the Seyfert 1 galaxy NGC 5548 are shown in Figure 11.5 (Clavel *et al.*, 1991; Peterson *et al.*, 1991). It is apparent from these results that the average time delays are different for atoms

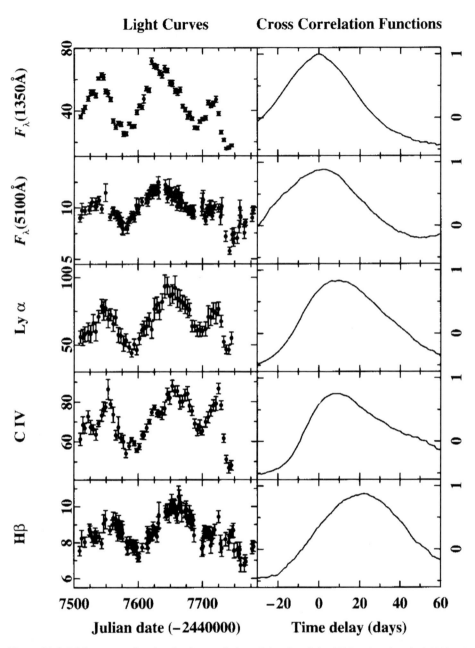

Figure 11.5: Light-curves showing the time variation of the ultraviolet (135 nm) and optical (510 nm) continuum intensity, as well as the fluxes of three prominent emission lines, Lyman-α (121.6 nm), CIV (154.9 nm) and Hβ (510 nm) in the Seyfert 1 galaxy NGC 5538. The ultraviolet observations were made with the IUE observatory and the optical observations are from ground-based telescopes (Clavel *et al.*, 1991; Peterson *et al.*, 1991). In the right-hand column, the cross-correlation functions of the time variation of the ultraviolet continuum intensity with the optical continuum and the different emission lines are shown, the first panel being the autocorrelation function of the ultraviolet continuum. The emission lines show the same pattern of variation as the ultraviolet and optical continuum, but with a time delay due to the light-travel time from the nucleus to the broad-line regions.

and ions in different states of ionisation. As expressed by Bradley Peterson (b. 1951), lines which are prominent in highly ionised gases, HeII, NV and CIV, have shorter lag times than those prominent at lower ionisation levels, for example the Balmer lines. There must therefore be stratification in the ionisation structure of the clouds about the active nucleus. For example, the CIV lines originate from a quite different region from those of CIII].

11.5   The masses of black holes in active galactic nuclei

Important aspects of the study of active galactic nuclei were the search for unambigious evidence for the presence of supermassive black holes and the accurate determination of their masses. As early as 1964, Zeldovich and Igor Novikov (b. 1935) pointed out that the masses of quasars had to be very large because their luminosities could not exceed the Eddington limiting luminosity, $L = 1.3 \times 10^{31}(M/M_\odot)$ W (Zeldovich and Novikov, 1964). If this value were exceeded, radiation pressure of the source itself would blow it apart (see Section A11.1.3). Since the luminosity of 3C 273 is at least 10^{40} W, it follows that the source of energy must have mass greater than $10^9 M_\odot$.

The general arguments in favour of the presence of supermassive black holes in active galactic nuclei were persuasive, but finding unambiguous evidence proved to be far from trivial.[16] There are two separate parts to the argument. First of all, reliable estimates are needed of the mass contained within the nuclear regions of galaxies. Useful quantities to be estimated are the *mass-to-luminosity ratio*, M/L, of the central regions, which can be compared with the values typically found in galaxies, and the mass density of the nuclear regions. Large values of M/L imply the presence of a dark massive object, while the mass densities can be compared with those of dense star clusters. The second step is to find convincing evidence that the dark massive object is a black hole, rather than, say, a dense cluster of stars.

In 1978, pioneering studies of the surface brightness distribution and velocity dispersion of stars in the nuclear regions of the nearby active galaxy M87 were carried out by Peter Young (1954–1981), Wallace Sargent (b. 1935) and their colleagues, who found that the mass-to-luminosity ratio increased towards the nucleus (Sargent et al., 1978; Young et al., 1978). Assuming the velocity dispersion, or more precisely the velocity ellipsoid, is isotropic, the presence of a dark central object with mass $3 \times 10^9 M_\odot$ was inferred. This proved to be a controversial result, however, because, in the case of M87, the velocity field of the stars showed negligible rotation and so the mass estimate depended strongly upon the assumption that the velocity distribution of the stars was isotropic throughout the central regions of the galaxy. Studies by Garth Illingworth, James Binney (b. 1950) and others had already shown that the flattening of elliptical galaxies could not be wholly attributed to rotation and hence the velocity ellipsoid must be anisotropic (Illingworth, 1977; Binney, 1978). The possibility could not be ruled out that, if the velocity ellipsoid in M87 were anisotropic, the radial velocity dispersion might become larger closer to the nucleus, thus removing the necessity of a massive central black hole.

A different approach was taken by Holland Ford (b. 1940) and his colleagues, who discovered a disc of ionised gas about the nucleus of M87 from imaging observations made

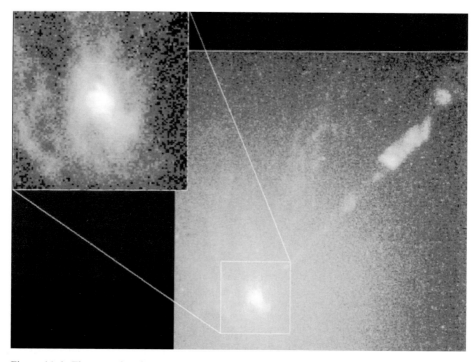

Figure 11.6: The central regions of the giant elliptical galaxy M87 (NGC 4486) in the nearby Virgo cluster of galaxies, as observed by the Hubble Space Telescope, showing the famous optical jet and the disc of ionised gas surrounding the luminous point-like source in the nucleus. The kinematics of the gaseous disc enabled the mass of the nuclear regions to be determined (Ford *et al.*, 1994). (Courtesy of Dr Holland Ford, NASA and the Space Telescope Science Institute.)

with the Hubble Space Telescope (Ford *et al.*, 1994). As can be seen from Figure 11.6, the famous optical jet lies along the axis of the nuclear disc. Spectroscopic observations by Richard Harms (b. 1946) and his colleagues established that the disc is in Keplerian rotation about the nucleus, the inferred mass of the nuclear regions being $3 \times 10^9 M_\odot$ (Harms *et al.*, 1994). The agreement between the mass inferred from the kinematics of the ionised disc and the estimates of Sargent and Young is reassuring. It would be surprising if it were fortuitous.

As more data on the physical properties of active galactic nuclei and the timescales of their variability became available in the 1980s, Amri Wandel (b. 1954) and Richard Mushotzky (b. 1947) analysed the spectra of quasars and Seyfert galaxies which were known to be intense variable X-ray sources (Wandel and Mushotzky, 1986). They estimated the masses of the central objects in two ways. Firstly, they used optical spectroscopic data to estimate the masses of the nuclei using the width of the emission lines as a measure of the typical velocities of the clouds and photoionisation models to find the distances of these clouds from the nuclei. The second approach was to assume that the X-ray variability originates from roughly the last stable orbit about the black hole, $r \approx 3r_g \approx 10(M/M_\odot)$ km, and so, assuming the timescale of variability is of the order of r/c, estimates of the mass of the black hole could be found (see Section A11.1.1). These two estimates were in good agreement, and the luminosities of the sources were all significantly less than the corresponding Eddington

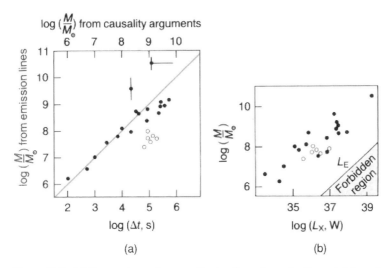

Figure 11.7: (a) Comparison of mass estimates for active galactic nuclei from the variability of their X-ray emission and from dynamical estimates. (b) Comparison of the inferred masses and luminosities with the Eddington limiting luminosity, $L_E = 1.3 \times 10^{31}(M/M_\odot)$. All the objects lie well below the Eddington limit (Wandel and Mushotzky, 1986).

luminosities (Figure 11.7). Thus, for these active galaxies and quasars, there was no problem, in principle, in accounting for their extreme luminosities and the short timescales of their variability, if it is assumed that their ultimate sources of energy were supermassive black holes in their nuclei.

While these observations were strongly suggestive of the presence of supermassive black holes in the nuclei of these active galaxies, conclusive evidence was provided by a beautiful set of radio observations of the nearby galaxy NGC 4258, or M106, in the water vapour (H_2O) maser line at a wavelength of 1.3 cm. Makoto Miyoshi (b. 1962) and his colleagues mapped the location and velocities of the H_2O masers with an angular resolution of better than one milliarcsec using the technique of very long baseline interferometry with the US Very Long Baseline Array (VLBA) (Miyoshi et al., 1995). The water vapour maser lines are very narrow, with typical velocity widths of about 1 km s^{-1}. Maser emission is only observed if the radiation passes through a sufficiently long path-length of the masing medium without the Doppler shift of the moving clouds shifting the wavelength of the radiation from the masing wavelength; that is, the maser emission must originate from regions in which the velocity gradient is zero. This occurs in regions in which the molecular ring is observed tangentially, and also along the line of sight directly towards the nucleus. The rotation curve derived from these observations is shown in Figure 11.8. It is important that, outside the central ±3 milliarcsec, the rotational velocities follow a Keplerian law and so there must be a compact massive object in the nucleus.

The mass of the central dark object was found to be about $3.7 \times 10^7 M_\odot$, and the mass density within the nuclear region was greater than $4 \times 10^9 M_\odot$ pc^{-3}. The latter figure is about 40 000 times greater than the density of the densest globular clusters known in our Galaxy, and it is so high that it is unlikely that the mass could be a cluster of any known

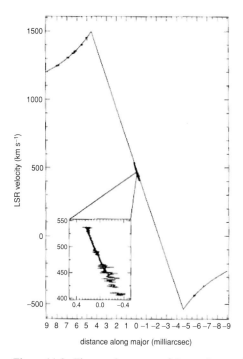

LSR velocity (km s^{-1})

distance along major (milliarcsec)

Figure 11.8: The rotation curve of the nuclear regions of the galaxy NGC 4258 from VLBI observations of the water vapour line at 1.3 cm. Note that outside the inner ±3 milliarcsec, the radial velocities follow a Keplerian rotation law, indicating the presence of a massive compact object in the nucleus (Miyoshi *et al.*, 1995).

type of stellar object. If the dark mass consisted of a cluster of normal stars, their number density would be so great that the cluster would evaporate by dynamical friction. Only if the objects were of low mass, $M \leq 0.03 M_\odot$, could the cluster survive for the age of the Galaxy. Such objects would have to be planets or brown dwarfs, in which case their collision times would be very short, roughly 10^5 years, and so their debris would rapidly fall towards the centre of the system, resulting in the formation of a massive black hole. It seems inevitable that there is a black hole with mass $3.7 \times 10^7 M_\odot$ in the nucleus of NGC 4258.

Equally spectacular have been the observations of the motions of stars about the centre of our own Galaxy in the infrared waveband at 2 µm with the Keck 10-metre telescope. Near-infrared observations were made by Andrea Ghez (b. 1965) and her colleagues with a very-high-speed infrared camera designed for speckle observations. Those frames which were diffraction-limited were selected and co-added. From observations over a period of four years, the orbits of stars within half an arcsec of compact radio source Sgr A*, which is assumed to define the Galactic nucleus, were measured (Figure 11.9(a)). These observations enabled the acceleration vectors of three of the stars to be determined, and, within the uncertainties with which these are determined, the centre of gravitational attraction coincides with the position of Sgr A* (Figure 11.9(b)) (Ghez *et al.*, 2000). These data implied that there is a dark object of mass $(2.61 \pm 0.35) \times 10^6 M_\odot$ within the central 0.015 pc of our Galaxy. The lower limit to the mass density within this region was $2.2 \times 10^{12} M_\odot \, \text{pc}^{-3}$.

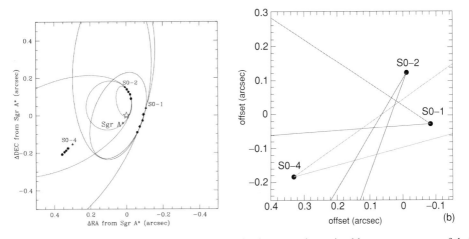

Figure 11.9: (a) The orbits of stars about the Galactic centre determined by measurements of their proper motions over a period of four years using near-infrared speckle techniques at 2 μm with the Keck 10-metre optical–infrared telescope. (b) Estimates of the directions of the acceleration vectors of the infrared stars S0-1, S0-2 and S0-4, the opening angles indicating the uncertainty in the measured directions. The most likely location of the central attractor includes the non-thermal radio source Sgr A*. (Ghez *et al.*, 2000.)

This value far exceeds the mass density inferred for the nucleus of NGC 4258, and the arguments given in the preceding paragraph apply with added force. For comparison, the Schwarzschild radius of a black hole of mass $2.6 \times 10^6 M_\odot$ is 2.6×10^{-7} pc – the size of the radio source Sgr A* measured by VLBI is only about 20 times the Schwarzschild radius of the central black hole.

A remarkable series of similar observations has been made using the facilities of the European Southern Observatory. Ten years of high-resolution astrometric imaging enabled Reinhard Genzel (b. 1952) and his colleagues to trace two-thirds of the orbit of the star labelled S0-2 in Figure 11.9(a) about Sgr A* (Figure 11.10) (Schödel *et al.*, 2002). These observations show that the star is moving in a bound, highly elliptical orbit with an orbital period of 15.2 years and that Sgr A* lies in one focus, as expected from Kepler's laws of planetary motion. The inferred mass of the black hole is $(3.7 \pm 1.5) \times 10^6 M_\odot$. Equally impressive is the fact that the distance of closest approach to the nucleus is only 17 light-hours, or 5×10^{-4} pc. To complete the story, the same set of infrared observations have revealed the very faint infrared counterpart of Sgr A*. In addition, these observations have discovered 'flares', probably associated with the process of accretion of mass onto the central black hole from the accretion disc (Genzel *et al.*, 2003).

Another, quite different, approach to demonstrating that supermassive black holes are present in the nuclei of active galaxies has been provided by X-ray spectral observations of asymmetric iron fluorescence lines in the spectra of Seyfert 1 galaxies. The most convincing case involves the X-ray properties of the Seyfert galaxy MGC-6-30-15, which was known to be a highly variable X-ray source, significant variations in its X-ray flux density being observed on the timescale of hours, indicating that the emission originates from very close to the nucleus itself.

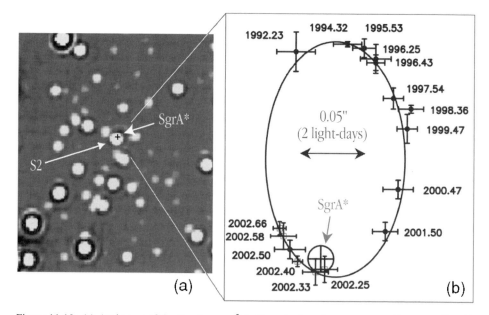

Figure 11.10: (a) An image of the 2×2 arcsec^2 field centred on the compact radio source Sgr A*, which is assumed to be located in the nucleus of our Galaxy. (b) The orbit of the infrared star S2 (the same as S0-2 in Fig. 11.9(a), as determined by infrared observations taken with the 3.5-metre New Technology Telescope at the ESO La Silla Observatory and the 8.2-metre YEPUN telescope of the ESO VLT at Cerro Paranal. (Schödel *et al.*, 2002.)

The most compelling model for the origin of the fluorescence lines involves a cool accretion disc embedded in an X-ray-emitting halo which is the source of the X-ray continuum radiation from the active galactic nucleus. Since the material of the disc is cooler than the surrounding X-ray emission, elements such as iron are not fully ionised, and, although they may be partially ionised, they possess filled K- and L-shells. The fluorescence process first involves the absorption of an X-ray photon by an iron ion, the greatest cross-section being for removal of one of the two K-shell electrons. The energy threshold for this process is 7.1 keV for neutral iron and it increases as the iron becomes more and more stripped of electrons. There is a 34% probability that the vacancy in the K-shell is filled by the emission of an X-ray photon of energy 6.4 keV in a permitted electromagnetic transition from the L-shell. Computations by Andrew Fabian and his colleagues have shown that the 6.4 keV fluorescent line of iron is by far the strongest of the fluorescent lines because of the large cross-section for photoelectric absorption by the K-shell of iron and iron's large cosmic abundance (Matt, Fabian and Reynolds, 1997). The 6.4 keV fluorescent iron line therefore acts as a tracer of the velocity field in the accretion disc, which extends into regions in which special and general relativistic effects are large and can strongly influence the shape of the profile of the 6.4 keV line.

The best example of this to date has been the observation of asymmetric broadening of the 6.4 keV line in the spectrum of the Seyfert 1 galaxy MCG-6-30-15. Figure 11.11

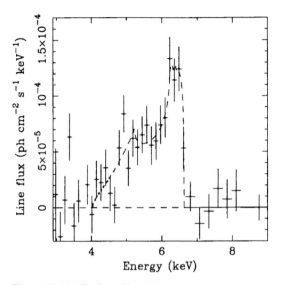

Figure 11.11: The broad iron line seen in a long ASCA observation of MCG-6-30-15 (Tanaka *et al.*, 1995). The dashed line shows a best fit of the data to the profile expected of an accretion disc about a Schwarzschild black hole. The inclination of the disc is $i = 30°$.

shows the spectrum of the 6.4 keV fluorescent line as observed in a long integration by the Japanese X-ray satellite ASCA (Tanaka *et al.*, 1995). The profile of this emission line has been determined by subtracting from the overall spectrum a smooth continuum spectrum. The key features of this remarkable observation are that the spectral line has an abrupt cut-off at energies greater than 6.7 keV, but extends to energies as low as about 4 keV to the low-energy side of the line. This type of asymmetry occurs very naturally if the thin accretion disc extends inwards towards the last stable orbit about the black hole. Two relativistic effects lead to asymmetries in the line. The first is the gravitational redshift, which shifts the spectrum to lower X-ray energies, and the second is the transverse Doppler effect, which appears in the expression for the Doppler shift of the observed energy of photons emitted from a source moving at an angle θ to the line of sight:

$$\epsilon_{obs} = \frac{\epsilon_0}{\gamma \left(1 - \frac{v}{c} \cos \theta\right)}. \tag{11.2}$$

Thus, if the plane of the accretion disc lies at a large angle to the line of sight, $\theta \to \pi/2$, the transverse Doppler shift associated with the Lorentz factor, γ, in the denominator dominates and shifts the line to lower energies. Therefore, the asymmetric profile of the 6.4 keV line has a natural interpretation if the largest redshifted parts of the line originate from close to the last stable orbit of an accretion disc about a massive black hole and the disc is observed more or less face-on.

  In 1995, John Kormendy and Douglas Richstone plotted their best estimates of the black-hole masses in the nuclei of nearby galaxies against the absolute magnitude of their spheroids

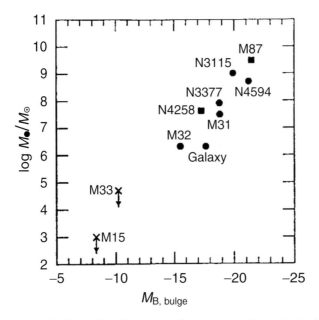

Figure 11.12: A plot of the masses of supermassive black holes in the nuclei of nearby galaxies against the absolute magnitude of their bulges, or spheroids (Kormendy and Richstone, 1995). The authors warn that the correlation may only be the upper envelope of a distribution which extends to lower values of $M_\bullet$.

for the cases in which they believed the masses were reasonably secure (Figure 11.12). It is a challenge to work out all the selection effects that went into the construction of Figure 11.12 (Kormendy and Richstone, 1995). In particular, good estimates are found in those elliptical galaxies which possess rapidly rotating nuclear discs. The black holes plotted may well form only the upper envelope of the distribution of black-hole masses at a given absolute magnitude, on the reasonable assumption that only those in which the evidence is compelling are included.

It is intriguing that, interpreted literally, Figure 11.12 suggests that the black-hole mass is proportional to the luminosity of the bulge of the galaxy. Adopting a standard mass-to-luminosity ratio for the bulge, Kormendy and Richstone suggest that the black-hole mass fraction is $M_\bullet/M_{\text{bulge}} \sim 0.002$–$0.003$. This conclusion is the subject of debate, in view of the many selection effects which go into the construction of Figure 11.12, particularly the absence of low-mass black holes in luminous elliptical galaxies.

11.6   Non-thermal phenomena in active galactic nuclei

From the very beginning, the non-thermal properties of active galactic nuclei posed intriguing astrophysical puzzles, many of them traceable to the very short timescales of their variability. In 1965, for example, Fred Hoyle, William Fowler and Wallace Sargent showed that the quasars are susceptible to what became known as the *inverse Compton catastrophe* (Hoyle, Burbidge and Sargent, 1966). Using the causality relation $r \leq c\tau$, an upper limit, r, to the size of the emitting region could be found from the timescale, τ, of variability of the source and hence a lower limit to the energy density of radiation could be determined. If the continuum optical emission were synchrotron radiation, the energy densities of radiation would be so great that the radiating relativistic electrons would lose much more energy by the inverse Compton scattering of the optical photons to higher energies than by the emission of the optical radiation itself, in the process creating intense fluxes of X- and γ-rays. The same argument can be applied to the variable radio emission of the cores of radio quasars and radio galaxies. It is found that there is a critical brightness temperature of about 10^{12} K, above which radiation is lost preferentially by inverse Compton scattering rather than by synchrotron radiation (see Section A11.2). In both cases, the electrons in the source regions would have very short lifetimes and would have to be replenished to maintain the sources' luminosities.

In a remarkably prescient paper of 1966, Martin Rees (b. 1942) showed how these problems could be overcome if the source components moved out of the nuclear regions at relativistic speeds (Rees, 1966, 1967). Several relativistic effects contribute to the alleviation of the problem of the short variability timescales, particularly aberration effects, which make the sources appear to be more luminous and larger than they are in their rest frames (see Section A11.4).

11.6.1 *Double radio sources and the acceleration of high-energy electrons*

During the late 1960s and early 1970s, the first high-resolution radio maps were made of extragalactic radio sources with the new generation of Earth-rotation aperture synthesis radio telescopes, and these began to reveal the details of the structures of these sources. The culmination of these studies were maps of superb quality, such as that of Cygnus A (Figure 11.1(b)) made with the US Very Large Array (Perley *et al.*, 1984). In 1974, Bernard Fanaroff (b. 1947) and Julia Riley (b. 1947) noticed that the morphologies of these radio structures depended very strongly upon their radio luminosities. In powerful double radio sources such as Cygnus A, the maximum radio surface brightness of the lobe structures is observed in *hot-spots* towards the outer ends of the diffuse radio lobes, and these are referred to as *Fanaroff–Riley Class 2*, or FR2, radio sources (Fanaroff and Riley, 1974). In contrast, those sources in which the maximum surface brightness occurs less than halfway from the active galactic nucleus to the edge of the diffuse radio lobe structure, the *Fanaroff–Riley Class 1*, or FR1, sources, have radio luminosities which are much lower than those of Class 2. Radio maps of the FR1 sources 3C 31 and 3C 66B are shown in Figure 11.13. The distinction between the two classes occurs rather abruptly at a well defined radio luminosity (Figure 11.14).

Many of the astrophysical problems associated with extragalactic radio sources were derived from the pioneering radio maps made by the Cambridge one-mile and 5-kilometre radio telescopes and the Westerbork Synthesis Radio Telescope. The features of particular interest in these maps were the 'hot-spots' observed towards the leading edges of the double

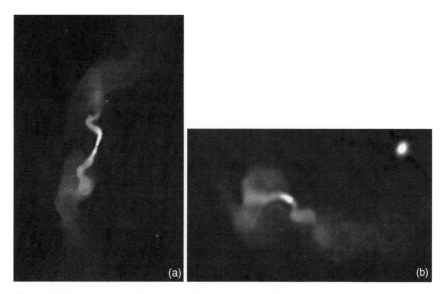

Figure 11.13: Examples of Fanaroff–Riley Class 1, or FR1, radio sources. (a) 3C 31 (courtesy of the US National Radio Astronomy Observatory); (b) 3C 66B (Hardcastle *et al.*, 1996), both observed by the US Very Large Array at 6 cm wavelength.

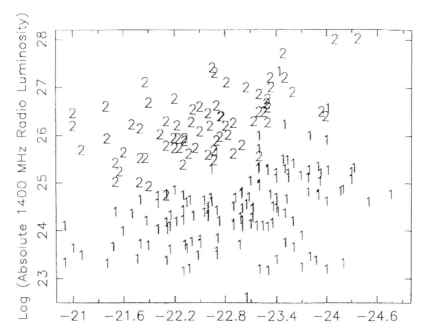

Figure 11.14: A plot of radio luminosity at 1.4 GHz against optical absolute magnitude, illustrating the distinction between FR1 and FR2 double radio sources (Owen and Ledlow, 1994).

radio structures. These components were inferred to have high energy densities in relativistic particles and magnetic fields, and there were two reasons why these regions had to be continuously replenished with relativistic material (Longair, Ryle and Scheuer, 1973). Firstly, the synchrotron lifetimes of the ultra-relativistic electrons in the most compact hot-spots were less than the age of the source and so they would have to be continuously accelerated or replenished. Secondly, the energy densities of the high-energy particles and magnetic fields were too great to be confined within the hot-spots of the radio sources. Peter Scheuer, Martin Ryle and I (b. 1941) inferred that the hot-spots are temporary phenomena which are continuously supplied with energy from the active galactic nucleus by beams or jets of particles.

Direct evidence for these jets was subsequently found in the radio maps of many radio galaxies and quasars, including Cygnus A. Variants of this model were proposed by Martin Rees in 1971 (Rees, 1971) and by Peter Scheuer (1930–2001) in 1974 (Scheuer, 1974), and the generic picture of continuous-flow models has remained the preferred picture to account for the radio emission of the extended radio sources. Numerical simulations during the 1980s demonstrated that many of the detailed features of the radio source structures could be accounted for by gas dynamical processes involving the interaction of gaseous jets with the ambient interstellar and intergalactic gas. The most elegant analytic model which describes many of the key features involved in these processes was presented by Christian Kaiser (b. 1970) and Paul Alexander (b. 1960), in which the gas dynamics of the double sources is described by a self-similar solution of the equations of gas dynamics (Kaiser and Alexander, 1997).

One of the difficulties associated with the presence of high-energy electrons in many different astronomical environments is the fact that, as a gas of relativistic electrons and its frozen-in magnetic field expand, energy is lost adiabatically. Thus, if the radio-emitting electrons had been accelerated in the initial explosion of a supernova, such as that which gave rise to the remnant Cassiopeia A, they would have lost all their energy in the adiabatic expansion of the sphere of hot gas as it does work against the ambient interstellar gas. Similar problems were found in accounting for the high energy densities of relativistic electrons observed in extended radio sources and the fluxes of energetic electrons in the interplanetary medium. Up till the mid 1970s, various acceleration mechanisms had been proposed, but none of them gained any general acceptance. One popular mechanism proposed by Enrico Fermi in 1949 involved stochastic acceleration of particles in elastic collisions with interstellar clouds (Fermi, 1949). The problem was that the *Fermi acceleration* process was second-order in the ratio of the velocity of the clouds to the speed of light and so was very slow. Electromagnetic acceleration in the magnetospheres of neutron stars was proposed by Gunn and Ostriker in 1969 (Ostriker and Gunn, 1969), but this did not overcome the problem of accelerating the particles in extended source regions.

In 1977 and 1978, a number of independent authors, including Ian Axford (b. 1933), Egil Leer (b. 1942), George Skadron (b. 1936), Germogen Krymsky (b. 1937), Anthony Bell (b. 1952), Roger Blandford (b. 1949) and Jeremiah Ostriker showed that *first-order Fermi acceleration* in strong shock-waves is a remarkably effective means of accelerating particles and creating a power-law energy spectrum (Axford, Leer and Skadron, 1977; Krymsky, 1977; Bell, 1978; Blandford and Ostriker, 1978). The idea is as follows: if high-energy particles are present in the vicinity of a shock-wave, they can be scattered back and forth across the shock-wave, and, in each crossing in either direction, the particles obtain a fractional increase in energy of order v/c, where v is the velocity of the shock-wave. Particles are lost from the accelerating region by the bulk motion of the material behind the shock downstream, and it is straightforward to show that these two competing effects lead to the formation of a power-law energy spectrum of the form $N(E)\,dE \propto E^{-2}\,dE$. The beauty of this process is that it depends only upon the particles encountering a strong shock-wave and being scattered in the interstellar media on either side of the shock. The energy acquired by the particles is derived from the kinetic energy of the matter behind the shock-wave.[17]

This mechanism had great attractions because it meant that the particles are accelerated wherever there are strong shocks. Thus, in supernova shells and at the interface between jets and the interstellar or intergalactic media, where there are strong shock-waves, this mechanism provides a means of accelerating the high-energy particles where they are needed. There remained a number of problems with the mechanism, including the fact that it has proved difficult to obtain spectra steeper than $N(E)\,dE \propto E^{-2}\,dE$. In addition, as applied to supernova remnants, particles could not be accelerated to energies much greater than 10^{15} eV, at which energy the gyroradii of the electrons become comparable to the size of the supernova remnant itself (Lagage and Cesarsky, 1983). Thus, some other mechanism is needed to account for the very highest energy cosmic rays. Nonetheless, the discovery of this mechanism has gone a long way to resolving the problem of accelerating high-energy particles in a wide range of astronomical environments.

The origin of the magnetic fields in supernova remnants was solved by Stephen Gull (b. 1949) in 1975 when he showed that an expanding supernova shell becomes unstable to Rayleigh–Taylor instabilities as it decelerates (Gull, 1975). The instability results in turbulent motions at the contact discontinuity between the expanding sphere of hot gas and the shocked interstellar material. The turbulent eddies can stretch and wind up the magnetic field until its energy density is roughly the same as that in the turbulent motions. Through numerical simulations, Gull showed that the magnetic field in supernova remnants could attain values up to about $B \sim 10\,\mathrm{nT}$, more or less the values observed. Similar instabilities are expected to take place at the interface between the jets and the intergalactic gas in extragalactic radio sources and can account for the presence of magnetic fields in the source components.

11.6.2 Superluminal radio sources

The continuous-flow model of double extragalactic radio sources laid responsibility for their extended structures on jets of material ejected from the quasar or radio galaxy nucleus. The radio jets observed in these sources were clear evidence for this process and were very close relatives of the optical jets seen in objects such as M87 and 3C 273. Even more remarkable was the evidence from very long baseline interferometric observations of structural changes in the compact radio cores of radio quasars and radio galaxies by Marshall Cohen (b. 1926) and his colleagues (Cohen *et al.*, 1971; Whitney *et al.*, 1971). By 1980, there was clear evidence that the structures observed in these sources on the scale of milliarcseconds were not stationary; the source components appeared to be moving apart at speeds exceeding the

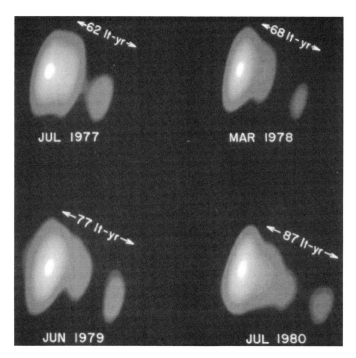

Figure 11.15: VLBI images of the nuclear regions of the quasar 3C 273 from 1977 to 1980. The radio component is observed to move a distance of 25 light-years in 3 years, implying a superluminal velocity of about ten times the speed of light (Pearson *et al.*, 1981).

speed of light. In the case of 3C 273 (Figure 11.15), the fainter component appeared to move a distance of about 25 light-years between 1977 and 1980, implying an apparent separation velocity of about ten times the speed of light (Pearson *et al.*, 1981, 1982). Observations of these sources since that time have shown that the phenomenon is remarkably common among radio sources with compact radio cores.[18]

There immediately followed a flurry of theoretical speculation about the origin of this phenomenon. The simplest, and still the most plausible, explanation had already been described by Martin Rees in his papers from the 1960s (Rees, 1966, 1967). If a source component is ejected at a relativistic speed from the nucleus at a small angle to the line of sight, the component can have an apparent velocity perpendicular to the line of sight of up to γv, where $\gamma = (1 - v^2/c^2)^{-1/2}$ is the Lorentz factor. This maximum value is found when the angle of ejection of the component to the line of sight is $\sin \theta = \gamma^{-1}$, that is, $\theta \approx \gamma^{-1}$. In addition, the luminosity of the approaching component is enhanced by the effects of aberration, and this may account for the one-sidedness of many of the jets observed in active galactic nuclei.[†] It is generally assumed that these superluminal motions are associated with the origin of the beams or jets responsible for fuelling the hot-spots and extended radio structures.

[†] Some of the more important results concerning the effects of aberration and Doppler shifting upon the observed intensities of sources of radiation are included in the Section A11.4.

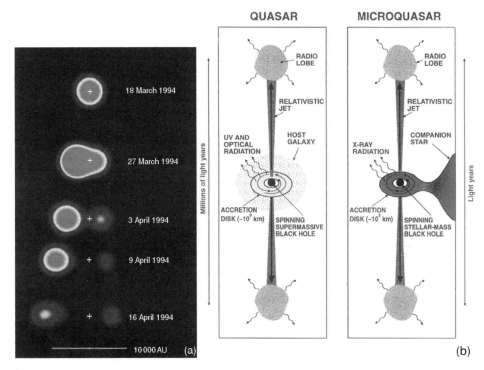

Figure 11.16: (a) The evolution of the radio structure of the binary X-ray source GRS 1915+105 over a three-week period in 1994 (Mirabel and Rodrigues, 1994). (b) A comparison between the properties of the radio jets in quasars and microquasars (Mirabel and Rodrigues, 1998).

The problems of understanding the physics of the double radio structures observed in extragalactic radio sources was, for a long time, the province of extragalactic radio astronomers. In 1994, however, evidence was found for relativistic jets in Galactic binary X-ray sources in which there is evidence for the presence of a black hole as the invisible companion. Felix Mirabel (b. 1944) and Luis Rodrigues (b. 1948) observed the intense hard X-ray source GRS 1915 + 105 during a radio outburst; they discovered that its radio structure was double and that the components were separating from the radio core at 1.25 times the speed of light (Mirabel and Rodrigues, 1994) (Figure 11.16(a)). The X-ray properties of the source strongly suggested that the energy source in the binary system is a black hole. Since this discovery in 1994, several other examples have been discovered. Mirabel and Rodrigues pointed out the remarkable similarities between the radio properties of these sources, which they term *microquasars*, and the radio galaxies and radio quasars (Mirabel and Rodrigues, 1998) (Figure 11.16(b)). The big advantage of studying the microquasars is that the timescales of the phenomena associated with their variability and energy production scale as the mass of the black hole, and so phenomena which would take thousands or millions of years for a $10^9 M_\odot$ black hole would be expected to take place on the timescale of minutes or days in a $10 M_\odot$ Galactic black hole. Such rapid changes in the radio and X-ray emission of these sources have been observed.

Relativistic beaming has also been invoked to solve the problems associated with the rapid variability of the intense γ-ray sources discovered by the Compton Gamma-ray Observatory. The Observatory carried out a complete survey of the γ-ray sky and among the most intense sources discovered were variable γ-ray sources associated with those extreme radio quasars which exhibited superluminal motions (Kniffen, Chipman and Gehrels, 1994). The γ-ray luminosities were so great and the timescales of variability so short that the energy density of γ-rays was very large, so large in fact that the γ-rays would be degraded into electron–positron pairs through the pair-production process. The importance of this process is determined by the *compactness factor*, C, defined to be the quantity

$$C = \frac{L_\gamma \sigma_T}{4\pi m_e c^4 t},$$ (11.3)

where L_γ is the γ-ray luminosity at 1 MeV, σ_T is the Thomson cross-section and t is the timescale of variability. Note that C is a measure of how far a 1 MeV γ-ray can propagate through the source before it is destroyed by electron–positron production (see Section A11.3). If the compactness factor is very much greater than unity, the γ-rays cannot escape from the source. In addition, the γ-ray luminosities are extreme, particularly when account is taken of the ultra-high-energy γ-radiation associated with some of the sources detected by the Compton Gamma-ray Observatory (see Figure 7.11). As demonstrated in Section A11.4, if it is assumed that the source components are relativistically beamed, the γ-ray luminosity is enhanced by a factor of roughly κ^5, where $\kappa = \gamma[1 + (v/c)\cos\theta]$ and θ is the angle between the axis of ejection and the line of sight, and so the γ-rays can survive without suffering degradation by electron–positron pair production, consistent with the observation of superluminal motion of the radio components.

11.6.3 Jets in active galactic nuclei

These advances enabled an empirical picture of the physics of active galaxies and extragalactic radio sources to be developed, but there are many pieces of the story which are uncertain. For example, the origin of the jets in active nuclei is unclear. One possibility is that they are associated with the funnels which are formed along the rotation axes of thick accretion discs, but these are thought to be unstable configurations. Other models ascribe the jets to electromagnetic processes occurring close to the black hole itself and, in the case of rotating black holes, there is a preferred axis along its rotation axis. In the type of model considered by Martin Rees and his colleagues in 1982, electromagnetic torques are used to extract the rotational energy of a rotating black hole, resulting in an outflow parallel to its rotation axis (Rees, 1976; Blandford and Znajek, 1977).

The most promising models associate the initial collimation of the jets with the winding up of magnetic fields lines frozen into the ionised gas in the accretion disc about a rotating black hole. The most detailed studies of these complex relativistic magnetohydrodynamic problems have been carried out by Richard Lovelace (b. 1941) and his colleagues, who have performed analytic and numerical simulations of the force-free magnetic field structures which arise as an initial dipolar magnetic field threading the accretion disc is wound up. They find solutions in which 'magnetic bubbles' are expelled at relativistic speeds along the

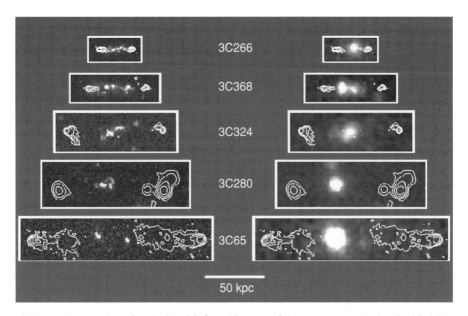

Figure 11.17: A series of optical and infrared images of the structures associated with 3CR radio galaxies at a redshift $z \sim 1$ on the same physical scale. The images are in order of increasing physical size. The images in the left-hand column were taken with the Hubble Space Telescope; those in the right-hand column were taken at 2.2 μm using the UK Infrared Telescope. The characteristic double radio structures of the radio sources as observed with the VLA are shown as white contour lines (Best *et al.*, 1996).

rotation axis of the disc (Lovelace and Romanova, 2003). Even the most extreme relativistic jets can be accounted for by these models.

The jets which power the outer radio lobes in the most luminous radio sources can have a dramatic influence upon any interstellar or intergalactic clouds surrounding the active galaxy as they are engulfed by the expanding cocoon of high-energy particles and magnetic field. This is most dramatically illustrated by the *alignment effect*, which was discovered by Patrick McCarthy (b. 1961), Kenneth Chambers (b. 1956) and their colleagues in 1987 (Chambers, Miley and van Breugel, 1987; McCarthy *et al.*, 1987). They found that the optical images of large-redshift 3CR radio galaxies were aligned with the radio axes of the double radio sources. The nature of the alignment effect was revealed by high-resolution images of these radio galaxies with the Hubble Space Telescope obtained by Philip Best (b. 1972), Huub Röttgering (b. 1963) and me (Best, Longair and Röttgering, 1996). In the montage shown in Figure 11.17, the optical, infrared and radio structures of five luminous 3CR radio galaxies are shown on the same linear scale. The right-hand panels show the infrared images of the galaxies at 2.2 μm and the associated radio sources, revealing the underlying old populations of the radio galaxies. The HST images in the left-hand panels show that the optical emission is aligned along the axis of the radio source, this emission vastly outshining the optical emission from the galaxies themselves. There is little evidence for the presence of the galaxy in the HST images, but rather the structures are dominated by emission regions aligned with the direction of passage of the radio jets which power the outer radio lobes. The spectroscopic evidence suggests that the optical structures observed in the radio sources with physical size less than about 120 kpc is associated with shock excitation of pre-existing cool clouds in the vicinity of the radio galaxy (Best, Longair and Röttgering, 2000). The natural interpretation of these observations is that the optical emission is excited with shock-waves associated with the highly supersonic passage of the radio jet through the interstellar and intergalactic medium in the vicinity of the host galaxy.

11.6.4 Unified models for active galaxies

One of the more remarkable industries of the last 20 years has been the attempt to 'unify' different types of active galactic nuclei into a coherent, unified picture. The concept of unification resulted from the realisation that projection effects must be important in determining the observed properties of active galactic nuclei, and the question is how much of their diversity can be accounted for in terms of a model in which a single class of active galaxy is viewed at different angles to the line of sight.[19]

The importance of projection effects in distinguishing between Seyfert 1 and Seyfert 2 galaxies was first convincingly demonstrated by Robert Antonucci (b. 1954) and Joseph Miller (b. 1941) in 1985 in their study of the nearby Seyfert 2 galaxy NGC 1068 (Antonucci and Miller, 1985). When observed spectropolarimetrically, the polarised line emission was found to be as broad as the broad permitted lines seen in Seyfert 1 galaxies. Antonucci and Miller interpreted these observations in terms of a picture in which there is an 'obscuring torus' about the nucleus so that, when the galaxy is observed at a small angle to the axis of the torus, the nuclear regions, which contain the active nucleus itself and the broad-line emitting regions, are observed and the galaxy is classified as a Seyfert 1 galaxy. The characteristic Seyfert 1 broad-line regions and strong blue and UV continuum are only observed along lines of sight close to the axis of the torus. If the axis of the torus is observed at a large angle to the line of sight, the nuclear regions are obscured and only the narrow-line regions, which are located much further from the nucleus than the broad-line regions, are observed in direct light. The obscured central regions can, however, be observed in the light reflected from clouds outside the region of the torus, perhaps by gas and dust in the narrow-line regions. The continuum and broad-line emission from the nuclear regions are polarised in the process of reflection by Thomson or Rayleigh scattering. In Antonucci's review of 1993, he provided a number of other compelling arguments for this unification picture (Antonucci, 1993).

Another unification scenario concerns those active galaxies which are strong radio sources: the radio quasars and the radio galaxies. The importance of projection effects in determining the observed properties of extragalactic radio sources was advocated by Peter Barthel (b. 1952). Barthel was principally concerned with the unification of radio galaxies with the radio quasars, and his analysis centred upon the properties of the classical double radio sources in the 3CR catalogue (Barthel, 1989, 1994). The principal arguments concerned the observation of superluminal motions in the radio quasars, suggesting that they are observed at small angles to the line of sight, the one-sided jet emission observed in the radio quasars, but not in the radio galaxies, the relative sizes of the radio structures of the quasars and radio galaxies, and so on. In the unification picture, radio quasars are observed

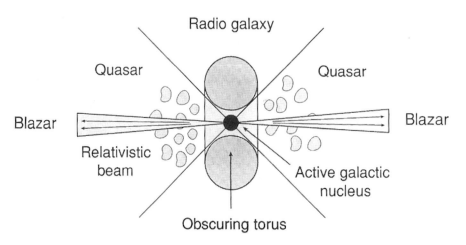

Figure 11.18: Illustrating the unified model for the radio galaxies and radio quasars observed in samples of bright radio sources. Radio quasars are observed when the axis of the radio source lies within about 45° of the line of sight. When observations are made almost along the axis of the radio jet, superluminal radio sources and blazars are observed.

when the nucleus is observed within a cone of half-angle roughly 45° and a radio galaxy is observed when the nuclear regions are hidden by the obscuring torus (Figure 11.18). Highly collimated radio jets are assumed to be emitted by the nucleus along the axis of the torus, and these jets are responsible for powering the outer radio hot-spots and extended radio lobes. If the radio jet is observed at an angle close to the line of sight, the emission of the relativistic jet is strongly enhanced by aberration and Doppler effects and so superluminal motions can be observed. It is natural to attribute the extreme *BL Lac* or *blazar* phenomena to radio sources, in which the relativistic jets point almost precisely along the line of sight to the observer. In a few extreme blazars, high-dynamic-range observations have enabled the underlying double radio source to be observed. Another unification scenario concerns the relation between the BL Lac objects and radio galaxies. A good case can be made for BL Lac objects being the relativistically beamed jets originating in the nuclei of FR1 radio galaxies (Urry and Padovani, 1994).

11.7   The γ-ray bursts

The discovery of γ-ray bursts was briefly recounted in Section 7.5. The paper by Ray Klebesadel (b. 1932), Ian Strong (b. 1930) and Roy Olson of the Los Alamos National Laboratory was published in 1973, six years after the first of these events had been discovered by the Vela surveillance satellites (Klebesadel *et al.*, 1973). The delay in publication was not caused by security considerations, but rather because the authors wished to be certain that the γ-ray bursts were genuine astronomical phenomena. The bursts lasted from a fraction of a second to several minutes (Figure 11.21), and, during that time, the γ-ray bursts were the most luminous objects in the γ-ray sky.

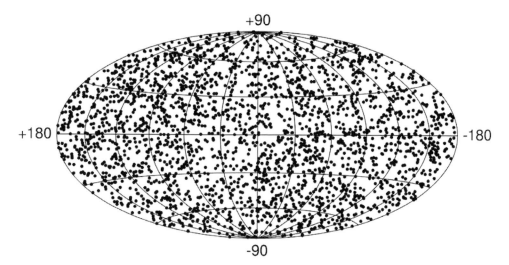

Figure 11.19: The locations in galactic coordinates of the 2704 γ-ray bursts recorded by the Burst and Transient Source Experiment of NASA's Compton Gamma-ray Observatory during its nine-year mission. (Courtesy of NASA, G. J. Fishman and the CGRO Science Team.)

The Compton Gamma-ray Observatory (CGRO), launched in 1991, was equipped with a detector system specifically designed to detect and locate γ-ray bursts, the Burst and Transient Source Experiment (BATSE). Over the following nine years, this instrument detected 2704 bursts and established beyond any question that they are uniformly distributed over the sky (Figure 11.19). Even early in the mission, it was apparent that the faint and bright bursts were uniformly distributed over the sky. Furthermore, the number counts of the bursts showed departures from those expected of a uniform Euclidean distribution of sources. Initially, it was thought that the uniformity of the distribution of the γ-ray bursts could be most naturally interpreted in terms of a distribution of nearby sources in the vicinity of the Solar System. The theorists had a field day, coming up with a very wide variety of exotic explanations,[20] the most plausible of which involved a nearby distribution of neutron stars. In this picture, it was difficult, however, to account for both the isotropy of their distribution and the uniform convergence of the numbers of events at low fluxes over the whole sky.

The big problem was that the γ-ray bursts were of very short duration and the angular resolution of the BATSE instrument was only a few degrees, so that there was not time or adequate positional accuracy to make secure identifications of the sources. The solution to the problem was stimulated by a theoretical paper by Peter Mészáros (b. 1943) and Martin Rees in 1993 (Mészáros and Rees, 1993). They argued that, if the γ-ray bursts were extragalactic phenomena, the energy densities during the event would be so extreme that a relativistic shock-wave would be created in which electrons would be accelerated to very high energies. These electrons would emit synchrotron radiation with a power-law spectrum, consistent with the observed spectra of the bursts. Although the intense γ-ray emission would last only a short time at γ-ray energies, the emission would remain observable at lower energies. In other words, there would be an *afterglow* which would appear progressively in the X-ray, then optical, infrared and radio wavebands.

The search for afterglows of γ-ray bursts became possible with the launch of the Italian–Dutch BeppoSAX satellite. In 1997, the X-ray telescope was pointed at the position of the γ-ray burst GRB 970228, within 8 hours of the event having taken place, and its X-ray afterglow was discovered (Costa *et al.*, 1997). Subsequently, afterglows were observed throughout the electromagnetic spectrum. These observations enabled a precise position for the burst to be determined and the association of this burst with a very faint galaxy established by observations with the Hubble Space Telescope (Sahu *et al.*, 1997). Within four years, over 40 γ-ray-burst afterglows had been detected, and 30 of these could be associated with distant host galaxies. Most of the galaxies associated with the γ-ray bursts show evidence for active star formation, implying the presence of young massive stars. These observations demonstrated that the γ-ray bursts are extreme events occurring in galaxies at cosmological distances – the most distant γ-ray burst detected to date has redshift $z = 4.5$ (Andersen *et al.*, 2000).

The γ-ray bursts are the most extreme high-energy astrophysical events yet discovered. The very short timescales of variability of their energy release indicate that the sources must originate from regions less than about 100 km in size on timescales much less than one second. Hence, the bursts must involve stellar-mass objects, and the huge energy releases, typically 10^{47} J if the emission is isotropic, would require the conversion of about one solar mass to be converted into high-energy particles. Furthermore, the energy density in γ-rays in the source region would be so great that the same problems afflicting the extreme γ-ray luminosities of blasars and quasars, the degradation of the γ-rays by photon–photon interactions, would apply equally severely. Simply from the energy and scale of the emitting region, it followed that the source region must expand relativistically, and therefore, just as in the extreme γ-ray sources, the γ-ray bursts must involve relativistic bulk motions.[21]

Many of the results derived by Martin Rees in his papers of 1966 and 1967 can be applied directly to the observed emission (Rees, 1966, 1967). In the simplest picture, the source region can be taken to be a *relativistic fireball*, meaning a relativistically expanding sphere which heats the surrounding gas and drives a relativistic shock-wave into it. If the sphere is optically thick, it is expected that the radiation would be thermalised and a thermal spectrum observed. However, the observed spectra of the γ-ray bursts are of power-law form with $N(\varepsilon) \propto \varepsilon^{-\alpha}$, where $\alpha \sim$ 2–3 at energies greater than about 0.1–1 MeV, and therefore the radiation must originate from optically thin regions later in the expansion. The most attractive model involves the acceleration of electrons in the relativistic shock-front by the *first-order Fermi acceleration mechanism* discussed in Section 11.6.1. Thus, the general picture involves a sphere of ultra-relativistic electrons and magnetic fields expanding relativistically. The characteristic evolution of the synchrotron spectrum of such a sphere has proved to be a good match to the time evolution of the afterglows throughout the electromagnetic spectrum (Figure 11.20).

Relativistic beaming alleviates the problems of $\gamma\gamma$ annihilation, but does not solve the energy problem. The fireball need not, however, be isotropic. In the case in which the energy of the γ-ray burst is emitted in a narrow collimated beam, the energy requirements

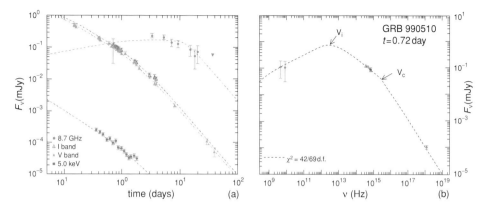

Figure 11.20: (a) Model light-curves (dotted lines) at X-ray, optical and radio wavelengths compared with the observed evolution for the γ-ray burst GRB 990510. (b) The spectrum of the γ-ray burst GRB 990510 at a wide range of wavelengths after 0.72 day. The dotted line shows the expectation of the standard synchrotron radiation model (Mészáros, 2002).

would be very significantly reduced. The same general picture described above would apply, provided the opening angle, θ, of the beam is greater than γ^{-1}, where γ is the Lorentz factor associated with the bulk motion of the emitting region. As the jet sweeps up the surrounding gas and decelerates, the Lorentz factor of the beam decreases, and the beaming associated with the collimation of the jet itself is more important than the relativistic beaming. This change is expected to be associated with a more rapid decline in the intensity of the γ-ray burst, and this has been observed in the γ-ray burst GRB 990510 Figure 11.20(a).

If this jet model is adopted, the total energy requirements of the γ-ray bursts can be evaluated, and it has been found that, rather than a dispersion in intrinsic luminosities from about 10^{44} to 10^{47} J, if the radiation is isotropic, there is only about an order of magnitude spread about the value 8×10^{43} J, a value typical of the energy released in the core collapse of a supernova explosion. In fact, in a number of cases, the bursts are roughly coincident with supernova explosions occurring in distant galaxies. The supernova $-\gamma$-ray burst association was established with certainty for the γ-ray burst GRB 030329 observed on 29 March 2003, for which the characteristic broad lines of an extremely energetic supernova were observed within days of the event (Hjorth *et al.*, 2003).

The distribution of the durations of the γ-ray bursts is bimodal, as illustrated in Figure 11.21. The results discussed above have been derived from studies of the longer duration bursts, and it is not so clear that the same considerations apply to the bursts with duration less than one second. While the relativistic fireball model involving relativistic jets can account for many of the features of the bursts, there is less agreement about the ultimate energy source for the γ-ray bursts. An appealing picture is to relate the phenomena to the collapse of a very massive stellar remnant to form a Kerr black hole and then to associate the relativistic jet with the presence of electric and magnetic fields in the vicinity of the hole and the axisymmetric collapse of the outer envelope onto the collapsed core. This picture is often referred to as a *collapsar* model for γ-ray bursts. The example of the γ-ray burst GRB 030329, which is definitely associated with a supernova explosion, may be an

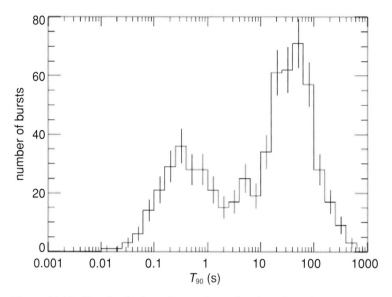

Figure 11.21: The distribution of γ-ray-burst durations from the bursts observed by the BATSE experiment of the CGRO. (Courtesy of NASA, G. J. Fishman and the CGRO Science Team.)

example of such an event. There are, however, many other important possibilities, including the merger of binary neutron stars, or of a neutron star and a black hole, or of two black holes, which might be associated with the very short γ-ray bursts. Many of these possibilities are discussed in the excellent review by Peter Mészáros (Mészáros, 2002).

Notes to Chapter 11

1 The luminosity, $L(\nu)$, of a source of synchrotron radiation depends upon the magnetic flux density, B, and the number density of ultra-relativistic electrons, defined by $N(E)\,\mathrm{d}E = \kappa V E^x\,\mathrm{d}E$, as $L_\nu \propto \kappa V B^{(x-1)/2}$, where V is the volume of the source region. Thus, a fixed luminosity can be produced by a small number of electrons in a strong magnetic field, or a large number of electrons gyrating in a weak magnetic field. Between these extremes, there is a minimum energy requirement which corresponds closely to the equipartition of energy between the magnetic field and the high-energy electrons. This minimum energy requirement is often referred to as the equipartition values of the magnetic field strength and total electron energy. For details, see Malcolm Longair, *High Energy Astrophysics*, vol. 2 (Cambridge: Cambridge University Press, 1997), Chapter 19.

2 The designation 3C refers to radio sources listed in the Third Cambridge Catalogue of Radio Sources (Shakeshaft *et al.*, 1955). An improved version of the catalogue was compiled by Andrew Bennett entitled the Revised 3C Catalogue, or 3CR (Bennett, 1962).

3 The proceedings of the First Texas Symposium contains many of the most important papers on radio sources, active galaxies and quasars published up to 1963: *Quasi-stellar Sources and Gravitational Collapse*, eds, I. Robinson, A. Schild and E. L. Schucking (Chicago: University of Chicago Press, 1965). A summary of the early observations of quasars is given by G. R. Burbidge and E. M. Burbidge, *Quasi-stellar Objects* (San Francisco: W. H. Freeman and Company, 1967).

4 See Gold's after-dinner speech in Robinson *et al.*, eds, *Quasar-stellar Sources and Gravitational Collapse*.

5 The arguments involved in the controversy concerning the origin of quasar redshifts are contained in the volume by G. B. Field, H. Arp and J. N. Bahcall, *The Redshift Controversy* (Reading, Massachusetts: W. A. Benjamin, 1973).

6 An excellent review of the history of black holes is presented by Werner Israel (b. 1931), Dark stars: the evolution of an idea, in *300 Years of Gravitation*, eds S. W. Hawking and W. Israel (Cambridge: Cambridge University Press, 1987), pp. 199–276.

7 Schwarzschild volunteered for military service at the outbreak of the First World War in 1914 and served in Belgium, France and Russia. He wrote the paper on his exact solution of Einstein's field equations while on service in Russia. Tragically, Schwarzschild died in May 1916 from the illness pemphigus contracted while on military service in Russia.

8 The Kerr metric is somewhat more complicated than the Schwarzschild metric and can be written in Boyer–Lundquist coordinates as follows:

$$ds^2 = \left(1 - \frac{2GMr}{\rho c^2}\right) dt^2 - \frac{1}{c^2}\left[\frac{4GMra \sin^2 \theta}{\rho c} dt \, d\phi + \frac{\rho}{\Delta} dr^2\right.$$
$$\left. + \rho \, d\theta^2 + \left(r^2 + a^2 + \frac{2GMra^2 \sin^2 \theta}{\rho c^2}\right) \sin^2 \theta \, d\phi^2\right],$$

where the black hole rotates in the positive ϕ direction; $a = (J/Mc)$ is the angular momentum of the black hole per unit mass, $\Delta = r^2 - (2GMr/c^2) + a^2$ and $\rho = r^2 + a^2 \cos^2 \theta$. If the black hole is non-rotating, $J = a = 0$ and the Kerr metric reduces to the Schwarzschild metric, equation (11.1).

9 John Archibald Wheeler had conjectured that all multipoles of a black hole except its mass, angular momentum and charge would be radiated away. It was only with the work of Carter and Hawking that this remarkable result was proved for black holes.

10 These results can be derived from simple analyses of the form of the Schwarzschild metric, as I demonstrate in Malcolm Longair, *Theoretical Concepts in Physics* (Cambridge: Cambridge University Press, 2003), Chapter 17.

11 This result can be deduced from the Kerr metric given in endnote 8, noting that, for a maximally corotating black hole, the event horizon occurs at $r_g = GM/c^2$.

12 The physics of black holes is described in C. W. Misner, K. S. Thorne and J. A. Wheeler, *Gravitation* (New York: W. H. Freeman and Company, 1973) and S. I. Shapiro and S. A. Teukolsky, *Black Holes, White Dwarfs and Neutron Stars: The Physics of Compact Objects* (New York: Wiley Interscience, 1983).

13 Many of the key features of accretion discs in astrophysics are described in the review by J. Pringle, Accretion discs in astrophysics, *Annual Review of Astronomy and Astrophysics*, **19**, 1981, 137–162, and in the excellent monograph by J. Frank, A. King and D. Raine, *Accretion Power in Astrophysics,* 3rd edn (Cambridge: Cambridge University Press, 2003).

14 Many astronomers contributed to the understanding of the optical spectra of active galactic nuclei. Many of the basic ideas are included in the excellent monograph by Donald E. Osterbrock, *Astrophysics of Gaseous Nebulae and Active Galaxies* (Mill Valley, California: University Science Books, 1989).

15 Excellent examples of the results of this type of study and the problems of interpretation are given by Bradley Peterson in *An Introduction to Active Galactic Nuclei* (Cambridge: Cambridge University Press, 1997).

16 The problems of finding definitive evidence for black holes in the nuclei of galaxies have been surveyed by Kormendy and Richstone (1995) and S. M. Faber, Black holes in galaxy centers, *Formation of Structure in the Universe*, eds A. Dekel and J. P. Ostriker (Cambridge: Cambridge University Press, 1999), pp. 337–359.

17 I have given an elementary derivation of these results in Chapter 21 of Longair, *High Energy Astrophysics*, Vol. 2.

18 Many important references to early studies of superluminal motions can be found in J. A. Zensus and T. J. Pearson, eds, *Superluminal Radio Sources* (Cambridge: Cambridge University Press, 1987).

19 Many different aspects of unification scenarios are discussed in G. V. Bicknell, M. A. Dopita and P. J. Quinn, eds, *First Stromlo Symposium: Physics of Active Galactic Nuclei*, ASP Conference Series, vol. 34 (San Francisco: ASP, 1994), and in the book by E. I. Robson, *Active Galactic Nuclei* (Chichester: John Wiley & Sons, in association with Praxis Publishing, 1996).

20 In his brief review of 1994, Robert Nemiroff (b. 1960) lists 100 models for γ-ray bursts. These include the merger of binary neutron stars or of a neutron star and a black hole, asteroids or comets falling onto neutron stars and black holes, and so on (Nemiroff, 1994).

21 An excellent survey of the physics of γ-ray bursts is given by Peter Mészáros in his review article (Mészáros, 2002).

A11   Explanatory supplement to Chapter 11

A11.1 Black holes as energy sources in high-energy astrophysics

The preferred model for the energy source in compact non-thermal sources involves accretion of matter onto black holes. The reasoning involves the following physical arguments.

A11.1.1 Time variability

One of the characteristic features of many luminous compact sources is their rapid time variability. For any mass M, the smallest physically meaningful dimension is the radius of the surface of infinite redshift, or event horizon, about a black hole of this mass. Radiation emitted from this radius is detected with zero frequency, or infinite wavelength, by an observer at infinity. For spherically symmetric, Schwarzschild black holes, this radius is at $r_g = 2GM/c^2 = 3(M/M_\odot)$ km, and for maximally rotating Kerr black holes it occurs at $r_g/2 = GM/c^2$. Therefore, to order of magnitude, causality implies that the shortest possible timescale associated with objects of mass M is given by

$$t_{min} \sim r_g/c = 10^{-5}(M/M_\odot) \text{ s.} \qquad (A11.1)$$

There are two points to be made about this relation. The first is that there is a last stable orbit about any black hole. For Schwarzschild black holes, this lies at $3r_g$, and for maximally rotating Kerr black holes it is at $r_g/2$. Nonetheless, to order of magnitude, equation (A11.1) is a useful lower limit to the timescale of variation associated with any mass M. The second point is that this calculation neglects the possibility that the sources of radiation might be moving relativistically. We discuss these effects in Section A11.4.

A11.1.2 The efficiency of accretion of matter onto black holes as an energy source

Accretion of mass onto compact objects is a very powerful source of energy for high-energy astrophysical objects. Suppose a mass element, m, is dropped from infinity onto a star with mass M and radius r. The kinetic energy acquired by the mass m increases as it falls towards the surface, and this is dissipated as heat when the material hits the surface. Thus, by conservation of energy, the energy released is GMm/r, and, if the material is continuously accreted at a rate $\dot{m}$, the luminosity due to accretion is given by $GM\dot{m}/r$. Expressing this result in terms of the Schwarzschild radius, $r_g = 2GM/c^2$, we obtain

$$L = \frac{GM\dot{m}}{r} = \frac{\dot{m}c^2}{2}\left(\frac{r_g}{r}\right). \tag{A11.2}$$

Thus, for matter accreting onto the surface of $1M_\odot$ neutron stars with radii 10 km, the luminosity is expected to be $0.15\dot{m}c^2$. In fact, this is an overestimate since this Newtonian calculation has neglected the effects of general relativity, which become important when r_g/r approaches unity. In the case of a spherically symmetric black hole, the maximum luminosity is $L = 0.057\dot{m}c^2$. Nonetheless, this efficiency of energy release is still about an order of magnitude greater than that associated with nuclear processes. For comparison, the conversion of hydrogen into helium only releases 0.7% of the rest-mass energy of the hydrogen nuclei. The corresponding energy release for a maximally rotating Kerr black hole is $0.42\dot{m}c^2$. Thus, in principle, the process of accretion onto black holes can release a substantial fraction of the rest-mass energy of the infalling material.

A11.1.3 The Eddington luminosity

From the considerations of Section A11.1.2, it might seem that, by increasing the mass accretion rate, $\dot{m}$, an arbitrarily large luminosity could be obtained, but this is not the case, because the radiation emitted by the compact object provides a radiation pressure which can halt the accretion process. The force of radiation pressure acts upon the electrons, which are strongly coupled electrostatically to the protons and nuclei in the plasma. Consequently, the inward force acting on 1 m³ of the infalling matter at radius r is $G(m_e + m_p)N_pM/r^2 \approx Gm_pN_pM/r^2$, where $N_p = N_e$ is the number density of electrons or protons. The radiation pressure acting on 1 m² of material at radius r is $p = \sigma_T N_e U_{rad}$, where σ_T is the Thomson cross-section and the energy density of radiation at radius r from a point source of luminosity L is $U_{rad} = L/4\pi r^2 c$. Equating the radiation pressure and gravitational forces, we find

$$\frac{\sigma_T N_e L}{4\pi r^2 c} = \frac{Gm_p N_p M}{r^2}. \tag{A11.3}$$

Note that both forces depend upon radius as $1/r^2$ and so the limiting luminosity, known as the *Eddington luminosity*, is independent of radius:

$$L_{Edd} = \frac{4\pi M m_p c}{\sigma_T} = 1.4 \times 10^{31}\left(\frac{M}{M_\odot}\right)\text{W}. \tag{A11.4}$$

Note that, in this calculation, we have adopted the minimum possible cross-section for the scattering of radiation by electrons. Thus, for isotropic radiation, the Eddington limit provides a rather firm upper limit to the luminosity of any source powered by accretion. Eddington's name is associated with this formula since the same result is found in considering the upper limit for the stability of massive stars which are radiation-dominated.

A11.2 Synchro-Compton radiation and the inverse Compton catastrophe

Wherever there are large number densities of soft photons, the presence of ultra-relativistic electrons in the same region must result in the production of high-energy photons, X-rays and γ-rays, by the process of *inverse Compton scattering*. The case of special interest in this chapter is that in which the same relativistic electrons which are the source of the soft photons are also responsible for scattering these photons to X-ray and γ-ray energies – this process is known as *synchro-Compton radiation*. One case of special importance is that in which the number density of low-energy photons is so great that most of the energy of the electrons is lost by synchro-Compton radiation rather than by synchotron radiation. This line of reasoning leads to what is known as the *inverse Compton catastrophe*.

We can derive the essential results from the formulae for synchrotron and inverse Compton radiation.[1] The mean energy loss rates for an electron of Lorentz factor γ in a magnetic field of flux density B and radiation field of energy density U_{rad} are, respectively, given by

$$-\left(\frac{dE}{dt}\right)_{sync} = \tfrac{4}{3}\sigma_T c \left(\frac{B^2}{2\mu_0}\right)\gamma^2, \tag{A11.5}$$

$$-\left(\frac{dE}{dt}\right)_{IC} = \tfrac{4}{3}\sigma_T c U_{rad}\gamma^2. \tag{A11.6}$$

Thus, the ratio, η, of the rates of loss of energy of an ultra-relativistic electron by inverse Compton and synchrotron radiation is given by

$$\eta = \frac{(dE/dt)_{IC}}{(dE/dt)_{sync}} = \frac{U_{photon}}{B^2/2\mu_0}. \tag{A11.7}$$

The synchro-Compton catastrophe occurs if this ratio is greater than unity. In that case, low-energy photons, say, radio photons produced by synchrotron radiation, are scattered to X-ray energies by the same flux of relativistic electrons. Since η is greater than unity, the energy density of the X-rays is greater than that of the radio photons and so the electrons suffer an even greater rate of loss of energy by scattering these X-rays to γ-ray energies. In turn, these γ-rays have a greater energy density than the X-rays . . . and so on. It can be seen that, as soon as the ratio given in equation (A11.7) becomes greater than unity, all the energy of the electrons is lost at the very highest energies and so the radio source should instead be a very powerful source of X-rays and γ-rays. Before considering the higher-order scatterings, let us study the first stage of the process for the case of a compact synchrotron self-absorbed radio source.

We need to determine the energy density of radiation within a synchrotron self-absorbed radio source. The flux density of such a source is given by

$$S_\nu = \frac{2kT_e}{\lambda^2}\Omega, \quad \text{where} \quad \Omega \approx \theta^2 = \frac{r^2}{D^2}; \tag{A11.8}$$

Ω is the solid angle subtended by the source, r is the size of the source and D is its distance. For a synchrotron self-absorbed source, the electron temperature of the relativistic electrons is the same as the brightness temperature of the source, $T_e = T_b$. The radio luminosity of the source is given by

$$L_\nu = 4\pi D^2 S_\nu = \frac{8\pi k T_e}{\lambda^2} r^2. \tag{A11.9}$$

Therefore, the energy density of the radio emission, U_{rad}, is given by

$$U_{\text{rad}} \sim \frac{L_\nu \nu}{4\pi r^2 c} = \frac{2k T_e \nu}{\lambda^2 c}. \tag{A11.10}$$

Note that we have used the fact that L_ν is the luminosity per unit bandwidth, and so the bolometric luminosity is roughly νL_ν. Therefore,

$$\eta = \frac{\left(\dfrac{2k T_e \nu}{\lambda^2 c}\right)}{\left(\dfrac{B^2}{2\mu_0}\right)} = \frac{4k T_e \nu \mu_0}{\lambda^2 c B^2}. \tag{A11.11}$$

We can now use the theory of self-absorbed radio sources to express the magnetic flux density, B, in terms of observables. The frequency of emission, ν, is related to the non-relativistic gyrofrequency, $\nu_g = eB/2\pi m_e$, by

$$\nu_g = \nu/\gamma^2, \tag{A11.12}$$

and the relation between temperature and Lorentz factor in the relativistic limit is given by

$$3k T_b = 3k T_e = \gamma m_e c^2, \tag{A11.13}$$

where T_b is the brightness temperature of the source. Reorganising these relations, we find

$$B = \frac{2\pi m_e}{e} \left(\frac{m_e c^2}{3k T_e}\right)^2 \nu. \tag{A11.14}$$

Therefore, the ratio of the loss rates, η, is given by

$$\eta = \frac{(dE/dt)_{\text{IC}}}{(dE/dt)_{\text{sync}}} = \left(\frac{81 e^2 \mu_0 k^5}{\pi^2 m_e^6 c^{11}}\right) \nu T_e^5. \tag{A11.15}$$

The important result is that the ratio of the loss rates depends very strongly upon the brightness temperature of the radio source. Putting in the values of the constants, the critical brightness temperature is given by

$$T_b = T_e = 10^{12} \nu_9^{-1/5} \text{ K}, \tag{A11.16}$$

where ν_9 is the frequency at which the brightness temperature is measured in units of 10^9 Hz, that is, in gigahertz. Thus, according to this calculation, no compact radio source should have brightness temperature greater than $T_B \approx 10^{12}$ K if the emission is incoherent synchrotron radiation.

The most compact sources, which have been studied by VLBI at centimetre wavelengths, have typical brightness temperatures $T_B \approx 10^{11}$ K which are less than the synchro-Compton limit. Note that this is direct evidence that the radiation is the emission of relativistic electrons, since the temperature of the emitting electrons must be at least 10^{11} K. This is not, however, the whole story. If the timescales of variability, τ, of the compact sources are used to estimate their physical sizes, $l \sim c\tau$, the source regions must be considerably smaller than those inferred from VLBI, and then values of T_B exceeding 10^{11} K are found. It is likely that relativistic beaming is the cause of this discrepancy.

It might appear that higher-order scatterings would result in a divergent situation in which the X-rays would be scattered to γ-ray energies. In fact, this does not occur because, at relativistic energies $h\nu \geq 0.5$ MeV, the Klein–Nishina cross-section rather than the Thomson cross-section should be used for photon–electron scattering. In the ultra-relativistic limit, the cross-section is given by

$$\sigma_{KN} = \frac{\pi^2 r_e^2}{h\nu} \left[\ln(2h\nu) + \frac{1}{2} \right], \qquad (A11.17)$$

and so the cross-section decreases as $(h\nu)^{-1}$ at high energies. Consequently, higher-order scatterings result in much reduced luminosities as compared with the non-relativistic calculation.

A11.3 The compactness parameter

The *compactness parameter* arises in considerations of whether or not a γ-ray source is opaque for $\gamma\gamma$ collisions because of pair production. Let us carry out a simple calculation which indicates how the compactness parameter arises. We carry out the calculation for the flux of γ-rays at the threshold for electron–positron pair production, $\varepsilon \sim m_e c^2$, for simplicity. The mean free path of the γ-ray for $\gamma\gamma$ collisions is $\lambda = (N_\gamma \sigma)^{-1}$, where N_γ is the number density of photons with energies $\varepsilon = h\nu \sim m_e c^2$. If the source has luminosity L_γ and radius r, the number density of photons within the source region is given by

$$N_\gamma = \frac{L_\gamma}{4\pi r^2 c\varepsilon}. \qquad (A11.18)$$

The condition for the source to be opaque is $r \approx \lambda$, that is

$$r \sim \frac{4\pi r^2 c m_e c^2}{L_\gamma \sigma}, \qquad \text{that is} \qquad \frac{L_\gamma \sigma}{4\pi m_e c^3 r} \sim 1. \qquad (A11.19)$$

The compactness factor, C, is defined to be the quantity

$$C = \frac{L_\gamma \sigma}{4\pi m_e c^3 r}. \qquad (A11.20)$$

Note that sometimes the compactness parameter is defined without the factor of 4π in the denominator. If the compactness parameter is very much greater than unity, the γ-rays are all destroyed by electron–positron pair production, resulting in a huge flux of electrons and positrons within the source region. Consequently, the source would no longer be a hard γ-ray source. The significance of the compactness parameter can be appreciated from observations of some of the intense γ-ray sources observed by the Compton Gamma-ray Observatory. These have enormous luminosities, $L_\gamma \sim 10^{41}$ W, and have been observed to vary significantly in intensity over timescales of the order of days. Inserting these values into equation (A11.20), it is found that $C \gg 1$, and so there is a problem in understanding why these sources exist. Fortunately, an answer is at hand since all the ultra-luminous

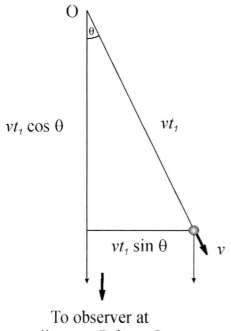

O

θ

$vt_1 \cos \theta$ vt_1

$vt_1 \sin \theta$ v

To observer at
distance D from O

Figure A11.1: The relativistic ballistic model of superluminal sources.

γ-ray sources are associated with compact radio sources, which exhibit synchrotron self-absorption, and many of which display superluminal motions. The inference is that the luminosities of the γ-ray sources and the timescales of variation have been significantly changed by the relativistic motion of the source region; we turn to this issue next.

A11.4 Superluminal motion and relativistic beaming

The most popular model for superluminal sources is known as the *relativistic ballistic model*. The simplest part of the calculation is the determination of the *kinematics* of relativistically moving source components. The aim is to determine the observed transverse speed of a component ejected at some angle, θ, to the line of sight at a high velocity v (Figure A11.1).

The observer is located at a distance D from the source. The source component is ejected from the origin, O, at some time t_0, and the signal from that event sets off towards the observer, where it arrives at time $t = D/c$ later. After time t_1, the component is located at a distance vt_1 from the origin and so is observed at a projected distance $vt_1 \sin \theta$ according to the distant observer. The light signal bearing this information arrives at the observer at time

$$t_2 = t_1 + \frac{D - vt_1 \cos \theta}{c}, \qquad (A11.21)$$

since the signals have to travel a slightly shorter distance $D - vt_1 \cos \theta$ to reach the observer. Therefore, according to the distant observer, the transverse speed of the component is given

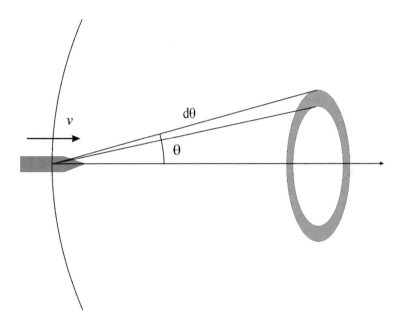

Figure A11.2: Illustrating the geometry of the propagation of light from a circular annulus on the surface of the Sun to an observer in a spaceship moving radially towards the Sun at speed v.

by

$$v_\perp = \frac{vt_1 \sin \theta}{t_2 - t} = \frac{vt_1 \sin \theta}{t_1 - \dfrac{vt_1 \cos \theta}{c}} = \frac{v \sin \theta}{1 - \dfrac{v \cos \theta}{c}}. \qquad (A11.22)$$

It is a simple sum to show that the maximum observed transverse speed occurs at an angle $\cos \theta = v/c$ and is given by $v_\perp = \gamma v$, where $\gamma = (1 - v^2/c^2)^{-1/2}$ is the Lorentz factor. Thus, provided the source component moves at a speed close enough to the speed of light, apparent motions on the sky, $v_\perp > c$, can be observed without violating causality and the postulates of special relativity. For example, if the source component were ejected at a speed $0.98c$, transverse velocities up to $\gamma c = 5c$ are perfectly feasible.

   This is the easy bit of the story. The trickier bit is to understand the effects of what is loosely referred to as 'relativistic beaming' upon the observed intensities of the source components. Let us consider first a classical undergraduate problem in relativity:

A rocket travels towards the Sun at speed $v = 0.8c$. Work out the luminosity, colour, angular size and brightness of the Sun as observed from the spaceship when it crosses the orbit of the Earth. It may be assumed that the Sun radiates like a uniform disc with a black-body spectrum at temperature T_0.

   This problem includes many of the effects found in relativistic beaming problems. Let us work out the separate effects involved in evaluating the intensity of radiation observed in the moving frame of reference. Consider the radiation from an annulus of angular width $\Delta\theta$ at angle θ with respect to the centre of the Sun (Figure A11.2).

The frequency shift of the radiation

It is simplest to use four-vectors to work out the frequency shifts and aberrations. The frequency four-vector in the frame of the Solar System, S, in Rindler's notation[2] is given by

$$\mathbf{K} = \left[\frac{\omega_0}{c}, -k_0 \cos \theta, -k_0 \sin \theta, 0 \right],$$
(A11.23)

where the light rays are assumed to propagate towards the observer at the orbit of the Earth, as illustrated in Figure A11.2. The frequency four-vector in the frame of reference of the spaceship, S', is given by

$$\mathbf{K}' = \left[\frac{\omega'}{c}, -k' \cos \theta', -k' \sin \theta', 0 \right].$$
(A11.24)

We use the time transform to relate the 'time' components of the four-vectors:

$$ct' = \gamma \left(ct - \frac{Vx}{c} \right),$$
(A11.25)

and so

$$\frac{\omega'}{c} = \gamma \left(\frac{\omega_0}{c} + \frac{V k_0 \cos \theta}{c} \right).$$
(A11.26)

Since $k_0 = \omega_0/c$,

$$\nu' = \gamma \nu_0 \left(1 + \frac{V}{c} \cos \theta \right) = \kappa \nu_0.$$
(A11.27)

This is the expression for the 'blueshift' of the frequency of the radiation due to the motion of the spacecraft.

The waveband $\Delta \nu$

This waveband, in which the radiation is observed, is blueshifted by the same factor,

$$\Delta \nu' = \kappa \Delta \nu_0.$$
(A11.28)

Time intervals

These are also different in the stationary and moving frames. This can be appreciated by comparing the periods of the waves as observed in S and S':

$$\nu' = \frac{1}{T'}; \quad \nu_0 = \frac{1}{T_0},$$
(A11.29)

and so

$$\frac{T'}{T} = \frac{\nu_0}{\nu'}.$$
(A11.30)

Since the periods T and T' can be considered to be the times measured on clocks, the radiation emitted in the time interval Δt is observed in the time interval $\Delta t'$ by the observer in S' such that

$$\Delta t' = \Delta t / \kappa.$$
(A11.31)

Solid angles

Finally, we need to work out how the solid angle subtended by the annulus shown in Figure A11.2 changes between the two frames of reference. It is simplest to begin with the cosine transform, which is derived from the '*x*' Lorentz transformation of the frequency four-vector:

$$\cos \theta' = \frac{\cos \theta + \dfrac{V}{c}}{1 + \dfrac{V}{c} \cos \theta}. \tag{A11.32}$$

Now, differentiating with respect to θ and θ' on both sides of this relation, we find

$$\sin \theta' \, \mathrm{d}\theta' = \frac{\sin \theta \, \mathrm{d}\theta}{\gamma^2 \left(1 + \dfrac{V}{c} \cos \theta\right)^2} = \frac{\sin \theta \, \mathrm{d}\theta}{\kappa^2}. \tag{A11.33}$$

This result has been derived for an annular solid angle with respect to the *x*-axis, but we can readily generalise to any solid angle since $\mathrm{d}\phi' = \mathrm{d}\phi$ and so

$$\sin \theta' \, \mathrm{d}\theta' \, \mathrm{d}\phi' = \frac{\sin \theta \, \mathrm{d}\theta \, \mathrm{d}\phi}{\kappa^2}; \quad \mathrm{d}\Omega' = \frac{\mathrm{d}\Omega}{\kappa^2}. \tag{A11.34}$$

Thus, the solid angle in S' is smaller by a factor κ^2 as compared with that observed in S. This is a key aspect of the derivation of the aberration formulae. Exactly the same form of beaming occurs in the derivation of the formulae for synchrotron radiation.

We can now put these results together to work out how the intensity of radiation from the region of the Sun within solid angle $\mathrm{d}\Omega$ changes between the two frames of reference. First of all, the intensity $I(\nu)$ is defined to be the power arriving at the observer per unit frequency interval per unit solid angle from the direction θ. The observer in the spacecraft observes the radiation arriving in the solid angle $\mathrm{d}\Omega'$ about the angle θ', and we need to transform its other properties to those observed in S'. Let us enumerate how the factors change the observed intensity. The energy, $h\nu N(\nu)$, received in S in the time interval Δt, in the frequency interval $\Delta \nu$ and in solid angle $\Delta \Omega$ is observed in S' as an energy $h\nu' N(\nu')$ in the time interval $\Delta t'$, in the frequency interval $\Delta \nu'$ and in solid angle $\Delta \Omega'$, where $N(\nu) = N(\nu')$ is the invariant number of photons. Therefore, the intensity observed in S' is given by

$$I(\nu') = I(\nu) \times \frac{\kappa \times \kappa \times \kappa^2}{\kappa} = I(\nu)\kappa^3. \tag{A11.35}$$

Now, let us apply this result to the spectrum of black-body radiation, for which

$$I(\nu) = \frac{2h\nu^3}{c^2} \left(e^{h\nu/kT} - 1\right)^{-1}. \tag{A11.36}$$

Then,

$$I(\nu') = \frac{2h\nu^3 \kappa^3}{c^2} \left(e^{h\nu/kT} - 1\right)^{-1} = \frac{2h\nu'^3}{c^2} \left(e^{h\nu'/kT'} - 1\right)^{-1}, \tag{A11.37}$$

where $T' = \kappa T$. In other words, the observer in S' observes a black-body radiation spectrum with temperature $T' = \kappa T$. A number of useful results follow from this analysis. For example, equation (A11.37) describes the temperature distribution of the cosmic microwave background radiation over the sky as observed from the Solar System, which is moving through the frame of reference in which the sky would be perfectly isotropic on the large scale at a velocity of about $600\,\mathrm{km\,s^{-1}}$. Since $V/c \approx 2 \times 10^{-3}$ and $\gamma \approx 1$, the temperature distribution is rather precisely a dipole distribution, $T = T_0[1 + (V/c)\cos\theta]$ with respect to the direction of motion of the Solar System through the cosmic microwave background radiation (see Figure 15.9(b)).

In the example of the spacecraft travelling at $v = 0.8c$ towards the Sun, we can illustrate a number of the features of relativistic beaming. In this case, $\gamma = 5/3$ and the angle at which there is no change of temperature, corresponding to $\gamma[1 + (V/c)\cos\theta] = 1$, is $\theta = 60°$.

Let us now turn to the case of relativistically moving source components. Evidently, all we need do is determine the value of κ for the source component moving at velocity V at an angle θ with respect to the line of sight from the observer to the distant radio source, as illustrated in Figure A11.1. In this case, a straightforward calculation shows that the value of κ is given by

$$\kappa = \frac{1}{\gamma\left(1 - \dfrac{V\cos\theta}{c}\right)}, \tag{A11.38}$$

where the source is moving towards the observer as illustrated in Figure A11.1. Just as in the above example, the observed flux density of the source is therefore

$$S(\nu_{obs}) = \frac{L(\nu_0)}{4\pi D^2} \times \kappa^3, \tag{A11.39}$$

where $\nu_{obs} = \kappa\nu_0$. In the case of superluminal sources, the spectra can often be described by a power law, $L(\nu_0) \propto \nu_0^{-\alpha}$, and so

$$S(\nu_0) = \frac{L(\nu_0)}{4\pi D^2} \times \kappa^{3+\alpha}. \tag{A11.40}$$

Thus, if the superluminal sources consisted of identical components ejected from the radio source at the same angle to the line of sight in opposite directions, the relative intensities of the two components would be in the ratio

$$\frac{S_1}{S_2} = \left(\frac{1 + \dfrac{v}{c}\cos\theta}{1 - \dfrac{v}{c}\cos\theta}\right)^{3+\alpha}. \tag{A11.41}$$

It is therefore expected that there should be large differences in the observed intensities of the jets. For example, if we adopt the largest observed velocities for a given value of γ, $\cos\theta = v/c$, then, in the limit $v \approx c$,

$$\frac{S_1}{S_2} = (2\gamma^2)^{3+\alpha}. \tag{A11.42}$$

Thus, since values of $\gamma \sim 10$ are quite plausible and $\alpha \sim 0\text{-}1$, it follows that the advancing component would be very much more luminous than the receding component. It is, therefore, not at all unexpected that the sources should be one-sided.

Another complication is the fact that the emission is often assumed to be associated with jets. Care has to be taken because, if the jet as a whole is moving at velocity v, then the time dilation formula, equation (A11.31), shows that the advancing component is observed in a different proper time interval as compared with the receding component, the time which has passed in the frame of the source being $\Delta t_1 = \kappa \Delta t_0$, where Δt_0 is the time measured in the observer's frame of reference. If the jet consisted of a stream of components ejected at a constant rate from the active galactic nucleus, the observed intensity of the jet would be enhanced by a factor of only $\kappa^{2+\alpha}$. Thus, the precise form of the relativistic beaming factor is model-dependent, and care needs to be taken about the assumptions made.

Let us first apply these considerations to the cases of sources exceeding the limiting surface brightness, $T_b = 10^{12}$ K, discussed in Section A11.2 and the *compactness parameter* discussed in Section A11.3. In the case of the inverse Compton catastrophe, equation (A11.15) shows that the ratio of the loss rates for inverse Compton scattering and synchrotron radiation depends upon the product $v T_b^5$. Since the brightness temperature $T_{obs} = \kappa T_0$ and $\nu_{obs} = \kappa \nu_0$, it follows that $\eta \propto \kappa^6$, and so the observed value of T_b can exceed 10^{12} K if the source is moving at such a high velocity that $\kappa \gg 1$.

In the case in which the compactness parameter,

$$C = \frac{L_\nu \sigma_T}{4\pi m_e c^3 \times ct},$$

(A11.43)

far exceeds unity, the relativistic beaming factors enable us to understand why these sources should exist. In equation (A11.36), it is assumed that the dimensions of the source are $l \approx ct$ from its rapid time variability. The observed luminosity is enhanced by a factor $\kappa^{3+\alpha}$, and, in addition, because the timescale of variability appears on the denominator of equation (A11.36), the observed value is shorter by a factor κ, and so the compactness parameter is increased by relativistic beaming by a factor of roughly $\kappa^{4+\alpha}$. Since $\alpha \approx 1$, it can be seen that $C \propto \kappa^5$, and so, in the frame of the source components themselves, the value of the compactness parameter can be reduced below the critical value.

Notes to Section A11

1 I have given detailed derivations of the formulae for synchrotron and inverse Compton radiation in Malcolm Longair, *High Energy Astrophysics*, vols I & II (Cambridge: Cambridge University Press, 1997, revised editions).

2 In Rindler's notation, the components of the four-vectors transform exactly as $[ct, x, y, z]$ according to the standard Lorentz transformation $ct' = \gamma(ct - Vx/c)$, $x' = \gamma(x - Vt)$, $y' = y$, $z' = z$. The invariant norm of the four-vector is $|R|^2 = c^2 t^2 - x^2 - y^2 - z^2$.

Part V

Astrophysical cosmology since 1945

The final five chapters of this book concern different aspects of cosmology. From a subject which barely existed before the Second World War, it developed into one of the central pillars of modern physics and astrophysics.

12 Astrophysical cosmology

This chapter concerns the development of astrophysical cosmology from 1945 to the early 1970s, by which time the success of the standard Big Bang models convinced the community at large that these provided the most satisfactory framework for the investigation of cosmological models. Then, in Chapter 13, we describe the endeavours to determine the values of the cosmological parameters and the problems which faced the observational cosmologists. It turned out that many of these endeavours encountered the problems of the evolution of the properties of the objects studied with cosmological epoch, and this is the subject of Chapter 14. In Chapter 15, we trace the development of ideas about the formation and evolution of galaxies and the large-scale structure of the Universe. These studies have provided many of the tools necessary to ask physical questions about the very early stages of the Universe, which is the subject of Chapter 16.

Many of the issues covered in this chapter on astrophysical cosmology up to the early 1970s are described in the book *Cosmology and Controversy* by Helge Kragh.[1]

12.1 Gamow and the Big Bang

During the 1930s, there were two reasons why the synthesis of the chemical elements in the early stages of evolutionary world models was taken seriously. Firstly, the studies of Cecilia Payne and Henry Norris Russell had shown that the abundances of the elements in stars were remarkably uniform, suggesting a common origin for the elements (see Section 3.3). The second consideration was that the interiors of stars seemed not to be hot enough for the nucleosynthesis of the chemical elements to take place. The starting point for studies of primordial nucleosynthesis was therefore to work out the equilibrium abundances of the elements at some high temperature and assume that, if the density and temperature decreased sufficiently rapidly, these abundances would remain 'frozen' as the Universe expanded and cooled.

Detailed calculations were carried out in 1942 by Chandrasekhar and Louis Henrich, who confirmed the expectation of equilibrium theory that, if the elements were in equilibrium at a high temperature, their abundances would be inversely correlated with their binding energies (Chandrasekhar and Henrich, 1942). The typical physical conditions under which this result was found involved densities of $\rho \approx 10^9 \, \mathrm{kg \, m^{-3}}$ and temperatures $T \approx 10^{10}$ K. There were, however, several gross discrepancies between their predictions and the observed abundances of the elements. The light elements, lithium, beryllium and boron, were predicted to be vastly overproduced relative to their cosmic abundances, and iron was predicted to be underproduced, as were all the heavier elements with mass numbers greater than about 70. This result was referred to as the 'heavy-element catastrophe'. It was concluded that all the chemical elements could not have been synthesised at a single density and temperature. Chandrasekhar and Henrich suggested that some non-equilibrium process was required.

In contrast to this equilibrium picture, Georges Lemaître proposed in 1931 that the Friedman models had evolved from an initial state which he termed a 'primaeval atom', consisting of vast numbers of protons, electrons and α-particles packed together at nuclear densities (Lemaître, 1931a). Such a huge 'atom' is necessarily unstable, and Lemaître proposed that the process of disintegration would give rise to the formation of the chemical elements. He also suggested that the energy released in the nuclear fission processes could account for the high energies of cosmic rays.

Lemaître's ideas provided the starting point for George Gamow's attack on the problem of the origin of the chemical elements.[2] In 1946, he accepted the conclusion that the synthesis of the chemical elements had to take place through non-equilibrium processes and he postulated that the early phases of the Friedman models were the most likely location where this might occur (Gamow, 1946). He extrapolated the Friedman models back to very early cosmological epochs, at which the densities were high enough for the nucleosynthesis to take place, and he found that the timescale of the Universe was then too short to establish an equilibrium distribution of the elements. In his original proposal, the initial state consisted of a sea of neutrons and subsequent β-decays and neutron capture processes would move nuclei towards the locus of nuclear stability.

Ralph Alpher (b. 1921) joined Gamow as a graduate student in 1946 and was given the task of working out the products of nucleosynthesis according to Gamow's prescription. Neutron capture cross-sections were available as a by-product of the nuclear physics programmes carried out during the Second World War, and these showed the encouraging result that there is an inverse correlation between the relative abundances of the chemical elements and their neutron capture cross-sections. In Alpher's first calculations, a smooth curve was fitted to the available data, and it was assumed that the initial conditions consisted of a sea of free neutrons. As protons became available as a result of the β-decay of the neutrons, heavier elements were synthesised by neutron capture. The nuclear reactions were assumed to begin only after the temperature had fallen below that corresponding to the binding energy of deuterium, $kT = 0.1$ MeV, and the Universe was assumed to be static. This theory was published in 1948 by Alpher, Bethe and Gamow, Bethe's name being added to complete the $\alpha\beta\gamma$ pun, and they found reasonable agreement with the observed abundances of the elements (Alpher, Bethe and Gamow, 1948). The importance of the paper was that it drew attention to the necessity of a hot, dense phase in the early Universe if the chemical elements were to be synthesised cosmologically.

In the same year, Alpher and Robert Herman (1914–1997) began improved calculations of primordial nucleosynthesis, but now including the dynamics of the expansion of the early Universe (Alpher and Herman, 1948). They realised that, at the necessary very high temperatures at early epochs, the Universe was then radiation- rather than matter-dominated, and they could then work out the subsequent thermodynamic history of the Universe. They found that the temperature history of the thermal background radiation corresponded closely

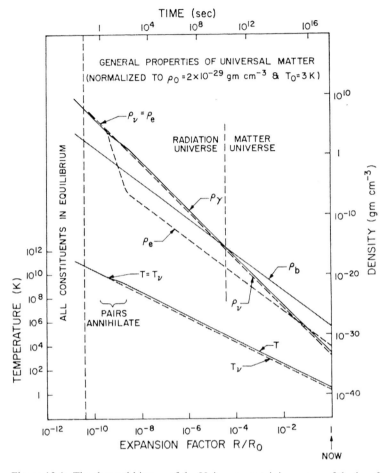

Figure 12.1: The thermal history of the Universe containing many of the key features described by Alpher and Herman (1948). This diagram was published by Wagoner, Fowler and Hoyle in 1967, following an earlier version by Robert Dicke and his colleagues (Dicke *et al.*, 1965).

to the adiabatic expansion of a photon gas, $T \propto R^{-1}$, where R is the scale factor of the Universe (Figure 12.1). From these results, they came to the far-reaching conclusion that the cooled remnant of these hot early phases should be present in the Universe today, and they estimated that the temperature of the thermal background should be about 5 K. This was the first prediction that there should exist diffuse background radiation in the centimetre and millimetre wavebands associated with what became known as the *Big Bang theory*[3] of the evolving Universe. Penzias and Wilson announced their discovery of the cosmic microwave background radiation in 1965 (Penzias and Wilson, 1965).

   There was, however, a major problem with this picture, which Gamow and his colleagues were well aware of – there are no stable nuclei with mass numbers 5 and 8, and hence it was difficult to understand how elements such as carbon, nitrogen and oxygen could have been created by the addition of further protons, neutrons or α-particles to helium nuclei. Enrico Fermi and Anthony Turkevich (1916–2002) carried out calculations of the evolution of the

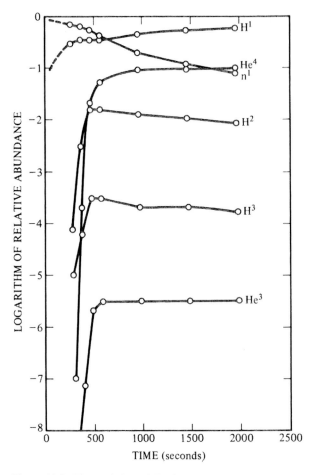

Figure 12.2: The evolution of the fraction (by number) of the light nuclei in a radiation-dominated Universe, according to calculations by Fermi and Turkevich, published by Alpher and Herman in 1950 (Alpher and Herman, 1950). The models began with 100% of the material in the form of neutrons.

abundances of the light elements, including 28 nuclear reactions for elements up to mass number 7 in a radiation-dominated expanding Universe, and their results were published by Alpher and Herman in 1950 (Alpher and Herman, 1950). These calculations showed that only about one part in 10^7 of the initial mass was converted into elements heavier than helium, far less than the cosmic abundances of the heavy elements (Figure 12.2).

In 1950, another key link in the chain was provided by Chushiro Hayashi, who pointed out that, in the early phases of the Universe at temperatures only ten times greater than those at which nucleosynthesis takes place, the neutrons and protons were maintained in thermodynamic equilibrium through the weak interactions

$$e^+ + n \leftrightarrow p + \bar{\nu}_e, \qquad \nu_e + n \leftrightarrow p + e^- \tag{12.1}$$

(Hayashi, 1950). Furthermore, at about the same temperature, electron–positron pair production ensures a plentiful supply of positrons and electrons. The result was that, rather than assume arbitrarily that the initial conditions consisted of a sea of neutrons, the equilibrium abundances of protons, neutrons, electrons and all the other constituents of the early Universe could be calculated exactly. In 1953, Alpher, James Follin (b. 1919) and Herman worked out the evolution of the proton–neutron ratio as the Universe expanded, and they obtained answers remarkably similar to modern calculations (Alpher, Follin and Herman, 1953). They left, however,

for future study to re-examine the formation of the elements by thermonuclear reactions as a subsequent part of the picture developed here.

They had come very close indeed to the modern picture of the thermal and nuclear evolution of the early Universe, including the important result that, in the standard Big Bang picture, about 25% of the primordial material by mass is converted into helium. Before this result became an established feature of astrophysical cosmology, however, steady state cosmology and the nucleosynthesis of the chemical elements in stars occupied centre stage.

12.2   Steady state cosmology

As Hermann Bondi[4] has remarked, the period immediately after the Second World War was one of considerable uncertainty concerning the physical basis of cosmology. Observationally, there was a timescale problem because Hubble's estimate of the rate of expansion of the Universe corresponded to a value of H_0^{-1} of only 2×10^9 years. This is the maximum age which any of the Friedman models can have, if the cosmological constant is set equal to zero, and it was known that the age of the Earth was greater than this value. The only solution within the standard picture was to introduce the cosmological constant so that the timescale of the Universe could be stretched out as illustrated in Figure 6.4. This model did not have much appeal for Bondi who felt that it was a contrived solution to suppose that the cosmological constant had precisely the value that would result in a Universe which almost reaches the Einstein stationary state at the present epoch, but not quite.

There were many new ideas in the air. Milne had developed his theory of kinematic relativity in which he supposed that there are two different times, one associated with dynamical phenomena and another with electromagnetic phenomena (Milne, 1948). Dirac had been profoundly impressed by coincidences between the very large numbers in physics and the properties of the Universe – for example, the square of the ratio of the strengths of electromagnetic and gravitational forces is roughly equal to the numbers of protons in the Universe. A consequence of his identification of these large numbers was his inference that the gravitational constant might change with time (Dirac, 1937). Eddington was completing his *Fundamental Theory*, in which he attempted to account for the values of the fundamental constants of physics and in which the cosmological constant appeared as a fundamental constant of nature (Eddington, 1946).[5] What was remarkable about these new ideas was that they were based upon concepts about what the underlying physics might be without any strong physical motivation. Perhaps the most extreme example was the remark of Eddington, in which he asserted

Generalisations that can be reached epistemologically have a security which is denied to those that can only be reached empirically.

These views can scarcely have appealed to a generation of physicists who were coming to terms with the completely new concepts of relativistic quantum mechanics, which accounted spectacularly for the experimentally determined properties of matter at the atomic level. Herbert Dingle (1890–1978) was particularly outspoken in condemning Eddington, Milne, Dirac and others for what he termed 'Modern Aristotelianism' (Dingle, 1937).

It was in this atmosphere that the idea of steady state cosmology was born. Fred Hoyle has recorded the delightful story of the flash of inspiration which gave rise to the concept of steady state cosmology. In his reminiscences of late 1946 or early 1947, he wrote:[6]

In a sense, the steady-state theory may be said to have begun on the night that Bondi, Gold and I patronised one of the cinemas in Cambridge. The picture, if I remember rightly, was called *The Dead of Night*. It was a sequence of four ghost stories, seemingly disconnected as told by the several characters in the film, but with the interesting property that the end of the fourth story connected unexpectedly with the beginning of the first, thereby setting-up the potential for a never-ending cycle. When the three of us returned that evening to Bondi's rooms in Trinity College, Gold suddenly said: 'What if the Universe is like that?'

In an earlier version of his reminiscences,[7] Hoyle remarked:

One tends to think of unchanging situations as being necessarily static. What the ghost-story film did sharply for all three of us was to remove this wrong notion. One can have unchanging situations that are dynamic, as for example a smoothly flowing river. The universe had to be dynamic, since Hubble's red-shift law proved it to be so ... From this position, it did not take us long to see that there would need to be a continuous creation of matter.

There the matter rested until 1948 when the three of them returned to the cosmological problem. Bondi and Gold took a quite different approach from Hoyle to the exposition of the theory. Bondi and Gold derived the theory by very general, almost philosophical, arguments (Bondi and Gold, 1948), whereas Hoyle built his theory up from a field-theoretical description of the process of continuous creation of matter, which he described by what he called the C-field (Hoyle, 1948). The creation rate of matter amounted to only one particle per cubic metre every 300 000 years. Both theories resulted in the same unique form for the metric of space-time.

In the approach of Bondi and Gold, the cosmological principle was extended to what they termed the *perfect cosmological principle*, according to which the Universe presents the same large-scale picture to all observers *at all times*. Hence, Hubble's constant becomes a fundamental constant of nature, on a par with the charge on the electron or the gravitational constant. This hypothesis also disposed of a problem, which concerned Bondi and Gold, namely the general question of whether or not it was safe to assume that the laws of physics are unchanging with time – by definition, all the laws were to be unchanging with time. They showed that the perfect cosmological principle leads to a unique metric for the dynamics of the Universe with zero spatial curvature[8] and a scale factor which changes with time as $R(t) = R_0 \exp(t/t_0)$, where $t_0 = H_0^{-1}$. Consequently, the Universe was infinite in age, but the age of typical objects observed in the local Universe is only $\frac{1}{3} H_0^{-1}$. Thus, our own Galaxy had to be rather older than the typical object in the Universe, but that was not too unreasonable because, if Hubble's constant were $500 \ \mathrm{km \ s^{-1} \ Mpc^{-1}}$, our own Galaxy would be exceptionally large, much larger than other spiral galaxies. Bondi and Gold went on to evaluate the counts of galaxies expected in the steady state model and found that they agreed remarkably well with Hubble's counts of galaxies.

The papers by Bondi and Gold and by Hoyle were published in 1948, and they immediately attracted considerable attention, both within the astronomical community and, through Hoyle's radio broadcasts, among the public at large. One consequence of the theory of the greatest importance for astrophysics was that Hoyle set about attempting to find an alternative means of understanding the formation of the chemical elements, and this was one of the motivations for his remarkable prediction of the carbon resonance (Hoyle, 1954) and the subsequent fundamental paper on the processes of nucleosynthesis in stars by Burbidge *et al.* (1957) (see Section 8.2). Bondi asked what evidence there was for any relics of the hot early phases of the Universe and, with these new insights, the abundances of the chemical elements disappeared as evidence.

The idea of the continuous creation of matter was a major stumbling block for many physicists and astronomers, but as early as 1951, William McCrea had a deep insight into the physics of Hoyle's proposal (McCrea, 1951). McCrea realised that there was a quite different interpretation of the metric of steady state cosmology, which evokes a resonance with contemporary cosmology. To quote McCrea,

The single admission that the zero of absolute stress may be set elsewhere than is currently assumed on somewhat arbitrary grounds permits all of Hoyle's results to be derived within the system of General Relativity theory. Also, this derivation gives the results an intellectual physical coherence.

McCrea wrote the physics of the steady state picture in terms of a negative energy equation of state, $p = -\rho c^2$, and recovered the three features of the theory, namely,

- the density of the Universe is a constant;
- the spatial geometry is flat;
- the scale factor varies as $\exp[H_0(t - t_0)]$.

It is intriguing that McCrea had realised that there is nothing intrinsically implausible about a negative energy equation of state. In fact, these ideas had already been foreshadowed by Lemaître in 1933. There is a close relation between the mathematics of a Universe expanding exponentially under the influence of the cosmological constant Λ and the steady state picture.[9] Lemaître had suggested that the Λ-term could be interpreted in terms of a negative energy equation of state (Lemaître, 1933). In his words,

Everything happens as though the energy *in vacuo* would be different from zero.

In the 1950s, two important results were reported of central importance for cosmology. The first concerned the value of Hubble's constant. At the meeting of the International Astronomical Union in Rome in 1952, Walter Baade announced that the distance of the Andromeda Nebula (M31) had been underestimated by a factor of 2 (Baade, 1952). The reason for this change was his discovery of the different types of stellar populations in galaxies (Section 5.7). The principal indicators used to determine the distance of M31 were the Cepheid variables, and Baade discovered that there is a difference in the period–luminosity relations for Cepheids of Populations I and II. By using the same type of Cepheid variable in our own Galaxy, in the Magellanic Clouds and in M31, the distance to M31 increased by a factor of 2. This result also eliminated the problems that the globular clusters in M31 appeared to be intrinsically fainter than those in our Galaxy and that our Galaxy appeared to be exceptionally large. From the point of view of cosmology, Hubble's constant was reduced to 250 km s^{-1} Mpc^{-1} and hence H_0^{-1} increased to 4×10^9 years. In 1956, Humason, Nicholas Mayall (1903–1993) and Sandage published their redshift–magnitude relation for 474 galaxies and Hubble's constant was revised downwards again to 180 km s^{-1} Mpc^{-1} and hence H_0^{-1} increased to 5.6×10^9 years (Humason, Mayall and Sandage, 1956). These revisions reduced, and quite possibly eliminated, the discrepancy between the age of the Earth and the age of the Universe according to the standard Friedman models if the cosmological constant Λ is set equal to zero.

12.3   The counts of radio sources

The second piece of evidence resulted from the surveys of extragalactic radio sources which began in the early 1950s. The central figure in this story was Martin Ryle, who was leading the initiatives in radio astronomy at the Cavendish Laboratory in Cambridge. Initially, he had been strongly wedded to the idea that most of the radio sources were radio stars belonging to our own Galaxy, but, by early 1954, he had been converted to the idea that most of the radio sources observed at high galactic latitudes are extragalactic. Ryle and Anthony Hewish designed and constructed a large four-element interferometer to carry out a new survey of the sky at 81.5 MHz, which, being an interferometer, would be sensitive to small angular diameter sources. The second Cambridge (2C) survey of radio sources was completed in 1954, and the first results were published in the following year (Shakeshaft *et al.*, 1955). Ryle and his colleagues found that the small-diameter radio sources were uniformly distributed over the sky and that the numbers of sources increased enormously as the survey extended to fainter and fainter flux densities. In any uniform Euclidean model, the numbers of sources brighter than a given limiting flux density, S, are expected to follow the relation $N(\geq S) \propto S^{-3/2}$ (see Section A5.1). In contrast, Ryle found a huge excess of faint radio sources, the slope of the source counts between 20 and 60 Jy being described by $N(\geq S) \propto S^{-3}$ (Figure 12.3). He concluded that the only reasonable interpretation of these data was that the sources were extragalactic, that they were objects similar in luminosity to the radio galaxy Cygnus A and that there was a much greater comoving number density of radio sources at large distances than there are nearby. As Ryle expressed it in his Halley Lecture in Oxford in 1955 (Ryle, 1955),

This is a most remarkable and important result, but if we accept the conclusion that most of the radio stars are external to the Galaxy, and this conclusion seems hard to avoid, then there seems no way in which the observations can be explained in terms of a Steady-State theory.

These remarkable conclusions came as a surprise to the astronomical community. There was enthusiasm, and also some scepticism, that such profound conclusions could be drawn

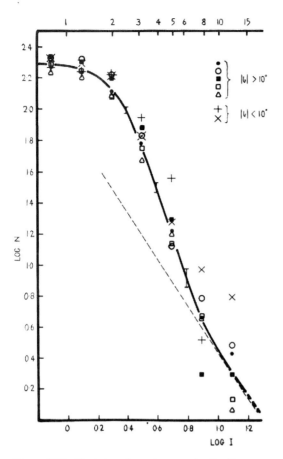

Figure 12.3: The integral number counts of radio sources from the 2C survey of radio sources; N is the number of radio sources brighter than flux density I, the units of I being 10^{-25} W m^{-2} sr^{-1}. The dashed line shows the 'Euclidean' number counts of radio sources. The observations show a very large excess of faint radio sources relative to the expectations of the Euclidean world model (Ryle, 1955).

from the counts of radio sources, particularly when their physical nature was not understood and only the brightest 20 or so objects had been associated with relatively nearby galaxies.

The Sydney group, led by Bernard Mills (b. 1920), were carrying out similar radio surveys of the southern sky at about the same time with the large cross-shaped radio telescope known as the Mills Cross, and they found that the source counts could be represented by the relation $N(\geq S) \propto S^{-1.65}$, which they argued was not significantly different from the expectation of uniform world models. In 1957 Mills and Bruce Slee stated (Mills and Slee, 1957):

We therefore conclude that discrepancies, in the main, reflect errors in the Cambridge catalogue, and accordingly deductions of cosmological interest derived from its analysis are without foundation. An analysis of our results shows that there is no clear evidence for any effect of cosmological importance in the source counts.

The problem with the Cambridge number counts was that they extended to surface densities of radio sources such that the flux densities of the faintest sources were overestimated because of the presence of faint sources in the beam of the telescope, a phenomenon known as *confusion*. Peter Scheuer, who was Martin Ryle's research student from 1951 to 1954, devised a statistical procedure for deriving the number counts of sources from the survey records themselves without the need to identify individual sources (Scheuer, 1957). The technique which he discovered, which he referred to as the $P(D)$ technique and which has since been adopted in many other astronomical contexts, showed that the slope of the source counts was actually -1.8. Ironically, this result, which is exactly the correct answer, was not trusted, partly because the mathematical techniques used by Scheuer were somewhat forbidding and also because his result differed from the prejudices of both Ryle and Mills. The dispute reached its climax at the Paris Symposium on Radio Astronomy in 1958, but the conflicting positions were not resolved (Bracewell, 1959). The strong feelings aroused by the contrasting views of the proponents of the evolutionary and steady state models are vividly described in Kragh's book *Cosmology and Controversy* (see endnote 1).

The resolution of the controversy only came with the construction of the next generation of radio telescopes, which had higher angular resolution and hence were less sensitive to the effects of source confusion. In the next Cambridge catalogues, the 3C catalogue (Edge *et al.*, 1959) and, particularly, the revised 3C catalogue (Bennett, 1962), much more care was taken to eliminate the effects of confusion, and the accuracy of the radio source positions improved so that identifications could be made with fainter galaxies. These showed that Ryle's conclusions of 1955 were basically correct, but that the magnitude of the excess had been considerably overestimated. More radio sources were identified with distant galaxies, and the optical identification programmes led to the discovery of quasars in the early 1960s (see Section 11.2). The lessons of the 1950s had been taken to heart, and detailed studies were made of the effects of confusion and the absence of sources of large angular size from the catalogues. The radio source counts derived from the 4C catalogues (Pilkington and Scott, 1965; Gower, Scott and Wills, 1967) showed an excess over the expectations of Euclidean world models (Gower, 1966). In fact, the discrepancies with the uniform Friedman and steady state models were much greater than this simple comparison suggested because the predicted radio source counts converge rapidly as soon as the source populations extended to significant redshifts (Longair, 1971; Scheuer, 1975). By the mid 1960s, the evidence was compelling that there was indeed an excess of sources at large redshifts and this was at variance with the expectations of the steady state theory.[10]

12.4   The helium problem

In 1964, while I was completing my first year of research at Cambridge, Fred Hoyle gave a course of lectures on the problems of extragalactic research. He would arrive with, at best, a scrap of paper with some notes and expound an area of current research. One week, the topic was the problem of the cosmic helium abundance. Helium is one of the more difficult elements to observe astronomically because of its high excitation potential, and so it can only be observed in very hot stars. Donald Osterbrock and John Rogerson had shown in

1961 that the abundance of helium seemed to be remarkably uniform wherever it could be observed and that it corresponded to about 25% by mass (Osterbrock and Rogerson, 1961). A further important observation was reported by O'Dell (b. 1937) in 1963 of the helium abundance in a planetary nebula in the old globular cluster M15 (O'Dell, Peimbert and Kinman, 1964). Despite the fact that the heavy elements were deficient relative to their cosmic abudances, the helium abundance was still about 25%.

The evidence on the cosmic helium abundance was reviewed by Hoyle, and he then described the work of Gamow, Alpher, Herman and Follin concerning the problems of synthesising the heavy elements in the early phases of the Big Bang (see Section 12.1). Although helium is synthesised in the central regions of stars during their long phases of evolution on the main sequence, it is most unlikely that this process could have created as much helium as 25% by mass of the baryonic matter in the Universe. Most of the luminosity of galaxies is associated with the burning of hydrogen into helium in main-sequence stars and so, if the luminosity of our Galaxy had remained more of less the same throughout its lifetime, an upper limit of about 1% of the mass of the Galaxy could have been converted into helium. Furthermore, the stars move off the main sequence when only about 10% of their mass has been converted into helium, and then the helium is burned into heavier elements. It was difficult to understand why there should be a universal abundance of about 25% by mass if the helium was created in stars.

By 1964, when Hoyle was delivering his lecture, it was possible to carry out primordial nucleosynthesis calculations more accurately. At that time, Roger Tayler (1929–1997) had just returned to Cambridge and was present in the audience. Hoyle and Tayler realised that they could undertake much more precise calculations and, over the following week, they and Tayler's research student, John Faulkner (b. 1937), worked out the details of the formation of helium in the early phases of the Big Bang. The audience had the privilege of being present as a key piece of modern astrophysics was created in real time in a graduate lecture course. Hoyle and Tayler obtained the result that about 25% helium by mass is synthesised in the Big Bang, in remarkable agreement with observation and essentially independent of the overall baryonic matter density in the Universe. The reason for the constancy of the cosmic helium abundance is that it is primarily determined by the thermodynamics of the early Universe, rather than by the microphysics involved in the nuclear reactions (see Section A12.1). Their paper was published in *Nature* in 1964 (Hoyle and Tayler, 1964).

12.5 The discovery of the cosmic microwave background radiation

One consequence of the Big Bang model which Hoyle and Tayler did not mention explicitly in their paper was that the cooled remnant of the thermal radiation present during the very hot early phases should be detectable at centimetre and millimetre wavelengths.[11] Alpher and Herman's prediction had been more or less forgotten when Gamow's theory of primordial nucleosynthesis had failed to account for the creation of the chemical elements. The idea of searching for thermal radiation from the Big Bang was revived in the early 1960s by Yakov Zeldovich and his colleagues in Moscow and by Robert Dicke (1916–1997) and his colleagues in Princeton.[12]

In 1964, Andrei Doroshkevich (b. 1937) and Igor Novikov reanalysed the physics of the Big Bang model and showed that the thermal background radiation with a Planck spectrum at radiation temperature between about 1 and 10 K should be present in the Universe at the present day (Doroshkevich and Novikov, 1964). They pointed out that this prediction provided a key test of the Big Bang scenario. They also noted that useful limits to the background radiation temperature could be obtained from the measurements of Edward Ohm (b. 1926) in 1961 of the radio background emission at centimetre wavelengths published in the reports of the Bell Telephone Laboratories (Ohm, 1961). In fact, Ohm had discovered an excess noise temperature of 3.3 K in his experiments, but believed that this figure was within the measurement errors of the total signal detected by his antenna and receiver system, which was 22.3 K. There were, however, earlier indications of a diffuse extragalactic component of the background radiation. In 1955, Émile Le Roux (b. c1930) detected a uniform background of 3 ± 2 K at a wavelength of 33 cm at the Nançay Radio Observatory, somewhat greater than expected from the population of discrete radio sources (Dennise, Le Roux and Steinberg, 1957). In 1957, Tigran Shmaonov (b. 1930) published a measurement of the radiation temperature of the background radiation at 3.2 cm of 4 ± 3 K, but it had been forgotten, and he only brought the result to the attention of Igor Novikov[13] much later, in 1983 (Shmoanov, 1957).

The very next year, in 1965, the microwave background radiation was discovered by Arnold Penzias and Robert Wilson, more or less by accident. They had joined the Bell Telephone Laboratories in the early 1960s with the intention of using the same 20-foot horn reflector used by Ohm, which had been built to test telecommunication with the Echo satellite, for radio astronomical observations. Penzias and Wilson had the responsibility of calibrating the antenna for use at these frequencies, for which they had built a 7.35 cm cooled maser receiver. The understanding was that the telescope could be used for astronomical observations for some fraction of the observing time. Wherever they pointed the telescope on the sky, they found an excess antenna temperature, which could not be accounted for by noise sources in the telescope or receiver system. A list of contributions to the total detected signal is given in Table 12.1. Having carefully calibrated all parts of the telescope and receiver system, they found that there remained about 3.5 ± 1 K excess noise contribution (Penzias and Wilson, 1965).

At almost exactly the same time, Robert Dicke's group in Princeton were preparing exactly the same type of experiment to detect the cooled remnant of the Big Bang. Discussions with the Princeton group ensued, and it became apparent that Penzias and Wilson had discovered the diffuse cosmic microwave background radiation, exactly what the Princeton physicists were searching for. Within a few months, the Princeton group had measured a background temperature of 3.0 ± 0.5 K at a wavelength of 3.2 cm, confirming the black-body nature of the background in the Raleigh–Jeans region of the spectrum (Roll and Wilkinson, 1966).

Remarkably, there was earlier evidence for a diffuse component of millimetre radiation with this radiation temperature from the study of several faint interstellar absorption lines associated with the molecules CH, CH^+ and CN. In the case of CN, for example, absorption was observed from the first rotationally excited state of the molecule as well as the ground state. In 1941, Andrew McKellar (1910–1960) had shown that the necessary excitation

Table 12.1. *Contributions to the total measured radio signal in Penzias and Wilson's experiments at* 4.08 GHz (7.35 cm)

Signal	Noise signal T/K
Total zenith noise temperature	6.7 ± 0.3
Atmospheric emission	2.3 ± 0.3
Ohmic losses	0.8 ± 0.4
Backlobe response	≤ 0.1
Cosmic background radiation	3.5 ± 1.0

temperature to populate the first excited state was 2.3 K, although the origin of the excitation was then unknown (McKellar, 1941).

Many measurements of the background radiation were made in the following years. At millimetre wavelengths, the observations were very difficult because of atmospheric absorption, and a number of balloon observations were made which were broadly consistent with a black-body radiation spectrum at a temperature of about 2.7 K. The best way of avoiding the problems of atmospheric absorption was to carry out the observations from space, and this was achieved with the launch of the Cosmic Background Explorer (COBE) in November 1990. This first results of this experiment showed that the spectrum of the cosmic background radiation is of black-body form, the radiation temperature being 2.725 ± 0.01 K (Mather *et al.*, 1990). The final results, reported in 1996, show that deviations from a perfect black-body spectrum amount to less than 0.03% of the maximum intensity over the waveband 2.5 to 0.5 mm, the most perfect naturally occurring black-body spectrum (Fixsen *et al.*, 1996) (Figure 12.4). The radiation temperature was 2.728 ± 0.004 K (95% confidence level).

12.6 The helium problem revisited

The appearance of the paper by Hoyle and Tayler (1964) and the discovery of the cosmic background radiation stimulated a number of detailed studies of the synthesis of the light elements during the period of primordial nucleosynthesis when the Universe was a few minutes old. After Dicke's group published their results confirming the discovery of the background radiation, James Peebles published two papers exploring the constraints on cosmological models which could be derived from the observed abundances of deuterium and helium in the Universe at the present time (Peebles, 1966a,b). The standard radiation-dominated Big Bang picture makes quite definite predictions about the abundance of the light elements, but it depends upon the dynamics of the Universe through the epochs when the neutrinos decouple from matter and the synthesis of the light elements begins. Peebles showed that if the early expansion were speeded up by a factor of 10 to 100 as compared with the standard picture, the neutrinos would decouple earlier and greater amounts of helium

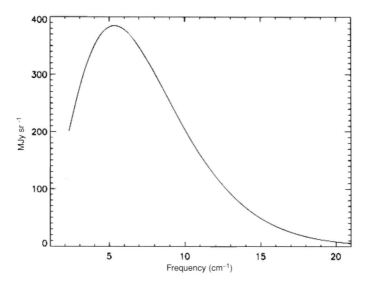

Figure 12.4: The final spectrum of the cosmic microwave background radiation as measured by the COBE satellite (Fixsen *et al.*, 1996). The units of the abscissa are inverse centimetres, so that ten units corresponds to 1 mm and five correspond to 2 mm. The experimental uncertainties are less than the thickness of the line.

would be produced. This might happen if the gravitational constant had changed with time, as in the Brans–Dicke cosmology (Brans and Dicke, 1961), or if the early expansion of the Universe had been anisotropic. Similar calculations were carried out independently by the Moscow group (Doroshkevich *et al.*, 1971).

In 1967, Robert Wagoner (b. 1938), William Fowler and Fred Hoyle repeated the analysis carried out by Hoyle and Tayler, but now using all the available cross-sections for many more nuclear interactions between light nuclei, and with the knowledge that the cosmic microwave background radiation had a temperature of about 2.7 K (Wagoner, Fowler and Hoyle, 1967).[14] Fowler's deep understanding of nuclear physics contributed greatly to all aspects of these computations[15] and enabled a very detailed network of the many nuclear interactions involved in the synthesis of the light elements to be created (Figure 12.5). These calculations confirmed that about 25% of helium by mass is created by primordial nucleosynthesis and that this figure is remarkably independent of the present density of baryonic matter in the Universe. The reason for this is that the amount of helium produced depends primarily upon the neutron–proton ratio when the neutrinos decouple from the nuclear reactions which maintain the abundances in equilibrium (see Section A12.1). Of particular importance was their demonstration that the abundances of other products of nucleosynthesis, deuterium, ^{3}He and ^{7}Li are sensitive to the mean baryon density in the Universe (Figure 12.6). The importance of observations of these elements is that they are very difficult to synthesise in stars because they have relatively small nuclear binding energies – deuterium and ^{3}He are destroyed rather than created in stars.

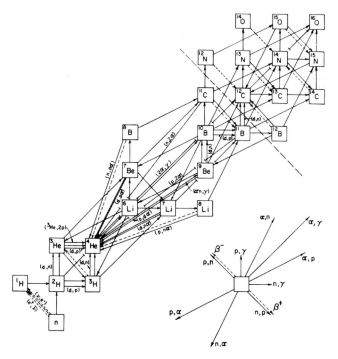

Figure 12.5: The network of reactions used by Wagoner in his determination of the primordial abundances of the elements (Wagoner, 1973). This network is an enhanced version of that used by Wagoner, Fowler and Hoyle in their paper of 1967 (Wagoner *et al.*, 1967).

Interstellar absorption lines of deuterium were discovered in the ultraviolet region of the spectrum by John Rogerson (b. 1922) and Donald York (b. 1944) in 1973 from observations made by the Copernicus ultraviolet satellite (Figure 7.12) (Rogerson and York, 1973). An interstellar deuterium abundance of 1.5×10^{-5} by mass relative to hydrogen was found. Subsequent observations showed that the same deuterium abundance is found along the line of sight to other stars which could be observed by the Copernicus satellite (Vidal-Madjar *et al.*, 1977). These observations enabled an upper limit to be placed upon the mean baryonic density of the Universe of 1.5×10^{-28} kg m^{-3}, corresponding to $\Omega_B h^2 \leq 10^{-2}$. If the mean baryonic density of the Universe were any greater, deuterium would be underproduced by primordial nucleosynthesis and no other way of creating deuterium astrophysically was known. This important upper limit to the baryon density in the Universe was at least an order of magnitude less than the critical cosmological density.

The story of the ^{7}Li abundance was more complicated. Observations of this isotope in Population I stars had shown that there is an upper limit to its abundance of about 1 part in 10^9 relative to hydrogen by mass. Stars like the Sun have much lower ^{7}Li abundances than this value because lithium is destroyed when it is convected or diffuses into regions with temperatures greater than about 2×10^6 K. It seemed natural to suppose that the primordial abundance of ^{7}Li was about 10^{-9}. In 1982, François Spite (b. 1930) and Monique Spite

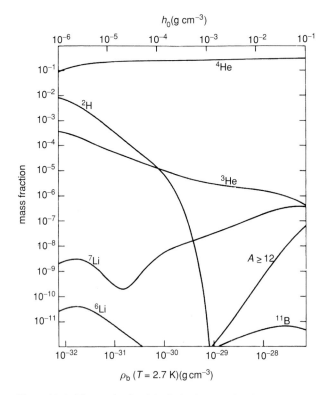

Figure 12.6: The synthesis of the light elements in Big Bang according to the calculations of Wagoner, Fowler and Hoyle, as revised by Wagoner in 1973 (Wagoner *et al.*, 1967; Wagoner, 1973). These computations demonstrated the sensitivity of the abundances of deuterium, ^{3}He and ^{7}Li to the present baryon density of the Universe.

(b. 1939) observed a remarkably constant value for the ^{7}Li abundance of 10^{-10} in old Population II stars (Spite and Spite, 1982). The abundance did not vary with the surface temperatures of the stars, suggesting that the ^{7}Li was primordial. It was, however, intriguing that an abundance of ^{7}Li in the range 10^{-9} to 10^{-10} would be entirely consistent with a cosmological origin (Figure 12.6).

It was appreciated by Hoyle, Tayler, Fowler, Wagoner and Peebles, and also by the Moscow group working with Zeldovich, that the synthesis of the heavy elements provided an important diagnostic tool for the dynamics of the Universe at the epoch of nucleosynthesis. If the Universe expanded too rapidly, the neutron–proton ratio would freeze out at a higher temperature and helium would be overproduced. This result enabled important constraints to be placed upon any variations of the gravitational constant with time as well as restricting the number of permissible neutrino species to three, a result subsequently confirmed by experiments with the Large Electron–Positron Collider at CERN from studies of the energy widths of the decay products of the Z^0 bosons.[16]

Thus, by the end of the 1960s, the overwhelming balance of opinion was that there was convincing evidence for two relics of the hot early phases of the Universe, the cosmic microwave background radiation and the cosmic abundances of the light elements. Since that time the Big Bang picture has become the preferred framework for astrophysical cosmology. But, equally important, rather than being the preserve of speculation, geometrical and astrophysical cosmology acquired a quite new status as the province of genuine astrophysical enquiry. It is not coincidental that from this period onwards particle physicists, and physicists in general, had to embrace the early Universe as an integral part of physics. The astronomers had provided the particle physicists with the ultimate particle physics laboratory.

Notes to Chapter 12

1 H. Kragh, *Cosmology and Controversy: The Historical Development of Two Theories of the Universe* (Princeton: Princeton University Press, 1996).

2 The history of Gamow's work on the Big Bang theory is described by R. A. Alpher and R. C. Herman in Early work on 'Big Bang' cosmology and the cosmic black body radiation, *Modern Cosmology in Retrospect*, eds B. Bertotti, R. Balbinot, S. Bergia and A. Messina (Cambridge: Cambridge University Press, 1990), pp. 129–157, and by R. V. Wagoner, Deciphering the nuclear ashes of the early Universe: a personal perspective, in *Modern Cosmology in Retrospect*, pp. 159–185.

3 There had been references to the expansion of the standard Friedman models as an explosion or 'bang', but the term *Big Bang* entered the cosmological literature with some force following a series of BBC radio broadcasts by Fred Hoyle in the spring of 1949. His use of the term has been interpreted as a pejorative remark, in contrast to his enthusiasm for the steady state theory that he, Bondi and Gold had developed over the preceding years (see Section 12.2). According to Simon Mitton, Hoyle did not intend the term to be derogatory (S. Mitton, *Conflict in the Cosmos: Fred Hoyle's Life in Science* (Washington D.C.: Joseph Henry Press, 2005)). The lectures were subsequently published, essentially unmodified, by Hoyle in his book *The Nature of the Universe* (Oxford: Basil Blackwell, 1950). The term is not a particularly happy one, since it conjures up a somewhat misleading impression of how the standard isotropic world models are constructed, but it is now firmly embedded in the literature.

4 Bondi's revealing thoughts about observational and theoretical cosmology during the pre- and post-War years are contained in H. Bondi, The cosmological scene 1945–1952, in *Modern Cosmology in Retrospect*, pp. 189–196.

5 Eddington (1946) was published posthumously under the editorship of E. Whittaker.

6 This reminiscence is recounted by Hoyle in An assessment of the evidence against the steady-state theory, in *Modern Cosmology in Retrospect*, pp. 221–231.

7 This reminiscence is contained in F. Hoyle, *Steady-State Cosmology Re-visited* (Cardiff: Cardiff University Press, 1980).

8 I have shown how these results may be simply derived from the Robertson–Walker metric in Section 19.5 of Malcolm Longair, *Theoretical Concepts in Physics* (Cambridge: Cambridge University Press, 2003).

9 There is a delightful sequel to this story. On the occasion of his 80th birthday in 1995, I invited Hoyle to lecture to the Cavendish Physical Society. He was delighted to accept this invitation because he had given his first lecture on steady state cosmology to the Cavendish Physical Society in 1948. Hoyle remarked wryly that his only mistake had been to call his creation field C rather than ψ. The exponential expansion of the early Universe according to the inflation picture involves a scalar field ψ which performs exactly the same function as Hoyle's C-field or Lemaître's cosmological constant Λ.

10 It was a great sadness that relations between Hoyle and Ryle were so soured by the controversy over the radio source counts. In 1965, Peter Scheuer and I attempted a reconciliation between them when the four of us got together in Hoyle's newly founded Institute of Theoretical Astronomy in

Cambridge to try to understand their different positions. Sadly, there was no longer any common ground, and each simply repeated their entrenched views. It was one of the saddest events of my career.

11 According to Roger Tayler (personal communication), he had included this result in his draft of their paper, but it did not appear in the published version.

12 Some appreciation of the scope of Zeldovich's contributions to astrophysics and cosmology can be found in *Selected Works of Yakob Borisovich Zeldovich*, vol. 2, eds J. P. Ostriker, G. I. Barenblatt and R. A. Sunyaev (Princeton: Princeton University Press, 1993). An account of Dicke's role in the revival of the observational study of Gamow's picture of the Big Bang can be found in P. J. E. Peebles, *Principles of Physical Cosmology* (Princeton: Princeton University Press, 1993).

13 Novikov describes the remarkable early history of attempts to detect the cosmic microwave background radiation in his paper Discovery of the CMB, Sakharov oscillations and polarization, in *Historical Development of Modern Cosmology*, ASP Conference Series, vol. 252, eds V. J. Martinez, V. Trimble, and M. J. Pons-Bordeia (San Francisco: ASP, 2001), pp. 43–53. This also includes the story of Shmaonov's measurements. See also, Kragh, *Cosmology and Controversy*, p. 343.

14 When Hoyle first presented these results in Cambridge in 1967, many of us were surprised that it seemed as though he had been converted to the Big Bang picture. As in the paper itself, however, equal weight was given to the idea that these computations could also be applied at very much higher baryonic densities to very massive stars which collapsed and 'bounced'. The nucleosynthesis of the expansion phase was exactly the same as a Universe of very high baryonic mass density. The densities were so high that heavy elements could be synthesised in these stars. The subsequent paper by Wagoner concentrated upon the primordial synthesis of the light elements (Wagoner, 1973).

15 Wagoner's review of the history of these computations provides a vivid picture of how Hoyle, Fowler and he interacted during these exciting years. See Deciphering the nuclear ashes of the early Universe: a personal perspective, Bertotti *et al.*, eds, *Modern Cosmology in Retrospect*, pp. 159–185.

16 See, for example, the Opal Collaboration, A combined analysis of the hadronic and leptonic decays of the Z^0, *Physics Letters*, **B240**, 1990, 497–512.

A12   Explanatory supplement to Chapter 12

A12.1 *The primordial abundances of the light elements*

The reason for the remarkable stability of the prediction of 25% helium by mass in the standard Big Bang picture can be understood from the following physical arguments.

Consider a particle of mass m at very high temperatures such that its total energy is much greater than its rest-mass energy, $kT \gg mc^2$. If the timescales of the interactions which maintain this species in thermal equilibrium with all the other species present at temperature T are shorter than the age of the Universe at that epoch, the equilibrium number densities of the particle and its antiparticle are given by the standard expression from statistical mechanics,

$$N = \overline{N} = \frac{4\pi g}{h^3} \int_0^\infty \frac{p^2 \, \mathrm{d}p}{\mathrm{e}^{E/kT} \pm 1}, \tag{A12.1}$$

where g is the statistical weight of the particle, p is its momentum and the $\pm$ sign depends upon whether the particles are fermions ($+$) or bosons ($-$). Now, photons are massless bosons for which $g = 2$, nucleons, antinucleons, electrons and positrons are fermions with $g = 2$ and the electron, muon and tau neutrinos are fermions with helicity for which $g = 1$. The equilibrium number densities, N, and energy densities, ϵ, can be found from this expression. For (i) photons, (ii) nucleons and electrons and (iii) neutrinos, respectively, these are given by:

$$\text{(i)} \quad g = 2, \quad N = 0.244 \left(\frac{2\pi kT}{hc} \right)^3 \text{m}^{-3}, \quad \epsilon = aT^4; \tag{A12.2}$$

$$\text{(ii)} \quad g = 2, \quad N = 0.183 \left(\frac{2\pi kT}{hc} \right)^3 \text{m}^{-3}, \quad \epsilon = \tfrac{7}{8}aT^4; \tag{A12.3}$$

$$\text{(iii)} \quad g = 1, \quad N = 0.091 \left(\frac{2\pi kT}{hc} \right)^3 \text{m}^{-3}, \quad \epsilon = \tfrac{7}{16}aT^4. \tag{A12.4}$$

To find the total energy density, we add all the equilibrium energy densities together, to obtain

$$\text{total energy density} = \epsilon = \chi(T)\,aT^4. \tag{A12.5}$$

When the particles become non-relativistic, $kT \ll mc^2$, and the abundances of the different species are still maintained by interactions between the particles, the non-relativistic limit of the integral gives an equilibrium number density,

$$N = g \left(\frac{mkT}{h^2} \right)^{3/2} \exp\left(-\frac{mc^2}{kT} \right). \tag{A12.6}$$

Thus, once the particles become non-relativistic, they no longer contribute to the inertial mass density which determines the rate of expansion of the Universe.

Now consider the abundances of protons and neutrons in the early Universe. At redshifts less than 10^{12}, the neutrons and protons are non-relativistic, $kT \ll mc^2$, and their abundances are maintained at their thermal equilibrium values by the electron–neutrino weak interactions

$$e^+ + n \rightarrow p + \bar{\nu}_e; \qquad \nu_e + n \rightarrow p + e^-. \tag{A12.7}$$

For the neutrons and protons, the values of g are the same, and so the relative abundances of neutrons to protons is given by

$$\left[\frac{n}{p} \right] = \exp\left[-\frac{(m_n - m_p)c^2}{kT} \right] = \exp\left(-\frac{\Delta mc^2}{kT} \right), \tag{A12.8}$$

where Δmc^2 is the mass difference between the neutron and the proton.

This abundance ratio freezes out when the neutrino interactions can no longer maintain the equilibrium abundances of neutrons and protons. The condition for 'freezing out' is that the timescale of the weak interactions becomes greater than the age of the Universe.

The variation of the energy density of radiation and temperature during the early radiation-dominated phases of the Universe are given by

$$\epsilon = \chi a T^4 = \frac{3c^2}{32\pi G} t^{-2},$$ (A12.9a)

$$T = \left(\frac{3c^2}{32\pi G \chi a}\right)^{1/4} t^{-1/2} = 10^{10} t^{-1/2} \text{ K},$$ (A12.9b)

where it is assumed that the number of neutrino species $N_\nu = 3$ and so $\chi = 43/8$. The time, t, is measured in seconds. Note that equations (A12.9 a,b) illustrate how the early expansion rate depends upon the gravitational constant G and the number of neutrino species through χ.

The processes which prevent the neutrinos escaping freely are:

$$e^- + e^+ \rightarrow \nu_e + \bar{\nu}_e; \quad e^\pm + \nu_e \rightarrow e^\pm + \nu_e; \quad e^\pm + \bar{\nu}_e \rightarrow e^\pm + \bar{\nu}_e.$$ (A12.10)

Straightforward calculations[1] show that the timescales for the expansion of the Universe and the decoupling of the neutrinos are the same when the Universe was almost precisely one second old at a temperature of 10^{10} K At that time, $t = 1$ s, the neutron fraction, as determined by equation (A12.8), was

$$\left[\frac{n}{n+p}\right] = 0.21.$$ (A12.11)

The neutron fraction decreases very slowly after this time. Detailed calculations show that after 300 s the neutron fraction has fallen to 0.123. It is at this epoch that the bulk of the formation of the light elements takes place. Almost all the neutrons are combined with protons to form ^{4}He nuclei, so that, for every pair of neutrons, a helium nucleus is formed. The reactions involved are

$$p + n \rightarrow {}^3\text{He} + \gamma; \quad n + D \rightarrow {}^3\text{H} + \gamma; \quad p + {}^3\text{H} \rightarrow {}^4\text{He} + \gamma;$$ (A12.12)
$$n + {}^3\text{He} \rightarrow {}^4\text{He} + \gamma; \quad d + d \rightarrow {}^4\text{He} + \gamma; \quad {}^3\text{He} + {}^3\text{He} \rightarrow {}^4\text{He} + 2p.$$ (A12.13)

Most of the nucleosynthesis takes place at a temperature less than about 1.2×10^9 K since, at greater temperatures, the deuterons are destroyed by the γ-rays of the background radiation. The binding energy of deuterium is $E_B = 2.23$ MeV and so this energy is equal to kT at $T = 2.6 \times 10^{10}$ K. However, the photons far outnumber the nucleons, and it is only when the temperature of the expanding gas has decreased to about 26 times less than this temperature that the number of dissociating photons is less than the number of nucleons. Although the neutrons begin to decay spontaneously by this time, the bulk of them survive and so, according to the above calculation, the predicted helium to hydrogen mass ratio is just twice the neutron fraction:

$$\left[\frac{{}^4\text{He}}{\text{H}}\right] \approx 0.25.$$ (A12.14)

The predicted abundance of deuterium is a strong function of the present density of the Universe, in contrast to the constant abundance of helium. The reasons for this can be understood as follows. The helium abundance results from the equilibrium ratio of protons to neutrons as the Universe cools down; that is, it is primarily determined by the thermodynamics of the expanding radiation-dominated Universe. On the other hand, the abundance of deuterium depends upon the number density of nucleons. If the Universe has a high baryon number density, then essentially all the deuterons are converted into helium, whereas if the Universe is of low density, not all the deuterium is converted into ^{4}He. The same argument applies to ^{3}He. Thus, the deuterium and ^{3}He abundances set an upper limit to the present baryon density of the Universe.

Note to Section A12

1 I have given a simple version of this calculation in Malcolm Longair, *Galaxy Formation* (Berlin: Springer-Verlag, 1998), Sections 10.2 and 10.3.

13 The determination of cosmological parameters

13.1 Sandage and the values of H_0 and q_0

In 1952, Walter Baade announced that the value of Hubble's constant, H_0, had been over-estimated because the distance to the Andromeda Nebula, M31, adopted by Hubble was about a factor of 2 too small (Baade, 1952). The cause of the discrepancy was that there is a difference in the period–luminosity relations for Cepheid variables of Populations I and II (see Section 12.2). By using the same type of Cepheid variable in our own Galaxy, in the Magellanic Clouds and in M31, the distance to M31 increased by a factor of 2. Consequently, Hubble's constant was reduced to $250 \, \mathrm{km \, s^{-1} \, Mpc^{-1}}$ and H_0^{-1} increased to 4×10^9 years.

In 1956, Humason, Mayall and Sandage showed that the expected redshift–magnitude relation, $m = 5 \log_{10} z + \text{constant}$, is observed for galaxies selected at random, but there is a large scatter about the mean relation because of the breadth of the luminosity function of galaxies (Humason *et al.*, 1956). It had been known since Hubble's pioneering studies of the 1930s, however, that the brightest galaxies in clusters of galaxies follow a very much tighter relation[1] which follows precisely Hubble's law $v = H_0 r$ (Figure 13.1). Thus, in order to estimate the value of H_0, it was only necessary to calibrate the observed relation by measuring the distance of the nearest rich cluster of galaxies, the Virgo cluster of galaxies, by techniques independent of its redshift. Humason, Mayall and Sandage estimated the distance of the giant spiral galaxy NGC 4321, one of the brightest galaxies in the Virgo cluster, assuming that the brightest stars and nebulae in that galaxy were the same as those in M31. Hubble's constant was revised downwards again to $180 \, \mathrm{km \, s^{-1} \, Mpc^{-1}}$. In 1958, Sandage's best estimate of H_0 was reduced yet again from 180 to $75 \, \mathrm{km \, s^{-1} \, Mpc^{-1}}$ (Sandage, 1958). The principal reason for this further downward revision was that what had been thought to be the brightest stars in some of the most distant galaxies studied turned out to be regions of ionised hydrogen and star clusters.

Immediately after the Second World War, the prime instrument for cosmological research was the Palomar 200-inch telescope, which was commissioned in 1948. In 1961, Allan Sandage published an influential paper entitled 'The ability of the 200-inch telescope to discriminate between selected world models', in which different approaches to the determination of cosmological parameters with this telescope were discussed critically (Sandage, 1961b). The observed properties of galaxies at large redshifts depend upon the geometry

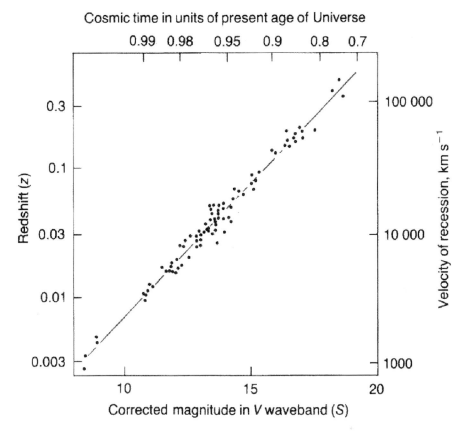

Figure 13.1: The redshift–magnitude relation in the visual (V) waveband for the brightest galaxies in clusters presented by Sandage in 1968 (Sandage, 1968). The straight line shows the expected relation if the galaxies all have the same intrinsic luminosity, $m = 5 \log_{10} z + $ constant. The sparsity of points at redshifts greater than 0.3 illustrates the difficulty of finding clusters of galaxies at large redshifts that would be suitable for cosmological tests.

of the world model and upon its kinematics between the epochs of emission and reception of the radiation. To repeat the list of parameters, these are:

- Hubble's constant, $H_0 = \dot{R}/R$, the present rate of expansion of the Universe;
- the deceleration parameter, $q_0 = -\ddot{R}/\dot{R}^2$, the present deceleration of the Universe;
- the curvature of space, $\kappa = R_0^{-2}$;
- the mean density of matter in the Universe at the present epoch, ρ_0, and its value relative to the critical density, $\rho_{\text{crit}} = 3H_0^2/8\pi G$;
- the present age of the Universe, T_0;
- the cosmological constant, λ.

As pointed out in Section 6.8, these are not independent, *provided* the Friedman models are a correct description of the large-scale dynamics of the Universe. Thus, for the Friedman models,

$$\kappa = \mathcal{R}^{-2} = \frac{(\Omega_0 - 1) + \frac{1}{3}(\lambda/H_0^2)}{(c/H_0)^2}; \quad q_0 = \frac{\Omega_0}{2} - \frac{1}{3}\frac{\lambda}{H_0^2}, \tag{13.1}$$

where $\Omega_0 = \rho_0/\rho_{crit}$ is the *density parameter*. If $\lambda = 0$, there is a simple one-to-one relation between the geometry of the world models, their densities and dynamics, $q_0 = \Omega_0/2$ and $\kappa = \mathcal{R}^{-2} = (\Omega_0 - 1)/(c/H_0)^2$. Thus, if the Friedman models with non-zero cosmological constant are adopted, three independent parameters need to be determined, for example H_0, q_0 and Ω_0. If the cosmological constant is zero, only two parameters need be determined, say H_0 and q_0. The steady state theory was uniquely defined by the single parameter H_0. Ideally, the three parameters should be determined independently, and then equations (13.1) provide a test of the general theory of relatively on the largest scales accessible to us.

Sandage fully recognised the magnitude of the task involved. He discussed the use of the redshift–magnitude relation for giant elliptical galaxies, the angular diameter–redshift test, the number counts of galaxies and the ages of the oldest stars as means of constraining, if not estimating, the values of the cosmological parameters. The differences between the world models only become significant at redshifts greater than about 0.3. For example, the difference in apparent magnitude of a galaxy at redshift 0.5 between the steady state model with $q_0 = -1$ and a Friedman model with $q_0 = 1$ and $\lambda = 0$ was only 0.9 magnitudes.[2]

Sandage discussed in detail the problems of using these different techniques to determine cosmological parameters, and he concluded that the most promising route was the use of the redshift–apparent magnitude relation for the brightest galaxies in clusters for which the dispersion in absolute magnitude about the mean relation was only about $\Delta m \approx 0.3$. His best estimate for q_0 was 1 ± 0.5, but it could have ranged between 0 and 3. He warned that the analysis involves a number of important selection effects which needed to be taken into account before a convincing estimate could be made. In particular, he emphasised the importance of the *Malmquist bias*, according to which intrinsically brighter objects are selected in studies which extend to the limit of observational capability (Malmquist, 1920).

Sandage also noted that there was a discrepancy between the ages of the oldest globular clusters, which were estimated to be about 15×10^9 years, and the age of the Universe, which, for the $q_0 = 1$, $\lambda = 0$ model with $H_0 = 75 \, \mathrm{km\,s^{-1}\,Mpc^{-1}}$, was $T_0 = 7.42 \times 10^9$ years. A solution to this problem would be to assume that the cosmological constant was positive, which would result in a negative value for q_0. Sandage took the view that there were probably too large uncertainties in the estimates of H_0, q_0 and T_0 for this result to be taken too seriously. He devoted an enormous effort to the determination of the basic cosmological parameters, Hubble's constant, H_0, the deceleration parameter, q_0, and the age of the Universe, T_0 using the telescopes at the Palomar Observatory, and, until the 1970s, his work dominated the field.[3] Indicative of his approach to observational cosmology during these years was the title of his paper, 'Cosmology – the search for two numbers' (Sandage, 1970).

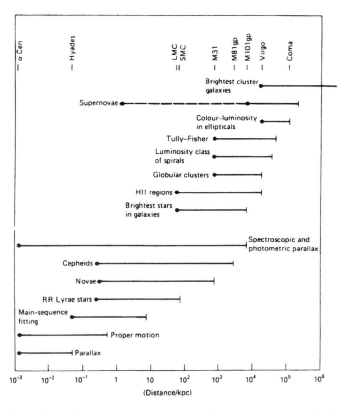

Figure 13.2: Illustrating the 'cosmological distance ladder'; after M. Rowan-Robinson, *The Cosmological Distance Ladder* (New York: W. H. Freeman and Company, 1985). The diagram shows roughly the range of distances over which different classes of object can be used to estimate astronomical distances.

13.2 Hubble's constant

Hubble's constant, H_0, appears ubiquitously in cosmological formulae, and its value was the subject of considerable controversy for many years. The use of the redshift–magnitude relation for the brightest cluster galaxies had the advantage that Hubble's law is defined well beyond distances at which there might have been deviations associated with the peculiar motions of clusters and superclusters of galaxies. Therefore, Hubble's constant could be found if the distances to the nearest rich clusters of galaxies could be estimated accurately.

The traditional approach to this calibration involved a hierarchy of distance indicators to extend the local distance scale from the vicinity of the Solar System to the nearest giant cluster of galaxies, the Virgo cluster. The only direct methods of distance measurement involve stellar parallaxes, and these can only be used for stars in the neighbourhood of the Sun. To extend the distance scale further, it is assumed that objects of the same intrinsic types can be identified at greater distances. Then, their relative brightnesses provide estimates of their distances. Examples of the different techniques used are summarised in Figure 13.2.

The period–luminosity relation for Cepheid variables, discovered by Henrietta Leavitt in 1912 (Leavitt, 1912), provides one of the best means of extending the distance scale from our own Galaxy to nearby galaxies, but, even using the 200-inch telescope, it was only possible to use this procedure to distances of about 1–2 Mpc. Other techniques were used to extend the distance scale from the neighbourhood of our Galaxy to the Virgo cluster, including the luminosity functions of globular clusters, the brightest stars in galaxies and the luminosities of Type I supernovae at maximum light. In 1977, Brent Tully and Richard Fisher discovered the relation between the absolute magnitudes of spiral galaxies and the velocity widths of their 21 cm line emission (Tully and Fisher, 1977). This relation could be determined for a number of spiral galaxies in a nearby group or cluster and then relative distances found by assuming that the same correlation between their intrinsic properties is found in more distant groups and clusters (see Section 10.1).

From the 1970s until the 1990s, there was an ongoing controversy concerning the value of Hubble's constant.[4] In a long series of papers, Sandage and Gustav Tammann (b. 1932) found values of Hubble's constant about $50 \, \text{km} \, \text{s}^{-1} \, \text{Mpc}^{-1}$, whereas de Vaucouleurs, Aaronson, Mould and their collaborators found values of about $80 \, \text{km} \, \text{s}^{-1} \, \text{Mpc}^{-1}$. The nature of the discrepancy can be appreciated from their estimates of the distance to the Virgo cluster. If its distance is 15 Mpc, the higher estimate of H_0 is found, whereas if the distance is 22 Mpc, values close to $50 \, \text{km} \, \text{s}^{-1} \, \text{Mpc}^{-1}$ are obtained. Sandage and Tammann repeatedly emphasised how sensitive the distance estimates are to observational selection effects, such as the Malmquist effect, and systematic errors.

During the 1990s, a major effort was made to resolve these differences, much of it stimulated by the capability of the Hubble Space Telescope (HST) to measure Cepheid variable stars in the Virgo cluster of galaxies. When the HST project was approved in 1977, one of its major scientific objectives was to use its superb sensitivity for faint star-like objects to enable the light curves of Cepheid variables in the Virgo cluster to be determined precisely and so estimate the value of Hubble's constant to 10% accuracy. This programme was raised to the status of an HST Key Project in the 1990s, with a guaranteed share of observing time to enable a reliable result to be obtained.

The Key Project team, led by Wendy Freedman (b. 1957), carried out an outstanding programme of observations and analysis of these data. Equally important was the fact that the team used, not only the HST data, but also all the other distance measurement techniques, to ensure internal self-consistency of the distance estimates. For example, the improved determination of the local distance scale in our own Galaxy from the parallax programmes of the *Hipparcos* astrometric satellite improved significantly the reliability of the calibration of the local Cepheid distance scale. The great advance of the 1990s was that the distances of many nearby galaxies became known very much more precisely than they were previously. As a result, by 2000, there was relatively little disagreement among the experts about the distances of those galaxies which had been studied out to the distance of the Virgo cluster. If there were differences, they arose from how the data were to be analysed once the distances were known, in particular in the elimination of systematic errors and biases in the observed samples of galaxies. The final result of the project, published in 2001, was $70 \pm 7 \, \text{km} \, \text{s}^{-1} \, \text{Mpc}^{-1}$, where the errors are one-sigma errors (Freedman *et al.*, 2001).

In addition to the traditional approach, new *physical methods* of measuring H_0 became available, which have the advantage of eliminating many of the steps involved in the cosmological distance ladder. They are based upon measuring a physical dimension l of a distant object, independent of its redshift, and its angular size θ, so that an angular diameter distance, D_A, can be found from $D_A = l/\theta$ at a known redshift z. A beautiful example of the use of this technique was described by Nino Panagia (b. 1943) and his colleagues, who combined IUE observations of the time-variability of the emission lines from the supernova SN 1987A in the Large Magellanic Cloud with Hubble Space Telescope observations of the emission-line ring observed about the site of the explosion to measure the physical size of the ring (Panagia *et al.*, 1991) (see Section 8.10.2). The distance found for the Large Magellanic Cloud was as accurate as that found by the traditional procedures.

Another promising method, suggested originally by Walter Baade in 1926 and modified by Adriaan Wesselink (1909–1995) in 1947, involves measuring the properties of an expanding stellar photosphere (Baade, 1926; Wesselink, 1947). If the velocity of expansion can be measured from the Doppler shifts of the spectral lines, and the increase in size estimated from the change in luminosity and temperature of the photosphere, the distance of the star can be found. The *Baade–Wesselink method* was first applied to supernovae by David Branch (b. 1942) and Bruce Patchett (1948–1996) in 1973 (Branch and Patchett, 1973) and by Robert Kirshner and John Kwan (b. 1947) in 1974 (Kirshner and Kwan, 1974). It was successfully applied to the supernovae SN 1987A in the Large Magellanic Cloud by Ronald Eastman (b. 1958) and Kirshner, resulting in a distance consistent with other precise distance measurement techniques (Eastman and Kirshner, 1989). Extending the Baade–Wesselink technique to ten Type II supernovae with distances ranging from 50 kpc to 120 Mpc, Brian Schmidt (b. 1967) and his colleagues found a value of H_0 of 60 ± 10 km s^{-1} Mpc^{-1} (Schmidt, Kirshner and Eastman, 1992).

Another approach which has produced promising results involves the use of the hot gaseous atmospheres in clusters of galaxies, the properties of which can be measured from their X-ray emission and from the Sunyaev–Zeldovich decrement in the cosmic microwave background radiation due to inverse Compton scattering (Figure 10.12) (Gunn, 1978). As discussed in Section 10.5.1, clusters of galaxies contain vast quantities of hot gas which is detected by its X-ray bremsstrahlung. The X-ray surface brightness depends upon the electron density, N_e, and the electron temperature, T_e, through the relation $I_\nu \propto \int N_e^2 T_e^{-1/2} \, dl$. The electron temperature, T_e, can be found from the shape of the bremsstrahlung spectrum. Furthermore, the decrement in the background due to the Sunyaev–Zeldovich effect is proportional to the Compton optical depth, $y = \int (kT_e/m_e c^2)\sigma_T N_e \, dl \propto \int N_e T_e \, dl$. Thus, the physical properties of the hot gas are over-determined and the physical dimensions of the X-ray emitting volume can be found. Steven Myers (b. 1962) and his colleagues estimated a value of $H_0 = 54 \pm 14$ km s^{-1} Mpc^{-1} from detailed studies of the Abell clusters A478, A2142 and A2256 (Myers *et al.*, 1997).

Another example of a physical method of measuring H_0 is to use gravitational lensing of distant objects by intervening galaxies or clusters. The first gravitationally lensed quasar, 0957+561 (Figure 13.3), was discovered by Dennis Walsh (1933–2005), Robert Carswell (b. 1940) and Ray Weymann (b. 1934) in 1979 (Walsh, Carswell and Weymann 1979). The gravitational deflection of the light from the quasar by the intervening galaxy splits its

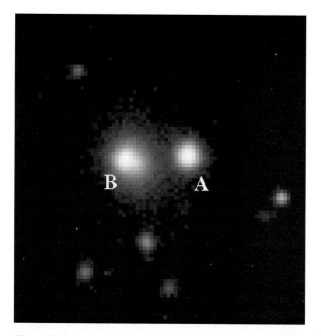

Figure 13.3: An optical image of the double quasar 0957+561 discovered by Walsh, Carswell and Weymann in 1979 (Walsh *et al.*, 1979). The spectra of the two quasars were identical. The small extension to the right of image B is the foreground galaxy responsible for the gravitational lensing. North is to the right of the image. (Image courtesy of Richard Ellis and Jean-Paul Kneib.)

image into a number of separate components. If the background quasar is variable, a time delay is observed between the variability of the different images because of the different path-lengths from the quasar to the observer on Earth. For example, a time delay of 418 days has been measured for the two components of the double quasar 0957+561 (Kundic *et al.*, 1997). This observation enables physical scales at the lensing galaxy to be determined, the main uncertainty resulting from the modelling of the mass distribution in the lensing galaxy. In the case of the double quasar 0957+561, Tomislav Kundic (b. 1968) and his colleagues claim that the mass distribution in the galaxy is sufficiently well constrained for the model-dependent uncertainties to be small, and they derived a value of Hubble's constant of $H_0 = 64 \pm 13$ km s^{-1} Mpc^{-1} at the 95% confidence level.

The estimates of Hubble's constant found by these physical methods are consistent with the value found by Freedman and her colleagues.

13.3   The age of the Universe, T_0

With Baade's revision of the value of Hubble's constant in 1952 and the further revision by Sandage in 1958, the discrepancy between the age of the Earth and H_0^{-1} was eliminated, but it was known that the ages of the oldest stars in globular clusters were considerably greater than that of the Solar System. Globular cluster ages were estimated by the method pioneered by Sandage and Schwarzschild in 1952 and involved the comparison of the Hertzsprung–Russell diagrams of the oldest, metal-poor, globular clusters with the expectations of the theory of stellar evolution from the main sequence onto the giant branch (Sandage and Schwarzschild, 1952).

The feature of these diagrams which is particularly sensitive to the age of the cluster is the *main-sequence termination point*. In the oldest globular clusters, the main-sequence termination point has reached a mass of about $0.9 M_\odot$, and in the most metal-poor, and presumably oldest, clusters the abundances of the elements with $Z \geq 3$ are about 150 times lower than their Solar System values. These facts made the determination of stellar ages much simpler than might be imagined. As Michael Bolte (b. 1955) pointed out, low-mass, metal-poor stars have radiative cores and so are unaffected by the convective mixing of unprocessed material from their envelopes into their cores (Bolte, 1997). Furthermore, the corrections to the perfect gas law equation of state are relatively small throughout most solar-mass stars. Finally, the surface temperatures of these stars are high enough for molecules to be rare in their atmospheres, simplifying the conversion of their effective temperatures into predicted colours. Taking account of the various sources of uncertainty, Brian Chaboyer (b. 1965) demonstrated that the absolute magnitude of the main-sequence termination point is the best indicator of the age of the cluster (Chaboyer, 1998).

As understanding of the theory of stellar evolution advanced, improved estimates of the ages of the oldest globular clusters became available. A good example of what could be achieved is illustrated in Figure 13.4, which shows a comparison of the Hertzsprung–Russell diagram for the old globular cluster 47 Tucanae with the predicted isochrones for various assumed ages for the cluster. In this case the abundance of the heavy elements is only 20% of the solar abundance, and the age of the cluster is estimated to be between 12×10^9 and 14×10^9 years (Hesser *et al.*, 1987).

In 1994, André Maeder (b. 1942) reported evidence that the ages of the oldest globular clusters are about 16×10^9 years (Maeder, 1994), and similar results were reported by Sandage in 1995 (Sandage, 1995). In 1997, Bolte argued that the ages of the oldest globular clusters were given by

$$T_0 = 15 \pm 2.4 \text{ (stat)} {}^{+4}_{-1} \text{ (syst) Gyears.} \tag{13.2}$$

The first results of the *Hipparcos* astrometric survey relating to determination of the local distance scale were announced in 1997, with the result that it increased by about 10% (Feast and Catchpole, 1997). This result meant that the stars in globular clusters were more luminous than previously thought and so their main-sequence lifetimes were reduced. In Chaboyer's review of 1998, the ages of globular clusters were estimated to be $T_0 = (11.5 \pm 1.3)$ Gyears (Chaboyer, 1998).

Constraints on the age of the Galaxy can also be obtained from estimates of the cooling times for white dwarfs. According to Chaboyer, these provide a firm lower limit of 8 Gyears. The numbers of white dwarfs observed in the vicinity of the Solar System enable an estimate of $\left(9.5 {}^{+1.1}_{-0.8} \right)$ Gyears to be made for the age of the disc of our Galaxy (Oswalt *et al.*, 1996).

Just as Rutherford had used the relative abundances of the radioactive species to set a lower limit to the age of the Earth in 1904 (see Section 3.1), so lower limits to the age of

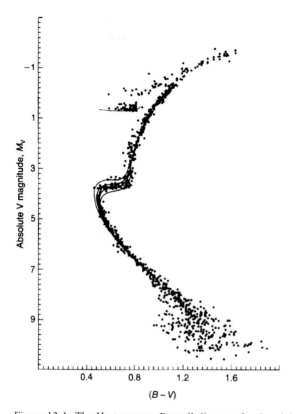

Figure 13.4: The Hertzsprung–Russell diagram for the globular cluster 47 Tucanae (Hesser *et al.*, 1987). The solid lines show fits to the data using theoretical models of the evolution of stars of different masses from the main sequence to the giant branch due to VandenBerg (b. 1947). The isochrones shown have ages of 10, 12, 14 and 16×10^9 years, the best fitting values lying in the range $(12–14) \times 10^9$ years. The cluster is metal-rich relative to other globular clusters, the metal abundance corresponding to about 20% of the solar value.

the Universe can be derived from the discipline of *nucleocosmochronology*. A secure lower limit to the age of the Universe can be derived from the abundances of long-lived radioactive species. In 1963, Edward Anders (b. 1926) used these to determine an accurate age for the Earth of 4.6×10^9 years (Anders, 1963). Some pairs of long-lived radioactive species, such as ^{232}Th–^{238}U, ^{235}U–^{238}U and ^{187}Re–^{187}Os, can provide information about nucleosynthetic timescales before the formation of the Solar System (Schramm and Wasserburg, 1970). These pairs of elements are all produced by the r-process, in which the timescale for neutron capture is less than the β-decay lifetime. The production abundances of these elements can be predicted and compared with their present observed ratios (Cowan, Thielemann and Truran, 1991).

   The best astronomical application of this technique has been carried out by Christopher Sneden (b. 1947), John Cowan (b. 1948) and their colleagues for the ultra-metal-poor K giant star CS 22892-052, in which the iron abundance is 1000 times less than the solar value (Sneden *et al.*, 1992). A number of species never previously observed in such metal-poor stars were detected, for example Tb (terbium, $Z = 65$), Ho (holmium, $Z = 67$),

Tm (thulium, $Z = 69$), Hf (hafnium, $Z = 72$) and Os (osmium, $Z = 76$), as well a single line of Th (thorium, $Z = 90$). The thorium abundance is significantly smaller than its scaled Solar System abundance, and so the star must have been formed much earlier than the Solar System. A lower limit to the age of CS 22892-052 of $(15.2 \pm 3.7) \times 10^9$ years was found.

A conservative lower bound to the cosmological timescale can be found by assuming that all the elements were formed promptly at the beginning of the Universe. From this line of reasoning, David Schramm (1945–1997) found a lower limit to the age of the Galaxy of 9.6×10^9 years (Schramm, 1997). The best estimates of the age of the Galaxy are somewhat model-dependent, but typically ages of about $(12 - 14) \times 10^9$ years are found (Cowan *et al.*, 1991).

13.4 The deceleration parameter, q_0

The hope of the pioneers of observational cosmology was that the value of q_0 could be found from studies of distant galaxies through the apparent magnitude–redshift relation, the angular diameter–redshift relation or the number counts of galaxies. This programme proved to be very much more difficult than the pioneers had expected for a number of reasons that are dealt with in this section. By the end of the twentieth century, real progress was made by two somewhat different routes, one involving the use of supernovae of Type Ia and the other involving observations of the spectrum of fluctuations in the cosmic microwave background radiation (see Section 15.4). The major problem encountered by many of the traditional approaches concerned the evolution with cosmic time of the properties of the objects studied. These issues are of key importance in understanding the processes involved in the formation of galaxies and larger scale structures and are dealt with in more detail in Chapter 15. The emphasis in this section is upon endeavours to estimate the value of q_0 from observations of distant objects.

13.4.1 *Redshift–magnitude relation for the brightest galaxies in clusters*

The redshift–magnitude relation for the brightest galaxies in clusters showed an impressive linear relation (Figure 13.1) (Sandage, 1968), but it only extended to redshifts $z \sim 0.5$ at which the differences between the world models are relatively small. Sandage was well aware of the many effects which needed to be considered before a convincing estimate of q_0 could be found. Some of these were straightforward, such as the need to determine the luminosities of galaxies within a given metric diameter, but others were more complex. For example, as discussed in Section 10.5, Sandage and Hardy discovered that the brightest galaxy in a cluster is more luminous, the greater the difference in magnitude between the brightest and next brightest galaxies in the cluster (Sandage and Hardy, 1973). In what they termed the *Bautz–Morgan effect*, the second- and third-ranked members of the cluster were intrinsically fainter than the corresponding galaxies in other clusters with less dominant first-ranked galaxies. It seemed as though the brightest galaxy became brighter at the expense of the next brightest members, a phenomenon which could plausibly be attributed to the effects of galactic cannibalism (Hausman and Ostriker, 1977). Sandage adopted an empirical correction to reduce the clusters to a standard Bautz–Morgan type.

Sandage was well aware of the need to take account of the evolution of the stellar populations of the galaxies with cosmic time. These corrections followed naturally from his work
on the Hertzsprung–Russell diagrams of globular clusters of different ages which mimic
the cosmic evolution of the old stellar populations of galaxies. He included evolutionary
corrections in the K-correction to the absolute magnitudes of the galaxies. There were,
however, other worrying pieces of evidence which did not fit easily into a picture of passive evolution of the galaxies in clusters. Dramatic evidence for the evolution of galaxies
in rich, regular clusters at relatively small redshifts was first described in the pioneering
analyses of Harvey Butcher (b. 1947) and Augustus Oemler (b. 1945) (Butcher and Oemler,
1978, 1984). They found that the fraction of blue galaxies in such clusters increased from
less than 5% in a nearby sample to percentages as large as 50% at redshift $z \sim 0.4$. The
Butcher–Oemler effect was the subject of a great deal of study and debate, the major observational problems concerning the contamination of the cluster populations by foreground
and background galaxies, as well as bias in the selection criteria for the clusters selected
for observation (Dressler, 1984).

The determination of q_0 might seem to be easier if the samples of galaxies extended to
larger redshifts, but it proved far from trivial to find suitable clusters at redshifts greater
than 0.5. Those in which the brightest galaxies were observed often turned out to be bluer
than expected. This finding reflects a basic problem with this approach to measuring the
deceleration parameter – the differences between the expectations of the world models only
become appreciable at large redshifts when the Universe was significantly younger than it
is now. Consequently, careful account has to be taken of the evolutionary changes of the
objects which are assumed to have 'standard' properties.

By the time of Sandage's review of the problem in 1993, the uncertainties in the value
of q_0 had not decreased, his estimate being $q_0 = 1 \pm 1$ (Sandage, 1995). In fact, by that
time, Alfonso Aragòn-Salamanca (b. 1962), Richard Ellis and their colleagues had extended
the infrared apparent magnitude–redshift relation for the brightest galaxies in clusters to
redshift $z = 0.9$ (Aragòn-Salamanca et al., 1993). They found evidence that the galaxies
were bluer at the larger redshifts, but, perhaps surprisingly, that their apparent magnitude–
redshift relation followed closely a model with $q_0 = 1$, with no corrections for the evolution of the stellar populations of the galaxies, for cluster richness or for Bautz–Morgan
type.

13.4.2 Redshift–magnitude relation for radio galaxies

Another approach to extending the redshift–magnitude relation to large redshifts became
possible in the early 1980s when the use of the first generation of CCD cameras enabled
complete samples of bright 3CR radio sources to be identified with very faint galaxies. These
galaxies turned out to have very strong, narrow emission-line spectra, and spectroscopy by
Hyron Spinrad (b. 1934) and his colleagues showed that many of these radio galaxies had
very large redshifts.[5] These observations showed that the 3CR radio galaxies are among the
most luminous galaxies known.

At about the same time, infrared photometry of these galaxies in the 1–2.2 μm waveband
became feasible with the development of sensitive indium antimonide detectors. There

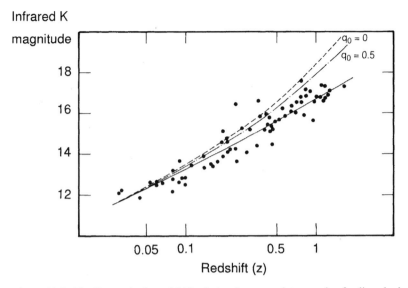

Figure 13.5: The K magnitude–redshift relation for a complete sample of radio galaxies from the 3CR catalogue. The infrared apparent magnitudes were measured at a wavelength of 2.2 µm. The dashed lines show the expectations of world models with $q_0 = 0$ and 0.5. The solid line is a best fitting line for standard world models which include the effects of stellar evolution on the old stellar population of the galaxies (Lilly and Longair, 1984).

were several advantages in defining the redshift–apparent magnitude relation in the infrared waveband, one of them being that dust becomes transparent in the near-infrared waveband and so extinction corrections to the luminosities of the galaxies are very small. A second advantage is that the stars that contribute most of the luminosity at these wavelengths belong to the old red giant population of the galaxy. As a result, the magnitudes are not affected by the bursts of star formation, which can profoundly influence the optical magnitudes of the galaxies and which are largely responsible for the fact that the galaxies at redshifts greater than 0.5 were found to be significantly bluer than those observed at lower redshifts (see Section A13.1).

In 1984, Simon Lilly (b. 1959) and I determined the redshift–apparent magnitude relation for a complete sample of 3CR radio galaxies at an infrared wavelength of 2.2 µm (Figure 13.5) (Lilly and Longair, 1984). We found that there is a remarkably well defined K magnitude–redshift relation which extended to redshifts of 1.5. It was also clear that the galaxies at large redshifts were more luminous than expected for world models with $q_0 \sim$ 0–0.5. When simple evolutionary corrections were made for the increased rate at which stars evolved onto the giant branch at earlier epochs (see Section A13.1), values of q_0 in the range 0 to 1 were found. This appeared to be evidence for the evolution of the stellar populations of these galaxies over cosmological timescales.

There were, however, problems even with this simple picture. In the late 1980s, as discussed in Section 11.6.3, Chambers, Miley, McCarthy and their collaborators discovered the alignment of the radio structures with the optical images of the 3CR galaxies, and this complicated the interpretation of these data (Chambers *et al.*, 1987; McCarthy *et al.*,

1987). Using a combination of surface photometry of these galaxies in the optical and infrared wavebands, we were able to show that the alignment effect does not have a strong influence upon the K magnitude–redshift relationships (Best, Longair and Röttgering, 1998). More serious was the fact that surveys of fainter samples of 6C radio galaxies by Stephen Eales (b. 1960), Steven Rawlings (b. 1961) and their colleagues found that, although the K magnitude redshift relation agreed with our relation at redshifts less than 0.6, their sample of radio galaxies at redshifts $z \sim 1$ were significantly less luminous than the 3CR galaxies by about 0.6 magnitudes (Eales *et al.*, 1997). Our most recent analysis of these data for a cosmological model with $\Omega_0 = 0.3$ and $\Omega_\Lambda = 0.7$, including corrections for the evolution of their stellar populations, have demonstrated clearly that the large-redshift 3CR radio galaxies are significantly more luminous than their nearby counterparts (Inskip *et al.*, 2002), and so the apparent success in accounting for the K magnitude-redshift relation for 3CR radio galaxies was an unfortunate cosmic conspiracy.

The lesson of this story is that the selection of galaxies as standard objects at large redshifts is a hazardous business – we generally learn more about the astrophysics of the most massive galaxies than about geometrical cosmology.

13.4.3 Redshift–magnitude relation for Type Ia supernovae

The discussion of Section 13.4.2 makes it clear that what is required is a set of standard objects which are not susceptible to poorly understood evolutionary changes with cosmic epoch. The use of supernovae of Type Ia to extend the apparent magnitude–redshift relation to redshifts $z > 0.5$ has a number of attractive features. First of all, it is found empirically that these supernovae have a very small dispersion in absolute luminosity at maximum light (Branch and Tammann, 1992). This dispersion can be further reduced if account is taken of the correlation between the maximum luminosity of Type Ia supernovae and the duration of the initial outburst (see Section 8.10.1). This correlation, referred to as the *luminosity–width relation*, is in the sense that the supernovae with the slower decline rates from maximum light are more luminous than those which decline more rapidly. Secondly, there are good astrophysical reasons to suppose that these objects are likely to be good standard candles, despite the fact that they are observed at earlier cosmological epochs. The preferred picture is that these supernovae result from the explosion of white dwarfs, which are members of binary systems that accrete mass from the other member of the binary. Although the precise mechanism that initiates the explosion has not been established, the favoured picture is that mass accreted onto the surface of the white dwarf raises the temperature of the surface layers to such a high temperature that nuclear burning is initiated and a deflagration front propagates into the interior of the star, causing the explosion which results in its destruction.

In 1995 Ariel Goobar (b. 1962) and Saul Perlmutter (b. 1959) discussed the feasibility of observing Type Ia supernova out to redshift $z \approx 1$ in order to estimate the values of the density parameter, Ω_0, and the cosmological constant (Goobar and Perlmutter, 1995). In 1996, they and their colleagues described the first results of systematic searches for Type Ia supernovae at redshifts $z \sim 0.5$ using an ingenious approach to detect them before they reach maximum light (Perlmutter *et al.*, 1996). Deep images of selected fields, including a number which contain distant clusters of galaxies, are taken during one period of the new moon and the fields are imaged in precisely the same way during the next new moon. Using rapid image analysis techniques, any supernovae which appeared between the first-and second-epoch observations were quickly identified and reobserved photometrically and spectroscopically over the succeeding weeks to determine their types and light curves.

Using this search technique, Perlmutter and his colleagues discovered 27 supernovae of Type Ia between redshifts 0.4 and 0.6 in three campaigns in 1995 and 1996 (Perlmutter *et al.*, 1996, 1997). The team used these data to demonstrate directly the effects of cosmological time dilation by comparing the light curves of Type Ia supernovae at redshifts $z \sim 0.4$–0.6 with those of the same type at the present epoch, thus testing directly the cosmological time dilation–redshift relation (Goldhaber *et al.*, 1996). The same peak luminosity–width relation was found as that observed at small redshifts. When account was taken of this relation, the intrinsic spread in the luminosities of the Type Ia supernovae was only 0.21 magnitudes.

This same technique has been used to discover Type Ia supernovae at redshifts greater than $z = 0.8$ as a result of observations with the Hubble Space Telescope. In two independent programmes, Peter Garnavich (b. 1958), Perlmutter and their colleagues discovered the Type Ia supernovae SN1997ck at redshift $z = 0.97$ and SN1997ap at redshift $z = 0.83$, respectively (Garnavich *et al.*, 1998; Perlmutter *et al.*, 1998). The great advantage of the HST observations is that their high angular resolution enables very accurate photometry to be carried out on stellar objects in distant galaxies.

The redshift–apparent magnitude relation presented by Perlmutter is shown in Figure 13.6 and is similar to that found by Garnavich and his colleagues. The major result of these observations, which was found by both groups independently, is that the data favour cosmological models in which the cosmological constant, λ, is non-zero. This was the first time in the history of observational cosmology that compelling evidence for a finite value of the cosmological constant had been found. Both groups have continued to extend this technique to large redshifts through the discovery of Type Ia supernovae at very large redshifts (Knop *et al.*, 2003; Tonry *et al.*, 2003). The best presentation of these results is in terms of a diagram in which the density parameter of the matter content of the Universe, Ω_0, is plotted against the value of the cosmological constant, written in terms of the energy density in the vacuum fields through the normalisation $\Omega_\Lambda = \lambda/3H_0^2$ – the reason for this form of presentation is that, if the global geometry of the Universe were flat, $\Omega_0 + \Omega_\Lambda = 1$. The results of the Supernova Cosmology Project are shown in Figure 13.7.

There are various ways of interpreting Figure 13.7, particularly when taken in conjunction with independent evidence on the mean mass density of the Universe and the evidence from the spectrum of fluctuations in the cosmic microwave background radiation. Perhaps the most conservative approach is to note that the matter density in the Universe must be greater than zero, and, as discussed in Section 13.5, all the data are consistent with values of $\Omega_0 \approx$ 0.25–0.3. Consequently, the cosmological constant must be non-zero. The data would be consistent with $\Omega_0 + \Omega_\Lambda = 1$ if $\Omega_0 \approx 0.25$–0.3. We will come back to these results in Chapter 15.

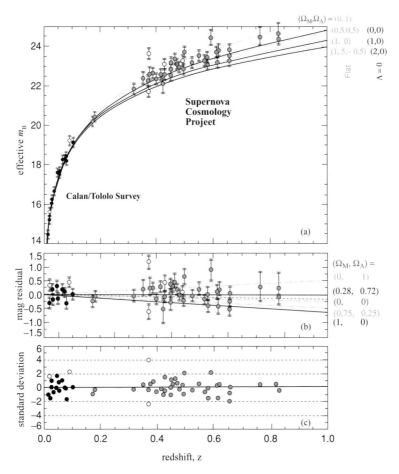

Figure 13.6: Results of the Supernova Cosmology Project (Perlmutter *et al.*, 1999). (a) The Hubble diagram for 42 high-redshift Type Ia supernovae and 19 low-redshift Type Ia super-novae from the Calan/Tololo Supernova survey plotted on a linear redshift scale. The theoretical lines are for $(\Omega_0, \Omega_\Lambda) = (0, 0); (1, 0); (2, 0)$ from top to bottom (solid lines) and $(\Omega_0, \Omega_\Lambda) = (0, 1); (0.5, 0.5); (1, 0); (1.5, -0.5)$ from top to bottom (dotted lines). (b) The magnitude residuals from the best-fit flat cosmological model with $\Omega_M = 0.28$ and $\Omega_\Lambda = 0.72$. (c) The uncertainty-normalised residuals from the best-fit cosmological model.

13.4.4 Number counts of galaxies

In his assessment of approaches to the determination of cosmological parameters using the number counts of galaxies, Sandage was not optimistic (Sandage, 1961a):

Galaxy counts are insensitive to the model ... There seems to be no hope of finding q_0 from the $N(m)$ counts because the predicted differences between the models are too small compared with the known fluctuations of the distribution.

These concerns have been fully justified by subsequent studies. The determination of precise counts of galaxies has proved to be one of the more difficult areas of observational cosmology.[6] The reasons for these complications are multifold. First of all, galaxies are

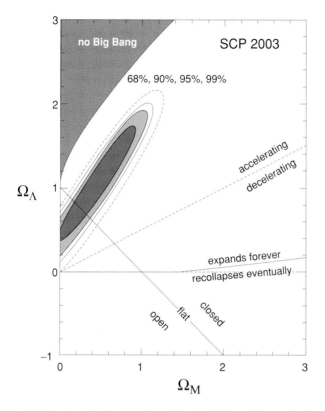

Figure 13.7: The 68%, 90%, 95% and 99% confidence limits for the values of Ω_0 and Ω_Λ determined by the Supernova Cosmology Project. The line labelled 'flat' on the diagram shows the condition $\Omega_0 + \Omega_\Lambda = 1$, which corresponds to flat geometry (Knop *et al.*, 2003).

extended objects, often with complex brightness distributions, and great care must be taken to ensure that the same types of object are compared at different magnitude limits and redshifts. Furthermore, the distribution of galaxies is far from uniform on scales less than about $50h^{-1}$ Mpc, as illustrated by the large voids and walls in the local distribution of galaxies (see Figures 10.7–10.9). Even at the faintest magnitudes, this 'cellular' structure in the distribution of galaxies results in fluctuations in the number counts of galaxies which exceed the statistical fluctuations expected in a random distribution. In addition, the probability of finding galaxies of different morphological types depends upon the galaxy environment. Finally, the luminosity function of galaxies is quite broad (Figure 10.2), and so the geometrical differences between models are masked by the convolution of the predictions of the world models with this function.

Up till about 1980, the deepest counts extended to apparent magnitudes of about 22 to 23 and, although there were disagreements between the results of different observers, there was no strong evidence that the counts of galaxies departed from the expectations of uniform world models. In the 1980s, much deeper counts became feasible with the use of CCD cameras on 4-metre class telescopes, the deepest surveys being carried out by Anthony Tyson (b. 1940) (Tyson, 1990). It was found that there is an excess of faint galaxies at blue magnitudes greater than about 22 (Figure 13.8). In contrast, the counts of galaxies at red and

b_j number counts

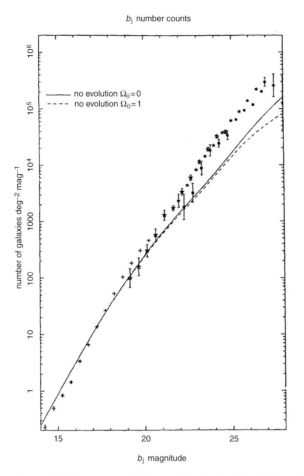

Figure 13.8: The counts of faint galaxies in the blue (B) waveband compared with the expectations of uniform world models with $\Omega_0 = 0$ and 1. For the references to the data points, see Ellis (1987).

infrared wavelengths showed little evidence of such strong evolution. While there is little prospect of using these observations to determine cosmological parameters, they are of the greatest interest in studying the astrophysical evolution of galaxies with cosmic epoch, and this topic will be taken up in Chapter 14.

13.4.5 Angular diameter–redshift test

The angular diameter–redshift relation provides an attractive route for the determination of cosmological parameters if accurate proper distances, l, of astronomical objects can be measured at large redshifts and their corresponding angular sizes, θ, measured. Then, the angular diameter distance $D_A = l/\theta$ can be determined as a function of redshift and compared with the predictions of the standard world models. The physical methods of measuring proper distances at large redshifts described in Section 13.2, involving the Sunyaev–Zeldovich effect in conjunction with X-ray observations of the hot gas in clusters, gravitational lenses and the various versions of the Baade–Wesselink method, all provide means of undertaking this test, but as yet the techniques are not sufficiently precise to provide useful results. A possible problem with this programme is the extent to which the predicted angular diameter–redshift relations are modified by inhomogeneities in the distribution of mass along the line of sight, which can significantly change the predicted relations (Dashevsky and Zeldovich, 1964; Zeldovich, 1964b; Dyer and Roeder, 1972).[7]

The alternative approach is to use objects which may be considered to be 'rigid rods', but the problem is to find suitable standard objects that can be used in the test. A distinctive feature of this test is that there is expected to be a minimum angular diameter, as the objects are observed at large redshifts. A good example is the use of the separation of the radio components of double radio sources, such as those illustrated in Figure 11.1(b). Large samples of these objects can be found spanning a wide range of redshifts. This version of the angular diameter–redshift test was first carried out by George Miley (b. 1942), who used the largest angular size of the radio structures of radio galaxies and quasars as a 'rigid rod' (Miley, 1968, 1971), but no minimum was found in the observed relation. Vijay Kapahi (1944–1999) confirmed this result, using instead the median angular separation, θ_m, of the radio source components as a function of redshift (Kapahi, 1987), but again no minimum was found (Figure 13.9(a)). The median angular separation of the source components is observed to be roughly inversely proportional to redshift, and this was interpreted as evidence that the median physical separation of the source components, l_m, was smaller at large redshifts. Examples of fits to the observational data using evolution functions of the form $l_m \propto (1 + z)^{-n}$ are shown in Figure 13.9(a) for world models with $q_0 = 0$ and 0.5 – values of $n \approx 1.5$–2.0 can provide good fits to the data. There are, of course, many reasons why the separation of the radio source components might be smaller in the past; for example the ambient interstellar and intergalactic gas may well have been greater in the past and so the source components could not penetrate so far through the surrounding gas. Again, we learn more about astrophysical changes with cosmic epoch than about geometrical cosmology.

Another version of the same test was described by Kenneth Kellermann (b. 1937) and involved using only compact double radio structures studied by very long baseline interferometry (Kellermann, 1993). He argued that these sources are likely to be less influenced by changes in the properties of the intergalactic and interstellar gas, since the components are deeply embedded within the central regions of the host galaxy. In his angular diameter–redshift relation, there is evidence for a minimum in the relation, which would be consistent with a value of $q_0 \sim 0.5$ (Figure 13.9(b)). The problem with this type of approach is that we cannot be certain that precisely the same types of double radio source are being selected at large and small redshifts.

13.5 The density parameter, Ω_0

The critical cosmological density, $\rho_{crit} = 3H_0^2/8\pi G$, depends upon the value of Hubble's constant, and so it is convenient to write the density parameter as

$$\Omega_0 = \frac{\rho_0}{\rho_{crit}} = \frac{\rho_0}{2 \times 10^{-26}h^2 \text{ kg m}^{-3}}, \quad \text{where} \quad h = \frac{H_0}{100 \text{ km s}^{-1} \text{ Mpc}^{-1}} \quad (13.3)$$

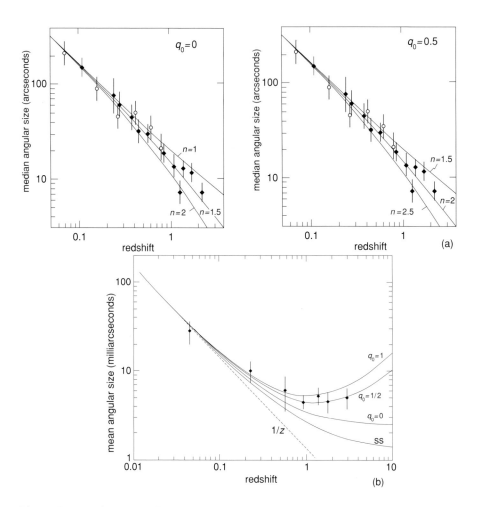

Figure 13.9: (a) The angular diameter–redshift relation for double radio sources, in which the median angular separation of the double radio source components, θ_m, is plotted against redshift (Kapahi, 1987). The observed relation follows closely the relation $\theta_m \propto z^{-1}$. The left-hand panel shows fits to the observations for a world model with $q_0 = 0$ and the right-hand panel is for a model with $q_0 = 0.5$; in both cases, the median separation of the components is assumed to change with redshift as $l_m \propto (1 + z)^{-n}$. (b) The mean angular diameter–redshift relation for 82 compact radio sources observed by VLBI (Kellermann, 1993). In addition to the standard Friedman models, the relation for steady state cosmology (SS) as well as the relation $\theta \propto z^{-1}$ (dashed line) are shown.

and ρ_0 is the present average mass density in the Universe. Estimates of ρ_0 for galaxies were included in Hubble's first paper on the extragalactic nature of the diffuse nebulae using his estimates of their average mass-to-light ratios. He found the value $\rho_0 = 1.5 \times 10^{-28}\ \mathrm{kg\,m^{-3}}$ (Hubble, 1926).

A convenient way of evaluating the average mass density of the Universe was first to work out the average luminosity density due to galaxies by integrating over the luminosity function of the galaxies, and then to convert this into a mass density by adopting a suitable average value for the mass-to-light ratio of the matter in galaxies. An analysis of this nature was carried out by Oort in 1958, who found that the average mass density was 3.1×10^{-28} kg m^{-3}, assuming that Hubble's constant was 180 km s^{-1} Mpc^{-1} (Oort, 1958).

In 1978, James Gunn expressed the same result in terms of the mass-to-light ratio which would be needed if the Universe were to attain the critical density (Gunn, 1978). He found $(M/L)_{\text{crit}} = 2600h$, very much greater than the values found in our vicinity in the plane of the Galaxy and in the visible parts of galaxies. As described in Sections 10.2 and 10.5, however, the mass of dark matter in galaxies and clusters of galaxies far exceeds that in the visible parts of galaxies. If account is taken of the dark matter, the overall mass-to-luminosity ratio attains values of $M/L \sim 100$–150. In well studied rich clusters, such as the Coma cluster, the value of M/L is of the order of 250, but this value is biassed towards elliptical and S0 galaxies, which have three times larger values of M/L than the spiral galaxies, the latter contributing most of the light per unit volume in the Universe at large. These values of M/L are significantly less than the value needed to close the Universe. Gunn's best estimate of the density parameter for bound systems such as galaxies, groups and clusters of galaxies was about 0.1 and was independent of the value of h.

Neta Bahcall (b. 1942) described many different approaches that can be taken to derive values of M/L for clusters of galaxies – cluster mass-to-light ratios, the baryon fraction in clusters and studies of cluster evolution. These have all found the same consistent result that the mass density of the Universe corresponds to $\Omega_0 \approx 0.25$, and furthermore that the mass approximately traces light on large scales (Bahcall, 2000). These results reflect the generally accepted view that, if mass densities are determined for bound systems, the total mass density in the Universe is about a factor of 4 less than that needed to close the Universe.

On scales greater than those of clusters of galaxies, estimates of the mass density in the general field can be found from the *cosmic virial theorem* (Peebles, 1976). In this procedure, the random velocities of galaxies with respect to the mean Hubble flow are compared with the varying component of the gravitational acceleration due to large-scale inhomogeneities in the distribution of galaxies. As in the other methods of mass determination, the mass density is found by comparing the random kinetic energy of galaxies with their gravitational potential energy, this comparison being carried out in terms of two-point correlation functions for both the velocities and positions of galaxies selected from the general field. Application to the random velocities of field galaxies suggested that Ω_0 might be larger than 0.2 (Davis, Geller and Huchra, 1978; Davis and Peebles, 1983).

A similar argument involves studies of the infall of galaxies into superclusters of galaxies. Galaxies in the vicinity of a supercluster are accelerated towards it, thus providing a measure of the mean density of gravitating matter within the system. The velocities induced by large-scale density perturbations depend upon the *density contrast*, $\Delta\rho/\rho$, between the system studied and the mean background density. A typical formula for the infall velocity, u, of test particles into a density perturbation is given by

$$u \propto H_0 r \Omega_0^{0.6} \left(\frac{\Delta\rho}{\rho} \right)_0 \tag{13.4}$$

(Gunn, 1978). In the case of small spherical perturbations, a result correct to second order in the density perturbation was presented by Alan Lightman (b. 1948) and Paul Schechter

POTENT   OPTICAL

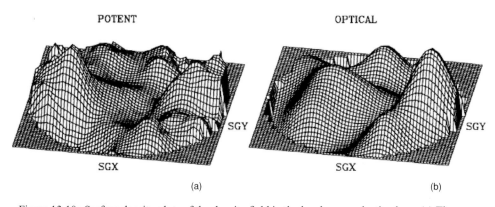

(a)   (b)

Figure 13.10: Surface density plots of the density field in the local supergalactic plane. (a) The mass distribution reconstructed from the peculiar velocity and distance information for the galaxies in this region using the POTENT numerical procedure. (b) The density field of optical galaxies. Both images have been smoothed with a Gaussian filter of radius 1200 km s^{-1}. The density contrast is proportional to the height of the surface above (or below) the plane of the plot. (Hudson *et al.*, 1995.)

(Lightman and Schechter, 1990):

$$\frac{\Delta v}{v} = -\frac{1}{3}\Omega_0^{4/7}\left(\frac{\Delta\rho}{\rho}\right)_0 + \frac{4}{63}\Omega_0^{13/21}\left(\frac{\Delta\rho}{\rho}\right)_0^2. \tag{13.5}$$

In Gunn's analysis, this method resulted in values of Ω_0 of about 0.2 to 0.3. In the 1990s, complete samples of IRAS galaxies became available, and they were used to define the local density and velocity fields. A mean density close to the critical density was found (Lynden-Bell *et al.*, 1988).

In an ambitious programme, Avishai Dekel (b. 1951) and his colleagues devised numerical procedures for deriving the distribution of mass in the local Universe entirely from the measured velocities and distances of complete samples of nearby galaxies, the objective being to determine a three-dimensional map of velocity deviations from the mean Hubble flow. Then, applying Poisson's equation, the mass distribution responsible for the observed peculiar velocity distribution can be reconstructed numerically. Figure 13.10 shows an example of a reconstruction of the local density distribution using this procedure (Hudson *et al.*, 1995). Despite using only the velocities and distances, and *not* their number densities, many of the familiar features of our local Universe are recovered – the Virgo supercluster and the 'Great Attractor' can be seen, as well as voids in the mean mass distribution. These procedures tended to produce somewhat larger values of Ω_0, Dekel stating that the density parameter is greater than 0.3 at the 95% confidence level.

The issue of the total amount of dark matter present in the Universe was the subject of heated debate throughout the 1990s. Some flavour for the points of contention among the experts in the field can be appreciated from the discussion in 1996 at the Princeton meeting *Critical Dialogues in Cosmology* between David Burstein (b. 1947) and Avishai Dekel, the discussion being moderated by Simon White (b. 1951) (Dekel, Burstein and White, 1997). The upshot of these considerations was that there was agreement that the value of Ω_0 is greater than 0.1, and that a value of 0.2 to 0.3 would probably be consistent with most of the data, the only concern being the somewhat larger values favoured by Dekel and his colleagues.

The infall test became feasible on very large scales following the completion of the two-degree-field (2dF) survey of galaxy redshifts carried out at the Anglo-Australian Telescope. The survey involved measuring redshifts for over 200 000 galaxies randomly selected from the Cambridge APM galaxy survey. A cut through that survey was shown in Figure 10.9. The concept behind the test was that superclusters of galaxies generate a systematic infall of other galaxies in their vicinities and this would be evident in the pattern of recessional velocities, resulting in anisotropy in the inferred spatial clustering of galaxies. Using the redshifts of more than 141 000 galaxies from the 2dF galaxy redshift survey, John Peacock and his colleagues discovered convincing statistical evidence for infall and estimated the overall density parameter to be $\Omega_0^{0.6} = 0.43b \pm 0.07$, where b is the bias parameter, the factor by which visible matter is more clustered than the dominant dark matter. When this result was combined with data on the anisotropy of the cosmic microwave background, their result favoured a low-density Universe with $\Omega_0 \approx 0.3$ (Peacock $et\ al.$, 2001).

When taken in conjunction with the results derived from the power spectrum of fluctuations in the cosmic microwave background radiation discussed in Section 15.4, the consensus view is that the best estimate of the overall density parameter for the Universe is given by $\Omega_0 \approx 0.25$–0.3. An immediate consequence of this result is that most of the mass cannot be in the form of baryonic matter, which is constrained by the production of the light elements in the early stages of the Big Bang. To rephrase the results discussed in Section 12.6, a conservative upper limit to the density parameter in baryons is $\Omega_{bar} \leq 0.0375h^{-2}$; otherwise, less than the observed abundance of deuterium is expected to be synthesised during the epoch of nucleosynthesis. Thus, adopting $h = 0.7$, there cannot be sufficient baryons to account for the observed mass density. Most of the mass in the Universe must be in some non-baryonic form.

13.6 Summary

From a subject dogged by controversy and strong feeling for most of the twentieth century, classical cosmology saw a dramatic change in perspective during the last decade of that century. New methods were developed which eliminated many of the problems of the pioneering efforts of previous decades. Whilst the emphasis in this chapter has been upon the traditional route to the determination of cosmological parameters, the consensus picture received a remarkable boost from analyses of the fluctuation spectrum of the cosmic microwave background radiation, a story which is told in the context of the understanding of the formation of large-scale structures in the Universe in Chapter 15. Indeed, many cosmologists would now look first to these observations as providing the key to unlocking many of the problems of classical cosmology.

Notes to Chapter 13

1 In Sandage's historical review of efforts to determine the redshift–apparent magnitude relation for galaxies, he reproduces many different versions of this relation, from Hubble's first example of 1929 to his preferred relation of 1993 (Sandage, 1995).

2 The reader may wonder why the pioneers of observational cosmology worked in terms of the deceleration parameter, q_0, rather than in terms of the density parameter, Ω_0, and the cosmological constant, λ. The reason is that the expression for the comoving radial distance coordinate, r, and what I call the 'effective distance', D, used to relate intrinsic properties to observables are independent of the density parameter, Ω_0, and the curvature of space, κ, to second order in the redshift, z. Explicitly, we find

$$D = r = \frac{c}{H_0} \left[z + \frac{z^2}{2}(1 + q_0) + \dots \right].$$

I have given details of this result in Section 8.2 of Malcom Longair, *Galaxy Formation* (Berlin: Springer Verlag, 1998). A briefer version is given by James Gunn in his review of 1978 (Gunn, 1978). Thus, provided the observations do not extend to redshifts greater than, say, 0.4–0.5, the various cosmological tests provide direct information about the present deceleration, or acceleration, of the Universe.

3 In his survey of his contributions to observational cosmology of 1995, Sandage provided an excellent overview of the history of the determination of cosmological parameters by the techniques listed above using optical observations (Sandage, 1995). The text includes many intriguing comments and footnotes about the problems of theory and observation which faced the pioneers of the subject from the 1920s to the 1990s. His delightful and touching remark about the comparison between his and my expositions of the foundations of cosmology is a measure of the differences in approach between the 1950s and the 1990s to the study of astrophysical cosmology (see Sandage (1995), footnote 8, p. 25).

4 For a detailed discussion of the different approaches to the determination of Hubble's constant during the 1970s and 1980s, see M. Rowan-Robinson, *The Cosmological Distance Ladder* (New York: W. H. Freeman and Company, 1985); Rowan-Robinson's conclusions were updated in M. Rowan-Robinson, The extragalactic distance scale, *Space Science Reviews*, **48**, 1988, 1–71.

5 An account of Spinrad's many contributions to the study of large-redshift radio galaxies are described in the volume celebrating his 65th birthday, *The Hy-redshift Universe*, APS Conference Series, vol. 193, eds. A. J. Bunker and W. J. K van Breugel (San Francisco: APS, 1999).

6 An excellent account of the problems of determining and interpreting the counts of galaxies has been given by Richard Ellis, Faint blue galaxies, *Annual Review of Astronomy and Astrophysics*, **35**, 1997, 389–443.

7 I have given simple derivations of the angular diameter–redshift relations for inhomogeneous world models in Section 7.4 of Longair, *Galaxy Formation*.

A13 Explanatory supplement to Chapter 13

A13.1 *Observing distant galaxies in the infrared waveband*

There are advantages in carrying out studies of the redshift–apparent magnitude relation in the 1–2.2 μm wavebands. To demonstrate these, a model has been developed for the spectral energy distribution of a giant elliptical galaxy at the present epoch, involving a wide range of stellar masses and detailed information about their spectra throughout the ultraviolet, optical and infrared wavebands. Then, models for the evolution of such a spectrum with cosmic epoch can be determined and these depend upon the assumed initial mass function of the stars in the galaxy and the star formation rate as a function of cosmic epoch.

The example shown in Figure A13.1 illustrates a number of important aspects of such computations (Inskip *et al.*, 2002). It is assumed that all the stars in the model giant elliptical

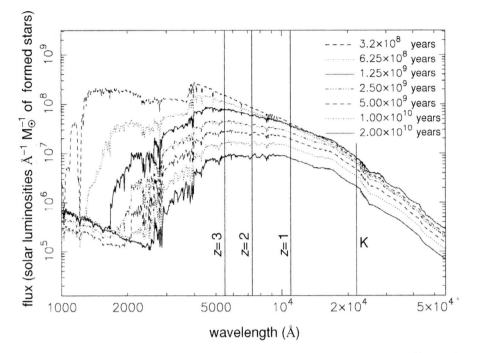

Figure A13.1: Illustrating the spectral evolution of a giant elliptical galaxy in which the stellar population is formed in an initial star-burst of duration 10^8 years. The computations used the stellar synthesis codes for the spectral energy distribution of galaxies developed by Gustavo Bruzual and Stefane Charlot (b. 1964) (Bruzual and Charlot, 2003). The different spectra show the predicted spectrum at later times, the intensity at infrared wavelengths decreasing monotonically with increasing age, the ages of the galaxies being shown by the key to the top right of the diagram. The computations were designed so that the spectrum of a gaint elliptical galaxy would be reproduced at the present epoch (Inskip *et al.*, 2002). The vertical line labelled K is the standard K waveband at 2.2 µm. The vertical lines labelled $z = 1, 2$ and 3 show the regions of the galaxy spectrum observed at the present epoch if the galaxy were observed at these redshifts.

galaxy were formed in the first 10^8 years of its history, during which time the star-formation rate was taken to be constant. The diagram shows the evolution of the spectral energy distribution of the galaxy at subsequent times in the galaxy's rest frame, resulting in the observed spectrum of a giant elliptical galaxy at the present epoch. There is a one-to-one relation between the mass of a star, its luminosity and the waveband in which it emits most of this luminosity, so that stars radiating in the ultraviolet waveband are massive and have short lifetimes. This is reflected in the rapid decrease with cosmic epoch of the spectral energy distribution in the ultraviolet regions of the spectrum. In contrast, in the infrared waveband, the stars are long-lived, and the gradual decrease in luminosity reflects the decreasing rate at which stars evolve from the main sequence onto the red giant branch.

The origin of the slow decline with cosmic epoch of the infrared luminosity of the old stellar populations of galaxies can be appreciated from the following simple calculation due to Gunn (1978). Most of the luminosity of the galaxy in the infrared waveband 1–3 µm is associated with red giant stars. Since the time stars spend on the giant branch, t_g, and their luminosities are relatively independent of their main-sequence masses, we need only determine the rate at which stars evolve from the main sequence onto the giant branch as a function of main-sequence mass in order to find the change in luminosity of a galaxy with redshift. In the *passive evolution model* of galactic evolution, it is assumed that all the stars were formed in an initial brief star-burst and that the subsequent luminosity evolution of the galaxy is due to the stellar evolution of this population. For illustration, let us assume that the initial mass function of the stars is of Salpeter form, $dN = N(M)\,dM \propto M^{-y}\,dM$, where $y = 2.35$. It is a straightforward calculation to show that the number of stars on the giant branch, N_g, is given by

$$N_g = t_g \frac{dN}{dt} = t_g \left(\frac{dN}{dM}\right)\left(\frac{dM}{dt}\right). \tag{A13.1}$$

Thus, using the relation between mass and main-sequence lifetime derived in Section A4.1, $t = t_{\odot}(M/M_{\odot})^{-(x-1)}$, we find

$$L(t) = L(t_0)\,t^{-(x-y)/(x-1)}. \tag{A13.2}$$

Inserting the values $x = 5$ and $y = 2.35$, we find $L \propto t^{-0.66}$. For the case of the critical world model, $t/t_0 = (1+z)^{-3/2}$, and so, to a good approximation, $L \propto (1+z)$. Thus, at a redshift of unity, the old stellar populations of galaxies should be about twice as luminous as they are at the present epoch, and at redshift $z = 3$, four times as luminous. This rough calculation explains the slow decrease in the luminosity of the model galaxy in the infrared (K) waveband seen in Figure A13.1.

The vertical line towards the right of Figure A13.1 shows the observing wavelength of 2.2 μm at zero redshift; the vertical lines to the left show the region of the spectrum observed at 2.2 μm at redshifts of 1, 2 and 3. It can be seen that, even at a redshift of 3, the evolution of the stellar energy distribution is quite smooth, and relatively stable corrections can be made for the effects of evolution of the stellar populations. This behaviour contrasts strongly with what would be observed if the same exercise had been carried out in the optical waveband. Then, the spectrum at large redshifts would originate from the strongly varying ultraviolet regions of the spectrum. Furthermore, the predicted spectrum at large redshifts is very sensitive to assumptions about the star-formation rate as a function of epoch. If there were bursts of star formation, for example, the ultraviolet region of the spectrum would be very strongly enhanced and the corrections for such effects would become difficult to predict.

Thus, from the point of view of studying the properties of the integrated stellar populations of galaxies, there are real advantages in carrying out these studies in the 1–2.2 μm waveband.

14 The evolution of galaxies and active galaxies with cosmic epoch

Evidence for strong evolutionary changes in the properties of extragalactic objects with cosmic epoch was first found in the 1950s and 1960s as a result of surveys of radio sources and quasars. An excess of faint sources was found in radio source and quasar surveys, as compared with the expectations of uniform world models. The inference was that there were many more of these classes of object at early cosmic epochs as compared with their number at the present epoch. During the 1980s, as the first deep counts of galaxies became available, a large excess of blue galaxies at faint apparent magnitudes was discovered. These studies culminated in the remarkable observations of the Hubble Deep Field in 1998 and the Hubble Ultra-Deep Field in 2004 by the Hubble Space Telescope.

In the 1990s, the first deep surveys of the X-ray sky were carried out by the ROSAT X-ray observatory, and evidence for an excess of faint X-ray sources was found, similar in many ways to the evolution inferred from studies of extragalactic radio sources and quasars. In the thermal infrared wavebands, the IRAS survey, although not extending to as large redshifts as the surveys mentioned above, also provided evidence for an excess of faint sources, which appear to be evolving in a manner similar to the active galaxies. Then, in the last few years of the century, evidence was found for a large population of submillimetre or far-infrared galaxies at large redshifts.

The primary evidence for these evolutionary changes came from counting the numbers of objects in well defined complete samples. In addition, since the 1970s, vast amounts of new data accumulated on many different aspects of galaxy formation and evolution – neutral hydrogen absorption-line systems, abundances of the elements in large-redshift absorption systems, star-formation rates as a function of cosmic epoch and so on.[1]

14.1 The cosmological evolution of active galaxies

14.1.1 Source counts for the standard world models

In his earliest studies of galaxies as extragalactic systems, Hubble realised that the number counts of galaxies potentially contain information about the large-scale structure of the Universe. In his monograph *The Realm of the Nebulae*, he used counts of galaxies to the limit of the Mount Wilson 100-inch telescope to demonstrate that, overall, the distribution

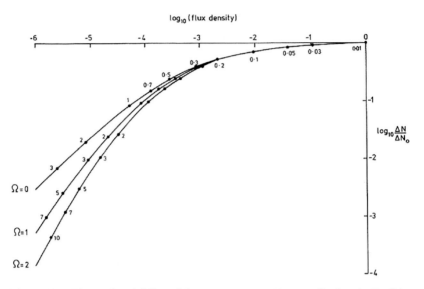

Figure 14.1: The predicted differential source counts, ΔN, normalised to the Euclidean prediction, $\Delta N_0 \propto S^{-5/2} \Delta S$, for a single luminosity class of source having spectral index $\alpha = 0.75$ for different cosmological models with $\lambda = 0$. The numbers opposite each point are the redshifts at which the sources are observed (Longair, 1978).

of galaxies is homogeneous on the large scale (Hubble, 1936). As illustrated in Figure 6.2, he compared the number counts with the expectations of a uniform Euclidean world model, which were derived in Section A5.1:

$$N(\geq S) \propto S^{-3/2} \quad \text{or} \quad \log N(\leq m) = 0.6m + \text{constant}, \tag{14.1}$$

both results being independent of the luminosity function, $N_0(L)$.

As the number counts extended to significant redshifts, however, departures from this relation are expected. These are most simply demonstrated by a plot of the *differential* number counts, normalised to the differential Euclidean number counts, $\Delta N_0 \propto S^{-2.5} \Delta S$, as a function of flux density, S, (Figure 14.1). The Euclidean prediction, $\Delta N / \Delta N_0 = \text{constant}$, is represented by the abscissa, $\log_{10}(\Delta N / \Delta N_0) = 0$. The numbers opposite each point on these relations are the redshifts at which the sources are observed. It can be seen that the predicted differential counts depart rapidly from the Euclidean prediction, even at relatively small redshifts. For example, for the case $\Omega_0 = 1$, the source counts at redshift $z = 0.5$ have differential slope -2.08 rather than -2.5, corresponding to a slope of the integral source counts of -1.08 rather than -1.5. In fact, the source populations cannot be represented by a single luminosity, but, rather, the counts shown in Figure 14.1 must be convolved with the luminosity function, $N_0(L)$, of the sources.

14.1.2 *Extragalactic radio sources and radio quasars*

Historically, the considerations of the Section 14.1.1 were important because the counts of radio sources at high flux densities had integral slope $\beta = 1.8$, much steeper than expected in the uniform world models. In the 1960s, this was strong evidence that there must have been many more radio sources at large redshifts than were predicted by the uniform, isotropic models, indicating that the source population must have evolved with cosmic epoch. Pioneering studies of how strong the evolution had to be to account for the observations were carried out by William Davidson (b. 1924), who showed how extreme the divergence was between the expectations of the steady state theory and the observed number counts (Davidson, 1962). He also constructed the first empirical expressions for the evolution of the population of radio sources (Davidson and Davies, 1964).

As increasing numbers of optical identifications became available, the local luminosity function of extragalactic radio sources became better defined, and so the nature of the cosmological evolutionary changes in the radio source population could be quantified in more detail. By 1966, I was able to impose important constraints on the form of the evolution, the key features of the successful models being strong evolution of either the luminosities or comoving number densities of radio sources out to redshifts $z \sim 2$–3, beyond which the strong evolution could no longer continue (Longair, 1966). Otherwise, the convergence of the counts indicated by the observations of Martin Ryle and Ann Neville (b. 1938) could not be reproduced (Ryle and Neville, 1962) and the isotropic radio background radiation would be exceeded. Typically, the luminosities of the most luminous radio sources had to change with cosmic epoch as $L(z) \propto (1 + z)^3$ out to redshifts $z \sim 2$–3. Because of the steepness of the radio luminosity function, the corresponding increase in the comoving space density of sources of a given luminosity was even more extreme, $n(z) \propto (1 + z)^{5.5}$.

The discovery of the radio quasars, and the large numbers of redshifts which rapidly became available for them, confirmed that there was an excess of these objects at large redshifts. In 1968, Michael Rowan-Robinson (b. 1942) and Maarten Schmidt independently developed a procedure, known as the luminosity-volume or V / V_{max} test, to show that the radio quasars were located towards the limits of their observable volumes, implying the same type of strong evolutionary effect inferred somewhat less directly from the counts themselves (Rowan-Robinson, 1968; Schmidt, 1968).

From the 1960s onwards, number counts of radio sources were derived at frequencies throughout the radio waveband, all of them displaying the excess of radio sources observed at high flux densities, followed by convergence at low flux densities. Examples of the differential number counts at frequencies from 150 MHz to 8.44 GHz compiled by Jasper Wall (b. 1942) are shown in Figure 14.2 (Wall, 1996). While the number counts were well established by the efforts of many radio astronomers, the determination of how the luminosity function of the sources changed with cosmic epoch required the redshifts of the sources. This meant identifying very faint galaxies and quasars reliably, which required very accurate radio positions, and preferably radio structures, and then measuring the redshifts of these extremely faint objects.

An example of an ambitious study using large complete samples of radio sources with redshifts was completed by James Dunlop (b. 1962) and John Peacock (b. 1956) in 1990 (Dunlop and Peacock, 1990). The inferred changes of the radio luminosity function with cosmic epoch are very strong (Figure 14.3) and can be well represented by a model in which the luminosity function was shifted to greater radio luminosities at large redshifts. This increase cannot continue, however, beyond redshifts $z \sim 2$–3. There needs to be some

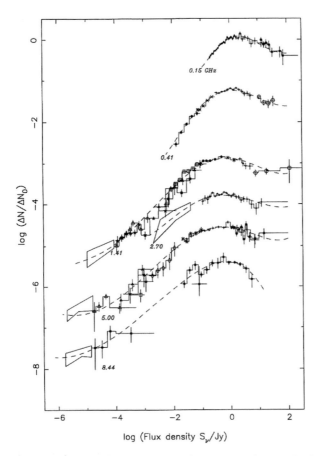

Figure 14.2: The differential, normalised counts of extragalactic radio sources at a wide range of frequencies throughout the radio waveband (Wall, 1996). The points show the number counts derived from surveys of complete samples of radio sources. The boxes indicate extrapolations of the source counts to very low flux densities using the $P(D)$ technique described in Section 12.3. (Diagram courtesy of Dr Jasper Wall.)

form of cut-off, and there has been considerable debate about how the luminosity function of the radio sources changes at redshifts greater than $z \sim 2$–3. Dunlop pointed out that, although the precise form of the evolution of the radio luminosity function at large redshifts is still uncertain, the integrated radio emissivity of all the radio sources as a function of redshift is rather well defined. The quantity $\int L N(L, z) \, dL$ attains a maximum at redshifts $z \approx 2$–3 and decreases rapidly at larger redshifts (Dunlop, 1994). Thus, the epoch corresponding to $z \sim 2$–3 was the era when the radio quasars and radio galaxies were most populous.

14.1.3 The radio-quiet quasars

Intensive searches for radio-quiet quasars[2] began in 1965 as soon as Sandage announced their discovery in that year (Sandage, 1965). In these early days, the clue to finding

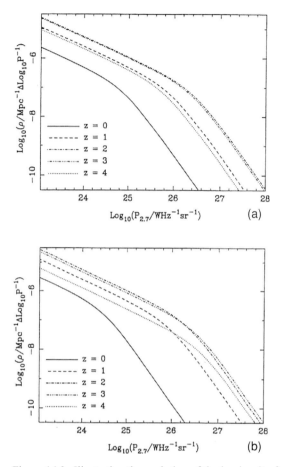

Figure 14.3: Illustrating the evolution of the luminosity function of extragalactic radio sources with (a) steep and (b) flat radio spectra as a function of redshift, z, or cosmic epoch. The luminosity functions shown describe the number of radio sources with different radio luminosities per unit comoving volume, that is in a coordinate system which expands with the Universe, so that the figure shows changes over and above the changing density due to the expansion of the Universe (Dunlop and Peacock, 1990).

radio-quiet quasars among the vastly more populous foreground stars was the fact that their optical spectra were non-thermal, their continua being of roughly power-law form. As a result, they displayed ultraviolet and infrared excesses in comparison with stellar spectra. Quasars were sought in catalogues of blue stellar objects, and surveys were made to search specifically for objects with ultraviolet excesses. The problem with these surveys was that the objects with ultraviolet excesses might be white dwarfs, and so spectroscopic confirmation that the objects had large redshifts with the characteristic quasar spectra was required. In 1970, Alessandro Braccesi (b. 1937) and his colleagues compiled a list of 175 quasar candidates which resulted in a complete sample of 19 quasars to apparent magnitude $B = 18$ with confirmed redshifts in an area of 36 square degrees (Braccesi, Formiggini and

Gandolfi, 1970). These were the first observations to show convincingly that the number counts of optically selected quasars had slope steeper than $\beta = 1.5$.

In 1972, Richard Green (b. 1949) and Maarten Schmidt began a major survey using the Palomar 18-inch Schmidt telescope to find all the bright quasars in a region of 10 700 square degrees of the northern sky away from the Galactic plane (Schmidt and Green, 1983). The survey was made in two colours, U and B, and several thousands of stellar objects with ultraviolet excesses were discovered. Of these, 108 turned out to be quasars which formed a complete sample with apparent magnitudes less than B = 16.2. Comparison of these numbers with those of fainter radio-quiet quasars again showed that the counts of optically selected quasars were significantly steeper than the Euclidean expectation.

The ultraviolet excess technique was only useful out to redshifts $z \leq 2.2$ because at that redshift the Lyman-α emission line is redshifted into the B filter and so the quasars no longer exhibit ultraviolet excesses. The extension of this technique involved the use of multicolour photometry to discriminate between stars and objects with the typical spectra of large redshift quasars. David Koo (b. 1951) and Richard Kron (b. 1951) used (U, J, F, N) photometry to find radio-quiet quasars with ultraviolet excesses to B = 23 (Koo and Kron, 1982). This survey provided the first evidence for the convergence of the number counts of radio-quiet quasars at redshifts $z \geq 2$.

The technology for undertaking quantitative surveys of large samples of stars and galaxies improved greatly with the availability of high-speed measuring machines, such as the APM facility in Cambridge and the COSMOS machine at the Royal Observatory, Edinburgh. These enabled scans to be made of the very large numbers of objects present on the sky survey plates taken with the UK 48-inch Schmidt telescope at the Siding Spring Observatory in Australia. In 1987, the multicolour technique was extended by Stephen Warren (b. 1957) and his colleagues to four-colour photometry using observations in the U, J, V, R and I wavebands, providing four colours (U–J, J–V, V–R, R–I) (Warren et al., 1987). Stars lie along a rather narrow locus in this four-dimensional colour space. By searching for objects which lay well away from that locus, they found the first quasar with redshift $z > 4$. In 1991, this technique was further refined by Michael Irwin (b. 1952) and his colleagues, who realised that quasars with redshifts greater than 4 could be found by means of two-colour photometry in the (B_J, R, I) wavebands because the colours of these very large redshift quasars are very different from those of stars (Irwin, McMahon and Hazard, 1991). The reason for the success of this approach was that, at these very large redshifts, the redshifted Lyman-α forest enters the B_J waveband and so strongly depresses their continuum intensities. The largest redshift found in their survey was $z = 4.8$. At the same time, Brian Boyle (b. 1960) and his colleagues used the ultraviolet-excess technique to derive a complete sample of over 400 quasars down to B = 21, all confirmed by spectroscopy (Boyle et al., 1991).

Other techniques were developed to search for quasars. One approach made use of the fact that the Lyman-α and CIV emission lines are always very strong in the spectra of quasars and are superimposed upon a roughly power-law continuum energy distribution. If a dispersion prism, or grating, is used in conjunction with a wide-field telescope, low-resolution spectra are obtained for each object in the field, and this proved to be a powerful means of discovering quasars with redshifts $z > 2$, since these strong lines are then redshifted into the optical waveband. Pioneering surveys were carried out at the Cerro Tololo Observatory in Chile

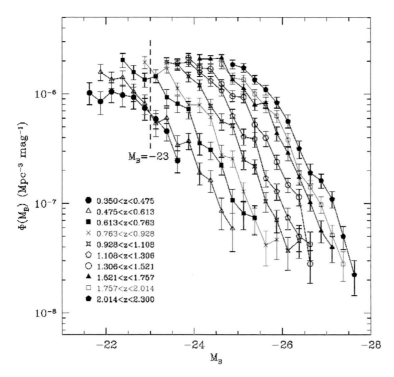

Figure 14.4: The evolution of the optical luminosity function for 6000 optically selected quasars in the redshift range $0.35 \leq z \leq 2.3$ observed in the 2dF survey carried out at the Anglo-Australian Telescope (Boyle *et al.*, 2000).

using the Curtis Schmidt telescope and the 4-metre telescope by Malcolm Smith (b. 1942), Art Hoag (1921–1999) and Patrick Osmer (b. 1943) in the late 1970s (Hoag and Smith, 1977). By the early 1980s, it was known that the optically selected quasars exhibited strong cosmological evolution, similar in character to the radio quasars. Osmer found that there seemed to be a lack of large-redshift objects as compared with the expectations of the evolutionary models, which could explain the redshift distributions and number counts of objects to a redshift of about 2 (Osmer, 1982). The problem with these surveys was the question of the completeness of the samples and the many complex selection effects involved in selecting the candidate objects.

One of the most successful applications of this technique was the survey of Maarten Schmidt and his colleagues, who used the Palomar 200-inch telescope as a fixed transit instrument in conjunction with a grism and a large-area CCD camera, which was clocked at the sidereal rate. In this way, six narrow bands across the sky were scanned both photometrically in the v and i wavebands, as well as spectroscopically, resulting in a total scanned area of 62 square degrees. Of 1660 candidate emission-line objects, 141 were found to be quasars in the redshift interval $2.0 < z < 4.7$ (Schmidt, Schneider and Gunn, 1986, Schneider *et al.*, 1994).

Perhaps the most ambitious programme to date to determine the optical luminosity function of radio-quiet quasars and its evolution with cosmic epoch has resulted from the 2dF quasar survey carried out at the Anglo-Australian telescope in conjunction with the 2dF Galaxy Redshift Survey. The quasar candidates were selected by the multicolour technique previously employed by Boyle and his colleagues and resulted in a sample of over 6000 quasars in the redshift interval $0.35 \leq z \leq 2.3$ (Boyle *et al.*, 2000). From these data, they reconstructed the evolving comoving luminosity function of quasars which can be conveniently described by luminosity evolution of the quasar luminosity function as $L(z) = L_0 \exp(7\tau)$ out to redshift $z = 2.3$, where τ is the *fractional look-back time*, $\tau = (t_0 - t)/t_0$, t_0 being the present age of the world model and t being the time when the light was emitted by the quasar (Figure 14.4).

Systematic surveys were made to determine the large-redshift evolution of the radio-quiet quasars by a number of authors (Warren, Hewett and Osmer, 1994; Kennefick, Djorgovski and de Carvalo, 1995; Schmidt *et al.*, 1995). Good agreement was found between the results of these surveys. There are fewer large-redshift optically selected quasars than expected if the comoving optical luminosity function had remained constant at all redshifts at $z \geq 2$: the comoving number densities of luminous quasars decreased by a factor of about 5 to 7 over the redshift interval $2 \leq z \leq 4$. The general consensus was that both the radio sources and the optically selected quasars show a maximum in their comoving space densities at redshifts $z \sim 2$–3 and decline steeply at both lower and higher redshifts.

14.1.4 X-ray sky surveys

Following the bright X-ray source survey carried out by the UHURU X-ray Observatory in the early 1970s, the first deep surveys of the X-ray sky were carried out by the Einstein X-ray Observatory in 1979 (Giacconi *et al.*, 1979). One of the prime motivations of these studies was to understand the origin of the X-ray background radiation, which was discovered in Giacconi's pioneering rocket flight in 1962 and which is remarkably bright (Figure 14.5(a)). Giacconi and his colleagues found that about 26% of the background intensity could be attributed to discrete X-ray sources, most of which were associated with active galaxies and quasars (Giacconi *et al.*, 1979).

Definitive evidence for cosmological evolution of the population of X-ray sources was derived from the deep surveys carried out by the German–American–British ROSAT satellite, which was launched in 1991. The principal objective of the mission was to carry out a complete survey of the sky in the X-ray energy band 0.1 to 2.4 keV, and this was successfully achieved, resulting in a catalogue of about 60 000 sources and information about their X-ray spectra in four X-ray 'colours'. In addition to the sky survey, very deep observations were carried out in a small region of sky to define the X-ray source counts to the faintest achievable flux densities. The deep survey was made in the 'Lockman Hole', a region of sky in which the neutral hydrogen column density has a very low value, $N_H = 5.7 \times 10^{19}$ cm^{-2}, so that there is minimum photoelectric X-ray absorption by the interstellar gas.

The X-ray source counts were derived by Günther Hasinger (b. 1954) and his colleagues in two ways from these observations (Hasinger *et al.*, 1993). For the medium deep survey, particular care was taken to understand the effects of source confusion at low X-ray flux densities. The very deep survey was analysed using a $P(D)$ analysis similar to that described in Section 12.3. The resulting differential, normalised X-ray source counts are shown in

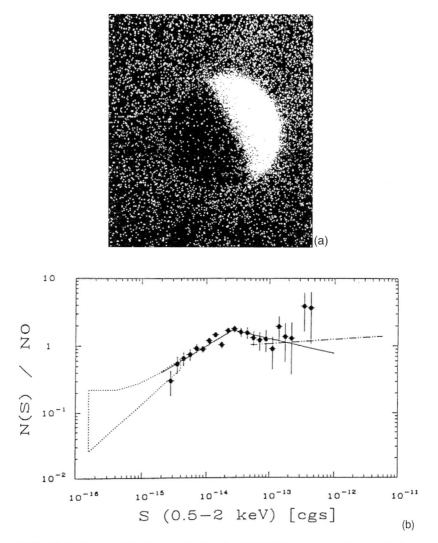

Figure 14.5: (a) An image of the Moon taken by the ROSAT Observatory showing the fluorescent X-ray emission from the sunlit side of the Moon. On the dark side, the Moon is seen occulting the diffuse X-ray background emission. (Courtesy of Professor J. Trümper, Max Planck Institute for Extraterrestrial Physics, Garching.) (b) The differential, normalised counts of faint X-ray sources observed by the ROSAT X-ray observatory. The dot–dash line is the best-fit source count from the Einstein Observatory surveys. The dotted area at faint flux densities shows the 90% confidence limits from the fluctuation analysis of the deepest ROSAT survey in the Lockman Hole (Hasinger *et al.*, 1993).

Figure 14.5, and these bear a strong resemblance to the differential counts of radio sources (Figure 14.2). At high X-ray flux densities, $S > 3 \times 10^{-14} \, \mathrm{erg \, s^{-1} \, cm^{-1}}$, the differential counts have slope $(\beta + 1) = 2.72 \pm 0.27$, and below this flux density the counts converge with source count slope $(\beta + 1) = 1.94 \pm 0.19$. The identifications of sources in the medium deep survey were consistent with a picture in which the X-ray sources follow the same type of cosmological evolutionary behaviour as the radio galaxies, radio quasars and optically selected quasars. The background intensity which could be attributed to discrete sources with flux densities greater than $2 \times 10^{-15} \, \text{erg cm}^{-2} \, \text{s}^{-1}$ amounted to 59% of the total background intensity in this energy range. Extending the counts to the limit of the deep survey, discrete X-ray sources can account for about 75% of the background intensity.

14.1.5 Infrared and submillimetre wavebands

The IRAS satellite carried out the first essentially complete sky survey in those infrared wavebands between 12.5 and 100 μm which can only be observed from space. Among the many important discoveries of the mission was the realisation that many galaxies are intense far-infrared emitters, the most intense being the *star-burst galaxies*. The catalogues of IRAS galaxies proved to be important cosmologically because they provided complete samples of galaxies unaffected by obscuration by interstellar dust. As a result of a major effort by many astronomers, the redshifts of complete samples of IRAS galaxies were measured and the local luminosity function at 60 μm determined.

Counts of IRAS galaxies were made at 60 μm from the IRAS Point Source Catalogue, the IRAS Faint Source Survey and a survey in the region of the ecliptic poles (Oliver, Rowan-Robinson and Saunders, 1992). The normalised differential counts showed that there are more faint IRAS sources than expected, in the same sense as the counts of radio sources, X-ray sources and quasars. The counts did not, however, extend deep enough to constrain the large-redshift behaviour of the source population.

An interesting consequence of the evolution of the IRAS galaxies resulted from the strong correlation between the radio emission of normal and star-burst galaxies and their far-infrared emission (Helou, Soifer and Rowan-Robinson, 1985). The proportionality extends over many orders of magnitude and can be written $S(60 \, \mu\text{m}) = 90 S(1.4 \, \text{GHz})$, where both flux densities are measured in janskies (Jy). As a result, it is straightforward to predict the counts of star-burst and normal galaxies in the radio waveband. It turns out that it is possible to account for the flattening of the radio source counts at radio flux densities $S \leq 10^{-3}$ Jy seen in Figure 14.2 in terms of the evolution of the population of IRAS galaxies (Rowan-Robinson *et al.*, 1993). This conclusion is supported by the identification content of the millijansky radio sources, many of which are blue and have spectra similar to those of star-burst galaxies (Windhorst, Dressler and Koo, 1987; Windhorst, 1995).

Galaxies undergoing bursts of star formation are not only sources of intense ultraviolet continuum radiation, but are also strong emitters in the far-infrared waveband because of the presence of dust in the star-forming regions. The ultraviolet and optical radiation are absorbed by the dust which reradiates the absorbed energy at the temperature to which the dust is heated, which is typically about 30–60 K, and so corresponds to emission in the far-infrared and submillimetre wavebands. In a study of star-forming galaxies in the Markarian catalogues of ultraviolet-excess galaxies, Joseph Mazzarella (b. 1961) and Vicki Balzano (b. 1961) found that star-forming galaxies are, on average, stronger emitters in the far-infrared than in the ultraviolet waveband (Mazzarella and Balzano, 1986). Similarly, in a sample of star-forming galaxies studied by the International Ultraviolet Explorer, Daniel Weedman found that most of the galaxies emit much more strongly in the far-infrared rather

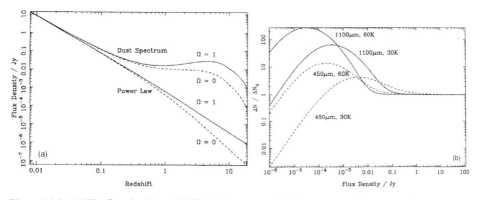

Figure 14.6: (a) The flux density–redshift relations expected for a source with a power-law spectrum of the form $S \propto \nu^{-1}$ and for a galaxy with an inverted dust spectrum at a temperature of 45 K, as observed at a submillimetre wavelength of 850 μm, for world models with $\Omega_0 = 1$ and 0. (b) The predicted differential normalised counts of sources at 450 and 1100 μm assuming the galaxies have typical dust spectra and that they have a far-infrared luminosity function given by that of IRAS galaxies at 60 μm (Saunders $et~al.$, 1990). No evolution of the source population is assumed. The predictions for dust temperatures of 30 and 60 K are shown (Blain and Longair, 1993).

than in the ultraviolet region of the spectrum (Weedman, 1994). These observations had two consequences. First of all, some of the star-forming galaxies may well be obscured by dust and so not be present in optical–ultraviolet multicolour surveys. Secondly, it is quite possible that a significant fraction of the radiation associated with the formation of stars and the heavy elements was not radiated in the ultraviolet–optical region of the spectrum, but at far-infrared wavelengths.

One way of estimating the significance of absorption by dust is to carry out observations in the millimetre and submillimetre wavebands. The dust emission in the millimetre/submillimetre waveband provides a complementary measure of the star-formation rate to that inferred from optical observations. What makes this approach feasible observationally is the fact that the spectrum of dust in the submillimetre waveband is strongly 'inverted' with spectra of the form $S_\nu \propto \nu^x$, where $x \sim 3$–4. As a consequence, the 'K-corrections' are so large and negative that a typical star-forming galaxy is expected to have essentially the same flux density, whatever its redshift in the range $1 < z < 10$ (Figure 14.6(a)) (Blain and Longair, 1993). A further consequence is that the number counts are expected to be inverted, even if there if no evolution of the source population (Figure 14.6(b)).

Deep surveys in the submillimetre waveband became feasible with the construction of bolometer array receivers, in particular the SCUBA instrument operating on the James Clerk Maxwell Telescope. Ian Smail (b. 1967), Robert Ivison (b. 1966) and Andrew Blain (b. 1970) made the first deep submillimetre surveys in the fields of two clusters of galaxies with SCUBA and discovered a large population of faint background submillimetre sources (Smail, Ivison and Blain, 1997). What was remarkable was that the excess of faint submillimetre sources was even greater than the largest excess predicted by our most extreme evolutionary models of 1993. All subsequent observations have confirmed the existence of this large population of faint submillimetre sources, which are almost all star-burst galaxies at redshifts $z \geq 1$. One of the more remarkable SCUBA observations was the submillimetre

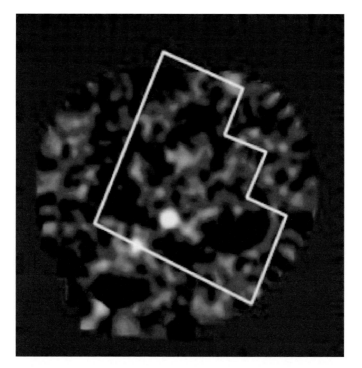

Figure 14.7: The submillimetre image of the Hubble Deep Field observed at a wavelength of 850 μm by the SCUBA submillimetre camera on the James Clerk Maxwell Telescope. Although only a few sources are detected, they contribute significantly to the star-formation rate at large redshifts (Hughes *et al.*, 1998; Serjeant *et al.*, 2003).

image of the Hubble Deep Field (Hughes *et al.*, 1998; Serjeant *et al.*, 2003), the central bright source being associated with a very faint extremely red object with an estimated redshift $z \approx 4$ (Dunlop *et al.*, 2004) (Figure 14.7). The galaxies associated with the SCUBA sources are very faint optically, but they make a major contribution to the global star-formation rate at large redshifts (see Section 14.6).

14.2 The counts of galaxies

The determination of precise counts of galaxies proved to be one of the more difficult areas of observational cosmology.[3] The reasons for these complications are multifold. First of all, galaxies are extended objects, often with complex brightness distributions, and great care must be taken to ensure that the same types of object are compared at different magnitude limits and redshifts. Secondly, the distribution of galaxies is far from uniform, as illustrated by the huge voids and walls in the local distribution of galaxies (Figure 10.8). Even at the faintest magnitudes, this 'cellular' structure in the distribution of galaxies results in fluctuations in the number counts of galaxies which exceed the statistical fluctuations expected of

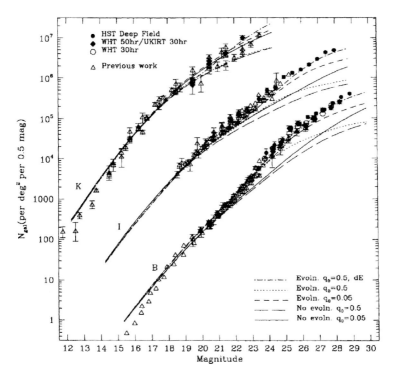

Figure 14.8: The counts of faint galaxies observed in the B, I and K wavebands compared with the expectations of various uniform world models, as well as other models in which various forms of the evolution of the luminosity function of galaxies with redshift are assumed (Metcalfe *et al.*, 1996). The galaxy counts follow closely the expectations of uniform world models at magnitudes less than about 21, but there is a excess of galaxies in the B and I wavebands at fainter magnitudes.

a random distribution (Figure 10.9). Finally, the probability of finding galaxies of different morphological types depends upon the galaxy environment.

A major complication concerns the K-corrections which should be used for galaxies of different types. There have been remarkably few systematic surveys of the ultraviolet spectra of normal galaxies, and only in a few cases have images of galaxies in the ultraviolet waveband been obtained (Giavalisco *et al.*, 1996). The problem is exacerbated by the fact that the ultraviolet spectra of galaxies can be dominated by bursts of star formation, and this fact alone makes the comparison of the optical images of galaxies at the present epoch with those at redshifts of one and greater problematic. On the other hand, as discussed in Section A13.1, counts of galaxies in the infrared K waveband at 2.2 μm have a number of advantages and enable the old, stable populations of galaxies to be studied out to redshifts $z \sim 3$.

Number counts in the B (440 nm), I (800 nm) and K (2.2 μm) wavebands from a number of independent determinations by ground-based optical and infrared telescopes, as well as deep number counts in the Hubble Deep Field, are shown in Figure 14.8 (Metcalfe *et al.*,

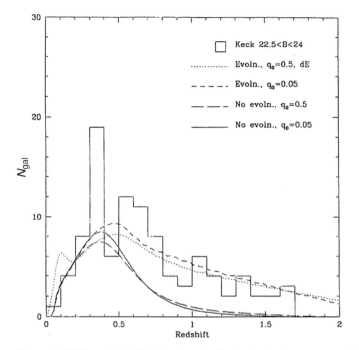

Figure 14.9: The redshift distribution of galaxies in the deep survey of Cowie and his colleagues in the magnitude interval 22.5 < B < 24 (Cowie *et al.*, 1996; Metcalfe *et al.*, 1996). The solid line and the long-dashed line show the expected redshift distributions if there were no evolution of the local luminosity function of galaxies. It can be seen that there is a long high-redshift tail of blue galaxies, which, from the presence of strong [OII] lines, are inferred to be undergoing rapid star formation.

1996). The lines labelled 'No evolution' show the expectations of uniform world models and include appropriate K-corrections for the types of galaxy observed in bright galaxy samples. In the infrared K waveband (2.2 μm), the counts follow reasonably closely the expectations of uniform world models with $q_0 \sim 0$–0.5. In contrast, in the B and I wavebands, there is a large excess of faint galaxies, particularly in the B waveband. The departure from the expectations of the uniform models sets in at about B = 23, and, at fainter magnitudes, there is a large excess of faint blue galaxies. The lines on the diagram illustrate the results of various modelling exercises to account for the observed counts on the basis of models for the evolution of the stellar populations of spiral and elliptical galaxies.

The nature of the excess of faint blue galaxies is a key cosmological problem. Redshift distributions for complete samples of galaxies at which the excess of blue galaxies is observed are required, but this was beyond the capabilities of the generation of 4-metre class telescopes. Richard Ellis (b. 1950) noted that the complete redshift surveys with such telescopes are effectively limited to B ≤ 24, I ≤ 22 and K ≤ 18 (Ellis, 1997). An indication of what is possible with the 8–10-metre optical–infrared telescopes was provided by the first surveys of faint galaxies carried out with the Keck 10-metre telescope (Cowie, Hu and Songaila, 1995; Cowie *et al.* 1996). Figure 14.9 shows the redshift distribution for an almost complete sample of galaxies in the magnitude interval 22.5 < B < 24, compared with the expectations of the 'no evolution' models. There is a significant excess of blue galaxies extending to redshifts $z = 1.7$. According to Lennox Cowie (b. 1950) and his colleagues, the distribution is composed of a mixture of normal galaxies at small redshifts plus galaxies undergoing rapid star formation from $z = 0.2$ to beyond $z = 1.7$ (Cowie *et al.*, 1996). There is unquestionably an increase in the numbers of star-forming galaxies with increasing redshift. At the same time, they found little change in the K-band luminosity function out to redshifts $z \approx 1$, suggesting that most of their stellar populations were already in place by a redshift of 1.

The nature of the excess of blue galaxies has been at least partially elucidated by studies with the Hubble Space Telescope. By combining observations of the HST Medium Deep Survey with those of the Hubble Deep Field, number counts have been determined for galaxies of different morphological types (Figure 14.10). The high-resolution images enabled the morphologies of galaxies to be classified into spheroidal/compact, spiral and irregular/peculiar/merger categories. The counts for the different morphological classes show that the spheroidal and spiral galaxies more or less follow the expectations of the uniform world models, while the objects classified as irregular/peculiar/merger systems show a distinct excess relative to their populations in bright galaxy samples (Abraham *et al.*, 1996). These results are consistent with visual impressions of the Hubble Deep Field, which suggests that about 25% of the galaxies seem to be irregular/interacting/merging systems. They are also consistent with the imaging results of Cowie, David Schade (b. 1953) and their colleagues, which indicate that about the same fraction of the blue galaxies in their surveys have peculiar morphologies (Cowie *et al.*, 1995; Schade *et al.*, 1995).

14.3 The Lyman-α clouds

As soon as the first quasar with redshift $z > 2$ was discovered in 1965, the radio quasar 3C 9, James Gunn and Bruce Peterson (b. 1941), and independently Peter Scheuer, realised that the continuum radiation to the short-wavelength side of the redshifted Lyman-α line was observable in the optical waveband and provided a sensitive test for the presence of intergalactic neutral hydrogen (Gunn and Peterson, 1965; Scheuer, 1965). The Gunn–Peterson test makes use of the fact that the cross-section for the Lyman-α transition at 121.6 nm is very large, and so, when the ultraviolet continuum of distant quasars is shifted to the redshift at which it has wavelength 121.6 nm, the radiation is scattered many times, so that, if sufficient neutral hydrogen is present at these redshifts, an absorption trough would be expected to the short-wavelength side of the redshifted Lyman-α line.

The predicted absorption trough was searched for in those quasars with redshifts $1 + z \geq (330\,\mathrm{nm})/\lambda_{\mathrm{Ly}}$, $z \geq 2$. No evidence for such a depression to the short-wavelength side of the Lyman-α line was observed in 3C 9 or in any of the other large-redshift quasars which were observed over the succeeding years. As larger redshift quasars were discovered, it was possible to search for a feature to the short-wavelength side of the corresponding line of neutral helium, HeI, which has rest wavelength 58.4 nm, but no evidence of an absorption trough was observed. A typical upper limit to the number density of neutral hydrogen atoms at a redshift $z = 3$ was $N_{\mathrm{H}} \leq 10^{-5}\,\mathrm{m}^{-3}$, which is very small indeed compared with typical

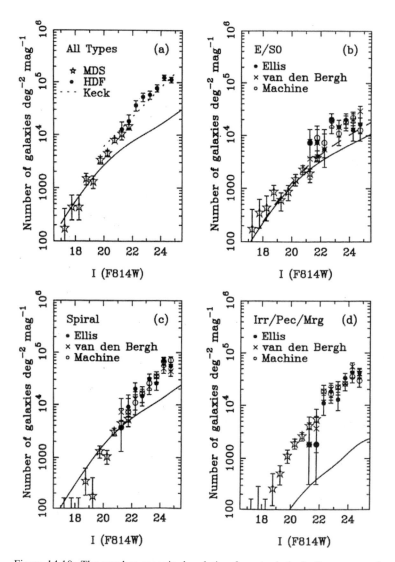

Figure 14.10: The number–magnitude relation for morphologically segregated samples of galaxies from the Medium Deep Survey (MDS) and the Hubble Deep Field (Abraham *et al.*, 1996). The morphological classifications were carried out independently by Richard Ellis, Sidney van den Bergh (b. 1929) and by an automated machine-based classification algorithm. The dotted line in (a) shows the total counts of galaxies from a field observed by the Keck Telescope. The solid lines show the expected counts of the different morphological classes, assuming their properties do not change with cosmic epoch (Glazebrook *et al.*, 1995).

cosmological baryonic densities. Therefore, if there were significant amounts of hydrogen in the intergalactic medium, it would have to be very highly ionised.

Although there was little evidence for continuum absorption due to diffuse neutral gas, as more large-redshift quasars were discovered, various types of absorption lines and absorption-line systems were observed.[4] The first quasar to be discovered with a rich absorption-line spectrum was 3C 191 with redshift $z = 1.953$, which was observed by the Burbidges, Roger Lynds (b. 1928) and Alan Stockton (b. 1942) (Burbidge, Lynds and Burbidge, 1966; Stockton and Lynds, 1966). The Burbidges found that the absorption-line systems were mostly observed in quasars with redshifts greater than about $z = 1.9$ (Burbidge and Burbidge, 1967). Some of the lines could be associated with strong reso-nance absorption lines, such as Lyman-α and CIV, which enabled an absorption redshift to be measured, but others remained unidentified. Soon, examples of multiple-absorption-line systems of the common elements were found in large-redshift quasars (Bahcall, Greenstein and Sargent, 1968; Burbidge, Lynds and Stockton, 1968).

These studies benefitted enormously from the electronic revolution in astronomical spec-troscopy during the 1970s. For example, the image photon counting system developed by Alexander Boksenberg revolutionised these studies and enabled superb absorption-line spectra to be obtained.[5] Similar spectrographs were eventually built for most of the 4-metre class telescopes and ultimately for the generation of 8- to 10-metre telescopes. These studies culminated in absorption-line spectra of the extraordinary quality seen in Figure 14.11. The study of such spectra became a major field in its own right, and these developments impacted a wide range of different astrophysical and cosmological topics, including the physics of the intergalactic gas, the evolution of galaxies and the history of metal enrichment in galaxies.[6]

While some of the absorption-line systems are associated with matter ejected from the quasar, there are two broad classes of absorption-line system which are of direct importance for cosmological studies. By far the most common are those belonging to the *Lyman-α forest*, which dominates the spectra of large-redshift quasars, such as that illustrated in Figure 14.11. These systems have neutral hydrogen column densities in the range $10^{16} \leq N_{HI} \leq 10^{21}$ m^{-2}. In those systems with $N_{HI} \geq 10^{19}$ m^{-2}, evidence for low abundances of the heavy elements, corresponding to about 1% of the solar value, is usually found (Boksenberg, 1997). The absorbers responsible for the Lyman-α forest are interpreted as intergalactic clouds, containing largely unprocessed primordial material. There must, however, be some mild enrichment of these primordial clouds, the chemical abundances of which are similar to those of halo stars in our Galaxy.

In contrast, the rarer *Lyman limit* and *damped Lyman-α systems* have much larger col-umn densities, $10^{21} \leq N_{HI} \leq 10^{26}$ m^{-2}, and have correspondingly larger optical depths for Lyman-α absorption. The Lyman-limit systems are those with column densities in the range $10^{21} \leq N_{HI} \leq 2 \times 10^{24}$ m^{-2}. A continuum break is observed at the redshifted wavelength of the Lyman limit at 91.2 nm and the spectra display the corresponding absorption lines of the common elements. Jacqueline Bergeron (b.1942) showed that these systems can be associated with the extended gaseous haloes of galaxies, similar to those observed about nearby galaxies (Bergeron, 1988). The abundances of the heavy elements are less than 10% of their cosmic abundances, consistent with the inference that the haloes are very extensive, $\sim$50–100 kpc, and so the gas in these regions is not expected to be as enriched as the gas in the disc of a galaxy. These systems probably contain most of the mass density of neutral gas in the Universe (Lanzetta, Wolfe and Turnshek, 1995).

The damped Lyman-α systems are those with the greatest column densities, $2 \times 10^{24} \leq N_{HI} \leq 10^{26}$ m^{-2}. In 1988, Arthur Wolfe (b. 1939) showed that these can be convincingly associated with galactic discs, and, in his pioneering paper, he identified them with the

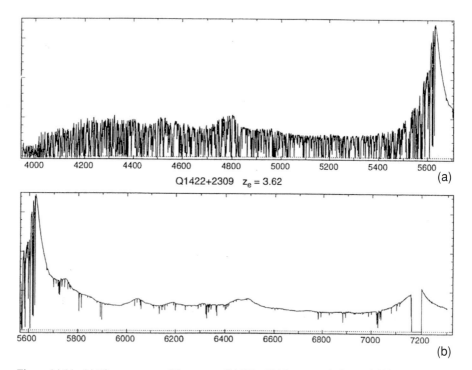

Figure 14.11: (a) The spectrum of the quasar Q1422+2309 at an emission redshift $z = 3.62$ showing the remarkable 'Lyman-α forest' to the short-wavelength side of the strong redshifted Lyman-α emission line, which has observed wavelength 560 nm. (Courtesy of Dr W. L. W. Sargent, from Boksenberg (1997).) (b) To the long-wavelength side of the Lyman-α line, the spectrum is very much smoother, only weak metal absorption lines being observed, most of them associated with the CIV absorption line.

progenitors of the stellar discs of present-day spiral galaxies (Wolfe, 1988). This is a natural assumption, since stars form in the coldest regions of the interstellar gas. It was reinforced by the important analysis of Robert Kennicutt (b. 1951), who made detailed observations of a number of nearby spiral galaxies and showed that active star formation only takes place in their discs if the column density of neutral gas exceeds 2×10^{24} m^{-2} (Kennicutt, 1989). It is striking that this criterion is identical to the lower limit at which absorbers are identified as damped Lyman-α systems. Kennicutt argued that this criterion is consistent with the stability criterion for rotating thin gaseous discs. Thus, the damped Lyman-α systems can provide information about the chemical evolution of disc galaxies.

An important aspect of the distribution of both classes of absorption system is their variation with redshift. It is convenient to parameterise the variation of the number density of absorbers with redshift by a power-law distribution of the form

$$N(z)\,dz = A(1+z)^\gamma\,dz. \tag{14.2}$$

Typically, it is found that, for the Lyman-α forest systems, $A \approx 10$ and $\gamma = 2$–3 whereas, for the Lyman-limit systems, $A \approx 1$ and $\gamma \sim 1$. These variations with redshift can be compared with the expected distribution if the properties of the absorbers were unchanging with cosmic epoch, that is if the absorbers had the same proper cross-sections and constant *comoving* number density.[7] If $\Omega_0 = 1$, $N(z) \propto (1+z)^{1/2}$ and, if $\Omega_0 = 0$, $N(z) \propto (1+z)$. Thus, the observed number density of Lyman-α forest absorbers changes more rapidly with increasing redshift than expected according to the uniform absorber model. The sense of the evolution is that there were more Lyman-α forest systems at large redshifts as compared with low redshifts. On the other hand, the Lyman-limit systems seem to show little variation with redshift other than that expected if their cross-sections and comoving number densities remained unchanged with cosmic epoch.

14.4 The abundances of elements in Lyman-α absorbers

Michael Fall (b. 1951) has emphasised that one of the great attractions of using absorption lines in Lyman-α absorbers to estimate the relative abundances of any species along the line of sight to distant quasars is that, provided the absorbers are randomly oriented, average relative abundances of different species, such as atoms, ions and molecules, can be found independent of the structures or clumpiness of the clouds (Fall, 1997). There will generally be a distribution of column densities, $N(N_X) \, dN_X$, for any species X, and this is expected to change with cosmic epoch. Gas is condensed into stars, and the interstellar gas in galaxies is enriched as a result of nucleosynthesis in stars and the subsequent recycling of processed material to the gas. The beauty of this result is that, if we average over many lines of sight, we take averages over all systems in all orientations which contribute to the mean density parameter of that species, $\Omega_X(z)$, at that redshift. Furthermore, we can determine how the global metallicity, $Z = \Omega_m / \Omega_g$, and other relative abundances change with redshift by taking appropriate ratios.

The damped Lyman-α systems are of particular importance in understanding the evolution of neutral gas and the build-up of the heavy elements with cosmic epoch (Lanzetta *et al.*, 1995; Storrie-Lombardi, McMahon and Irwin, 1996; Pettini *et al.*, 1997). The picture which emerges is that, at redshifts $z \sim 3$, the comoving density parameter in neutral hydrogen is $\Omega_{HI} \approx (1-2) \times 10^{-3} h^{-1}$ and this decreases with decreasing redshift until, at $z = 0$, its value is only about 2×10^{-4}, a value consistent with independent measures of the amount of neutral hydrogen present in galaxies and their environs at the present epoch (Figure 14.12). It is striking that the density parameter in neutral gas at a redshift $z = 2.5$ is of the same order of magnitude as the density parameter corresponding to the visible mass of galaxies at the present epoch, which is indicated by the hatched area in Figure 14.12.

The big advantage of studying the damped Lyman-α systems is that they have such large column densities that relatively rare species can be used to probe the chemical abundances of the elements. Max Pettini (b. 1949) and his colleagues, for example, used observations of singly ionised zinc, Zn^+ or ZnII, which has a number of advantages as a tracer of the overall abundance of the heavy elements (Pettini *et al.*, 1997). Zinc shows little affinity for dust and is predominantly in the form of Zn^+ in HI regions. Although zinc is a relatively rare species, the Solar System value corresponding to [Zn/H] $= 3.8 \times 10^{-8}$, this is an advantage since the absorption lines are optically thin and so accurate column densities can be determined.

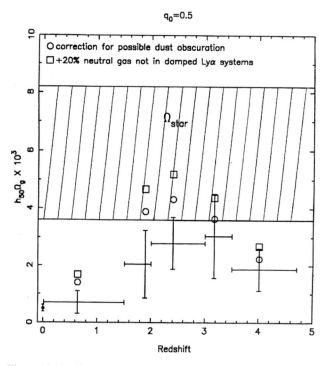

Figure 14.12: The evolution of the mass density of neutral gas as a function of redshift as determined by the mass density of damped Lyman-α absorbers in the spectra of distant quasars (Storrie-Lombardi *et al.*, 1996). The circles show the estimates of $\Omega_{\rm HI}$ corrected for the effects of dust extinction and the boxes show the corrected values which take account of neutral hydrogen not associated with damped Lyman-α systems. The hatched area indicates estimates of the density parameter in stars, that is the mass associated with the visible light of galaxies, at the present epoch.

In Figure 14.13, the abundance of zinc relative to hydrogen is shown for a number of large-redshift damped Lyman-α clouds. Recalling that we should take averages over all systems at a given redshift, it can be seen that the heavy element abundances have built up from values of only about 10% of the present solar abundances at a redshift $z \approx 2$ to $Z \approx Z_\odot$ at the present epoch. Over the same redshift interval, the mean dust-to-gas ratio has increased by roughly the same factor, while the mean dust-to-metals ratio has remained roughly constant (Fall, 1997). These results strongly suggest that a large fraction of the build-up of the heavy elements in galaxies took place over the redshift interval $1 < z < 3$.

14.5 The Lyman-break galaxies

In an important paper of 1987, Simon Lilly and Lennox Cowie first showed how the rate of formation of heavy elements could be inferred from the flat blue continuum spectra of star-forming galaxies and that these estimates are independent of the choice of cosmological model (Lilly and Cowie, 1987). A prolonged burst of star formation has a flat intensity

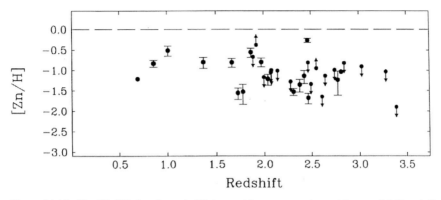

Figure 14.13: The [Zn/H] abundance in 24 damped Lyman-α systems at large redshifts relative to the solar value, which is indicated by the dashed line (Pettini *et al.*, 1997). The usual convention is used of plotting the logarithm of the abundance ratio [Zn/H] on the ordinate.

spectrum at wavelengths longer than the Lyman limit at 91.2 nm, as illustrated by the model star-bursts synthesised by Gustavo Bruzual (b. 1949) using his spectral synthesis codes to predict the spectra of galaxies at different phases of their evolution. Figure 14.14 shows the spectrum of a star-burst galaxy which lasts 12 Gyears, as observed at different ages, assuming that the star-formation rate is constant with the same Salpeter initial mass function. The flatness of the spectrum is associated with the fact that, although the most luminous blue stars have short lifetimes, they are continuously being replaced by new stars. To a good approximation, it can be assumed that the spectrum of the star-forming galaxy can be described by a power law, $I(\nu) \propto \nu^{-\alpha}$, with $\alpha = 0$ at wavelengths $\lambda > 91.2$ nm, and zero intensity at shorter wavelengths. The intensity of the flat part of the spectrum is directly proportional to the rate of formation of heavy elements in the star-burst since the conversion of hydrogen into helium is the essential first stage in the synthesis of the heavy elements in the central regions of the massive stars which are primarily responsible for the ultraviolet continuum. The important result found by Cowie and Lilly was that the background intensity associated with flat-spectrum star-forming galaxies is directly related to the rate at which elements are formed, and this rate, as a function of redshift, is *independent* of the cosmological model.[8]

Cowie, Lilly and their colleagues undertook deep multicolour surveys to discover flat-spectrum star-forming galaxies at large redshifts. Their survey was successful in finding such galaxies, which have roughly equal intensities in the U, B and V wavebands, the background due to such objects amounting to about 10^{-24} W m^{-2} Hz^{-1} sr^{-1}. They interpreted this result as meaning that a significant fraction of the heavy elements, about 1.5×10^{-31} kg m^{-3}, must have been synthesised at redshifts of about unity (Cowie *et al.*, 1988).

In a remarkable pioneering set of observations, Charles Steidel (b. 1962) and his colleagues extended the multicolour technique to find star-forming galaxies at redshifts $z > 3$. The objective was to search for star-forming galaxies in which the Lyman limit, clearly seen in the models of star-burst galaxies in Figure 14.14, is redshifted into the optical waveband. The technique is illustrated in Figure 14.15(a), in which the spectrum of a star-forming galaxy at redshift $z = 3.15$ is observed through carefully chosen filters in the ultraviolet,

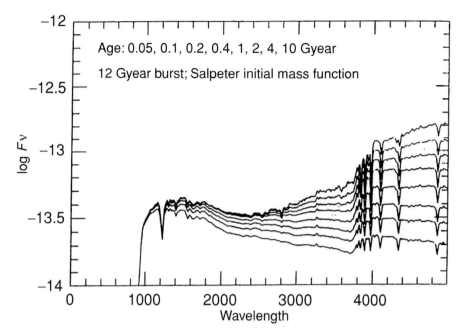

Figure 14.14: Synthetic spectra of a 12 Gyear star-burst with constant star-formation rate as observed at the ages indicated. A Salpeter initial stellar mass function, $N(M)\,\mathrm{d}M \propto M^{-2.35}\,\mathrm{d}M$, is assumed with cut-offs at $75\,M_\odot$ and $0.08\,M_\odot$. The spectra were generated by Gustavo Bruzual using his evolutionary synthesis programmes (White, 1989).

blue and red spectral regions. The signature of such a star-burst galaxy is that its image should be roughly equally bright in the two longer wavebands, but should not be present in the ultraviolet waveband. Steidel's original intention was to use this technique to identify the galaxies responsible for the Lyman-limit absorption systems in the spectra of distant quasars, and this programme turned out to be remarkably successful (Steidel and Hamilton, 1992). It was soon found, however, that the technique was also a remarkably effective means of discovering star-forming galaxies in the general field at $z > 3$ (Steidel, 1998). Observations of the Hubble Deep Field have proved to be ideal for exploiting this approach because, in addition to very precise photometry in four wavebands spanning the wavelength range $300 < \lambda < 900$ nm, high-resolution optical images have enabled the morphologies of these galaxies to be studied. HST images in four wavebands of one of the Lyman-break galaxies in the Hubble Deep Field are shown in Figure 14.15(b).

This technique has been successful in identifying star-forming galaxies at redshifts $z \geq 3$. Spectroscopic confirmation of their large redshifts has been obtained for almost 200 of these galaxies as a result of observations with the Keck telescope. An example of the redshift distribution found for the galaxies in Steidel's surveys is shown in Figure 14.16, in which the observed redshift distribution more or less follows that expected according to the colour selection criteria. There is a large 'spike' at $z = 3.09$, which is probably associated with the large-scale clustering of galaxies at that redshift. Thus, not only the properties of these star-forming galaxies, but also their three-dimensional spatial distribution, can be studied.

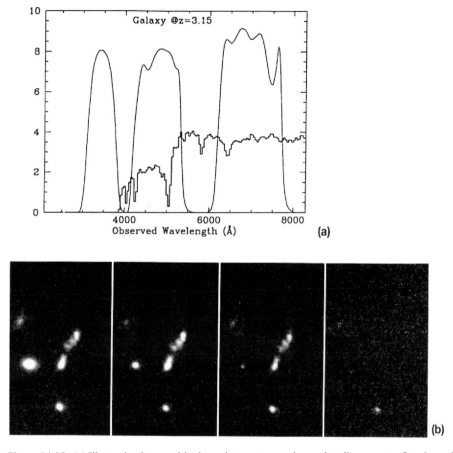

Figure 14.15: (a) Illustrating how multicolour photometry can be used to discover star-forming galaxies at redshifts $z > 3$, at which the Lyman limit is redshifted into the optical region of the spectrum (Steidel, 1998). (b) Images of a distant star-forming galaxy present in the Hubble Deep Field. From left to right, the images were taken at red (I), green (V), blue (B) and ultraviolet (U) wavelengths. Because the Lyman limit has been redshifted beyond the U waveband, no image of the galaxy appears on the U image (Macchetto and Dickinson, 1997).

These data can be analysed to determine the average star-formation rates within these galaxies, and they have turned out to be not so different from the typical overall star-formation rate in galaxies at the present epoch (see Section 14.6 and Figure 14.17). In addition, it has been possible to learn a great deal about the astrophysical nature of these objects from spectroscopic observations with the Keck telescope, for example their internal kinematics, their stellar populations and so on.

14.6 The global star-formation rate

In an important paper of 1996, Piero Madau (b. 1958) and his colleagues attempted to synthesise all the information discussed in Sections 14.2 to 14.5 into a unified picture for

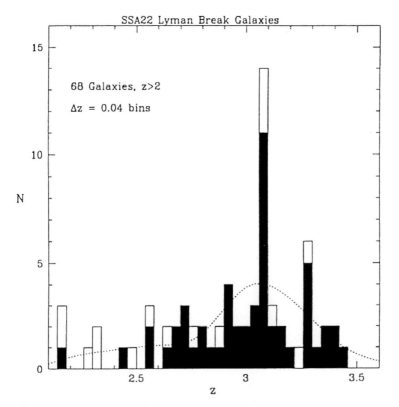

Figure 14.16: The redshift distribution of Lyman-break galaxies in a single $9' \times 18'$ area of sky, all of which have spectroscopically confirmed redshifts. The differently shaded histograms reflect slightly different selection criteria. The dotted curve shows the expected redshift distribution defined by the colour selection criteria for the complete sample of objects. The 'spike' at $z = 3.09$ is significant at the 99.9% confidence level (Steidel, 1998).

the evolution of the global star-formation rate with cosmic epoch (Madau *et al.*, 1996). By global star-formation rate, we mean the rate of star formation integrated over all types of galaxies as a function of cosmic epoch, or redshift. In Figure 14.17, the results are presented as the number of solar masses per year per cubic megaparasec, derived from an analysis of a large body of data. The sources of these include the global star-formation rate at the present epoch (Gallego *et al.*, 1995), the 280 nm continuum emission of star-forming galaxies at $z < 1$ (Lilly *et al.*, 1995), the Hα emission of star-forming galaxies in the redshift interval $1 < z < 2$ (Connolly *et al.*, 1997) and the Lyman-limit galaxies, including those observed in the Hubble Deep Field, which set limits to the star-formation rate at redshifts $z > 3$ (Madau *et al.*, 1996; Madau, Pozzetti and Dickinson, 1998).

A literal interpretation of Figure 14.17 suggests that the cosmic rate of star formation peaked in the redshift interval $1 \leq z \leq 2$, at which time it was about an order of magnitude greater than it is at the present epoch. At larger redshifts, $z \geq 3$, the star-formation rate was less than this maximum, and not so different from that occurring at the present epoch, consistent with the observations of Steidel and his colleagues.

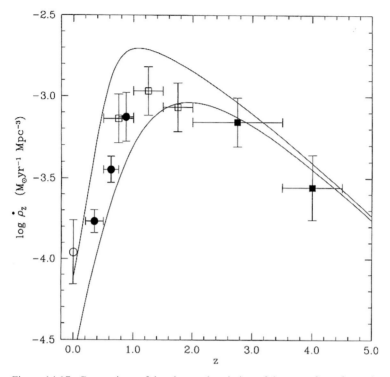

Figure 14.17: Comparison of the observed variation of the rate of star formation with cosmic epoch with that predicted on the basis of the absorption history of the interstellar gas in galaxies (Madau *et al.*, 1998). Both the observed rates and the model predictions have been corrected in a self-consistent manner for dust extinction. The model predictions are due to Pei and Fall (1995).

In 1995, Michael Fall and Yichuan Pei (b. 1962) developed a simple formalism for interpreting the emission and absorption histories of star formation and the build-up of the chemical elements with cosmic epoch (Pei and Fall, 1995). This approach was based upon the *equations of cosmic chemical evolution*, which were originally derived by Beatrice Tinsley (1941–1981) for the chemical evolution of galaxies (Tinsley, 1980). Many more details of these types of calculation and of the physics of the chemical evolution of stars and galaxies are given by Bernard Pagel (b. 1930) (Pagel, 1997).

In the approach adopted by Pei and Fall, all mention of galaxies disappears, all that remains being the global averages of the density parameters for the different constituents of the Universe. The density parameter in stars, Ω_s, is zero when star formation begins and builds up to the mean value $\Omega_s(t_0) \approx (4\text{–}8) \times 10^{-3}$ by the present epoch. At the same time, the density parameter in gas, Ω_g, initially comprised 100% of the baryonic matter in galaxies and has decreased to only about 5% of its initial value by the present epoch. The density parameter in heavy elements, Ω_m, was initially zero and has built up to about 1% of the baryonic mass by the present epoch. Finally, dust was also formed as the abundance of the heavy elements built up. A typical figure would be that about 50% of the heavy elements in the interstellar media of galaxies is in the form of dust.

The objective of the approach is to relate the absorption history of the Universe, as defined by the build-up of the heavy elements and the decrease in the abundance of neutral hydrogen, to the emission history, which provides direct information about the global star-formation rate. To give a flavour for this approach, the first equation describes the conservation of mass:

$$\frac{d}{dt}\left(\Omega_g + \Omega_s\right) = \dot{\Omega}_f,$$ (14.3)

where $\dot{\Omega}_f$ is the rate of infall into, or of the expulsion of baryonic matter from, galaxies. If this term is zero, equation (14.3) states that $\dot{\Omega}_g = -\dot{\Omega}_s$, that is the rate at which gas is depleted is equal to the rate at which mass is condensed into stars.

The second equation describes the rate at which the mass of heavy elements in diffuse gas changes with time. The metallicity of the gas at any epoch is defined to be $Z = \Omega_m/\Omega_g$, and so the rate of change of Ω_m is given by

$$\frac{d\Omega_m}{dt} = \frac{d}{dt}\left(Z\Omega_g\right) = y\frac{d\Omega_s}{dt} - Z\frac{d\Omega_s}{dt} + Z_f\dot{\Omega}_f.$$ (14.4)

The first term on the right-hand side, $y\,d\Omega_s/dt$, describes the rate of increase of the mass of heavy elements associated with the rate of star formation. The quantity y is called the *yield*. This term means that, in the time dt, a mass of stars $d\Omega_s$ is formed per unit comoving volume and these eventually return a mass of heavy elements, $y\,d\Omega_s$, to the interstellar medium. In the spirit of this analysis, the yield is assumed to be independent of cosmic epoch. The second term on the right-hand side of equation (14.4) describes the loss of heavy elements because of the formation of stars from gas which has already attained a metallicity Z. In the time dt, the loss of heavy elements per unit comoving volume is given by $-d\Omega_m = -Z\,d\Omega_s$. The third term on the right-hand side represents enhancement of the heavy element abundance by the infall of baryonic material from intergalactic space, assumed to have metallicity Z_f. This term takes account of the fact that the primordial gas might have been enriched by early generations of star formation.

Pei and Fall use these equations to predict the emission history from the absorption history, taking account of the effect of dust in a self-consistent manner. The results of their calculations are shown in Figure 14.17 by the pair of continuous lines which bound the observed star-formation rate determined by Madau and his colleagues. It is notable that their analysis is sensitive to dust obscuration of the quasars, which are the prime source of information on the absorption-line history of metal enhancement, as Pei and Fall (1995) demonstrate.

Another concern is the completeness of the information on the global star-formation rate, which, in Figure 14.17, was entirely determined by optical observations. Fortunately, the advances in submillimetre astronomy described in Section 14.1.5 have enabled estimates to be made of the incompleteness of the optical surveys. The observations of the Hubble Deep Field and other areas by the SCUBA array have shown that a population of obscured star-forming galaxies exists at large redshifts and that these sources increase the global star-formation rate significantly above the rates shown in Figure 14.17 (Hughes *et al.*, 1998). Subsequent analyses of the large-redshift global star-formation rate, including improved corrections for obscured star formation, have suggested that the rate may well be roughly constant at redshifts beyond the peak at $z \sim 1$–2.

14.7 Conclusion

The developments described in this chapter have been among the most important in astrophysical cosmology in the second half of the twentieth century. Most of the developments described above have concerned the evolution of discrete objects once they have formed. The major challenge is to relate these to theories of the origin of galaxies and larger-scale structures, which are the subject of Chapter 15.

Notes to Chapter 14

1 I have given many more details of the observations and data involved in the studies described in this chapter in Part IV of Malcolm Longair, *Galaxy Formation* (Berlin: Springer-Verlag, 1998).

2 Lodewijk Woltjer (b. 1930) has presented a concise summary of the relative merits of the different selection procedures for finding radio-quiet quasars and other types of active galaxy in Woltjer, Phenomenology of active galactic nuclei, in *Active Galactic Nuclei* by Blandford, R. D. Netzer, H. and Woltjer, L., eds T. Courvoisier and M. Mayor (Berlin: Springer-Verlag, 1990).

3 The review of counts of faint galaxies by Richard Ellis (Ellis, 1997) describes vividly the complications in determining precisely and in interpreting the counts of galaxies.

4 A contemporary account of the early years of quasar studies is given in the book by Geoffrey and Margaret Burbidge (Burbidge and Burbidge, 1967). The book includes much of the early history of quasar studies, both the observations and interpretations of the data. The puzzles presented by the absorption-line spectra are surveyed as they were understood in 1967.

5 Boksenberg's image photon counting system was a superb instrument which was particularly effective in astronomical spectroscopy. During the 1970s, it was so superior to all other instruments that he was awarded very large amounts of time on the premier telescopes available at that time. In consequence, he and his support team, affectionately known as Boksenberg's Flying Circus, were involved in many of the most important programmes in the spectroscopy of faint objects.

6 The nature and properties of these absorption-line systems are vast subjects; for many more details, reference should be made to the following volumes: J. C. Blades, D. Turnshek and C. A. Norman, eds, *QSO Absorption Lines: Probing the Universe* (Cambridge: Cambridge University Press, 1988); G. Meylan, ed., *QSO Absorption Lines* (Berlin: Springer-Verlag, 1995); and N. R. Tanvir, A. Aragón-Salamanca and J. V. Wall, eds, *The Hubble Space Telescope and the High Redshift Universe* (Singapore: World Scientific Publishing Company, 1997).

7 I have derived these relations in Section 18.3.2 of Longair, *Galaxy Formation*.

8 I have derived this result in more detail in Section 18.4.1 of Longair, *Galaxy Formation*.

15 The origin of galaxies and the large-scale structure of the Universe

Galaxies and clusters of galaxies are complex systems, but the aim of the cosmologist is not to explain all their detailed features. Rather, it is to explain how large-scale structures formed in the expanding Universe in the sense that, if $\delta\rho$ is the enhancement in density of some region over the average background density ρ, the *density contrast* $\delta\rho/\rho$ reached amplitude 1 from initial conditions which must have been remarkably isotropic and homogeneous. Once the initial perturbations have grown in amplitude to $\delta\rho/\rho \sim 1$, their growth becomes non-linear and they rapidly evolve towards bound structures in which star formation and other astrophysical process lead to the formation of galaxies and clusters of galaxies as we know them. The cosmologist's objectives are therefore twofold – to understand how density perturbations evolve in the expanding Universe and to derive the initial conditions necessary for the formation of structure in the Universe.[1]

Galaxies, clusters of galaxies and other large-scale structures of our local Universe must have formed relatively late in the history of the Universe. The average density of matter in the Universe today corresponds to a density parameter $\Omega_0 \sim 0.3$. The average densities of gravitationally bound systems, such as galaxies and clusters of galaxies, are much greater than this value, typically their densities being about 10^6 and 1000 times greater than the mean background density, respectively. Superclusters have mean densities a few times the background density. Therefore, the density contrasts $\delta\rho/\rho$ for galaxies, clusters of galaxies and superclusters at the present day are about $\sim 10^6$, 1000 and a few, respectively. Since the average density of matter in the Universe, ρ, changes as $R^{-3} = (1+z)^3$, where R is the scale factor and z is redshift, it follows that typical galaxies must have had $\delta\rho/\rho \sim 1$ at a redshift $z \approx 100$. They could not have separated out as discrete objects at larger redshifts, or else their mean densities would be greater than those observed at the present epoch. The same argument applied to clusters and superclusters indicates that they could not have separated out from the expanding background at redshifts greater than $z \sim 10$ and 1, respectively.

Therefore galaxies and larger-scale structures must have separated out from the expanding Universe at redshifts significantly less than 100. This epoch occurred long after the epoch of recombination at $z \approx 1000$ when the primordial plasma recombined (see Section 15.2.1) and well into the matter-dominated phase of the standard Big Bang. Thus, these structures were not formed in the inaccessibly remote past, but at redshifts which are accessible to observation.

15.1 Gravitational collapse and the formation of structure in the expanding Universe

The standard Friedman world models are isotropic and homogeneous, and so the diversity of structure we observe in the Universe about us is absent. The next step is to include density perturbations of small amplitude, $\delta\rho/\rho$, into the Friedman models and study their development under gravity. This problem was solved by James Jeans in 1902 for the case of a stationary medium (Jeans, 1902). He derived the *dispersion relation*, the relation between wavenumber, k, and angular frequency, ω, for perturbations in a medium of density ρ_0 and pressure p_0:

$$\omega^2 = c_s^2 k^2 - 4\pi G \rho_0, \tag{15.1}$$

where c_s is the speed of sound in the medium and $k = 2\pi/\lambda$, λ being the wavelength of the perturbation.[2] Thus, for small wavelengths and large wavenumbers, the right-hand side is positive and the perturbations behave as propagating sound waves. If the right-hand side is negative, gravitational collapse occurs and the density perturbations grow exponentially with characteristic timescale $\tau \sim (G\rho_0)^{-1/2}$ in the limit of long wavelengths. The criterion for collapse is thus that the size of the perturbation should exceed the *Jeans' length*, $\lambda_J = c_s/(G\rho_0/\pi)^{1/2}$, meaning that the gravitational force of attraction by the matter of the perturbation exceeds the pressure gradient which resists collapse. This criterion is important in studies of star formation.

The analysis was repeated for the case of an expanding medium in the 1930s by Georges Lemaître and by Richard Tolman (1881–1948) for the case of spherically symmetric perturbations (Lemaître, 1933; Tolman, 1934), and the solution for the general case was found by Evgenii Lifshitz (1915–1985) in 1946 (Lifshitz, 1946). Lifshitz found that the condition for gravitational collapse is exactly the same as the Jeans' criterion at any epoch, but, crucially, the growth of the density contrast is no longer exponential but only algebraic (see Section A15.1). In the case of a matter-dominated Universe with the critical density $\Omega_0 = 1$, the density contrast grows linearly with the scale factor R, that is

$$\delta\rho/\rho \propto R = (1+z)^{-1} \propto t^{2/3}. \tag{15.2}$$

This is one of the most important equations in astrophysical cosmology. A similar result is found for radiation-dominated universes, $\delta\rho/\rho \propto R^2 = (1+z)^{-2}$. For other Friedman world models, the growth rate given by equation (15.2) is a good approximation for redshifts $z > \Omega_0^{-1}$, but, at smaller redshifts, the perturbations are stabilised. The implication of these results is that the fluctuations from which the large-scale structure of the Universe formed cannot have grown from infinitesimal statistical perturbations in the number density of particles – they must have developed from perturbations of finite amplitude. For this reason, Lemaître, Tolman and Lifshitz inferred that galaxies were not formed by gravitational collapse.

Other authors took the view that finite perturbations should be included in the initial conditions from which the Universe evolved and then the evolution of the perturbation spectrum with cosmic time should be studied in detail. The Moscow school, led by Yakov Zeldovich, Igor Novikov and their colleagues, and James Peebles at Princeton pioneered the study of the development of structure in the Universe in the 1960s. If perturbations on a particular physical scale were tracked backwards into the past, it was found that, at some large redshift, the scale of the perturbation was equal to the horizon scale of the Universe at that time, that is $r = ct$, where t is the age of the Universe. In 1964, Novikov showed that, to form structures on the scales of galaxies and clusters of galaxies, density perturbations on the scale of the horizon had to have amplitude $\delta\rho/\rho \sim 10^{-4}$ in order to guarantee the formation of galaxies by the present epoch (Novikov, 1964). These perturbations are certainly *not* infinitesimal, and their origin had to be ascribed to processes occurring in the very early Universe.

15.2 The thermal history of the Universe

The discovery of the cosmic microwave background radiation by Penzias and Wilson in 1965 had an immediate impact upon these studies, since the thermal history of the matter and radiation content of the Universe could be determined in detail. In turn, this enabled the variation of the speed of sound with cosmic epoch to be determined; this is needed to apply the Jeans' stability criterion on different physical scales at different cosmic epochs. The thermal history of the Big Bang established by Alpher and Herman (Alpher and Herman, 1948) could now be placed on a firm observational foundation. In the simplest picture, the temperature of the background radiation changes with scale factor as $T = T_0/R = T_0(1 + z)$, exactly the same as the adiabatic expansion of a photon gas, but there are some important elaborations (Figure 15.1).

15.2.1 The epoch of recombination and the last scattering surface

At a redshift $z = 1500$, the background radiation attained a temperature $T = T_0(1 + z) \approx 4000$ K, at which there were sufficient photons in the Wien region of the Planck distribution to ionise all the intergalactic hydrogen.[3] This epoch is referred to as the *epoch of recombination* since the hydrogen was fully ionised at earlier cosmic epochs.

The details of the process of recombination of the primordial plasma as the Universe expands and cools are important in understanding the origin of the temperature fluctuations in the cosmic microwave background radiation. These recombination calculations were first carried out by James Peebles and by Yakov Zeldovich, Vladimir Kurt (b. 1933) and Rashid Sunyaev independently in the late 1960s (Peebles, 1968; Zeldovich, Kurt and Sunyaev, 1968). Recombinations to the ground state of hydrogen release Lyman continuum photons which can immediately reionise any neutral hydrogen atoms which have recombined. If the recombination takes place to an excited state, the liberated photon excites a neutral hydrogen atom to an excited state from which it can be readily ionised. Therefore, the recombination rate is determined by the rate at which Lyman-α photons are destroyed by the rare two-photon process. The result is that the process of recombination takes place over a finite redshift range. Detailed calculations show that the pre-galactic gas was 50% ionised at a redshift $z_r \approx 1500$. At earlier epochs, $z \approx 6000$, helium was 50% ionised and rapidly became fully ionised before that time.

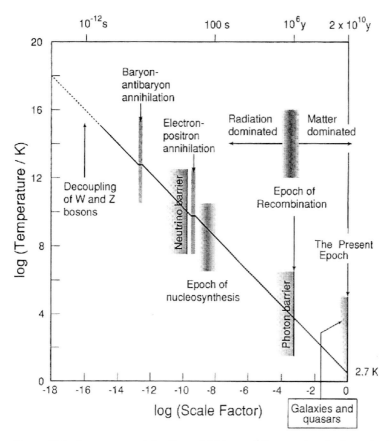

Figure 15.1: A summary of the thermal history of the cosmic microwave background radiation, showing some of the important events at different cosmic epochs. This diagram is a simplified version of Figure 12.1, which was first derived by Alpher and Herman in 1948 (Alpher and Herman, 1948; Wagoner *et al.*, 1967).

The most important consequence is that, at redshifts greater than about 1000, the Universe became opaque to *Thomson scattering*. This is the simplest scattering process, which impedes the propagation of photons from their sources to the Earth through an ionised plasma, the photons being scattered without loss of energy by free electrons. The intergalactic gas was essentially fully ionised at $z > 1000$, and so the optical depth at larger redshifts is given by

$$\tau_T = 0.035 \frac{\Omega_B}{\Omega_0^{1/2}} h z^{3/2}. \tag{15.3}$$

For reasonable values of Ω_B, Ω_0 and h, $\tau_T \gg 1$. Therefore, the Universe beyond a redshift of 1000 is unobservable. Any photons originating from larger redshifts were scattered many times before they propagated to the Earth, and consequently all the information they carry about their origin is lost. There is therefore a *photon barrier* or *last scattering surface* at a redshift of 1000, beyond which we cannot obtain information directly using photons. This is the surface on which the ripples in the cosmic microwave background radiation were imprinted by matter perturbations.

15.2.2 The radiation-dominated era

If the matter and radiation were not thermally coupled, they would cool independently, the hot gas having a ratio of specific heat capacities $\gamma = 5/3$ and the radiation $\gamma = 4/3$, corresponding to variations of the temperature of the matter and radiation with scale factor R as $T_m \propto R^{-2}$ and $T_r \propto R^{-1}$, respectively. The cosmic microwave background radiation provides by far the greatest contribution to the energy density of radiation in intergalactic space. Therefore, comparing the inertial mass density in the radiation and the matter as a function of redshift, we obtain

$$\frac{\rho_r}{\rho_m} = \frac{aT^4(z)}{\Omega_0 \rho_c (1+z)^3 c^2} = \frac{2.48 \times 10^{-5}(1+z)}{\Omega_0 h^2}. \tag{15.4}$$

Hence, the Universe became radiation- rather than matter-dominated at redshifts $z \gg 4 \times 10^4 \Omega_0 h^2$. It would be expected that the matter would cool more rapidly than the radiation, and this is indeed what is expected to take place during the post-recombination era. This is not the case, however, during the pre-recombination and immediate post-recombination eras because the matter and radiation are strongly coupled by *Compton scattering*. The optical depth of the pre-recombination plasma for Thomson scattering is so large that the small energy transfers which take place between the photons and the electrons in Compton collisions cannot be ignored.

The details of this process were worked out in pioneering papers by Raymond Weymann (Weymann, 1966) and in much more detail by Zeldovich and Sunyaev (Zeldovich and Sunyaev, 1969).[4] The analyses of Zeldovich and Sunyaev were based upon the theory of induced Compton scattering developed by Alexander Kompaneets (1914–1974), which had been published in 1956, long after this remarkable classified work had been completed in 1950 (Kompaneets, 1956).[5] What these papers showed was that, during the radiation-dominated epochs, the matter and radiation were maintained in very close thermal contact by Compton scattering so long as the intergalactic gas remained ionised. Therefore, since the radiation had much greater heat capacity than the matter, the matter cooled at the same rate as the radiation during the radiation-dominated epochs, $T_m \propto R^{-1}$.

Zeldovich and Sunyaev showed how significant distortions of the spectrum of the microwave background radiation could take place if the electrons were heated to a temperature greater than the radiation temperature by some process. These might involve, for example, the dissipation of primordial sound waves or turbulence, matter–antimatter annihilation, the evaporation of primordial black holes by the Hawking mechanism or the decay of heavy unstable leptons. If no photons were created, the spectrum of the radiation would be distorted from its Planckian form to a Bose–Einstein spectrum with a finite dimensionless chemical potential μ:

$$I_\nu = \frac{2h\nu^3}{c^2} \left[\exp\left(\frac{h\nu}{kT_r} + \mu\right) - 1 \right]^{-1}. \tag{15.5}$$

This is the form of equilibrium spectrum expected when there is a mismatch between the number of photons and the energy to be distributed among them in statistical equilibrium. Very strong upper limits to the value of μ were derived from the COBE spectral observations of the cosmic microwave background radiation. As illustrated in Figure 12.4, the microwave background radiation has a perfect black-body spectrum, the upper limit to μ being $|\mu| \leq 10^{-4}$ (Page, 1997). In general terms, this means that there cannot have been major injections of energy into the pre-galactic gas in the redshift interval $10^7 \geq z \geq 2 \times 10^4$.

As noted above, during the post-recombination era, $z \leq 1000$, matter and radiation were decoupled and so the matter cooled more rapidly with redshift than the radiation. Since $T_m \propto R^{-2}$ and $T_r \propto R^{-1}$, it would be expected that the matter would now be roughly a factor of 1000 colder than the background radiation. In fact, as indicated by the Gunn–Peterson test (Section 14.3), the intergalactic gas must be highly ionised at redshifts $z \leq 6$ because of the absence of absorption troughs to the short-wavelength sides of the redshifted Lyman-α emission lines observed in the spectra of large-redshift quasars. Therefore, at some time between the epoch of recombination and a redshift certainly greater than 5, the intergalactic gas must have been ionised and reheated, presumably by some form of activity associated with the early formation of stars and galaxies. We will return to this topic in Section 15.13.

15.2.3 Earlier epochs

Extrapolating back to redshifts $z \approx 3 \times 10^8$, $T = 10^9$ K, the radiation temperature was sufficiently high for the background photons to attain γ-ray energies, $\varepsilon = kT = 100$ keV. At this high temperature, the high-energy photons in the Wien region of the Planck distribution were energetic enough to dissociate light nuclei such as helium and deuterium. At earlier epochs, all nuclei were dissociated into protons and neutrons. When the clocks were run forward, this was the epoch when primordial nucleosynthesis of the light elements took place, a topic discussed in Sections 12.4 to 12.6.

At redshift $z \approx 10^9$, electron–positron pair production from the thermal background radiation took place and the Universe was flooded with electron–positron pairs, one pair for every pair of photons present in the Universe now. When the clocks were run forward from an earlier epoch, the electrons and positrons annihilated at about this epoch and their energy transferred to the photon field – this accounts for the little discontinuity in the temperature history when the electrons and positrons were annihilated (Figure 15.1). At a slightly earlier epoch, the opacity of the Universe for weak interactions became unity, resulting in a *neutrino barrier*, similar to the photon barrier at $z \approx 1000$.

We can extrapolate even further back in time to $z \approx 10^{12}$ when the temperature of the background radiation was sufficiently high for baryon–antibaryon pair production to take place from the thermal background. Just as in the case of the epoch of electron–positron pair production, the Universe was flooded with baryons and antibaryons, one pair for every pair of photons present in the Universe now. Again, there is a little discontinuity in the temperature history at this epoch.

These considerations lead to one of the great cosmological problems, the *baryon asymmetry problem*. In order to produce a matter-dominated Universe at the present epoch, there must have been a tiny asymmetry between matter and antimatter in the very early Universe.

Table 15.1. *Planck units*

Unit	Defining expression	SI value
Time	$t_P = (Gh/c^5)^{1/2}$	10^{-43} s
Length	$l_P = (Gh/c^3)^{1/2}$	4×10^{-35} m
Mass–energy	$m_P = (hc/G)^{1/2}$	5.4×18^{-8} kg $\equiv 3 \times 10^{19}$ GeV

Roughly, for every 10^9 antibaryons, there must have been $10^9 + 1$ baryons. When the clocks were run forward, the 10^9 baryons annihilated with the 10^9 antibaryons, leaving one baryon, which became the Universe as we know it with the correct photon-to-baryon ratio.

In 1965, Zeldovich showed that, if the Universe were completely symmetric with respect to matter and antimatter, the present-day photon-to-baryon/antibaryon ratio would be about 10^{18} (Zeldovich, 1965), very much greater than the observed value of about 10^9. Baryon-symmetric models of the Universe were proposed by Hannes Alfvén (1908–1995) and Oskar Klein (1894–1977) in 1962 (Alfvén and Klein, 1962) and by Roland Omnes in 1969 (Omnes, 1969), but none of these convincingly demonstrated how the matter and antimatter could be separated in the early Universe. The baryon asymmetry must have originated in the very early Universe. Fortunately, it is known that there is a slight asymmetry between matter and antimatter because of CP violation observed in the decays of K^0 mesons. We take this subject up in Section 16.4.

The process of extrapolation can be carried further and further back into the mists of the early Universe, as far as we believe we understand high-energy particle physics. Probably most particle physicists would agree that the standard model of elementary particles has been tried and tested to energies of at least 100 GeV, and so we can probably trust laboratory physics back to epochs as early as 10^{-6} s, although more conservative cosmologists might be happier to accept 10^{-3} s. The most ambitious theorists have no hesitation in extrapolating back to the very earliest Planck eras, $t_P \sim (Gh/c^5)^{1/2} = 10^{-43}$ s, when the relevant physics was certainly very different from the physics of the Universe from redshifts of about 10^{12} to the present day (see Table 15.1).[6] Some aspects of these ideas are taken up in Chapter 16.

15.3 The development of small perturbations with cosmic epoch

15.3.1 The speed of sound as a function of cosmic epoch

An important quantity for understanding the physics of the formation of structure in the Universe is the variation with cosmic epoch of the speed of sound, c_s,

$$c_s^2 = \left(\frac{\partial p}{\partial \rho}\right)_S, \tag{15.6}$$

where the subscript S means 'at constant entropy', that is adiabatic sound waves. From the epoch when the energy densities of matter and radiation were equal, to beyond the epoch of recombination, the dominant contributors to p and ρ changed dramatically as the Universe changed from being radiation- to matter-dominated. Since the matter and radiation were closely coupled throughout the pre-recombination era, the square of the sound speed can be written as

$$c_s^2 = \frac{c^2}{3} \frac{4\rho_r}{4\rho_r + 3\rho_m},$$ (15.7)

where ρ_r and ρ_m are the inertial mass densities in radiation and matter, respectively. Thus, in the radiation-dominated era, $z \gg 4 \times 10^4 \Omega_0 h^2$, $\rho_r \gg \rho_m$, and the speed of sound tended to the relativistic sound speed, $c_s = c/\sqrt{3}$. At smaller redshifts, the sound speed decreased as the contribution of the inertial mass density of the matter became more important. Specifically, between the epoch of equality of the matter and radiation energy densities and the epoch of recombination, the pressure of the sound waves was provided by the radiation, but the inertia was due to the matter.

After the decoupling of matter and radiation, the sound speed became the thermal sound speed of the matter which, because of the close coupling between the matter and the radiation, had temperature $T_r = T_m$ at redshifts $z \geq 550\,h^{2/5}\Omega_0^{1/5}$. Thus, at a redshift of 500, the temperature of the gas was about 1300 K.

In 1968, Joseph Silk (b. 1942) realised that, during the pre-recombination epochs, sound waves in the radiation-dominated plasma were damped by the diffusion of radiation out of the perturbation by repeated electron scatterings (Silk, 1968). The effect of damping was to dissipate fluctuations with masses less than $M = M_D = 10^{12}(\Omega_B h^2)^{-5/4} M_\odot$ by the epoch of recombination. Thus, for adiabatic perturbations, all fine-scale structure was wiped out and only objects with masses greater than those of large galaxies or clusters of galaxies survived to the epoch of recombination.

15.3.2 The evolution of perturbations on different physical scales

During the 1960s and 1970s, it was generally assumed that the principal sources of inertial mass in the Universe were baryonic matter and the cosmic microwave background radiation. The dark matter problem was fully appreciated, but within the limits of observational uncertainty at that time, the dark matter could well have been in some dark baryonic form. Consequently, the development of the spectrum of initial perturbations could be worked out assuming that the principal constituents of the Universe were baryonic matter and radiation and so, once the the variation of the speed of sound with cosmic epoch was established, the evolution of the primordial perturbation spectrum could be worked out. Of particular importance was the variation with cosmic epoch of the *Jeans' mass*, the mass which is just stable against collapse under gravity, in other words the mass contained within a region of dimensions the Jeans' length, $\lambda_J = c_s/(G\rho_0/\pi)^{1/2}$.

In the *adiabatic picture* developed by Zeldovich and his colleagues, it was assumed that a spectrum of small adiabatic perturbations was set up in the early Universe and their evolution followed according to the physical rules developed above. Figure 15.2(a) shows how perturbations on different mass scales evolve with cosmic epoch in the standard Big Bang (Sunyaev and Zeldovich, 1970b). Since the speed of sound in the radiation-dominated phases was close to the speed of light, the Jeans' mass was roughly equal to the mass

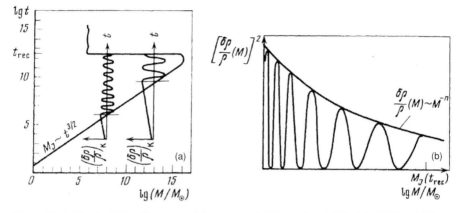

Figure 15.2: The 'stability diagram' of Sunyaev and Zeldovich published in 1970. (a) The region of stability is to the left of the solid line. The two superimposed graphs illustrate the evolution of adiabatic perturbations with different masses from early times, through the times when they enter the horizon up to the epoch of recombination. (b) Perturbations corresponding to different masses arrive at the epoch of recombination with different phases, resulting in a periodic dependence of the amplitude of the perturbations upon mass (Sunyaev and Zeldovich, 1970b).

contained within the horizon scale, $r_{\rm H} = ct$, during these epochs. As soon as masses on these scales came through the horizon, they were stabilised by the internal pressure of the photon gas and became sound waves. Specifically, during the pre-recombination era, after the perturbations came through the horizon, those perturbations with masses less than $M_{\rm J} = 3.75 \times 10^{15}/(\Omega_{\rm B}h^2)^2 M_\odot$ were sound waves.

The sound speed decreased after the epoch of equality of the energy densities in the matter and radiation until the epoch of recombination was approached. Then, after the decoupling of matter and radiation, the speed of sound dropped dramatically to the thermal sound speed in the baryonic matter, with the result that all masses greater than about $M_{\rm J} = 1.6 \times 10^5 (\Omega_0 h^2)^{-1/2} M_\odot \sim 10^6 M_\odot$ became unstable and began to grow in amplitude as $\delta\rho/\rho \propto (1+z)^{-1}$. It is therefore apparent why the adiabatic perturbations had to be of finite amplitude when they came through the horizon since they could only grow after the epoch of recombination and then only as $(1+z)^{-1}$. Figure 15.2(a), due to Sunyaev and Zeldovich, shows diagrammatically the oscillations of perturbations on different mass scales in the pre-recombination era.

Remarkably, Andrei Sakharov (1921–1989) studied the evolution of density perturbations in a cold universe in the years *before* the discovery of the cosmic microwave background radiation and showed that there would be preferred mass scales in the mass spectrum of large-scale structures in the Universe (Sakharov, 1965). Zeldovich and his colleagues applied these ideas to the evolution of adiabatic perturbations in the standard hot Big Bang model and determined the amplitudes and scales of the preferred scales which survived to the epoch of recombination, as illustrated in Figure 15.2(b) (Sunyaev and Zeldovich, 1970b). The oscillations seen in Figure 15.2(b) result from the fact that the fluctuations which develop into bound structures at late epochs are those with large amplitudes when they came through the horizon. Figure 15.2(a) shows examples of two perturbations coming through their particle horizons and oscillating as sound waves until the epoch of recombination. The amplitude of the oscillations at the epoch of recombination depended upon the phase of oscillation of the sound waves at that time. Those oscillations which completed an integral number of oscillations would have maximum amplitude as they began to collapse under gravity after the decoupling of matter and radiation. In contrast, those oscillations which had phases such that they had zero amplitude at the decoupling epoch did not form objects at all. The mass spectrum of perturbations at the decoupling epoch is shown in Figure 15.2(b). This spectrum of oscillations as a function of mass is sometimes called the spectrum of *acoustic* or, more appropriately, *Sakharov oscillations*, and is a general prediction of theories of structure formation which involve primordial adiabatic sound waves.

In the early 1970s, Zeldovich and Edward Harrison (b. 1919) independently put together information about the spectrum of the initial fluctuations on different physical scales and showed that observed structures in the Universe could be accounted for if the mass fluctuation spectrum had the form $\delta M/M \propto M^{-2/3}$ in the very early Universe, corresponding to a power spectrum of initial fluctuations of the form $P(k) \propto k^n$, with $n = 1$ (Harrison, 1970; Zeldovich, 1972).[7] The amplitude of the power spectrum had to be $\sim 10^{-4}$. Such a spectrum has the attractive feature that fluctuations on different mass scales had the same amplitude when they came through the horizon; in other words, it results in a fractal universe. This spectrum is known as the *Harrison–Zeldovich spectrum* of initial perturbations.

A key test of these models was provided by the fact that the presence of density fluctuations at the epoch of recombination should leave some imprint upon the cosmic microwave background radiation. In the simplest picture, if the process of recombination were instantaneous, the adiabatic perturbations would be expected to result in temperature fluctuations of $\Delta T/T = \frac{1}{3}\Delta\rho/\rho$ on the last scattering surface (Silk, 1968). In fact, the problem is much more complicated than this because the process of recombination was not instantaneous. The fluctuations which were imprinted upon the background radiation depended upon their sizes and optical depths relative to the thickness of this last scattering surface. The principal sources of temperature fluctuations on small scales were expected to be associated with first-order Doppler scattering due to the collapse of the perturbations (Sunyaev and Zeldovich, 1970b). These predictions provided a challenge for the observers since the predicted amplitudes of the fluctuations in these early theories were of the order $\Delta T/T \geq 10^{-3}$–10^{-4}.

At the opposite extreme from the adiabatic perturbations were the *isothermal perturbations*. In the radiation-dominated phase of the standard Big Bang, these fluctuations in the baryon density took place against the uniform cosmic background radiation. In the case of perfect gases, any pressure and density distribution in the radiation-dominated phases can be represented as the superposition of a distribution of adiabatic and isothermal perturbations. The perturbations were isothermal in the sense that they caused no fluctuations in the background radiation temperature during the radiation-dominated phases. Their internal temperature was the same as that of the uniform radiation background, and they were frozen into the radiation-dominated plasma. Throughout the radiation-dominated era, the timescale for the expansion of the radiation-dominated Universe was much shorter than the collapse timescale and so the isothermal perturbations scarcely grew at all. A simple calculation shows that the amplitude of these perturbations grew by only a factor of about 2.5 from the time they entered the horizon to the epoch of equality of matter and radiation energy densities. Subsequently, the perturbations grew according to the usual result $\delta\rho/\rho \propto R$. This important effect will reappear in a slightly different guise in considerations of the evolution of isocurvature perturbations involving collisionless cold dark-matter particles. This phenomenon was first described by Peter Mészáros, and is known as the *Mészáros effect* (Mészáros, 1974).

15.4 The adiabatic and isothermal scenarios for galaxy formation

In the 1970s, the concepts described in Section 15.3 gave rise to two different scenarios for the origin of structure in the Universe. In the *adiabatic* scenario, the initial perturbations were adiabatic sound waves before recombination and structure in the Universe formed by the fragmentation of the large-scale structures which reached amplitude $\delta\rho/\rho \sim 1$ at relatively late epochs. A realisation of this scenario was described by Doroshkevich and his colleagues in 1974 (Doroshkevich, Sunyaev and Zeldovich, 1974).

The alternative picture, favoured by James Peebles and his colleagues at Princeton, was one in which the perturbations were not sound waves but simply *isothermal* perturbations in the pre-recombination plasma which were in pressure balance with the background radiation. Low-mass perturbations were not damped out and so masses on all scales survived to the recombination epoch. Galaxies and clusters of galaxies then formed by the process of hierarchical clustering of low-mass objects under the gravitational influence of perturbations on larger scales.

Both models predicted similar amplitudes for the temperature fluctuations imprinted on the cosmic microwave background radiation at the epoch of recombination as the perturbations began to collapse to form bound objects. Their subsequent behaviours were, however, entirely different.

15.4.1 The adiabatic 'pancake' model

In the adiabatic picture developed by Zeldovich and his colleagues, only large-scale perturbations with masses $M \geq 10^{14} M_\odot$ survived to the epoch of recombination, all fluctuations on smaller mass scales being damped out by photon diffusion. During the pre-recombination era, after the perturbations came through their particle horizons, those with masses less than $M_J = 10^{16}–10^{17} M_\odot$ were sound waves, which oscillated until the epoch of recombination, when their internal pressure support vanished and the Jeans' mass dropped to $M_J \sim 10^6 M_\odot$.

Following recombination, all the surviving perturbations grew in amplitude as $\delta\rho/\rho \propto (1 + z)^{-1}$ until the epoch at which $\Omega_0 z \sim 1$ (see Section A15.1). In the early 1970s, the density parameter in baryons, Ω_B, was known to be less than about $0.05 h^{-2}$ from the constraints provided by primordial nucleosynthesis (Section 10.4), and so, even if $h = 0.5$, the perturbations would grow slowly at redshifts $z \leq 5$. In order to ensure the formation of galaxies and larger-scale structures, the amplitudes of the perturbations must have attained $\delta\rho/\rho = 1$ by $z \sim 5$. This was a satisfactory result, since quasars were known to exist at redshifts greater than 2, and the number counts of quasars and radio sources indicated that these objects had flourished at these early epochs (see Section 14.1). Zeldovich and his colleagues inferred that galaxies and the large-scale structure of the Universe began to form at relatively late epochs, $z \sim 3$–5. Since the fluctuations had attained amplitude $\delta\rho/\rho \sim 1$ at $z \sim 5$ and $\delta\rho/\rho \propto (1+z)^{-1}$, the amplitude of the density perturbations at the epoch of recombination must have been at least $\delta\rho/\rho \geq 3 \times 10^{-3}$.

The structures which survived on the scale of clusters and superclusters of galaxies were unlikely to be perfectly spherical and, in a simple approximation, could be described by ellipsoids with three unequal axes. In 1970, Zeldovich derived an analytic solution for the non-linear collapse of these structures and showed that such ellipsoids collapsed most rapidly along their shortest axis, with the result that flattened structures, which Zeldovich called 'pancakes', were formed (Zeldovich, 1970). The density became large in the plane of the pancake, and the infalling matter was heated to a high temperature as the matter collapsed into the pancake, a process sometimes called the 'burning of the pancakes'. Galaxies were assumed to form by fragmentation or thermal instabilities within the pancakes. In this picture, all galaxies formed late in the Universe, once the large-scale structures had collapsed. This baryonic pancake theory was developed in some detail by Zeldovich and his colleagues in the 1970s and can be thought of as a 'top-down' scenario for galaxy formation (Doroshkevich *et al.*, 1974). Among the successes of the theory was the fact that it accounted naturally for the large-scale structure in the distribution of galaxies. In three dimensions, the pancakes formed interconnected, flattened, stringy structures, not unlike the great holes and sheets of galaxies observed in the local Universe.

15.4.2 *The isothermal model and hierarchical clustering*

In contrast, Peebles and his Princeton colleagues favoured the isothermal picture, in which masses began to collapse on all scales greater than $M = M_J \sim 10^6 M_\odot$ immediately after the epoch of recombination. This scenario had the attractive feature that the first objects to form would have masses similar to those of globular clusters, which are the oldest known objects in our Galaxy. The process of galaxy and structure formation was ascribed to the *hierarchical clustering* of these small-scale structures under the influence of the power spectrum of perturbations, which extended up to the largest scales. One of the attractive features of this picture was that there would be early enrichment of the chemical abundances of the elements as a result of nucleosynthesis in the first generations of massive stars. This process could account for the fact that, even in the largest redshift quasars, the abundances of the elements were not so different from those observed locally. Many of these ideas were developed by Peebles in his important monograph, *The Large-Scale Structure of the Universe* (Peebles, 1980).[8]

The process of structure formation by hierarchical clustering was put on a formal basis by William Press (b. 1948) and Paul Schechter (b. 1948) in a remarkable paper of 1974 (Press and Schechter, 1974). Their objective was to provide an analytic formalism for the process of structure formation once the density perturbations had reached amplitude $\delta\rho/\rho \sim 1$. Their analysis started from the assumptions that the power spectrum of the primordial density perturbations was of power-law form, $P(k) \propto k^n$, and that the phases of the waves were random, what are known as *Gaussian fluctuations*. When the amplitude of the perturbations reached a critical value δ_c, it was assumed that they formed bound systems with mass M.

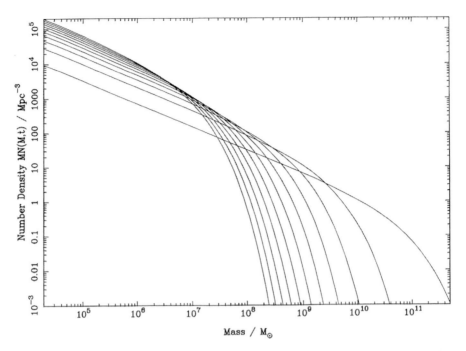

Figure 15.3: The variation of the form of the Press–Schechter mass function as a function of cosmic time in the Einstein–de Sitter world model, $\Omega_0 = 1$, according to equation (15.8). (Courtesy of Dr Andrew Blain.)

With these assumptions, they showed that the evolution of the spectrum of bound objects with cosmic time could be written in the following remarkably simple form:

$$N(M) = \frac{\bar{\varrho}}{\sqrt{\pi}} \frac{\gamma}{M^2} \left(\frac{M}{M^*} \right)^{\gamma/2} \exp\left[-\left(\frac{M}{M^*} \right)^{\gamma} \right], \qquad (15.8)$$

where $\gamma = 1 + (n/3)$ and $M^* = M^*(t_0)(t/t_0)^{4/3\gamma}$. The variation of this mass function with time is shown in Figure 15.3. Press and Schechter were well aware of the limitations of their approach, but it turned out that their mass function and its evolution with cosmic epoch were in good agreement with more detailed analyses and with the results of subsequent supercomputer simulations.[9] The Press–Schechter formalism has proved to be a very useful tool for studying the development of galaxies and clusters of galaxies in hierarchical scenarios for galaxy formation. In contrast to the adiabatic picture, the isothermal scenario is a 'bottom-up' picture, in which galaxies and larger-scale structures were assembled out of smaller objects by clustering and coalescence.

15.4.3 Confrontation with the observations

Despite these advances, there were major problems with both scenarios. First of all, it gradually became apparent that the dominant form of matter in the Universe was unlikely to be baryonic. The constraints from primordial nucleosynthesis of the light elements strongly suggested that the mean baryonic mass density of the Universe was about an order of magnitude less than the mean total mass density, $\Omega_0 \approx 0.2$–0.3. In addition, in 1980, the concept of the inflationary expansion of the early Universe, pioneered by Alan Guth and his colleagues, caught the imagination of theorists (Guth, 1981). One of the consequences of that picture, which could resolve a number of the fundamental cosmological problems, was that the Universe should have flat spatial geometry, and so, if $\Omega_\Lambda = 0$, it followed that $\Omega_0 = 1$. In this case, there was no question but that most of the mass in the Universe had to be in some non-baryonic form.

In addition, there was observational conflict with the expected amplitude of the temperature fluctuations in the cosmic microwave background radiation. As discussed above, after the epoch of recombination, both adiabatic and isothermal perturbations began to collapse and, for masses on the scales of clusters of galaxies and greater, their behaviour is similar. In purely baryonic theories, these fluctuations were expected to have large amplitudes on the last scattering surface, and these would cause observable fluctuations in the radiation temperature of the cosmic microwave background radiation.

The theory of these processes for both adiabatic and isothermal baryonic perturbations was worked out by Sunyaev and Zeldovich (Sunyaev and Zeldovich, 1970b), who found the important result that, for both types of perturbation, the root-mean-square temperature fluctuations were predicted to be

$$\left\langle \left(\frac{\delta T}{T}\right)^2 \right\rangle^{1/2} = 2 \times 10^{-5} \left(\frac{M\Omega_0^{1/2}}{10^{15} M_\odot}\right)^{1/2} (1 + z_0), \tag{15.9}$$

for masses $M \geq 10^{15} \Omega_0^{-1/2} M_\odot$, where z_0 is the redshift at which $\delta\varrho/\varrho = 1$.

Throughout the 1970s, increasingly sensitive searches were made for fluctuations in the cosmic microwave background radiation, these observations being analysed critically by Bruce Partridge (b. 1940) in his review of 1980 (Partridge, 1980a). His own observations had reached sensitivities of $\Delta T/T \approx 10^{-4}$ or slightly better by that time (Partridge, 1980b). Thus, by the early 1980s, the upper limits to the intensity fluctuations in the cosmic background radiation were beginning to constrain quite severely purely baryonic theories of structure formation.

15.5 Hot dark matter – neutrinos with finite rest mass

A potential solution to these problems appeared in 1980, when Valentin Lyubimov (b. 1929) and his collaborators reported that the electron neutrino had a finite rest-mass of about 30 eV (Lyubimov *et al.*, 1980). As early as 1966, Semion Gershtein (b. 1929) and Zeldovich had noted that relic neutrinos of finite rest mass could make an appreciable contribution to the mass density of the Universe (Gershtein and Zeldovich, 1966). In the 1970s, Györgi (George) Marx (1927–2002) and Alexander Szalay (b. 1949) had considered the role of neutrinos of finite rest mass as candidates for the dark matter, as well as studying their role in galaxy formation (Marx and Szalay, 1972; Szalay and Marx, 1976). The intriguing aspect of Lyubimov's result was that, if the relic neutrinos had this rest mass, this Universe would just be closed, $\Omega_0 = 1$. Zeldovich and his colleagues developed a new version of the adiabatic model in which the Universe was dominated by neutrinos with finite rest mass (Doroshkevich *et al.*, 1980a,b; Zeldovich and Sunyaev, 1980).

In the new picture, most of the inertial mass of the Universe was in the form of neutrinos of rest-mass energy 30 eV. The neutrinos were therefore highly relativistic during the epoch of nucleosynthesis and so none of the predictions of the standard Big Bang were changed. The differences began to appear at later epochs when the neutrinos became non-relativistic. This occurred at about the same redshift that the Universe changed from being radiation- to matter-dominated. This was not a coincidence. Prior to this epoch the energy densities in the photons and neutrinos were roughly the same, and so, when the neutrinos became non-relativistic, their inertial masses no longer decreased as the Universe expanded, unlike the photons, and so the Universe became matter-dominated. The neutrino fluctuations began to grow under gravity as soon as they became non-relativistic, but, since the neutrinos are weakly interacting, they streamed freely out of the perturbations and so the small-scale perturbations were damped out. Because of the decoupling of the matter and radiation from the neutrinos, except through their gravitational influence, the amplitudes of the sound waves in the matter and radiation remained at the same level they had when they came through the horizon. After recombination, the baryonic matter collapsed into the surviving, larger-amplitude neutrino fluctuations.

Only neutrino perturbations on the very largest scale with masses $\geq 10^{16} M_\odot$ survived to the epoch of recombination, and so, just as in the old adiabatic model, the largest-scale structures formed first and then the smaller-scale structures formed by a process of fragmentation. This model had the advantage of reducing very significantly the expected amplitude of the fluctuations in the microwave background radiation since the fluctuations in the baryonic matter were of low amplitude during the critical phases when the background photons were last scattered. The subsequent evolution of the perturbations was not so different from the adiabatic model. This scenario for galaxy formation became known as the *hot dark matter* picture of galaxy formation since the neutrinos were highly relativistic when they decoupled from thermal equilibrium.

There were, however, concerns about this picture. First of all, there were grave reservations about the experiments that claimed to have measured the rest mass of the electron neutrino, and it is now believed that the result was erroneous – the upper limit to the rest mass of the electron neutrino is now found to be $m_\nu \leq 3$ eV (Weinheimer, 2001).[10] Secondly, constraints could be set to the mass of the neutrinos if they were to constitute the dark matter in galaxies, groups and clusters of galaxies, as discussed in Section 10.2. Gunn and Tremaine showed that, while 30 eV neutrinos could bind clusters and the haloes of giant galaxies, those needed to bind dwarf galaxies would have to have masses much greater than 30 eV (Tremaine and Gunn, 1979).

15.6 Cold dark matter and structure formation

Once it was appreciated that non-baryonic dark matter had to be taken really seriously, many possibilities were proposed by the particle physicists. Examples included the axions, supersymmetric particles such as the gravitino or photino and ultraweakly interacting neutrino-like particles, all of which might be relics of the very early Universe. From roughly 1980 onwards, the particle physicists began to take the early Universe seriously as a laboratory for particle physics at energies which could not be achieved in terrestrial laboratories. According to James Peebles, Richard Bond (b. 1950) introduced the term *cold dark matter* in 1982 to encompass many of these exotic types of particle suggested by the particle physicists (Peebles, 1993). The matter was 'cold' in the sense that these particles decoupled from the thermal background after they had become non-relativistic. In the same year, Peebles demonstrated how the presence of such particles could reduce the amplitude of the predicted fluctuations in the cosmic microwave background radiation to levels consistent with the observational upper limits (Peebles, 1982).

A change of terminology was also introduced about this time. In the purely baryonic picture, the perturbations in the early Universe could be decomposed into isothermal and adiabatic modes. Now, another independent component, the cold dark matter, was added to the picture. In the three-component case, the decomposition can be made into similar modes, but the names 'isothermal' and 'adiabatic' were scarcely appropriate for fluids containing collisionless dark-matter particles. The corresponding modes were referred to as *curvature* and *isocurvature* modes.

- The *curvature modes* were the equivalent of the adiabatic modes in that, when these perturbations entered the horizon during the radiation-dominated era, the amplitudes of the perturbations in the radiation, the baryonic matter and the dark matter were all more or less the same. As result, there were variations in the local mass–energy density from point to point in the Universe, resulting in local perturbations to the curvature of space.
- In the *isocurvature modes*, the mass–energy density is constant throughout space and so there were no perturbations to the spatial curvature of the background world model, despite the fact that there might be fluctuations in the mass–energy density of each of the three components from point to point in the Universe.

The cold dark matter scenario was similar in many ways to the isothermal model (Davis *et al.*, 1992a). Since the matter was very cold, the perturbations were not damped by free streaming. Fluctuations on all scales survived, and so, when the pre-recombination Universe became matter-dominated, these perturbations begin to grow, decoupled from the matter and radiation. As in the hot dark matter scenario, after the epoch of recombination, the baryonic matter collapsed into the growing potential wells in the dark matter. Galaxies, groups and clusters formed by a process of hierarchical clustering, which can be modelled by the Press–Schechter formalism (Press and Schechter, 1974).

An important aspect of these models for the formation of large-scale structures was the fact that the initial power spectrum of the perturbations, taken to be of Harrison–Zeldovich form when they entered the horizon, was modified by various physical processes, and this modified spectrum became the input spectrum of perturbations for simulations of the subsequent post-recombination evolution. Thus, in the hot dark matter model, the free streaming of neutrinos damped out all perturbations on all scales up to about $10^{16} M_\odot$. In the adiabatic cold dark matter model, the evolution of the perturbations was driven by the perturbations in the radiation until the epoch at which the Universe became matter-dominated. In the isocurvature cold dark matter model, the growth of the perturbations

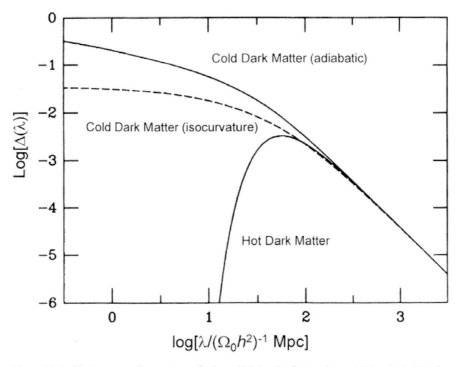

Figure 15.4: The 'processed' spectrum of primordial density fluctuations, $\Delta(\lambda) = \delta\rho/\rho(\lambda)$, observed at some time after the epoch of equality of matter and radiation energy densities (Kolb and Turner, 1990). Rather than mass, the abscissa is plotted in terms of the wavelength associated with the perturbations. The three models show the processed fluctuation spectrum for hot and cold dark matter models, assuming the fluctuations are adiabatic, and for an isocurvature cold dark matter model. In all three cases, it is assumed that the input power spectrum is of Harrison–Zeldovich form, $P(k) \propto k$, which is only preserved on the very largest physical scales.

under gravity was retarded by the Mészáros effect until the matter-dominated era. The net result is that each model involves a different transfer function which modifies the primordial Harrison–Zeldovich form of the spectrum (Figure 15.4). In an important paper, James Bardeen and his colleagues worked out the transfer functions for a number of different scenarios for structure formation and provided a convenient set of analytic formulae for astrophysical applications (Bardeen *et al.*, 1986).

By the early 1980s, the power of digital computers had developed to such an extent that simulations of the non-linear evolution of the various scenarios for galaxy and large-scale structure formation could be carried out successfully. Figure 15.5 shows a sample of the results of computer simulations of the hot and cold dark matter models carried out by Marc Davis, Carlos Frenk (b. 1951) and their colleagues (Frenk, 1986). These simulations represented the state of the art in numerical simulations of non-linear gravitational clustering in the mid 1980s, their N-body codes involving periodic boundary conditions and 32 768 particles. The models were run for different cosmological models and showed the evolution of the spectrum of perturbations during the post-recombination eras over a factor of 16 in length scale. In the *hot dark matter* picture, flattened structures like pancakes were produced very

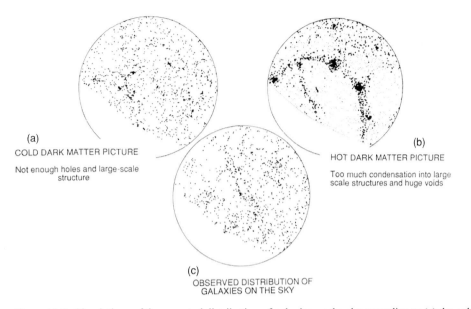

(a)
COLD DARK MATTER PICTURE

Not enough holes and large-scale structure

(b)
HOT DARK MATTER PICTURE

Too much condensation into large scale structures and huge voids

(c)
OBSERVED DISTRIBUTION OF GALAXIES ON THE SKY

Figure 15.5: Simulations of the expected distribution of galaxies on the sky according to (a) the cold dark matter scenario and (b) the hot dark matter scenario of galaxy formation including biassing compared with (c) the observed distribution of galaxies in the Harvard Center for Astrophysics northern sky survey (Frenk, 1986). The outer circle represents Galactic latitude $+40°$ and the empty regions lie at declinations less than $0°$.

effectively. The model was, in fact, far too effective in producing flattened, stringy structures. Essentially everything collapsed into thin pancakes, resulting in a much more highly structured Universe than is actually observed. Furthermore, it was difficult to understand how stars and galaxies could be formed before the structures on scales $M \geq 4 \times 10^{15} M_\odot$.

In the *cold dark matter* picture, masses on all scales began to collapse soon after recombination, and star clusters and the first generations of stars could be old in this picture. In the post-recombination epoch, large-scale systems, such as galaxies and clusters of galaxies, were assembled from their component parts by non-linear clustering. Figure 15.5 shows that structure indeed developed, but in the high-density models with $\Omega_0 = 1$, it was not as pronounced on the large scale as is observed in the local Universe. The physical reason for this was that it was difficult to produce elongated structures by gravitational clustering alone, which tended to make more symmetrical structures than the sheets and filaments of galaxies found in the Universe on the largest scales.

One of the important successes of the cold dark matter picture was that it could account for the observed two-point correlation function of galaxies.[11] A simulation in which the initial spectrum was of standard Harrison–Zeldovich form in a $\Omega_0 = 1$ universe and the phases of the waves were random, showed the two-point correlation function, $\xi(r)$, evolving towards a power law (Figure 15.6). The process of non-linear gravitational clustering converts the modified input power spectrum into one much more closely resembling the observed two-point correlation function, $\xi(r)$, with the observed slope of $\gamma = 1.8$ over a wide range of scales (Davis *et al.*, 1985). As the model evolved, the correlation function became steeper.

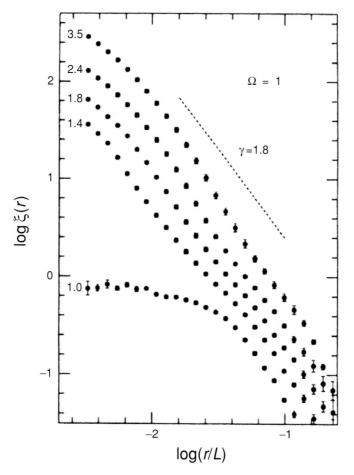

Figure 15.6: The two-point correlation functions, $\xi(r)$, at different scale factors. The error bars show the uncertainties in the correlation functions from five independent runs of the computational model. The dashed line shows the slope, $\gamma = 1.8$, of the observed two-point correlation function for galaxies (Davis *et al.*, 1985; Efstathiou, 1990).

As explained by George Efstathiou (b. 1955), this was very far from the end of the story (Efstathiou, 1990). In particular, in this realisation of the model, the velocity dispersion of galaxies chosen at random from the field was too large, but the match to observation could be improved if it was assumed that galaxy formation occurred preferentially in overdense regions, as proposed by Kaiser (1986). This led to the issue of the extent to which the visible matter traces the distribution of the dominant dark matter.

15.7 Biassing

So far, it has been assumed that the visible parts of galaxies trace the distribution of the dark matter, but one can imagine many reasons why this might not be so. The generic term for this phenomenon is *biassing*, meaning the preferential formation of galaxies in certain regions of space rather than in others. Part of the motivation behind the introduction of biassing was to improve the agreement between the predictions of the preferred cold dark matter scenario and the observed distribution of galaxies. In the hot dark matter picture, *anti-biassing* would be needed so that the formation of galaxies is not so highly concentrated into sheets and filaments.

Was there any observational evidence for biassing in the Universe? In the Coma Cluster and in other large-scale systems, for example, the mass of the dark matter amounts to about a factor of 10 greater than that in the visible matter, but this factor was only about one-quarter or one-third of the value necessary to attain a density parameter $\Omega_0 = 1$. If the Universe really had the critical density $\Omega_0 = 1$, there must be biassing by a factor of 3 towards the formation of galaxies on the scale of clusters and superclusters as opposed to the general field. Many possible biassing and anti-biassing mechanisms were described by Dekel and Rees (Dekel, 1986; Dekel and Rees, 1987).

One of the most important process for biassing was described by Nicholas Kaiser (b. 1954), who realised that inherent in the notion of the power spectrum of the perturbations is the fact that the perturbations have a Gaussian distribution of amplitudes about the root-mean-square value. If we write $\Delta = \delta\rho/\rho$, then, on any scale, Δ has mean value, $\overline{\Delta}$, with variance $\overline{\Delta^2}$, so that the probability of encountering a density contrast, Δ, at some point in space is proportional to $\exp(-\Delta^2/\overline{\Delta^2})$. Kaiser argued that galaxies are most likely to form in the highest peaks of the density distribution. Thus, if the density perturbations had to exceed some value, Δ_{crit}, in order that structures form, galaxy formation would be biassed towards the highest density perturbations over the mean background density (Kaiser, 1986). This picture can account for the fact that the clusters of galaxies are more strongly clustered than galaxies in general. If structure only forms when the density contrast exceeds a certain value Δ_{crit}, then galaxy formation within a large-scale density perturbation, which will eventually form a cluster of galaxies, is strongly favoured. This model was worked out in detail by John Peacock and Alan Heavens (b. 1959) and by James Bardeen and his colleagues (Peacock and Heavens, 1985; Bardeen *et al.*, 1986). The numerical simulations described by Efstathiou illustrated clearly how the density peaks of a Gaussian random field result in a much more highly structured distribution of galaxies as compared with the underlying mass distribution (Efstathiou, 1990). He showed that a standard cold dark matter model with $\Omega_0 = 1$ and $b = 2.5$ could be reconciled with a number of independent aspects of the large-scale distribution of galaxies, including the amplitude and slope of the two-point correlation function and the mean velocity dispersion of galaxies in the general field.

15.8 Reconstructing the initial power spectrum

Granted that biassing plays a role in determining the amplitudes of the correlation functions on different scales, was it possible to produce self-consistent models for the formation of structure from a single initial power spectrum? In a pioneering analysis, Andrew Hamilton (b. 1951) and his colleagues showed how it was possible to relate analytically the observed spectrum of perturbations well into the non-linear regime, $\xi(r) \gg 1$, to the initial spectrum

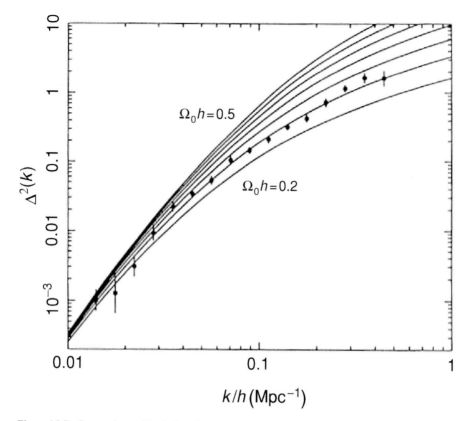

Figure 15.7: Comparison of the inferred power spectrum of large-scale structures at the present epoch with different variants of the cold dark matter model. The models have scale-invariant input spectra which have been modified by the cold dark matter transfer functions. Different values of the fitting parameter $\Omega_0 h$ are shown, $\Omega_0 h = 0.5, 0.45, \ldots, 0.25, 0.2$, in decreasing order of power at short wavelengths (Peacock and Dodds, 1994).

in the linear regime (Hamilton *et al.*, 1991). John Peacock and Stephen Dodds (b. 1970) extended this analysis to eight separate determinations of the present power spectrum of perturbations in the non-linear regime and then determined biassing factors for different samples of galaxies and clusters in order to generate a smooth power spectrum (Peacock and Dodds, 1994). This power spectrum could then be compared with the expectations of different versions of the cold dark matter picture for different values of the density parameter, Ω_0.

Peacock and Dodds found that there must be significant bias present such that $\Omega_0^{0.6}/b = 1.0 \pm 0.2$, where the bias factor, b, is defined by

$$\xi_{\mathrm{gal}}(r) = b^2 \xi_{\mathrm{D}}(r) \quad \text{or} \quad \left(\frac{\delta\rho}{\rho}\right)_{\mathrm{gal}} = b \left(\frac{\delta\rho}{\rho}\right)_{\mathrm{D}}, \tag{15.10}$$

the subscripts referring to galaxies (gal) and dark matter (D). The shapes of the curves were derived assuming that the initial power spectrum was of scale-invariant form, $P(k) \propto k^n$, and was modified by the transfer function for different values of the parameter $\Omega_0 h$. The best-fitting value of $\Omega_0 h$ was

$$\Omega_0 h = 0.255 \pm 0.017 + 0.32(n^{-1} - 1). \qquad (15.11)$$

Thus, for the standard Harrison–Zeldovich spectrum, $n = 1$, this analysis suggested that the 'standard' cold dark matter model, with $\Omega_0 = 1$, was a poor fit to the data. Even if $h = 0.5$, the best-fitting models would have Ω_0 significantly less than unity. This result was found by a number of workers, namely that the simplest cold dark matter models predict too much power on scales corresponding to $0.1 < k/h < 1 \, \text{Mpc}^{-1}$ if the large-scale power spectrum in the linear region is to be accommodated. The other key constraint, which has not yet been built into this picture, is the observation of intensity fluctuations in the cosmic microwave background radiation. The predicted power spectra shown in Figure 15.7 could be extrapolated to the scale of the COBE observations and reasonable consistency found with the observed fluctuation spectrum.

Although the 'standard' cold dark matter picture with $\Omega_0 = 1$, $\Omega_\Lambda = 0$ and $n = 1$ had some success in reproducing a number of features of the observed structures in the Universe on large scales, the consensus of opinion among cosmologists was that it is probably not good enough. These concerns were sufficiently worrying for Marc Davis and his colleagues to entitle their *Nature* review paper of 1992 'The end of cold dark matter?' (Davis *et al.*, 1992a). The outcome of these studies was the development of a number of variants of the standard picture.

15.9 Variations on a theme of cold dark matter

Many variants of the standard cold dark matter picture were proposed, all involving the introduction of additional parameters. In parallel with these developments, the ability to carry out large-scale simulations of the origin of large-scale structure increased dramatically, mirroring the exponential increase in computing power. These enabled many of these possibilities to be tested by supercomputer simulations. Some examples of the outcome of these simulations are shown in Figure 15.8, which shows some of the outputs of the programmes of the Virgo consortium (Kauffmann *et al.*, 1999). In these examples, the models were evolved from similar initial conditions with a power-law input spectrum of Harrison–Zeldovich form and were constrained to reproduce the same large-scale structure at the present epoch.

These models were debated in some detail in 1996 at the Princeton meeting *Critical Dialogues in Cosmology*, the proceedings of which convey the atmosphere of excitement aroused by these debates (Turok, 1997). Some of the more promising alternatives were the following.

Open cold dark matter (OCDM)

As the second row of Figure 15.8 shows, OCDM models could account satisfactorily for the observations and produce the type structures observed in large-scale redshift surveys. The differences as compared with the standard cold dark matter picture were, firstly, that the

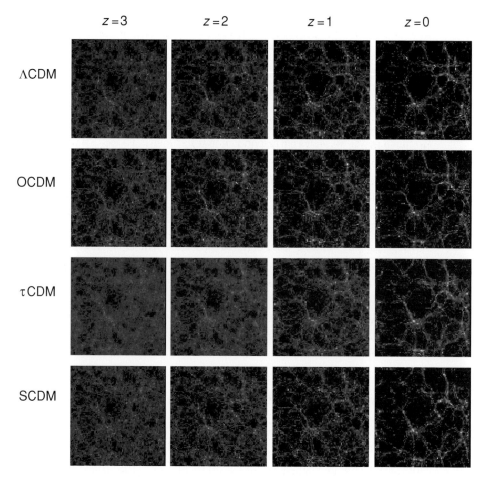

Figure 15.8: Some examples of the predicted large-scale structure in the distribution of galaxies from supercomputer simulations by the Virgo consortium. Each panel has side $240h^{-1}$ Mpc and the gravitational interactions of $256^3 = 1.7 \times 10^7$ particles were followed. The four models shown involve the standard cold dark matter (SCDM), open cold dark matter (OCDM), cold dark matter with a finite cosmological constant (ΛCDM) and cold dark matter with decaying neutrinos (τCDM). The parameters of the models have been chosen to reproduce the observed large-scale structure in the distribution of galaxies at the present epoch (Kauffmann *et al.*, 1999).

epoch of equality of matter and radiation occurs rather later and, secondly, that the growth of structure proceeds over a somewhat smaller range of redshifts, only until the epoch at which $\Omega_0 z \approx 1$. Consequently, the break in the power spectrum takes place at greater masses, resulting in less power at short wavelengths. This would be consistent with the dynamical evidence that the overall density parameter was about $\Omega_0 = 0.3$.

Cold dark matter with a finite cosmological constant (ΛCDM)

A way of preserving the flat geometry of space is to include the cosmological constant into the model so that $\Omega_0 + \Omega_\Lambda = 1$. There is not a great deal of difference in the dynamics of the underlying model as compared with the open cold dark matter model (see the top row of Figure 15.8). This is because the dynamics only differ from the $\Omega_\Lambda = 0$ case at redshifts $(1 + z) \leq \Omega_0^{1/3}$. Thus, the differences occur in the late stages of evolution when the effect of the cosmological constant is to stretch out the timescale of the model, allowing some further development of the perturbations.

Cold dark matter with decaying neutrinos (τCDM)

In this scenario, the ratio of radiation to matter energy densities is enhanced so that the epoch of equality of matter and radiation energy densities is shifted to lower redshifts, as in the case of the open cold dark matter picture. It is essential to ensure that the predictions of primordial nucleosynthesis are not violated in that, if there were additional relativistic components present in the Universe during the epoch of nucleosynthesis, the expansion rate would be increased and excessive amounts of helium would be produced. The trick was to suppose that there existed particles which decay after the epoch of nucleosynthesis, thus enhancing the radiation relative to the matter energy densities and so delaying the epoch of equality of matter and radiation energy densities.

By the end of the conference, the four scenarios illustrated in Figure 15.8 seemed equally plausible, but the whole picture was about to change with the new, highly suggestive, evidence that the cosmological constant was non-zero (see Section 13.4.3).

15.10 Fluctuations in the cosmic microwave background radiation

The study of the power spectrum of spatial fluctuations in the cosmic microwave background radiation has provided one of the most important means of confronting theories of the origin of the large-scale structure of the Universe with observation. Limits to the amplitude of the fluctuations continued to improve until, in 1992, a positive detection was made in the whole sky survey carried out by the Cosmic Background Explorer (COBE) led by George Smoot (b. 1945) (Smoot et al., 1992). The objectives of the project were to measure precisely the spectrum and spatial distribution of the radiation over the whole sky, and in both cases these were achieved with outstanding success. The maps of the whole sky demonstrated how uniform the background radiation is on large angular scales (Figure 15.9(a)). The dipole distribution of intensity expected due to the motion of the Earth through the frame of reference in which the radiation would have been isotropic on a large scale was clearly defined (Figure 15.9(b)). Finally, at the limiting sensitivity of the survey, fluctuations of cosmological origin were discovered on an angular scale of $10°$ with an amplitude of $\Delta I / I \approx 10^{-5}$ in directions away from the Galactic plane (Figure 15.9(c)).

These observations marked the beginning of a new era in astrophysical cosmology since they provided a direct link to the processes of formation of large-scale structures in the Universe when they were still in their linear stage of development. Many new ground-based experiments were developed, and approval was given for further space missions, specifically the Wilkinson Microwave Anisotropy Probe (WMAP) of NASA, which was launched in

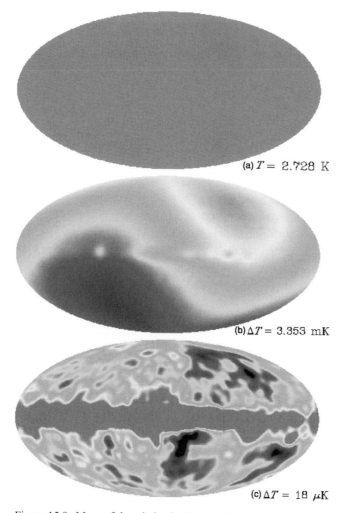

(a) $T = 2.728$ K

(b) $\Delta T = 3.353$ mK

(c) $\Delta T = 18$ μK

Figure 15.9: Maps of the whole sky in galactic coordinates as observed at a wavelength of 5.7 mm by the COBE satellite (Smoot *et al.*, 1992). (a) The distribution of total intensity over the sky. (b) Once the uniform component was removed, the dipole component associated with the motion of the Earth through the background radiation was observed, as was the emission from the Galactic plane. (c) Once the dipole component was removed, radiation from the plane of the Galaxy was seen as a bright band across the centre of the picture. The fluctuations seen at high galactic latitudes were largely noise from the telescope and the instruments, the root-mean-square value at each point being 36 μK. When averaged statistically over the whole sky at high latitudes, an excess sky noise signal of cosmological origin amounting to 30 $\pm$ 5 μK was observed.

2001 and which was named after David Wilkinson (1935–2002), and the Planck project of the European Space Agency, scheduled for launch in 2007.

From the early pioneering efforts of Peebles, Zeldovich and their colleagues, studies of the cosmic microwave background radiation have developed into one of the most important fields of cosmology. The subject is very rich astrophysically, and many different aspects of physics contribute to the interpretation of the observations.

15.10.1 The ionisation of the intergalactic gas through the epoch of recombination

As discussed in Section 15.2.1, primordial intensity fluctuations in the cosmic microwave background radiation originate in a rather narrow-redshift range at $z \approx 1000$. Following the pioneering studies of Zeldovich, Kurt and Sunyaev and Peebles (see Section 15.2.1), detailed calculations of the degree of ionisation through the critical redshift range were carried out by Bernard Jones (b. 1946) and Rosemary Wyse (b. 1957). These studies enabled the range of redshifts from which the photons of the background radiation observed today were last scattered to be determined. This probability distribution could be closely approximated by a Gaussian distribution with mean redshift 1070 and standard deviation $\sigma = 80$ in redshift. Thus, half the photons observed today were last scattered between redshifts of 1010 and 1130.[12]

The thickness of this last scattering layer corresponds to a physical scale of $10(\Omega_0 h^2)^{-1/2}$ Mpc at the present epoch, where Ω_0 is the density parameter in matter for the Universe as a whole. The mass contained within this scale is given by $M \approx 3 \times 10^{14}(\Omega_0 h^2)^{1/2} M_{\odot}$, corresponding roughly to the mass of a cluster of galaxies. The corresponding angular scale is $\theta = 6\Omega_0^{1/2}$ arcmin. For scales smaller than this value, a number of independent fluctuations are expected to be present along the line of sight through the last scattering layer. The random superposition of these perturbations leads to a statistical reduction in their amplitude by a factor of roughly $N^{-1/2}$, where N is the number of fluctuations along the line of sight.

This angular scale can be compared with the horizon scale, $r = 3ct$, at a redshift of 1000 which is $1.8\Omega_0^{1/2}$ degrees. Perturbations on this scale are the largest which can be in causal contact at that epoch and correspond to comoving scales of about $200(\Omega_0 h^2)^{-1/2}$ Mpc at the present day. Thus, the smallest angular scales observed by the COBE satellite, $\theta = 10°$, corresponded to structures on scales about five to ten times the horizon scale on the last scattering surface. The next task is to relate the density perturbations and their associated velocities to temperature, or intensity, fluctuations in the last scattering layer.

15.10.2 Large angular scales

On the very largest scales, the dominant source of intensity fluctuations results from the gravitational redshift associated with density perturbations at the last scattering layer. The fluctuations on these scales far exceeded the horizon scale at the epoch of recombination, and so they are more appropriately referred to as *metric perturbations*. Although the discussion of this subsection is concerned with perturbations on the very largest scales, *super-horizon* perturbations present a general problem for the formation of structure in the expanding Universe. Perturbations on all scales of astrophysical interest exceeded their particle horizons[13] early enough in the Universe.

Super-horizon perturbations need careful treatment because of the problems of defining the appropriate hypersurfaces corresponding to the same cosmic time. For sub-horizon scales, this is not a problem because the background metric can be taken to be of Robertson–Walker form and a *synchronous time* coordinate used to defined hypersurfaces of constant cosmic time. In principle, radar methods could be used to ensure the synchronisation of clocks on scales less than the particle horizon. This is not possible on super-horizon scales since the scale of the perturbation exceeds the particle horizon. These problems were elucidated in 1980 in an important paper by James Bardeen, in which he showed how these difficulties could be solved by introducing a *gauge-invariant* formalism (Bardeen, 1980). These issues are also discussed by Efstathiou and Hu (Efstathiou, 1990; Hu, 1996). As Efstathiou emphasises, there are no ambiguities so long as a well defined background model on super-horizon scales in the very early Universe is adopted.

To evaluate the expected temperature fluctuations in the last scattering layer, a general relativistic treatment was needed, and this was first performed by Raymond Sachs (b. 1932) and Arthur Wolfe (Sachs and Wolfe, 1967). The result was that the temperature fluctuation $\Delta T / T = (1/3)\Delta \phi / c^2$ was expected, recalling that the Newtonian gravitational potential, $\Delta \phi$, is a negative quantity – this source of temperature fluctuations is known as the *Sachs–Wolfe* effect.[14]

Although the photons pass through gravitational potential fluctuations during their subsequent propagation to the Earth, what they gain by falling into them is compensated by the gravitational redshift coming out, so long as the perturbations continue to grow linearly with redshift.[15] The amplitudes of the temperature fluctuations as a function of angular scale due to the Sachs–Wolfe effect depended only upon the spectral index, n, of the initial power spectrum of the fluctuations; specifically, in a simple analysis, $\Delta T / T \propto \theta^{(1-n)/2}$. Thus, if $n = 1$, the amplitudes were expected to be independent of angular scale. According to the COBE team and Partridge, the power spectrum at small multipoles, $l < 30$, determined from the four-year data set from the COBE experiment, provided an estimate of $n = 1.1 \pm 0.3$ (Bennett *et al.*, 1996; Partridge, 1999). This value was consistent with the expectations of the scale-free Harrison–Zeldovich spectrum, $n = 1$.

Whilst this agreement was encouraging, another source of fluctuations on these very large angular scales was associated with primordial gravitational waves. According to some inflationary theories of the early Universe, the quantum fluctuations which are responsible for the density perturbations from which the large-scale structure of the Universe developed might well be accompanied by a background of gravitational waves (Starobinsky, 1985; Davis *et al.*, 1992b; Crittenden *et al.*, 1993). This is claimed to be a general feature of a wide class of inflationary models for the formation of structure. According to these ideas, to a good approximation the spectral indices of the tensor (gravitational wave) and scalar (density perturbation) modes, n_t and n_s, respectively, are related by $n_t \approx 1 - n_s$, and the ratio of the amplitudes of their quadrupole power spectra, $r = C_2^t / C_2^s$, depends upon the spectral index of the scalar perturbations as $r = 7(1 - n_s)$. Although there is no direct evidence for such modes being present in the COBE observations, the importance of these ideas is that they provide a possible probe of the very early Universe indeed.

15.10.3 *Intermediate angular scales – the acoustic or Sakharov oscillations*

The COBE results acted as a spur to search for the *acoustic*, or *Sakharov, oscillations* in the power spectrum of the temperature fluctuations expected on angular scales $\theta \approx 1°$. The different variants of the cold dark matter model discussed in Section 15.9 made different predictions about the amplitude of the power spectrum of the fluctuations as a function of angular scale. The temperature fluctuations are distributed over the surface of a sphere, and so the spherical polar equivalents of Fourier transforms are needed to define their power spectra.[16] The appropriate complete sets of orthonormal functions are the spherical harmonics $Y_{lm}(\theta, \phi)$, each characterised by the amplitude, a_{lm}, the square of which, when averaged over the azimuthal harmonic, m, is a measure of the power, C_l, on angular scale $\theta \approx \pi/l$. In the simplest picture, the density perturbations are assumed to be Gaussian, meaning that the phases of the waves which make up the spherical harmonic decomposition over the sky are random, and the coefficients a_{lm} follow a Gaussian probability distribution with phases uniformly distributed between 0 and 2π (Kogut *et al.*, 1996). However, the seeds for the formation of structure need not necessarily be Gaussian. The temperature fluctuations on the sky might display non-Gaussian features, such as abrupt temperature discontinuities, intense hot spots, linear structures and so on. These types of feature are predicted by theories in which large-scale structures were seeded by topological defects, cosmic strings, or by cosmic textures (see Section 16.3) (Shellard, 2003). The non-Gaussian features would result in strongly correlated values of the coefficients a_{lm}.

To predict the power spectrum of temperature fluctuations, the evolution of the power spectrum of density perturbations on all scales is followed from the time they entered the horizon through the epoch of recombination. This involved using the collisional Boltzmann equation to follow the evolution of the independent Fourier modes of the perturbations in the dark matter, the baryonic matter and the radiation field. These computations were described by James Peebles and Jer Tsang Yu (b. 1942) for baryonic perturbations (Peebles and Yu, 1970) and by George Efstathiou and Wayne Hu (b. 1968) and Naoshi Sugiyama (b. 1961) for dark-matter cosmologies (Efstathiou, 1990; Hu and Sugiyama, 1995).[17] A good example of the predicted power spectrum of temperature fluctuations for such models is shown in Figure 15.10 (Bersanelli *et al.*, 1995). Let us disentangle some of the physical processes contributing to the temperature fluctuations expected on scales $0.1° \leq \theta \leq 1°$.

The abundances of the light elements created by primordial nucleosynthesis indicated that the density parameter in baryonic matter is low, $\Omega_B h^2 \approx 10^{-2}$ (see Section 12.6). In this case, according to equation (15.7), the sound speed throughout most of the pre-recombination era was $c/\sqrt{3}$ and so the Jeans' length was of the same order as the horizon scale. Therefore, as soon as they came through the horizon, the coupled photon–baryon perturbations were stabilised and oscillated as sound waves. At the same time, the perturbations in the decoupled dark matter continued to grow in amplitude from the time the Universe became matter-dominated. A good analogy is that the photon–baryon perturbations can be considered to be forced oscillations within the growing potential wells in the dark-matter perturbations. Thus, essentially *all* baryonic perturbations with wavelengths less than the horizon scale were *acoustic waves*. As a result, they gave rise to Sakharov oscillations on the last scattering surface.

The longest-wavelength Sakharov oscillations had wavelengths roughly equal to the *sound horizon*, $\lambda_s = c_s t$, on the last scattering layer, where c_s is the speed of sound and t is the age of the Universe. This is the maximum distance over which coherent oscillations could have existed at the epoch of recombination and sets an upper limit to the wavelengths which acoustic waves could have at that epoch. Adopting the relativistic sound speed $c_s = c/\sqrt{3}$

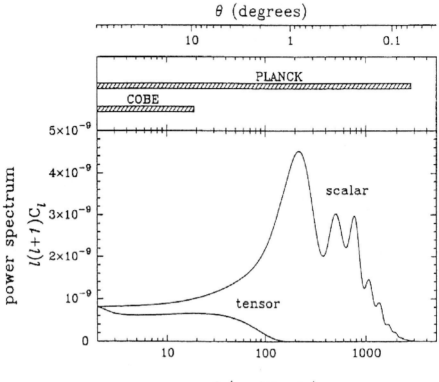

Figure 15.10: The predicted power spectrum of temperature fluctuations in the cosmic microwave background radiation plotted as a function of the multipole, l, for an inflationary cold dark matter cosmology (Bersanelli et al., 1995). The quantity plotted on the ordinate is $l(l + 1)C_l$, where C_l is the power spectrum defined in endnote 15. The multipole, l, corresponds to an angular scale $\theta = \pi/l$. The curve labelled *scalar* shows the contribution of small density perturbations and that labelled *tensor* shows the contribution of gravitational waves. The relative amplitudes of these contributions depend upon the specific inflationary model. In the case of the scale-invariant spectrum, $n = 1$, there is expected to be no contribution of gravitational waves and the spectrum of density perturbations would be flat, corresponding to $n = 1$, at multipoles $l \leq 30$. The estimates shown in the diagram are for $n_s = 0.86$. The bars along the top of the diagram show the ranges of multipoles probed by COBE and by the *Planck* mission of the European Space Agency.

as an upper limit to c_s, the sound horizon corresponds to a comoving distance scale of $32(\Omega_0 h^2)^{-1}$ Mpc at the present epoch and to an angular scale of

$$\theta \approx \frac{\lambda_s(\text{comoving})}{D} = 0.3\,\Omega_0^{1/2} \text{ degrees.} \tag{15.12}$$

Detailed solutions of the coupled Boltzmann equations for baryonic matter, dark matter and radiation showed that the first acoustic peak should occur at $l = 200$ if $\Omega_0 = 1$ and at $l = 430$ if $\Omega_0 = 0.2$. Note that the wavelength of the first Sakharov oscillation acts as a 'rigid rod' and so can be used to determine cosmological parameters. This is the beginning of a long story which involves using the details of the observed angular power spectrum of the fluctuations in the cosmic microwave background radiation to estimate many different cosmological parameters.

As discussed in Section 15.3.2, the amplitudes of the Sakharov oscillations depend upon the phase difference from the time they came through the horizon to last scattering surface, that is they depend upon

$$\int d\phi = \int c_s \, dt. \tag{15.13}$$

The first peak in the temperature spectrum corresponds to waves with wavelength equal to the sound horizon at the last scattering layer. Oscillations which are $n\pi$ out of phase with the first acoustic peak also correspond to maxima in the temperature power spectrum at the epoch of recombination. Note that the odd harmonics correspond to maximum compression of the waves and so to increases in the temperature, whereas the even harmonics correspond to rarefactions of the acoustic waves and so to temperature minima. Those with phase differences $\pi(n + \frac{1}{2})$ relative to that of the first acoustic peak have zero amplitude at the last scattering layer and correspond to minima in the power spectra. Thus, the maxima correspond to frequencies

$$\omega t_{\mathrm{rec}} = n\pi. \tag{15.14}$$

Adopting the short-wavelength dispersion relation for the oscillations, the condition becomes

$$c_s k_n t_{\mathrm{rec}} = n\pi; \qquad k_n = \frac{n\pi}{\lambda_s} = nk_1. \tag{15.15}$$

Thus, the acoustic peaks are expected to be evenly spaced in wavenumber.

If the contribution of baryons to the speed of sound at the last scattering surface is neglected, that is $\mathcal{R} = 3\rho_B/4\rho_{\mathrm{rad}} \ll 1$, the amplitudes of the maxima in the power spectrum shown in Figure 15.10 would change smoothly with increasing wavenumber. When the inertia of the baryons can no longer be neglected, however, the amplitude of the temperature fluctuations at maximum compression is $(1 + 6\mathcal{R})$ times that of the Sachs–Wolfe effect. Furthermore, the amplitudes of the oscillations are asymmetric, the temperature excursions varying between $-(\Delta\phi/c^2)(1 + 6\mathcal{R})$ for $k\lambda_s = (2n + 1)\pi$ and $(\Delta\phi/c^2)$ for $k\lambda_s = 2n\pi$. These results account for some of the features of the temperature fluctuation spectrum shown in Figure 15.10. The temperature perturbations associated with the acoustic peaks are expected to be much larger than the Sachs–Wolfe fluctuations. The asymmetry between the even and odd peaks in the fluctuation spectrum is associated with the extra compression at the bottom of the gravitational potential wells when account is taken of the inertia of the baryonic matter.

The damping of the oscillations seen in Figure 15.10 at multipoles $l \sim 1000$ are principally associated with the effects of photon diffusion, or Silk damping, discussed in Section 15.3.1. The *suppression factor* for waves of wavenumber k can be written

$$\int \frac{d\tau}{dz} e^{-\tau(z)} e^{-k/k_D(z)} \, dz, \tag{15.16}$$

where $k_D(z) = 2\pi/\lambda_D(z)$ and $\lambda_D(z)$ is the scale on which fluctuations were damped out at redshift z. The effects of damping and the random superposition of the perturbations lead to a strong suppression of all primaeval temperature fluctuations with wavenumbers greater than about 2000, that is on the scale of a few arcminutes.

The upshot of this discussion is that measurements of the details of the power spectra of fluctuations in the cosmic microwave background radiation contain a wealth of information about many key cosmological parameters, in addition to being a key probe of the very early Universe. As a result, throughout the 1990s and the early years of the twenty-first century, many ground-based and balloon-borne experiments were carried out to search for the Sakharov oscillations, these efforts culminating in the observations of the NASA WMAP space observatory.

15.11 The discovery of Sakharov oscillations

The race was on to discover whether or not Sakharov oscillations were present in the power spectrum of the fluctuations in the cosmic microwave background radiation. Strong hints that they were present were suggested by experiments during the 1990s, but the definitive results were published from 2000 onwards. The first results of the Boomerang experiment, which involved a balloon flight around the Antarctic continent, were published in April 2000 (de Bernardis *et al.*, 2000); the first results of the Maxima experiment, another balloon-borne project launched from Palestine in Texas, appeared in November 2000 (Hanany *et al.*, 2000); the power spectrum of the DASI experiment, a ground-based interferometer based at the South Pole, was published in March 2002 (Halverson *et al.*, 2002); the Archeops team, who developed a balloon-borne detector which was flown across northern Europe and Russia containing prototype detectors for the Planck project (Benôit *et al.*, 2003a), published their power spectrum in March 2003, while the Very Small Array (VSA), located on a high site on Tenerife, resulted in a power spectrum which was published in June 2003 (Scott *et al.*, 2003); the CBI experiment, another interferometer located in the Atacama Plateau at 5000 m in Chile, resulted in a power spectrum for wavenumbers up to about 2000, which was published in July 2003 (Pearson *et al.*, 2003). The results of these experiments are summarised in Figure 15.11.

The Wilkinson Microwave Anisotropy Probe (WMAP), the successor to COBE, was launched in August 2001, and the results of the first year's observations were published in September 2003. Both the intensity and polarisation power spectra are shown in Figure 15.12, which shows the extraordinary quality of the WMAP data (Bennett *et al.*, 2003). These data have defined the angular power spectrum and the temperature–polarisation cross-spectrum with very high precision to multipole moments, l, beyond the second peak of the power spectrum, clearly showing the Sakharov oscillations. Both the scalar power spectrum and the linear polarisation of the perturbations are in excellent agreement with the expectations of an adiabatic ΛCDM model, as illustrated by the solid lines in Figure 15.12. Particularly striking are the peak and trough in the cross-polarisation spectrum at $l = 300$ and $l = 150$ (Figure 15.12(b)), respectively. The polarisation signal is associated with Thomson scattering of the background radiation in the last scattering layer. The anti-correlation of

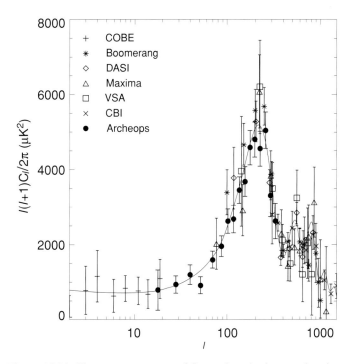

Figure 15.11: The power spectrum of fluctuations in the cosmic microwave background radiation published by the Archeops team, comparing their results with those of the other experiments listed on the diagram (Benôit *et al.*, 2003b).

these features with the first maximum of the power spectrum is consistent with the adiabatic ΛCDM model.

A surprise was the large polarisation signal observed at small multipoles. This signal must arise long after the epoch of recombination because it occurs on very large angular scales. It has been interpreted as polarisation induced during the re-ionisation epochs when the intergalactic gas was reheated and ionised.

15.12 The determination of cosmological parameters

The discussion of Section 15.10 indicated how various features of the power spectrum of fluctuations in the cosmic microwave background radiation enable cosmological parameters to be determined. On their own, the WMAP observations provide remarkable limits to many cosmological parameters, but they become even more impressive when taken in combination with independent data on the large-scale structure of the Universe. Some appreciation of what became possible by combining the WMAP observations with the two-point correlation function for galaxies determined by the Sloan Digital Sky Survey (SDSS) was provided by the analysis of Max Tegmark (b. 1967) and his colleagues (Tegmark *et al.*, 2004) (Figure 15.13).

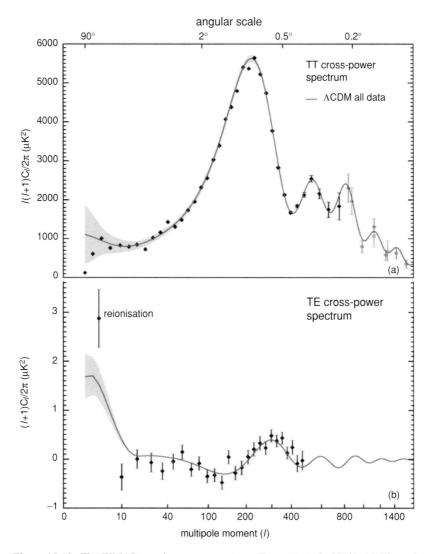

Figure 15.12: The WMAP angular power spectrum (Bennett *et al.*, 2003). (a) The points show the measured power spectrum from the first year of observations. The grey band shows the cosmic variance associated with the observations. The solid line shows the fit of a ΛCDM model to the observations. The grey points at $l > 800$ are derived from the CBI and ACBAR experiments. (b) The temperature–polarisation cross-spectrum. The peak in the spectrum near $l = 300$ is out of phase with the total power spectrum, as predicted for adiabatic initial conditions.

The procedure involves first defining the range of parameters to be included in the simulations in order to reproduce the observed spatial correlation function for galaxies and the power spectrum of fluctuations in the cosmic background radiation. A list of 13 parameters which span a very wide range of possible models is shown in Table 15.2. Many of these parameters have been discussed in previous sections, for example Ω_B, Ω_D, h, Ω_Λ, A_s, n_s, b, and the definitions are summarised in Table 15.2. The new parameters include the density

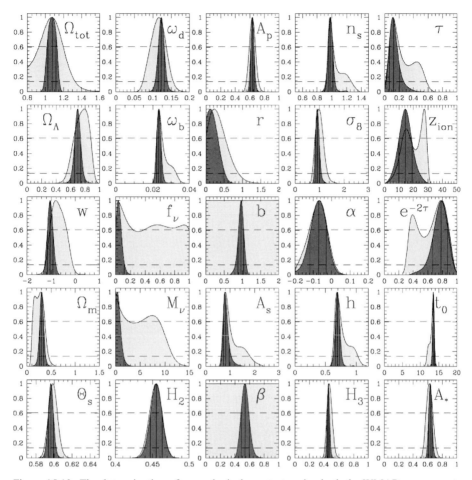

Figure 15.13: The determination of cosmological constants using both the WMAP power spectrum and the three-dimensional power spectrum for galaxies from the Sloan Digital Sky Survey (SDSS) (Tegmark *et al.*, 2004). The light shaded areas use the information from WMAP alone, while the darker shaded areas show the result of including the data from the SDSS. Each distribution has been marginalised over all other quantities in the six-parameter family (τ, Ω_Λ, ω_d, ω_b, A_s, n_s) of the 'vanilla' ΛCDM model, as well as the bias factor, b, when the SDSS data are included. The horizontal dashed lines show the 1σ and 2σ limits for each parameter.

parameter associated with space curvature, Ω_k, which is zero in the simplest inflationary model of the early Universe, the curvature of the initial power spectrum as parameterised by $\alpha = \mathrm{d}(\ln n_s)/\mathrm{d}(\ln k)$, the ratio of gravitational wave to scalar perturbations, r, the spectral index of the gravitational waves, n_t, and the ratio of mass densities in neutrinos to dark matter, f_ν.

Figure 15.13 shows how cosmological information can be extracted from these modelling procedures. Tegmark and his colleagues used Monte Carlo Markov Chain methods to determine the probability distributions with which the parameters are determined. The panels in Figure 15.13 include a large number of parameters, 13 of which are variables; the others are

Table 15.2. *The parameters involved in the construction of cosmological models for the origin of structure in the Universe (Tegmark et al., 2004)*

Parameter	Definition	Status	WMAP alone	WMAP + SDSS
$\omega_B = \Omega_B h^2$	baryon density parameter	not optional	$0.0245\,^{+0.0050}_{-0.0019}$	$0.0232\,^{+0.0013}_{-0.0010}$
$\omega_D = \Omega_D h^2$	dark-matter density parameter	not optional	$0.115\,^{+0.020}_{-0.021}$	$0.1222\,^{+0.0090}_{-0.0082}$
Ω_Λ	dark-energy density parameter	not optional	$0.75\,^{+0.10}_{-0.10}$	$0.699\,^{+0.042}_{-0.045}$
w	dark-energy equation of state			
τ	reionisation optical depth	not optional	$0.21\,^{+0.24}_{-0.11}$	$0.124\,^{+0.083}_{-0.057}$
$\Omega_k\,^a$	space curvature			
A_s	amplitude of scalar power spectrum	not optional	$0.98\,^{+0.56}_{-0.21}$	$0.81\,^{+0.15}_{-0.09}$
n_s	scalar spectral index[b]		$1.02\,^{+0.16}_{-0.06}$	$0.977\,^{+0.039}_{-0.025}$
α	running of scalar spectral index			
r	tensor–scalar ratio			
n_t	tensor spectral index			
b	bias factor	not optional	no constraint	$1.009\,^{+0.073}_{-0.083}$
$f_\nu = \rho_\nu/\rho_D$	neutrino fraction			

[a] $\Omega_B + \Omega_D + \Omega_\Lambda + \Omega_k = 1$.
[b] $n_s = 1$ is the preferred value according to the standard inflation picture.

derived from them. The light grey regions of the diagrams show the inferred values of the parameters using only the WMAP data. It can be seen that these strongly constrain many of the most important cosmological parameters. For example, there is unambiguous evidence for the presence of a non-zero cosmological constant, as parameterised by $\Omega_\Lambda = \Lambda/3H_0^2$. Likewise, there are strong constraints on ω_D and ω_B.

When the evidence of the two-point correlation function for galaxies from the Sloan Digital Sky Survey is included, as shown by the dark grey areas of Figure 15.13, the parameters are determined with roughly a factor of 2 greater precision. This arises because the two-point correlation function for galaxies provides a direct measure of the matter power spectrum with very much smaller error bars than could be inferred from the WMAP data alone.

There are many ways of interpreting the results of this remarkable analysis, as explained by Tegmark *et al.* (2004). One conservative approach is to derive the simplest set of parameters needed to account satisfactorily for all the WMAP and SDSS data. For illustration, such sets of parameters are shown in Table 15.2. Of the 13 parameters listed, six are absolutely necessary in order to account for the observations, and these have been labelled 'not optional'. In this 'vanilla' model, it is assumed that there are no primordial gravitational waves or relic neutrinos, the equation of state of the dark energy is $p = -\rho c^2$, the initial power spectrum is defined by a constant spectral index, n_s, and the spatial geometry of the Universe is flat, $\Omega_k = 0$. It turns out that satisfactory and self-consistent estimates of the six parameters can be found for this model. The estimates of the parameters for WMAP alone are listed in column 4 and for WMAP plus SDSS in column 5 of Table 15.2. Particularly striking results are the following.

- The spectral index of the primordial power spectrum is very close to $n_s = 1$, as expected for inflation models of the early Universe.
- The dark-energy density parameter $\Omega_\Lambda \approx 0.7$, while the dark matter has density parameter $\Omega_D h^2 \approx 0.122$.
- The baryonic matter has density parameter $\Omega_B h^2 \approx 0.023$.
- Hubble's constant can be inferred from combining the parameters in appropriate ways; it is found to be $h = 0.695^{+0.039}_{-0.031}$.
- When the SDSS data are included, the bias parameter, b, is very close to unity, meaning that the baryonic matter in galaxies traces the distribution of the cold dark matter.

What is spectacular about these results is that they are entirely consistent with the independent estimates of these parameters discussed in some detail in Chapter 13. For example, the estimates of Hubble's constant from the HST key project (Section 13.2) are in excellent agreement with that given in the above list, the estimates of Ω_Λ from the Type Ia supernovae projects agree very well with those listed above, as do the estimates of the overall density parameter in matter, $\Omega_0 = \Omega_B + \Omega_D$, and the density parameter in baryonic matter, Ω_B. Including the data derived from the Type Ia supernovae with the WMAP and SDSS data leads to an estimate of the age of the Universe of $t_0 = 14.1^{+1.0}_{-0.9}$ Gyears.

As expected, if additional parameters are included in the model-fits, the uncertainty in some of these parameters increases, but many of the results are robust. It is remarkable, however, how well the simplest models can account for many diverse data sets. As Tegmark *et al.* (2004) remark

Readers wishing to choose a concordance model for calculational purposes using Ockham's razor can adopt the best fit 'vanilla lite' model

$$(\tau, \Omega_\Lambda, \omega_d, \omega_b, A_s) = (0.17, 0.72, 0.12, 0.024, 0.89). \tag{15.17}$$

Note that this is even simpler than 6-parameter vanilla models, since it has $n_s = 1$ and only 5 free parameters.

It is significant that these results are in excellent agreement with the independent conclusions of the WMAP team, who used the data on the two-point correlation function for galaxies from the 2dF Galaxy Redshift Survey (Spergel *et al.*, 2003). The data in equation (15.17) correspond to the set of concordance parameters listed in Table 15.3.

15.13 The post-recombination Universe

15.13.1 *The epoch of re-ionisation*

Among the important results of the above analysis is the estimate of the optical depth of the intergalactic gas to Thomson scattering, $\tau \approx 0.17$, which is associated with the reheating and reionisation of the intergalactic gas, presumably by the earliest generations of stars or active galactic nuclei in galaxies. If the intergalactic gas were instantly 100% reionised at some redshift z_{ion}, the reionisation redshift was found to be $z_{ion} = 14.4^{+5.2}_{-4.7}$ according to the analysis of Tegmark and his colleagues. Note that this result is strongly dependent upon the polarisation information shown in Figure 15.12.

Table 15.3. *A concordance set of cosmological parameters for calculational purposes using Ockham's razor (Tegmark et al., 2004)*

Parameter	Definition	Value
H_0	Hubble's constant	$72 \text{ km s}^{-1} \text{ Mpc}^{-1}$
Ω_k	space curvature	0
Ω_Λ	dark-energy density parameter	0.72
$\Omega_0 = \Omega_B + \Omega_D$	total matter density parameter	0.28
Ω_B	baryon density parameter	0.047
Ω_D	dark-matter density parameter	0.233
n_s	scalar spectral index	1
A_s	amplitude of scalar power spectrum	0.89
τ	reionisation optical depth	0.17

This result can be contrasted with the information on the absence of a Gunn–Peterson trough in the spectra of large-redshift quasars (see Section 14.3). The quasars with redshifts up to $z \sim 5$ showed no evidence for neutral hydrogen absorption by diffuse intergalactic gas, but a Gunn–Peterson trough has been found in the spectrum of the largest-redshift quasar discovered as part of the Sloan Digital Sky Survey. Four quasars with redshifts 5.80, 5.82, 5.99 and 6.28 were discovered by the technique of searching for i-band drop-outs, similar to the technique illustrated in Figure 14.15, but now optimised for very-large-redshift quasars. The spectra of these four quasars observed with the Keck-2 telescope are shown in Figure 15.14 (Becker *et al.*, 2001).

As the redshift of the quasar increases, the Lyman-α forest, shown in Figure 14.11, depresses the continuum to the short-wavelength side of Lyman-α. As shown by Palle Møller (b. 1954) and Peter Jakobsen (b. 1953), the effect increases dramatically with increasing redshifts, particularly when account is taken of the evolution of the number of absorption systems per unit redshift, as discussed in Section 14.3 (Møller and Jakobsen, 1990). The first three quasars show increasing absorption to the short-wavelength side of the redshifted Lyman-α line, as expected if the absorption were due to an increasing number of discrete absorbing clouds. At the very largest redshifts, $5.95 \leq z \leq 6.15$, however, the continuum flux drops dramatically to zero to the short-wavelength side of the Lyman-α line in the largest-redshift quasar at $z = 6.28$. Robert Becker (b. 1946) and his colleagues interpreted this result as showing that the Gunn–Peterson trough has at last been observed, implying that the fractional abundance of diffuse neutral hydrogen begins to increase with increasing redshift beyond $z \sim 6$. This result was reinforced by observations of the trough to the short-wavelength side of the Lyman-β line. In the words of Becker *et al.* (2001),

the Universe is approaching the reionisation epoch at $z \sim 6$.

The study of the Universe at redshifts $z \geq 6$, often referred to as the *dark ages*, is one of the great challenges for twenty-first-century astrophysical cosmology. The first generations of stars in galaxies must have formed between redshifts $30 \geq z \geq 6$. The ultraviolet radiation of these newly formed stars, and any black holes which had formed in the nuclei of galaxies, must have resulted in heating and reionisation of the intergalactic gas. It is a great

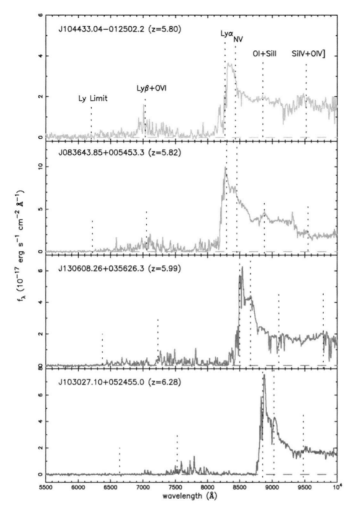

Figure 15.14: The optical spectra of four very-large-redshift quasars ($z \geq 5.8$) observed with the Keck-2 telescope. In each spectrum, the wavelength of prominent emission lines, as well as the Lyman limit, are indicated by dotted vertical lines (Becker *et al.*, 2001). The key observation is the zero continuum flux to the short-wavelength side of the Lyman-α line in the spectrum of the largest-redshift quasar, which can be contrasted with the residual Lyman-α forest in the lower-redshift quasars.

observational challenge to determine observationally the history of structure formation and the evolution of the intergalactic gas during these critical epochs.

15.13.2 *The effects of dissipation*

Most of the discussion of the formation of galaxies and large-scale structures has focussed upon the dynamical evolution of the potential wells defined by the dark matter, in other words the linear and non-linear evolution of perturbations under the influence of gravity

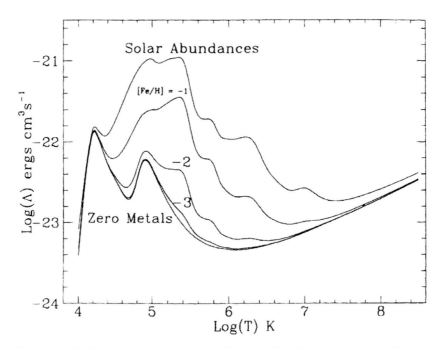

Figure 15.15: The cooling rate per unit volume, $\Lambda(T)$, of an astrophysical plasma of number density 1 nucleus cm^{-3} by radiation for different cosmic abundances of the heavy elements, ranging from zero metals to the present abundance of heavy elements as a function of temperature, T (Silk and Wyse, 1993). In the zero-metal case, the two maxima in the cooling curve are associated with the recombination of hydrogen ions and doubly ionised helium.

alone. In addition, the role of *dissipation*, meaning energy loss by radiation and which results in the loss of thermal energy from the system, needs to be taken into account. If the radiation process is effective in removing pressure support from the system, this can result in a runaway situation, known as a *thermal instability* (Field, 1965).

Dissipative processes play a dominant role in the formation of stars, and this leads to the question of whether or not similar processes are important in the formation of larger-scale structures. In 1977, the role of dissipative processes in galaxy formation was elegantly described by Martin Rees and Jeremiah Ostriker (Rees and Ostriker, 1977), who considered the cooling of the primordial plasma consisting of the primaeval abundances of hydrogen and helium. In 1993, Joseph Silk and Rosemary Wyse included cooling by heavy elements at different levels of enrichment from the primordial values into their cooling curves. The key relation is the energy loss rate of the plasma by radiation as a function of temperature (Figure 15.15) (Silk and Wyse, 1993). The cooling rate is presented in the form $dE/dt = -N^2 \Lambda(T)$, where N is the number density of hydrogen ions. In the absence of heavy metals, the dominant loss mechanism at high temperatures, $T > 10^6$ K, is thermal bremsstrahlung, the energy loss rate being proportional to $N^2 T^{1/2}$. At lower temperatures, the main loss mechanisms are free–bound and bound–bound transitions of hydrogen and ionised helium, corresponding to the two maxima in the cooling curve. As the abundance of the heavy

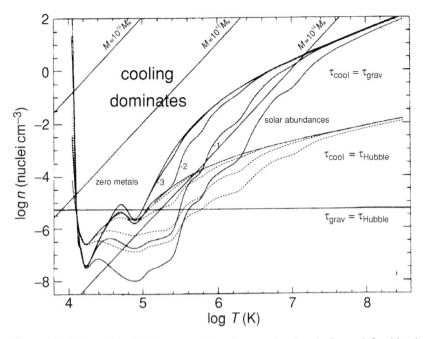

Figure 15.16: A number density–temperature diagram showing the locus defined by the condition that the collapse time of a region, t_{dyn}, should be equal to the cooling time of the plasma by radiation, t_{cool}, for different abundances of the heavy elements (Silk and Wyse, 1993). Also shown are lines of constant mass, a cooling time of 10^{10} years (dotted lines) and the density at which the perturbations are of such low density that they do not collapse in the age of the Universe.

elements increases, the overall energy loss rate can be more than an order of magnitude greater than that of the primordial plasma at temperatures $T \leq 10^6$ K.

For the case of a fully ionised plasma, the cooling time is defined to be the time it takes the plasma to radiate away its thermal energy, that is

$$t_{cool} = \frac{E}{|dE/dt|} = \frac{3NkT}{N^2 \Lambda(T)}. \tag{15.18}$$

This timescale can be compared with the timescale for gravitational collapse, $t_{dyn} \approx (G\rho)^{-1/2} \propto N^{-1/2}$. The significance of these timescales is best appreciated by inspecting the locus of the equality, $t_{cool} = t_{dyn}$, in a temperature–number density diagram (Figure 15.16). The locus $t_{cool} = t_{dyn}$ is a mapping of the cooling curve of the hydrogen–helium plasma loss rate into the (T, N) plane. Inside this locus, the cooling time is shorter than the collapse time, and so it is expected that dissipative processes are more important than dynamical processes in the collapse of the baryonic matter. Also shown in Figure 15.16 are lines of constant mass, as well as loci corresponding to the radiation loss time being equal to the age of the Universe, and to the perturbations having such low density that they do not collapse gravitationally in 10^{10} years. It can be seen that the range of masses which lie within the critical locus, and which can cool in 10^{10} years, corresponds to $10^{10} \leq M/M_\odot \leq 10^{13}$ – this is the key conclusion of this analysis. The fact that the masses lie naturally in the range of observed galaxy masses suggests that the typical masses of galaxies are not only determined by the initial fluctuation spectrum, but by astrophysical processes as well. As can be seen in Figure 15.16, the greater the abundance of the heavy elements, the shorter the timescale for cooling of a region of a given temperature and density.

Figure 15.16 can be used astrophysically in the following way. For any theory of the origin of the large-scale structure, the density and temperature of the gas can be worked out at each epoch. The diagram can then be used to determine whether or not cooling by radiative losses is important. A good example was found in the various versions of the adiabatic pancake theory. When the gas cloud collapses to form a pancake, the matter falls into a singular plane and, as a result, a shock-wave passes out through the infalling matter, heating it to a high temperature. In this picture, galaxies can form by thermal instabilities in the heated gas. Inspection of Figure 15.16 shows that, if the gas is heated above 10^4 K, there is no stable region for masses in the range 10^{10} to $10^{13} M_\odot$.

A second exercise, carried out by George Blumenthal (b. 1945) and his colleagues, was to plot the observed location of galaxies on the temperature–number density diagram (Blumenthal et al., 1984). The effective temperature associated with the velocity dispersion of the stars in a galaxy or the galaxies in a cluster, $\frac{1}{2}kT_{\text{eff}} \approx \frac{1}{2}mv^2$, was plotted, rather than the thermal temperature of the gas. The irregular galaxies fell well within the cooling locus, and the spirals, S0, and elliptical galaxies all lay close to the critical line. On the other hand, the clusters of galaxies lay outside the cooling locus. Thus, cooling is expected to be an important factor in certain scenarios for the formation of galaxies.

The success of the cold dark matter models in accounting for the large-scale distribution of galaxies encouraged the numerical cosmologists to tackle the next step of including diffuse baryonic matter into the computations and following the evolution of the gas using the procedures of single-particle hydrodynamics and similar computational algorithms (Hernquist et al., 1996; Katz et al., 1996; Miranda-Escudé et al., 1996). These programmes involved incorporating the role of dissipative processes in enabling the baryonic components of galaxies to form the types of structures observed in the real Universe. Once star formation got underway, the intense ultraviolet radiation of young massive stars contributed to the ionisation of the diffuse intergalactic gas. In addition, quasars were formed, and their ionising ultraviolet radiation had an important influence upon the physical state of the gas.

Figure 15.17 shows an example of the resulting structure of neutral hydrogen clouds at redshift $z = 2$. The simulations result in a network of filaments, with dense knots of neutral hydrogen forming in the vicinity of galaxies. The column densities found in these simulations span the range from about 10^{18} to 10^{26} m^{-2}, and their number density distribution follows closely the observed power-law relation $N(N_H) \propto N_H^{-1.5}$ over this range. According to the simulations, the high column density knots, shown as white blobs in Figure 15.17, arise from radiatively cooling gas associated with galaxies which form in high-density regions. In contrast, the low-density systems are associated with a wide range of different types of structure. To quote Lars Hernquist (b. 1954) and his colleagues (Hernquist et al., 1996),

the low column density absorbers are physically diverse: they include filaments of warm gas; caustics in frequency space produced by converging velocity flows; high-density halos of hot collisionally ionised gas; layers of cool gas sandwiched between shocks; and modest local undulations in undistinguished regions of the intergalactic medium. Temperatures of the absorbing gas range from below 10^4 K to above 10^6 K.

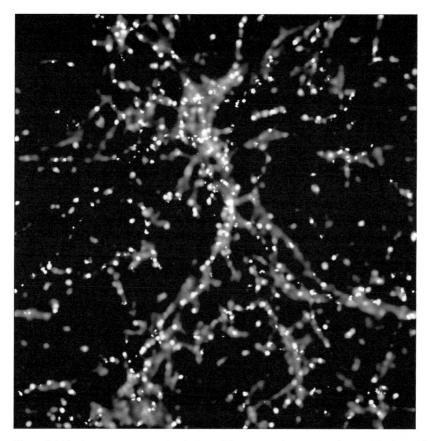

Figure 15.17: A supercomputer simulation of the expected structure of neutral hydrogen in the inter-galactic medium at a redshift $z = 2$ in a standard cold dark matter cosmology with $\Omega_0 = 1$. The size of the box corresponds to a comoving scale of 22.22 Mpc. The simulation includes self-shielding of the neutral hydrogen from the background ultraviolet ionising radiation. The grey-scale is such that the white blobs correspond to column depths $N(\text{HI}) \geq 10^{20.5} \text{ m}^{-2}$. The faint filamentary structures corre-spond to column densities $10^{19.5} \geq N(\text{HI}) \geq 10^{18.5} \text{ m}^{-2}$ and the black 'voids' correspond to regions with $N(\text{HI}) \leq 10^{18.5} \text{ m}^{-2}$ (Katz *et al.*, 1996).

Notes to Chapter 15

1 Part III of Malcolm Longair, *Galaxy Formation* (Berlin: Springer-Verlag, 1998) is devoted to an exposition of the detailed physics of the processes involved in structure formation in the Universe, and covers much of the content of this chapter.

2 The corresponding equation for the electrostatic case was only derived after the discovery of plasma oscillations by Irving Langmuir (1881–1957) and Lewi Tonks (1897–1971) in the 1920s, and describes the dispersion relation for longitudinal plasma oscillations, or Langmuir waves:

$$\omega^2 = c_s^2 k^2 + \frac{N_e e^2}{m_e \varepsilon_0},$$

where N_e is the electron density and m_e is the mass of the electron (Tonks and Langmuir, 1929). The formal similarity of the physics may be appreciated by comparing the attractive gravitational acceleration of a region of mass density ρ_0 and the repulsive electrostatic acceleration of a region of electron charge density $N_e e$. The equivalence of $-G\rho_0$ and $N_e e^2/4\pi\epsilon_0 m_e$ is apparent.

3 It may appear strange at first that the temperature at which the intergalactic gas was fully ionised is not closer to 150 000 K, the temperature at which $\langle h\nu \rangle = kT = 13.6\,\mathrm{eV}$, the ionisation potential of neutral hydrogen. The important point is that the photons far outnumber the baryons in the intergalactic medium by a factor of $3.6 \times 10^7/\Omega_B h^2$, and there is a broad range of photon energies present in the Planck distribution. Roughly speaking, the intergalactic gas will be ionised, provided there are as many ionising photons with $h\nu \geq 13.6\,\mathrm{eV}$ as there are hydrogen atoms, and this occurs when the temperature is only about 1/25 of the ionisation potential of hydrogen. This type of calculation appears in a number of different guises in astrophysics: the nuclear reactions which power the Sun take place at a much lower temperature than expected; the temperature at which regions of ionised hydrogen become fully ionised is only about 10 000 K; light nuclei are destroyed in the early Universe at much lower temperatures than would be expected. In all these cases, the tails of the Planck and Maxwell distributions contain large numbers of photons and particles, respectively, with energies very much greater than the mean.

4 The exchange of energy between photons and electrons is an enormous subject and has been treated in more detail by Sunyaev, Zeldovich and their colleagues (Sunyaev and Zeldovich, 1980; Pozdnyakov, Sobol and Sunyaev, 1983).

5 The topic of induced Compton scattering and the physics of radiation transfer had been the subject of detailed studies by Zeldovich, who led the team which developed the Soviet hydrogen bomb in the 1950s. Some of the very best Soviet physicists, including Lev Landau, contributed to the development of what is now called the Kompaneets equation. Kompaneets was the first physicist to receive security clearance so that the work could be published in the open literature.

6 It is intriguing that, as soon as Planck had introduced what he called the 'unit of action', h, into the expression for black-body radiation, he appreciated that it became possible for the first time to define a set of 'natural units' by combining the gravitational constant, G, and the speed of light, c, with what we now call Planck's constant, h, as listed in Table 15.1 (Planck, 1900).

7 The power spectrum $P(k)$, of the perturbations is defined in the following way. First of all, we define the Fourier transform pair for the density perturbations on scale r, $\delta\rho/\rho(r) = \Delta(r)$:

$$\Delta(r) = \frac{V}{(2\pi)^3} \int \Delta_k\, e^{-i k \cdot r}\, d^3 k;$$

$$\Delta_k = \frac{1}{V} \int \Delta(r)\, e^{i k \cdot r}\, d^3 x.$$

We now use Parseval's theorem to relate the integrals of the squares of $\Delta(r)$ and its Fourier transform Δ_k:

$$\frac{1}{V} \int \Delta^2(r)\, d^3 x = \frac{V}{(2\pi)^3} \int |\Delta_k|^2\, d^3 k.$$

The quantity on the left-hand side of this relation is the mean square amplitude of the fluctuation per unit volume, and $|\Delta_k|^2$ is the *power spectrum* of the fluctuations, which is often written as $P(k)$. Therefore, we can write

$$\langle \Delta^2 \rangle = \frac{V}{(2\pi)^3} \int |\Delta_k|^2\, d^3 k = \frac{V}{(2\pi)^3} \int P(k)\, d^3 k.$$

8 Peebles recounts the history of the development of many of these ideas in his major book, *The Principles of Physical Cosmology* (Peebles, 1993).

9 I have given a simple derivation of the Press–Schechter formula in Section 16.3 of Longair, *Galaxy Formation*.

10 The mass of the electron neutrino/antineutrino was measured in tritium β-decay experiments. The decay results in a ^{3}He nucleus, an electron and an electron antineutrino. If the antineutrinos had non-zero mass, the spectrum of the electrons would be deformed at high energies, that is the antineutrino mass determines the maximum energy of emitted electrons (Weinheimer, 2001). The results of the Mainz experiments provide an upper limit to the mass of the electron antineutrino of 2 eV. The particle data book suggests a conservative upper limit of 3 eV (see http://www-pdg.lbl.gov/pdg.html).

11 The review by George Efstathiou can be thoroughly recommended for those who wish to enter much more deeply into the details of these remarkable simulations (Efstathiou, 1990).

12 Further detailed calculations for the probability distribution of last scattering of photons through the epoch of recombination were presented by Hu and Sugiyama (1995).

13 Various types of horizon appear in cosmological discussions. The term *particle horizon* refers to the maximum separation which points could have had and could still be causally connected at some epoch t. In a simple approximation, the particle horizon is given by $r_{\mathrm{H}} \approx ct$. When account is taken of the fact that the Universe was expanding more rapidly in the past, the scale becomes $r_{\mathrm{H}} = 3ct$ in a matter-dominated Universe and $r_{\mathrm{H}} = 2ct$ in a radiation-dominated Universe. Another term found in the literature is the *event horizon*, which was introduced by Wolfgang Rindler (b. 1924) in 1956 (Rindler, 1956). This is defined to be the greatest comoving radial distance coordinate which an object can have at a particular epoch if it is ever to be observable, however long the observer waits.

14 Thanu Padmanabhan (b. 1957) shows how this result can be derived by perturbing the Friedman metric and relating the temperature fluctuation to the perturbation in the Newtonian gravitational potential $\Delta\phi$ (see also his even better solution of 1996) (Padmanabhan, 1993, 1996). Peter Coles (b. 1963) and Francesco Lucchin (1944–2002) rationalised how the Sachs–Wolfe answer can be derived (Coles and Lucchin, 1995). In addition to the Newtonian gravitational redshift, because of the perturbation of the metric, the cosmic time, and hence the scale factor R, at which the fluctuations are observed, are shifted to slightly earlier cosmic times. Temperature and scale factor change as $\Delta T/T = -\Delta R/R$. For all the standard models in the matter-dominated phase $R \propto t^{2/3}$, and so the increment of cosmic time changes as $\Delta R/R = (2/3)\Delta t/t$. But $\Delta \nu/\nu = -\Delta t/t$ is just the Newtonian gravitational redshift, with the net result that there is a positive contribution to $\Delta T/T$ of $-(2/3)\Delta\phi/c^2$. The net temperature fluctuation is given by $\Delta T/T = \frac{1}{3}\Delta\phi/c^2$.

15 There are variants on the theme of the Sachs–Wolfe effect. If the time evolution of the perturbations does not grow linearly with time, the integrated effect of the gravitational perturbation along the line of sight needs to be evaluated. This is known as the *integrated Sachs–Wolfe effect*. Another effect, noted by Rees and Sciama (1926–1999), is that, if the background radiation passes through large-scale density perturbations and the depth of the potential well increases during propagation through the perturbation, temperature fluctuations are induced in the cosmic background radiation (Rees and Sciama, 1968).

16 The first step is to make a spherical harmonic expansion of the temperature distribution over the whole sky:

$$\frac{\Delta T}{T}(\theta, \phi) = \frac{T(\theta, \phi) - T_0}{T_0} = \sum_{l=0}^{\infty} \sum_{m=-l}^{m=l} a_{lm} Y_{lm}(\theta, \phi), \tag{1}$$

where the normalised functions, Y_{lm}, are given by the expression

$$Y_{lm}(\theta, \phi) = \left[\frac{2l+1}{4\pi} \frac{(l - |m|)!}{(l + |m|)!} \right]^{1/2} P_{lm}(\cos\theta)\, e^{im\phi} \tag{2}$$

$$\times \begin{cases} (-1)^m & \text{for } m \geq 0 \\ 1 & \text{for } m < 0 \end{cases}, \tag{3}$$

and $P_{lm}(\cos\theta)$ are the associated Legendre polynomials of order l. The values of a_{lm} are found by multiplying the temperature distribution over the sphere by Y_{lm}^* and integrating over the sphere:

$$a_{lm} = \int_{4\pi} \frac{\Delta T}{T}(\theta, \phi) Y_{lm}^* \, d\Omega . \tag{4}$$

If the fluctuations can be represented by a superposition of waves of random phase, each of the $(2l + 1)$ coefficients of a_{lm} associated with the multipole, l, provides an independent estimate of the amplitude of the temperature fluctuations associated with that multipole. If the power spectrum is assumed to be circularly symmetric about each point in the sky, the mean value of $a_{lm}a_{lm}^*$, averaged over the whole sky, provides an estimate of the power associated with the multipole, l:

$$C_l = \frac{1}{2l + 1} \sum_m a_{lm}a_{lm}^* = \langle|a_{lm}|^2\rangle . \tag{5}$$

Because of the assumption that the fluctuations are Gaussian, the power spectrum, C_l, provides a *complete* statistical description of the temperature fluctuations.

17 Excellent surveys of the physics of temperature fluctuations in the cosmic microwave background radiation have been given by Hu and Sugiyama (1995), Hu (1996) and by Hu, Sugiyama and Silk (1997).

A15 Explanatory supplement to Chapter 15

A15.1 *The development of small perturbations in the expanding universe*

The results concerning the development of small perturbations in the expanding Universe are so important that it is worthwhile giving a simple demonstration of the origin of the fundamental results.[1]

The development of a spherical perturbation in the expanding Universe can be modelled by embedding a spherical region of density $\rho + \delta\rho$ in an otherwise uniform Universe of density ρ (Figure A15.1). According to Gauss's law for gravity, the spherical region behaves dynamically like a Universe of slightly higher density than the background model. It is simplest to begin with the parametric solutions for the dynamics of the Friedman world models, which can be written

$$R = a(1 - \cos\theta), \qquad t = b(\theta - \sin\theta), \tag{A15.1}$$

$$a = \frac{\Omega_0}{2(\Omega_0 - 1)}, \qquad b = \frac{\Omega_0}{2H_0(\Omega_0 - 1)^{3/2}}. \tag{A15.2}$$

Firstly, we find the solutions for small values of θ, corresponding to early stages of the matter-dominated era. Expanding to third-order in θ, $\cos\theta = 1 - \frac{1}{2}\theta^2$, $\sin\theta = \theta - \frac{1}{6}\theta^3$, we find the solution

$$R = \Omega_0^{1/3} \left(\frac{3H_0 t}{2}\right)^{2/3}. \tag{A15.3}$$

This solution shows that, in the early stages, the dynamics of all world models tend towards those of the Einstein–de Sitter model, $\Omega_0 = 1$, $\lambda = 0$, that is $R = (3H_0 t/2)^{2/3}$, but with a different constant of proportionality.

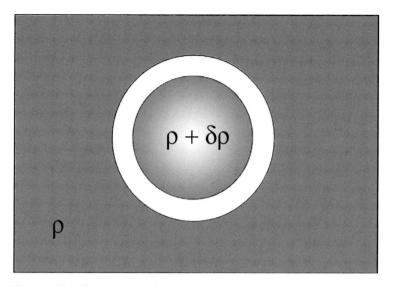

Figure A15.1: Illustrating a spherical perturbation with slightly greater density than the average in a uniformly expanding Universe. The region with slightly greater density behaves dynamically, exactly like a model Universe with density $\rho + \delta\rho$.

Now consider the region of slightly greater density embedded within the background model. We expand the expressions for R and t to fifth-order in θ: $\cos\theta = 1 - \frac{1}{2}\theta^2 + \frac{1}{24}\theta^4 - \cdots$, $\sin\theta = \theta - \frac{1}{6}\theta^3 + \frac{1}{120}\theta^5 - \cdots$. The solution is then given by

$$R = \Omega_0^{1/3}\left(\frac{3H_0 t}{2}\right)^{2/3}\left[1 - \frac{1}{20}\left(\frac{6t}{b}\right)^{2/3}\right].\tag{A15.4}$$

We can now write down an expression for the change of density of the spherical perturbation with cosmic epoch:

$$\rho(R) = \rho_0 R^{-3}\left[1 + \frac{3}{5}\frac{(\Omega_0 - 1)}{\Omega_0}R\right].\tag{A15.5}$$

Note that, if $\Omega_0 = 1$, there is no growth of the perturbation. The density perturbation may be considered to be a mini-Universe of slightly higher density than $\Omega_0 = 1$ embedded in an $\Omega_0 = 1$ model. Therefore, the density contrast changes with scale factor as

$$\Delta = \frac{\delta\rho}{\rho} = \frac{\varrho(R) - \varrho_0(R)}{\varrho_0(R)} = \frac{3}{5}\frac{(\Omega_0 - 1)}{\Omega_0}R.\tag{A15.6}$$

This result indicates why density perturbations grow only linearly with cosmic epoch. The instability corresponds to the slow divergence between the variation of the scale factors with cosmic epoch of the model with $\Omega_0 = 1$ and one with slightly greater density. This behaviour is illustrated in Figure A15.2. This is the essence of the argument developed by Tolman and Lemaître in the 1930s and, more generally, by Lifshitz in 1946 to the effect that, because the instability develops only algebraically, galaxies could not form by gravitational collapse.

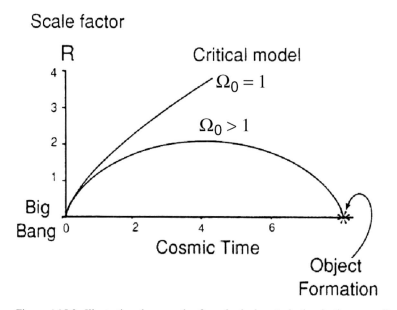

Figure A15.2: Illustrating the growth of a spherical perturbation in the expanding Universe as the divergence between two Friedman models with slightly different densities.

It can also be appreciated why the perturbations only grow at redshifts $z \geq 1/\Omega_0$. This is because at redshifts $z \geq 1/\Omega_0$ the dynamics of the world model tend to those of the critical Einstein–de Sitter model, and so the arguments given above apply to the divergence of the slightly denser model, which will then follow the behaviour shown in Figure A15.2. At redshifts $z \leq 1/\Omega_0$, the background model follows the 'open' trajectory, and it would need a very large density perturbation to move the perturbed region onto the collapsing branch in Figure A15.2.

Note to Section A15

1 I have given details of the full derivations in Chapter 11 of Malcolm Longair, *Galaxy Formation* (Berlin: Springer-Verlag, 1998).

16 The very early Universe

16.1 The big problems

The history recounted in the preceding four chapters represents quite extraordinary progress in understanding the astrophysical origins and evolution of our Universe. The contrast between the apparently insuperable problems of determining precise values of cosmological parameters up till the 1990s and the era of *precision cosmology* of the early years of the twenty-first century is startling.

Yet, despite the undoubted success of the concordance model, it raises as many problems as it solves. The picture is incomplete in the sense that, within the context of the standard world models, the initial conditions listed in Tables 15.2 and 15.3 have to be put in by hand in order to create the Universe as we observe it today. How did these initial conditions arise? As the quality of the observations improved, a number of fundamental issues for astrophysical cosmology became apparent. The resolution of these problems will undoubtedly provide insight into the laws of physics under physical conditions which at the moment can only be studied by cosmological observations.[1]

16.1.1 *The horizon problem*

This problem, clearly recognised by Robert Dicke in 1961, can be restated, 'Why is the Universe so isotropic?' (Dicke, 1961). At earlier cosmological epochs, the particle horizon $r \sim ct$ encompassed less and less mass and so the scale over which particles could be causally connected became smaller and smaller. A vivid example of this problem is to work out how far light could have travelled along the last scattering layer at $z = 1000$ since the Big Bang. In matter-dominated models, this distance is $r = 3ct$, corresponding to an angle of $\theta_H = 1.8\Omega_0^{1/2}$ ° on the sky. Thus, regions of the sky separated by greater angular distances could not have been in causal communication. Why then is the cosmic microwave background radiation so isotropic? How did causally separated regions 'know' that they had to have the same temperature to better than one part in 10^5?

16.1.2 *The flatness problem*

Why is the Universe so close to the critical density, $\Omega_0 = 1$? The flatness problem was also recognised by Dicke in his paper of 1961 and was reiterated by Dicke and Peebles in 1979 (Dicke, 1961; Dicke and Peebles, 1979). The problem arises from the fact that, according to the standard world models, if the Universe were set up with a value of the density parameter differing even slightly from the critical value $\Omega_0 = 1$, it would diverge very rapidly from $\Omega_0 = 1$ at later epochs. It is straightforward to show that, if the Universe has density parameter Ω_0 today, at redshift z, $\Omega(z)$ would have been given by

$$\left[1 - \frac{1}{\Omega(z)}\right] = f(z)\left[1 - \frac{1}{\Omega_0}\right], \tag{16.1}$$

where $f(z) = (1+z)^{-1}$ for the matter-dominated era and $f(z) \propto (1+z)^{-2}$ during the radiation-dominated era. Thus, since $\Omega_0 \sim 1$ at the present epoch, it must have been extremely close to the critical value in the remote past. Alternatively, if $\Omega(z)$ had departed from $\Omega(z) = 1$ at a very large redshift, Ω_0 would be very far from $\Omega_0 = 1$ today. Thus, the only 'stable' value of Ω_0 is $\Omega_0 = 1$. There is nothing in the standard world models that would lead us to prefer any particular value of Ω_0. This is sometimes referred to as the *fine-tuning problem*.

16.1.3 The baryon-asymmetry problem

The baryon-asymmetry problem arises from the fact that the photon-to-baryon ratio today is given by

$$\frac{N_\gamma}{N_B} = \frac{4 \times 10^7}{\Omega_B h^2} = 1.6 \times 10^9, \tag{16.2}$$

where Ω_B is the density parameter in baryons and the values of Ω_B and h have been taken from Table 15.3. If photons are neither created or destroyed, this ratio is conserved as the Universe expands. At temperature $T \approx 10^{10}$ K, electron–positron pair production takes place from the photon field. At a correspondingly higher temperature, baryon–antibaryon pair production takes place, with the result that there must have been a very small asymmetry in the baryon–antibaryon ratio in the very early Universe if we are to end up with the correct photon-to-baryon ratio at the present day. As explained in Section 15.2.3, at these very early epochs there must have been roughly $10^9 + 1$ baryons for every 10^9 antibaryons to guarantee the observed ratio at the present epoch. If the Universe had been symmetric with respect to matter and antimatter, the photon-to-baryon ratio would now be about 10^{18}, in gross contradiction with the observed value (Zeldovich, 1965). Therefore, there must be some mechanism in the early Universe which results in a slight asymmetry between matter and antimatter.

16.1.4 The primordial fluctuation problem

What was the origin of the density fluctuations from which galaxies and large-scale structures formed? According to the analyses of Chapter 15, the amplitudes of the density perturbations when they came through the horizon had to be of finite amplitude, $\delta\rho/\rho \sim 10^{-5}$, over a very wide range of mass scales. These cannot have originated as statistical fluctuations in the numbers of particles on, say, the scales of superclusters of galaxies. There must have been some physical mechanism which generated finite-amplitude perturbations with a power spectrum close to $P(k) \propto k$ in the early Universe.

16.1.5 The values of the cosmological parameters

The horizon and flatness problems were recognised before compelling evidence was found for the finite value of the cosmological constant, or, in modern parlance, the density parameter in the vacuum fields Ω_Λ, but these problems remain unchanged. The concordance values for the cosmological parameters create their own problems. The Universe seems to be geometrically flat, $\Omega_k = 0$, and so the sum of the density parameters in the matter and the dark energy must sum to unity, $\Omega_\Lambda + \Omega_m = 0.72 + 0.28 = 1$. Even if the sum of these two parameters were not precisely unity, it is a surprise that the two parameters are of the same order of magnitude at the present epoch. The matter density evolves with redshift as $(1 + z)^3$, while the dark-energy density parameter is unchanging with cosmic epoch. Why then do we live at an epoch when they have more or less the same values?

Inspection of the Friedman equation (6.13) shows that if Λ is positive, the term may be thought of as representing, in the words of Yakov Zeldovich, the 'repulsive force of a vacuum', the repulsion being relative to an absolute geometrical frame of reference (Zeldovich, 1968). There was no obvious interpretation of this term in the context of classical physics. There is, however, a natural interpretation according to quantum field theory.

The key insight was the introduction of the Higgs fields into the theory of weak interactions (Higgs, 1964). These and other ideas of quantum field theory were clearly described by Zeldovich (1968). The Higgs fields were introduced into the electro-weak theory of elementary particles in order to eliminate singularities in that theory and to endow the $W^\pm$ and Z^0 bosons with mass. Precise measurements of the masses of these particles at CERN confirmed that theory very precisely, although the Higgs particles themselves have not yet been detected – this is one of the major goals of the Large Hadron Collider (LHC) experiment at CERN, due to begin taking data in 2007. The Higgs fields are *scalar* fields, which have negative energy equations of state, $p = -\rho c^2$. Fields of this nature, associated with phase transitions when the strong force decoupled from the electro-weak force in the early Universe, is a possible candidate for a cosmological negative-energy equation of state.[2]

In their review of the problem of the cosmological constant, Sean Carroll (b. 1966), William Press and Edwin Turner (b. 1949) described how a theoretical value of Ω_Λ could be estimated using simple concepts from quantum field theory. They found the mass density of the repulsive field to be $\rho_v = 10^{95}$ kg m^{-3}, about 10^{120} times greater than permissable values at the present epoch which correspond to $\rho_v \le 10^{-27}$ kg m^{-3} (Carroll, Press and Turner, 1992).[3] This is quite a problem, but it should not be passed over lightly. If the inflationary picture of the very early Universe is taken seriously (see Section 16.4), this is exactly the type of force which drove the inflationary expansion. Then, we have to explain why ρ_v decreased by a factor of about 10^{120} at the end of the inflationary era. In this context, 10^{-120} looks remarkably close to zero, which would correspond to the standard Friedman picture with $\Omega_\Lambda = 0$, but there is now clear evidence that Ω_Λ is finite with value 0.72.

As if these problems were not serious enough, they are compounded by the fact that the natures of the dark matter and the dark energy are unknown. One of the consequences of precision cosmology is the troubling result that we do not understand the nature of about 95% of the material which drives the large-scale dynamics of the Universe. The concordance values for the cosmological parameters listed in Table 15.3 really are extraordinary – many of my colleagues regard them as crazy. Rather than being causes for despair, however, these problems should be seen as the great challenges for the astrophysicists and cosmologists of the twenty-first century. It is not too far-fetched to see an analogy with Bohr's theory of the hydrogen atom, which was an uneasy mix of classical and primitive quantum ideas, but which was ultimately to lead to completely new insights with the development of quantum mechanics.

16.1.6 The way ahead

In the standard Big Bang model, the problems are solved by assuming that the Universe was endowed with appropriate initial conditions in its very early phases. It is postulated that our Universe evolved from an initial state which was isotropic with flat geometry, was slightly matter–antimatter asymmetric, contained fluctuations with a Harrison–Zeldovich spectrum and had initial values of Ω_m and Ω_Λ such that they ended up being roughly equal at the present day. To put it crudely, we get out at the end what we put in at the beginning.

I have suggested five possible approaches to solving these problems (Longair, 1997).

- That is just how the Universe is – the initial conditions were set up that way.
- There are only certain classes of Universe in which intelligent life could have evolved. The Universe has to have the appropriate initial conditions, and the fundamental constants of nature should not be too different from their measured values or else there would be no chance of life forming as we know it. This approach involves the *anthropic cosmological principle*, according to which it is asserted that the Universe is as it is because we are here to observe it.
- The inflationary scenario for the early Universe can be adopted and its consequences studied.
- We should seek clues from particle physics and extrapolate that understanding beyond what has been confirmed by experiment to the earliest phases of the Universe.
- The solution may well turn out to be something else we have not yet thought of. This would certainly involve new physical concepts.

Let us consider each of these approaches.

16.2   The limits of observation

Even the first, somewhat defeatist, approach might be the only way forward if it turned out to be just too difficult to disentangle convincingly the physics responsible for setting up the initial conditions from which our Universe evolved. In 1970, William McCrea considered the fundamental limitations involved in asking questions about the very early Universe, his conclusion being that we can obtain less and less information the further back in time one asks questions about the early Universe (McCrea, 1970). A modern version of this argument would be framed in terms of the limitations imposed by the existence of a last scattering surface for electromagnetic radiation at $z \approx 1000$ and those imposed on the accuracy of observations of the cosmic microwave background radiation and the large-scale structure of the Universe because of their cosmic variances.

In the case of the cosmic microwave background radiation, the observations are already cosmic variance limited for values of spherical harmonic $l \leq 354$ – we will never be able to do any better than what we already know on these scales. Observations by the Planck Satellite will extend the cosmic variance limit to $l \approx 1500$. In these studies, the search for new physics will depend upon discovering discrepancies between the standard concordance model and future observations. The optimists, of whom the present author is one, would argue that the advances will come through extending our technological capabilities so that these classes of observation become cosmic variance limited and new approaches are adopted. For example, the detection of primordial gravitational waves, dark-matter particles and even the nature of the vacuum energy are likely to become the cutting edge of astrophysical cosmology during the twenty-first century. These approaches will be accompanied by discoveries in particle physics with the coming generations of ultra-high-energy particle experiments. There will also be surprises which open up completely new ways of tackling these apparently insuperable problems – for example, what will be discovered in ultra-high-energy cosmic-ray experiments, such as those to be carried out with the Auger array?

It is folly to attempt to predict what will be discovered over the coming years, but we might run out of luck. How would we then be able to check that the theoretical ideas proposed to account for the properties of the very early Universe are correct? Can we do better than boot-strapped self-consistency? The great achievement of modern observational and theoretical cosmology has been that we have made enormous strides in defining a convincing framework for astrophysical cosmology which has been observationally validated, and the big problems can now be addressed as areas of genuine physical enquiry.

16.3 The anthropic cosmological principle

There is certainly some truth in the fact that our ability to ask questions about the origin of the Universe says something about the sort of Universe we live in. The 'cosmological principle' asserts that we do not live at any special location in the Universe, and yet we are certainly privileged in being able to make this statement at all. In this line of reasoning, there are only certain types of Universe in which life as we know it could have formed. For example, the stars must live long enough for there to be time for biological life to form and evolve into sentient beings. This line of reasoning is embodied in the *anthropic cosmological principle*, first expounded by Brandon Carter in 1974 (Carter, 1974) and dealt with *in extenso* in the books by John Barrow (b. 1952) and Frank Tipler (b. 1947) and by John Gribbin (b. 1946) and Martin Rees (Barrow and Tipler, 1986; Gribben and Rees, 1989). Part of the problem stems from the fact that we have only one Universe to study – we cannot go out and investigate other Universes to see if they have evolved in the same way as ours. There are a number of versions of the principle, some of them stronger than others. In extreme interpretations, it leads to statements such as the strong form of the principle enunciated by John Wheeler (Wheeler, 1977),

Observers are necessary to bring the Universe into being.

It is a matter of taste how seriously one wishes to take this line of reasoning. To many cosmologists, it is not particularly appealing because it suggests that it will never be possible to find physical reasons for the initial conditions from which the Universe evolved, or for the values of the fundamental constants of nature. On the other hand, Steven Weinberg (b. 1933) found it such a puzzle that the vacuum energy density, Ω_Λ, is so very much smaller than the values expected according to current theories of elementary particles, that he invoked anthropic reasoning to account for its smallness (Weinberg, 1989, 1997). I prefer to regard the anthropic cosmological principle as the very last resort if all other physical approaches fail.

16.4   The inflationary Universe and clues from particle physics

The most important conceptual development for studies of the very early Universe can be dated to 1980 and the proposal by Alan Guth (b. 1947) of the *inflationary model* for the very early Universe (Guth, 1981).[4] There had been earlier suggestions foreshadowing his proposal. Zeldovich had noted in 1968 that there is a physical interpretation of the cosmological constant, Λ, associated with zero-point fluctuations in a vacuum (Zeldovich, 1968). Andrei Linde (b. 1948) in 1974 and Sidney Bludman (b. 1927) and Malvin Ruderman in 1977 showed that the scalar Higgs fields have similar properties to those which would result in a positive cosmological constant (Linde, 1974; Bludman and Ruderman, 1977).

Guth realised that, if there were an early exponential expansion of the Universe, this could solve the horizon problem and drive the Universe towards a flat spatial geometry, solving the flatness problem at the same time. Suppose the scale factor, R, increased exponentially with time as $R \propto e^{t/T}$. Such exponentially expanding models were found in some of the earliest solutions of the Friedman equations, in the guise of empty de Sitter models, driven by what is now termed the vacuum energy density, Ω_Λ (see Section 6.3 (Lanczos, 1922)). Consider a tiny region of the early Universe expanding under the influence of the exponential expansion. Particles within the region were initially very close together and in causal communication with each other. Before the inflationary expansion began, the region had physical scale less than the particle horizon, and so there was time for it to attain a uniform, homogeneous state. The region then expanded exponentially so that neighbouring points were driven to such large distances that they could no longer communicate by light signals – the causally connected regions were swept beyond their particle horizons by the inflationary expansion. At the end of the inflationary epoch, the Universe transformed into the standard radiation-dominated Universe and the inflated region continued to expand as $R \propto t^{1/2}$. In Guth's original inflationary scenario, the exponential expansion was associated with the symmetry breaking of 'grand unified theories' of elementary particles at very high energies through a first-order phase transition. At high enough energies, the strong and electroweak forces were unified, and only at lower energies did they appear as distinct forces. The grand unification phase transition was expected to take place at a characteristic energy $E \sim 10^{14}$ GeV, known as the GUT scale, only about 10^{-34} s after the Big Bang.

To order of magnitude, the argument ran as follows. The timescale 10^{-34} s was also the characteristic e-folding time for the exponential expansion. Over the interval from 10^{-34} s to 10^{-32} s, the radius of curvature of the Universe increased exponentially by a factor of about $e^{100} \approx 10^{43}$. The horizon scale at the beginning of this period was only $r \approx ct \approx 3 \times 10^{-26}$ m, and this was inflated to a dimension of 3×10^{17} m by the end of the period of inflation. This dimension then scaled as $t^{1/2}$, as in the standard radiation-dominated Universe, so that the region would have expanded to a size of 3×10^{42} m by the present day – this dimension far exceeds the present particle horizon, $r \approx cT_0$, of the Universe, which is about 10^{26} m. Thus, our present Universe would have arisen from a tiny region in the very early Universe which was much smaller than the horizon scale at that time. This guaranteed that our present Universe would be isotropic on the large scale, resolving the horizon problem. At the end of the inflationary era, there was an enormous release of energy associated with the 'latent heat' of the phase transition, and this reheated the Universe to a very high temperature indeed (Figure 16.1).

The exponential expansion also had the effect of straightening out the geometry of the early Universe, however complicated it may have been to begin with. Suppose the tiny region of the early Universe had some complex geometry. The radius of curvature of the geometry $R_c(t)$ scales as $R_c(t) = R_c(t_0)R(t)$, where $R_c(t_0)$ is the radius of curvature of the geometry at the present epoch t_0, and so radius of curvature of the geometry is inflated to dimensions vastly greater than the present size of the Universe, driving the geometry of the inflated region towards flat Euclidean geometry, $\Omega_k = 0$, and consequently the Universe must have $\Omega_0 + \Omega_\Lambda = 1$. It is important that these two aspects of the case for the inflationary picture can be made independently of a detailed understanding of the physics of the inflation.

Guth demonstrated how concepts from theories of elementary particles could lead to a physical realisation of this picture. Although the particles associated with any scalar fields have not yet been detected, scalar fields are now common in theories of elementary particles. The most important of these are the Higgs bosons associated with electroweak unification and which are expected to have mass about 100–200 GeV and which should be discovered in the Large Hadron Collider experiments at CERN. In Guth's picture, the masses of the scalar particles were of the order of 10^{14} GeV and the symmetry breaking took place at a temperature of about $T \sim 10^{27}$ K. An important feature of this picture was that, although occurring at energies far exceeding those accessible by laboratory experiments, the inflation took place long after the Planck era, $t_P \sim 10^{-43}$ s, and so the physics of inflation does not require a fully developed theory of quantum gravity before progress can be made.

In Guth's original proposal, the Universe was in a symmetric state, referred to as a false vacuum state, at a very high temperature before the inflationary phase took place. As the temperature fell, spontaneous symmetry breaking took place through the process of barrier penetration from the false vacuum state and the Universe attained a lower-energy state, the true vacuum. At the end of this period of exponential expansion, the phase transition took place, releasing a huge amount of energy. The problem with this realisation was that

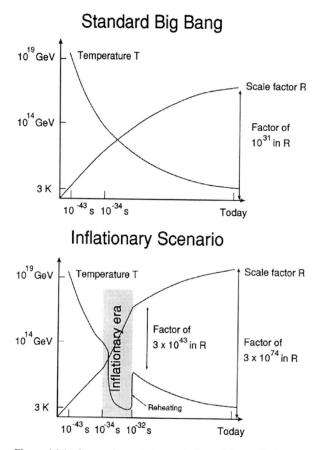

Figure 16.1: Comparison of the evolution of the scale factor and temperature in the standard Big Bang and inflationary cosmologies.

it predicted 'bubbles' of true vacuum embedded in the false vacuum, with the result that huge inhomogeneities were predicted.

Another concern about Guth's original proposal was that an excessive number of monopoles were created during the GUT phase transition. Thomas Kibble (b. 1923) showed that, when this phase transition took place, a variety of topological defects are expected to be created, including point defects (or monopoles), line defects (or cosmic strings) and sheet defects (or domain walls) (Kibble, 1976). Kibble showed that one monopole is created for each correlation scale at that epoch. Since that scale cannot be greater than the particle horizon at the GUT phase transition, it is expected that huge numbers of monopoles were created. According to the simplest picture of the GUT phase transition, the mass density in these monopoles in the standard Big Bang picture would vastly exceed $\Omega_0 = 1$ at the present epoch (Kolb and Turner, 1990).

The model was revised in 1982 by Linde and by Andreas Albrecht (b. 1957), and Paul Steinhardt (b. 1952), who proposed instead that, rather than through the process of barrier penetration, the transition took place through a second-order phase transition, which did not result in the formation of 'bubbles' and so excessive inhomogeneities (Albrecht and Steinhardt, 1982; Linde, 1982, 1983). This picture, often referred to as *new inflation*, also eliminated the monopole problem, since the likelihood of even one being present in the observable Universe was very small.

The original hope that a physical realisation for the inflationary expansion could be found within the context of particle physics beyond the standard model has not been achieved, but the underlying concepts of the inflationary picture have been used to define the necessary properties of the *inflaton* potential needed to create the Universe as we know it. The successful realisations are similar to those involved in the new inflationary picture. Once the inflationary expansion began at some stage in the early Universe, the change from the false to true vacuums states took place through a process of *slow roll-over*, meaning that the inflationary expansion took place over many e-folding times before the huge energy release takes place. An excellent introduction to these concepts and the changing perspective on the inflationary picture of the early Universe is contained in the book *Cosmological Inflation and Large-Scale Structure* by Andrew Liddle (b. 1965) and David Lyth (b. 1940) (Liddle and Lyth, 2000). Many different versions of the inflationary picture of the early Universe have emerged, an amusing table of over 100 possibilities being presented by Paul Shellard (b. 1959) (Shellard, 2003).

16.5 The origin of the spectrum of primordial perturbations

As discussed in the preceding section, there were hopes that an origin for the perturbations from which the large-scale structure of the Universe formed would be found in the phase transitions associated with the end of the inflationary era when the Universe was reheated. Whenever a first-order phase change occurs in nature, such as when water freezes or boils, large fluctuations are found as nucleation takes place. It was soon established that the amplitudes of the perturbations expected according to the old inflation model were huge, far exceeding the amplitude $\delta\rho/\rho \sim 10^{-5}$ necessary for the formation of cosmic structure.

The situation changed dramatically with the development of the new inflationary scenario, in which the phase transition is second-order. The history of the development of the idea that quantum fluctuations in the early Universe could create the observed scale-free spectrum of Harrison–Zeldovich type has been recounted by Guth (Guth, 1997, 2003). Stephen Hawking and Gary Gibbons (b. 1946) worked out the important result that quantum fluctuations in expanding de Sitter space produce thermal radiation with a well defined temperature (Gibbons and Hawking, 1977). This acted as a stimulus to apply similar ideas to the new inflationary picture with a view to estimating the perturbation spectrum. These ideas were thrashed out at the 1982 Nuffield Workshop held in Cambridge (Gibbons, Hawking and Siklos, 1983). The key result of these discussions was the agreement that the spectrum of quantum fluctuations of the vacuum Higgs fields were scale-free and result naturally in adiabatic curvature perturbations with a spectrum strikingly similar to the Harrison–Zeldovich spectrum with $n \approx 1$. For any particular model of the inflaton potential, the vacuum fluctuation spectrum can be determined exactly. This field has become one of the major growth areas of theoretical cosmology.[5] The success of these theories in suggesting a physical origin for the spectrum of initial perturbations is impressive, although the predicted amplitude of the perturbations is model-dependent. According to Liddle and Lyth (2000),

> Although introduced to resolve problems associated with the initial conditions needed for the Big Bang cosmology, inflation's lasting prominence is owed to a property discovered soon after its introduction. It provides a possible explanation for the initial inhomogeneities in the Universe that are believed to have led to all the structures we see, from the earliest objects formed to the clustering of galaxies to the observed irregularities in the microwave background.

At the same time, we should not neglect the possibility that there are other sources of perturbations resulting from various types of topological defect, such as cosmic strings, domain walls, textures and so on (Shellard, 2003). However, the startling success of inflationary ideas in accounting for the observed spectrum of fluctuations in the cosmic microwave background radiation has made it the model of choice for studies of the early Universe.

16.6 Baryogenesis

A key contribution of particle physics to the physics of the early Universe concerns the baryon-asymmetry problem, a subject referred to as *baryogenesis*. In a prescient paper of 1967, Andrei Sakharov enunciated the three conditions necessary to account for the baryon–antibaryon asymmetry of the Universe (Sakharov, 1967). *Sakharov's rules* for the creation of non-zero baryon number from an initially baryon-symmetric state are:

- *baryon number* must be violated;
- C (charge conjugation) and CP (charge conjugation combined with parity) must be violated;
- the asymmetry must be created under *non-equilibrium conditions*.

The reasons for these rules can be readily appreciated from simple arguments (Kolb and Turner, 1990). Concerning the first rule, it is evident that, if the baryon asymmetry developed from a symmetric high-temperature state, baryon number must have been violated at some stage – otherwise, the baryon asymmetry would have to be built into the model from the very beginning. The second rule is necessary in order to ensure that a net baryon number is created, even in the presence of interactions which violate baryon conservation. The third rule is necessary because baryons and antibaryons have the same mass and so, thermodynamically, they would have the same abundances in thermodynamic equilibrium, despite the violation of baryon number and C and CP invariance.

There is evidence that all three rules can be satisfied in the early Universe from a combination of theoretical ideas and experimental evidence from particle physics. Thus, baryon number violation is a generic feature of grand unified theories, which unify the strong and electroweak interactions – the same process is responsible for the predicted instability of the proton. C and CP violation have been observed in the decay of the neutral K^0 and $\bar{K}^0$ mesons. The K^0 meson should decay symmetrically into equal numbers of particles and antiparticles, but, in fact, there is a slight preference for matter over antimatter, at the level 10^{-3}, very much greater than the degree of asymmetry necessary for baryogenesis, $\sim 10^{-8}$. The need for departure from thermal equilibrium follows from the same type of reasoning that led to the primordial synthesis of the light elements. As in that case, so long as the timescales of the interactions which maintained the various constituents in thermal equilibrium were less than the expansion timescale, the number densities of particles and antiparticles of the same mass would be the same. In thermodynamic equilibrium, the number densities of different species did not depend upon the cross-sections for the interactions which maintain the equilibrium. It is only after decoupling, when non-equilibrium abundances were established, that the number densities depended upon the specific values of the cross-sections for the production of different species.

In a typical baryogenesis scenario, the asymmetry is associated with some very massive boson and its antiparticle, X, $\overline{X}$, which are involved in the unification of the strong and electroweak forces and which can decay into final states which have different baryon numbers. Edward (Rocky) Kolb (b. 1951) and Michael Turner (b. 1949) provide a clear description of the principles by which the observed baryon asymmetry can be generated at about the epoch of grand unification, or soon afterwards, when the very massive bosons could no longer be maintained in equilibrium. Although the principles of the calculations are well defined, the details are not understood, partly because the energies at which they are likely to be important are not attainable in laboratory experiments, and partly because predicted effects, such as the decay of the proton, have not been observed. Thus, although there is no definitive evidence that this line of reasoning is secure, well understood physical processes of the type necessary for the creation of the baryon–antibaryon asymmetry exist. The importance of these studies goes well beyond their immediate significance for astrophysical cosmology. As Kolb and Turner (1990) remark,

in the absence of direct evidence for proton decay, baryogenesis may provide the strongest, albeit indirect, evidence for some kind of unification of the quarks and the leptons.

16.7 The Planck era

It should be borne in mind that there is no evidence for the inflationary picture beyond the need to solve the big problems listed in Section 16.1. Enormous progress has been made in understanding the types of physical process necessary to resolve the four great problems, but it is not clear how independent evidence for them can be found. The methodological problem with these ideas is that they are based upon extrapolations to energies vastly exceeding those which can possibly be tested in terrestrial laboratories. Cosmology and particle physics come together in the early Universe and they boot-strap their way to a self-consistent solution. This may be the best that we can hope for, but it would be preferable to have independent constraints upon the theories.

A representation of the evolution of the Universe from the Planck era to the present day is shown in Figure 16.2. The *Planck era* is that time in the very remote past when the energy

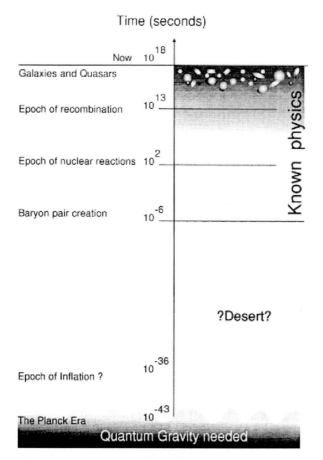

Figure 16.2: A schematic diagram illustrating the evolution of the Universe from the Planck era to the present time. The shaded area to the right of the diagram indicates the regions of known physics.

densities were so great that a quantum theory of gravity is needed. On dimensional grounds, this era must have occurred when the Universe was only about $t_P \sim (hG/c^5)^{1/2} \sim 10^{-43}$ s old. Despite enormous efforts on the part of theorists, there is no quantum theory of gravity, and so we can only speculate about the physics of these extraordinary eras.[6]

Being drawn on a logarithmic scale, Figure 16.2 encompasses the evolution of the whole of the Universe, from the Planck area at $t \sim 10^{-43}$ s to the present age of the Universe, which is about 4×10^{17} s, or 13×10^9 years, old. Halfway up the diagram, from the time when the Universe was only about one millisecond old, to the present epoch, we can be reasonably confident that we have the correct picture for the Big Bang, despite the basic problems discussed above.

At times earlier than about one millisecond, we quickly run out of known physics. This has not discouraged theorists from making bold extrapolations across the huge gap from 10^{-3} s to 10^{-43} s using the current understanding of particle physics and concepts from string theories. Some impression of the types of thinking involved in these studies can be found in the ideas expounded in Gibbons et al. (2003). Maybe many of these ideas will turn out to be correct, but there must be some concern that some fundamentally new physics may emerge at higher and higher energies before we reach the GUT era at $t \sim 10^{-36}$ s and the Planck era at $t \sim 10^{-43}$ s. This is why the particle physics experiments to be carried with the Large Hadron Collider at CERN are of such importance for astrophysics and cosmology, as well as for particle physics. It is fully expected that definite evidence will be found for the Higgs boson. In addition, there is the possibility of discovering new types of particles, such as the lightest supersymmetric particle or new massive ultra-weakly interacting particles, as the accessible range of particle energies increases from about 100 GeV to 1 TeV. These experiments should provide clues to the nature of physics beyond the standard model of particle physics and will undoubtedly feed into the understanding of the physics of the early Universe.

It is certain that at some stage a quantum theory of gravity is needed which may help resolve the problems of singularities in the early Universe. The singularity theorems of Roger Penrose and Stephen Hawking show that, according to classical theories of gravity under very general conditions, there is inevitably a physical singularity at the origin of the Big Bang, that is as $t \rightarrow 0$, the energy density of the Universe tends to infinity. However, it is not clear that the actual Universe satisfies the various energy conditions required by the singularity theorems. All these considerations show that new physics is needed if we are to develop a convincing physical picture of the very early Universe.

Notes to Chapter 16

1 The physics of the early Universe is splendidly described in Kolb and Turner (1990).
2 In the modern picture of the vacuum, there are zero-point fluctuations associated with the zero-point energies of all quantum fields. The stress–energy tensor of a vacuum has a negative-energy equation of state, $p = -\rho c^2$. This pressure may be thought of as a 'tension' rather than a pressure. When such a vacuum expands, the work done, p dV, in expanding from V to $V + $ dV is given by p d$V = -\rho c^2$ dV, so that, during the expansion, the mass density of the negative-energy field remains constant. The same result can be found from application of the first law of thermodynamics for a relativistic fluid in a cosmological setting,

$$\frac{\mathrm{d}\rho}{\mathrm{d}R} + 3\frac{\left(\rho + \dfrac{p}{c^2}\right)}{R} = 0.$$

(See Malcolm Longair, *Galaxy Formation* (Berlin: Springer-Verlag, 1998).) If the vacuum energy density is to remain constant, $\rho_{\mathrm{vac}} = $ constant, it follows from this equation that $p = -\rho c^2$.
3 John Peacock sets this as example 7.5 in his book *Cosmological Physics* (Cambridge: Cambridge University Press, 1999). Heisenberg's uncertainty principle states that a virtual pair of particles of mass m can exist for a time $t \sim \hbar/mc^2$, corresponding to a maximum separation $x \sim \hbar/mc$. Hence, the typical density of the vacuum fields is given by $\rho \sim m/x^3 \approx c^3 m^4/\hbar^3$. The mass density in the vacuum fields is unchanging with cosmic epoch and so, adopting the Planck mass for m (see Table 15.1), the mass density corresponds to about 10^{97} kg m^{-3}.
4 A popular account of the history of the development of ideas about the inflationary picture of the early Universe is contained in Alan Guth's book *The Inflationary Universe: The Quest for a New Theory of Cosmic Origins* (Guth, 1997). The pedagogical review published by Charles Lineweaver (b. 1954) is very readable and can also be strongly recommended. He adopts a healthily sceptical

attitude to the concept of inflation and our ability to test inflationary models through confrontation with observations (Lineweaver, 2005).

5 An excellent survey of these important ideas is contained in the book *Cosmological Inflation and Large-Scale Structure* by Liddle and Lyth (2000).

6 An excellent survey of many of these ideas is contained within the volume celebrating Stephen Hawking's 60th birthday, *The Future of Theoretical Physics and Cosmology* (Gibbons, Shellard and Rankin, 2003).

References

Aaronson, M. and Mould, J. (1983). A distance scale from the infrared magnitude/H I velocity–width relation. IV – The morphological type dependence and scatter in the relation; the distances to nearby groups, *Astrophysical Journal*, **265**, 1–17.

Abbot, C. G. (1924). Radiometer observations of stellar energy spectra, *Astrophysical Journal*, **60**, 87–107.

Abdurashitov, J. N., Bowles, T. J., Cleveland, B. T. *et al.* (2002). Results of the SAGE experiment, *Journal of Experimental and Theoretical Physics*, **95**, 181–193.

Abdurashitov, J. N., Bowles, T. J., Cleveland, B. T. *et al.* (2003). Measurement of the solar neutrino capture rate, *Nuclear Physics B Proceedings Supplements*, **118**, 39–46.

Abell, G. O. (1958). The distribution of rich clusters of galaxies, *Astrophysical Journal Supplement*, **3**, 221–288.

(1962). Membership of clusters of galaxies, in *Problems of Extragalactic Research*, ed. McVittie, G. C. (New York: Macmillan), pp. 213–238.

Abell, G. O., Corwin Jr, H. G. and Olowin, R. P. (1989). A catalogue of rich clusters of galaxies, *Astrophysical Journal Supplement*, **70**, 1–138.

Abraham, R. G., Tanvir, N. R., Santiago, B., Ellis, R. S., Glazebrook, K. and van den Bergh, S. (1996). Galaxy morphology to $I = 25$ mag in the Hubble Deep Field, *Monthly Notices of the Royal Astronomical Society*, **279**, L47–L52.

Abramowicz, M. A., Jaroszyński, M. and Sikora, M. (1978). Relativistic, accreting disks, *Astronomy and Astrophysics*, **63**, 221–224.

Adams, F. C., Lada, C. J. and Shu, F. H. (1987). Spectral evolution of young stellar objects, *Astrophysical Journal*, **312**, 788–806.

Adams, W. S. (1914). An A-type star of very low luminosity, *Publications of the Astronomical Society of the Pacific*, **26**, 198.

(1915). The spectrum of the companion of Sirius, *Publications of the Astronomical Society of the Pacific*, **27**, 236–237.

(1925a). The relativity displacement of the spectral lines in the companion of Sirius, *Proceedings of the National Academy of Sciences*, **11**, 382–387.

(1925b). The relativity displacement of the spectral lines in the companion of Sirius, *Observatory*, **48**, 336–342.

Adams, W. S. and Joy, A. H. (1919). The motions in space of some stars of high radial velocity, *Astrophysical Journal*, **49**, 179–185.

Adams, W. S. and Kohlschütter, A. (1914a). The radial velocities of one hundred stars with measured parallaxes, *Astrophysical Journal*, **39**, 341–349.

Adams, W. S. and Kohlschütter, A. (1914b). Some spectral criteria for the determination of absolute stellar magnitudes, *Astrophysical Journal*, **40**, 385–398.

Afonso, C., Albert, J. N., Andersen, J. *et al.* (2003). Limits on Galactic dark matter with 5 years of EROS SMC data, *Astronomy and Astrophysics*, **400**, 951–956.

Ahmad, Q. R., Allen, R. C., Andersen, T. C. *et al.* (2001). Measurement of the rate of $\nu_e + d \to p + p + e^-$ interactions produced by ^8B solar neutrinos at the Sudbury Neutrino Observatory, *Physical Review Letters*, **87**, 071301(1–6).

Aitken, D. K., Smith, C. H., James, S. D., Roche, P. F., Hyland, A. R. and McGregor, P. J. (1988). 10 micron spectral observations of SN 1987A – the first year, *Monthly Notices of the Royal Astronomical Society*, **235**, 19P–31P.

Akerib, D. S., Alvaro-Dean, J., Armel-Funkhouser *et al.* (2004). First results from the cryogenic dark matter search in the Soudan Underground Lab, *Physical Review Letters*, **93**, 211301(1–5).

Albrecht, A. and Steinhardt, P. J. (1982). Cosmology for grand unified theories with radiatively induced symmetry breaking, *Physical Review Letters*, **48**, 1220–1223.

Alcock, C., Akerlof, C. W., Allsman, R. A. *et al.* (1993a). Possible gravitational microlensing of a star in the Large Magellanic Cloud, *Nature*, **365**, 621–623.

Alcock, C., Allsman, R. A., Axelrod, T. S. *et al.* (1993b). The MACHO project – a search for the dark matter in the Milky-Way, in *Sky Surveys: Protostars to Protogalaxies*, ed. Soifer, T. (San Francisco: Astronomical Society of the Pacific Conference Series), pp. 291–296.

Alcock, C., Allsman, R. A., Alves, D. R. *et al.* (2000). The MACHO project: microlensing results from 5.7 years of Large Magellanic Cloud observations, *Astrophysical Journal*, **542**, 281–307.

Alfvén, H. and Herlofson, N. (1950). Cosmic radiation and radio stars, *Physical Review*, **78**, 616.

Alfvén, H. and Klein, O. (1962). Matter–antimatter annihilation and cosmology, *Arkiv für Fyzik*, **23**, 187–194.

Alpher, R. A. and Herman, R. C. (1948). Evolution of the Universe, *Nature*, **162**, 774–775. (1950). Theory of the origin and relative distribution of the elements, *Reviews of Modern Physics*, **22**, 153–212.

Alpher, R. A., Bethe, H. and Gamow, G. (1948). The origin of the chemical elements, *Physical Review*, **73**, 803–804.

Alpher, R. A., Follin, J. W. and Herman, R. C. (1953). Physical conditions in the initial stages of the expanding Universe, *Physical Review*, **92**, 1347–1361.

Ambartsumian, V. A. (1947). *Stellar Evolution and Astrophysics* (Yerevan: Armenian Academy of Sciences).

Anders, E. (1963). Meteorite ages, in *The Moon, Meteorites and Comets – The Solar System IV*, eds Middlehurst, B. M. and Kuiper, G. P. (Chicago: University of Chicago Press), pp. 402–495.

Andersen, M. I., Hjorth, J., Pedersen, H. *et al.* (2000). VLT identification of the optical afterglow of the gamma-ray burst GRB 000131 at $z = 4.50$, *Astronomy and Astrophysics*, **364**, L54–L61.

Anderson, C. D. (1932). The apparent existence of easily deflected positives, *Science*, **76**, 238–239.

Anderson, C. D. and Neddermeyer, S. H. (1936). Cloud chamber observations of cosmic rays at 4300 metres elevation and near sea-level, *Physical Review*, **50**, 263–271.

Anderson, W. (1929). Gewöhnliche Materie und Strahlende Energie als Verschiedene 'Phasen' eines und Desselben Grundstoffes (Ordinary matter and radiation energy as different phases of the same underlying matter), *Zeitschrift für Physik*, **54**, 433–444.

Ando, H. and Osaki, Y. (1975). Nonadiabatic nonradial oscillations: an application to the five-minute oscillation of the Sun, *Publications of the Astronomical Society of Japan*, **27**, 581–603.

Antonucci, R. R. (1993). Unified models for active galactic nuclei and quasars, *Annual Review of Astronomy and Astrophysics*, **31**, 473–521.

Antonucci, R. R. and Miller, J. S. (1985). Spectropolarimetry and the nature of NGC 1068, *Astrophysical Journal*, **297**, 621–632.

Aragòn-Salamanca, A., Ellis, R. S., Couch, W. J. and Carter, D. (1993). Evidence for systematic evolution in the properties of galaxies in distant clusters, *Monthly Notices of the Royal Astronomical Society*, **262**, 764–794.

Archer, F. S. (1851). On the use of collodion in photography, *The Chemist*, **2** (March) 257–258.

Argelander, H. (1838). Ueber die eigene Bewegung des Sonnensystems (On the proper motion of the Solar System), *Astronomische Nachrichten*, **16**, 45–48.

Arnett, W. D. and Clayton, D. D. (1970). Explosive nucleosynthesis in stars, *Nature*, **227**, 780–784.

Arp, H. C. (1966). *Atlas of Peculiar Galaxies* (Pasadena: California Institute of Technology).

Arp, H. C., Baum, W. A. and Sandage, A. R. (1952). The HR diagrams for the globular clusters M92 and M3, *Astronomical Journal*, **57**, 4–5.

Arp, H. C., Madore, B. F. and Roberton, W. E. (1987). *A Catalogue of Southern Peculiar Galaxies and Associations* (Cambridge: Cambridge University Press).

Atkinson, R. d'E. (1931a). Atomic synthesis and stellar energy I, *Astrophysical Journal*, **73**, 250–295.

   (1931b). Atomic synthesis and stellar energy II, *Astrophysical Journal*, **73**, 308–347.

   (1936). Atomic synthesis and stellar energy III, *Astrophysical Journal*, **84**, 73–84.

Atkinson, R. d'E. and Houtermans, F. G. (1929). Zur Frage der Aufbaumöglichkeit der Elemente in Sternen (On the possible synthesis of the elements in stars), *Zeitschrift für Physik*, **54**, 656–665.

Auger, P., Ehrenfest Jr, P., Maze, R., Daudin, J., Robley, X. and Fréon, A. (1939). Extensive air showers, *Reviews of Modern Physics*, **11**, 288–291.

Avery, L. W., Broten, N. W., Macleod, J. M., Oka, T. and Kroto, H. W. (1976). Detection of the heavy interstellar molecule cyanodiacetylene, *Astrophysical Journal*, **205**, L173–L175.

Axford, W. I., Leer, E. and Skadron, G. (1977). The acceleration of cosmic rays by shock waves, *Proceedings of the 15th International Cosmic Ray Conference*, **11**, 132–135.

Baade, W. A. (1926). Über eine Möglichkeit, die Pulsationstheorie der δ-Cephei-Veränderlichen zu Prüfen (On a possible method of testing the pulsation theory of the variations of δ-Cephei), *Astronomische Nachrichten*, **228**, 359–362.

(1944). The resolution of Messier 32, NGC 205, and the central region of the Andromeda Nebula, *Astrophysical Journal*, **100**, 137–146.

(1951). Galaxies – present day problems, *Publications of the Observatory of the University of Michigan*, **10**, 7–17.

(1952). A revision of the extra-galactic distance scale, *Transactions of the International Astronomical Union*, **8**, 397–398.

(1956). Polarization in the jet of Messier 87, *Astrophysical Journal*, **123**, 550–551.

Baade, W. A. and Minkowski, R. (1954). Identification of the radio sources in Cassiopeia, Cygnus A, and Puppis A, *Astrophysical Journal*, **119**, 206–214.

Baade, W. A. and Zwicky, F. (1934a). On super-novae, *Proceedings of the National Academy of Sciences*, **20**, 254–259.

(1934b). Cosmic rays from super-novae, *Proceedings of the National Academy of Sciences*, **20**, 259–263.

(1938). Photographic light-curves of the two supernovae in IC 4182 and NGC 1003, *Astrophysical Journal*, **88**, 411–421.

Bahcall, J. N. (1964). Solar neutrinos. I. Theoretical, *Physical Review Letters*, **12**, 300–302.

(1989). *Neutrino Astrophysics* (Cambridge: Cambridge University Press).

Bahcall, J. N. and Bethe, H. (1990). A solution of the solar neutrino problem, *Physical Review Letters*, **65**, 2233–2235.

Bahcall, J. N. and Davis Jr, R. (1976). Solar neutrinos: a scientific puzzle, *Science*, **191**, 264–267.

Bahcall, J. N. and Ulrich, R. (1988). Solar models, neutrino experiments and helioseismology, *Reviews of Modern Physics*, **60**, 297–372.

Bahcall, J. N., Greenstein, J. L. and Sargent, W. L. W. (1968). The absorption-line spectrum of the quasi-stellar radio source PKS 0237-23, *Astrophysical Journal*, **153**, 689–698.

Bahcall, J. N., Pinsonneault, M. H., Basu, S. and Christensen-Dalsgaard, J. (1997). Are standard solar models reliable?, *Physical Review Letters*, **78**, 171–174.

Bahcall, N. A. (2000). Clusters and cosmology, *Physics Reports*, **333**, 233–244.

Balbus, S. A. and Hawley, J. F. (1991). A powerful local shear instability in weakly magnetized Disks. I – Linear analysis. II – Nonlinear evolution, *Astrophysical Journal*, **376**, 214–233.

Baldwin, J. E., Beckett, M. G., Boysen, R. C. *et al.* (1996). The first images from an optical aperture synthesis array: mapping of Capella with COAST at two epochs, *Astronomy and Astrophysics*, **306**, L13–L16.

Balmer, J. J. (1885). Note on the spectral lines of hydrogen, *Annalen der Physik und Chemie*, **25**, 80–87.

Bardeen, J. M. (1970). Kerr metric black holes, *Nature*, **226**, 64–65.

(1980). Gauge-invariant cosmological perturbations, *Physical Review D*, **22**, 1882–1905.

Bardeen, J. M., Bond, J. R., Kaiser, N. and Szalay, A. S. (1986). The statistics of peaks of gaussian random fields, *Astrophysical Journal*, **304**, 15–61.

Barkla, C. G. (1906). Polarisation of secondary Röntgen radiation, *Proceedings of the Royal Society of London*, **A77**, 247–255.

Barnard, E. E. (1919). On the dark markings of the sky. With a catalogue of 182 such objects, *Astrophysical Journal*, **49**, 1–24.

Barrow, J. D. and Tipler, F. J. (1986). *The Anthropic Cosmological Principle* (Oxford: Oxford University Press).

Barthel, P. D. (1989). Is every quasar beamed?, *Astrophysical Journal*, **336**, 606–611.

(1994). Unified schemes of FR2 radio galaxies and quasars, in *First Stromlo Symposium: Physics of Active Galactic Nuclei*, ASP Conference Series, vol. 54, eds Bicknell, G. V., Dopita, M. A. and Quinn, P. J. (San Francisco: ASP), pp. 175–186.

Baum, W. A., Johnson, F. S., Oberly, J. J., Rockwood, C. C., Strain, C. V. and Tousey, R. (1946). Solar ultraviolet spectrum to 88 kilometers, *Physical Review*, **70**, 781–782.

Bautz, L. and Morgan, W. W. (1970). On the classification of the forms of clusters of galaxies, *Astrophysical Journal Letters*, **162**, L149–L153.

Baym, G., Pethick, C. and Pines, D. (1969a). Superfluidity in neutron stars, *Nature*, **224**, 673–674.

Baym, G., Pethick, C., Pines, D. and Ruderman, M. (1969b). Spin up in neutron stars: the future of the Vela pulsar, *Nature*, **224**, 872–874.

Baym, G., Bethe, H. A. and Pethick, C. J. (1971a). Neutron star matter, *Nuclear Physics A*, **175**, 225–271.

Baym, G., Pethick, C. J. and Sutherland, P. (1971b). The ground state of matter at high densities: equation of state and stellar models, *Astrophysical Journal*, **170**, 299–317.

Becker, R. H., Fan, X., White, R. L. *et al.* (2001). Evidence for reionisation at $z \sim 6$: detection of a Gunn–Peterson trough in a $z = 6.28$ quasar, *Astronomical Journal*, **122**, 2850–2857.

Becklin, E. E. and Neugebauer, G. (1967). Observations of an infrared star in the Orion Nebula, *Astrophysical Journal*, **147**, 799–802.

(1968). Infrared observations of the Galactic center, *Astrophysical Journal*, **151**, 145–161.

Becquerel, A. E. (1842). Mémoire sur la constitution du spectre solaire (Memoir concerning the constitution of the solar spectrum), *Bibliothéque Universelle de Gèneve*, **40**, 341–367.

Becquerel, H. (1896). Sur les radiations invisibles émises par les corps phosphorescents (On the invisible radiation emitted by phosphorescent bodies), *Comptes Rendus de l'Academie des Sciences*, **122**, 501–503.

Bell, A. R. (1978). The acceleration of cosmic rays in shock fronts. I, *Monthly Notices of the Royal Astronomical Society*, **182**, 147–156.

Bennett, A. S. (1962). The Revised 3C Catalogue of radio sources, *Memoirs of the Royal Astronomical Society*, **67**, 163–172.

Bennett, C. L., Banday, A. J., Górski, K. M. *et al.* (1996). Four-year COBE DMR cosmic microwave background observations: maps and basic results, *Astrophysical Journal*, **464**, L1–L4.

Bennett, C. L., Halpern, M., Hinshaw, G. *et al.* (2003). First-year Wilkinson Microwave Anisotropy Probe (WMAP) observations: preliminary maps and basic results, *Astrophysical Journal Supplement Series*, **148**, 1–27.

Benôit, A., Ade, P., Amblard, A. *et al.* (2003a). The cosmic microwave background anisotropy power spectrum measured by Archeops, *Astronomy and Astrophysics*, **399**, L19–L23.

(2003b). Cosmological constraints from Archeops, *Astronomy and Astrophysics*, **399**, L25–L30.

Benson, B. A., Church, S. E., Ade, P. A. R. *et al.* (2003). Peculiar velocity limits from measurements of the spectrum of the Sunyaev–Zel'dovich effect in six clusters of galaxies, *Astrophysical Journal*, **592**, 674–691.

Bergeron, J. (1988). Properties of the heavy-element absorption systems, in *QSO Absorption Lines: Probing the Universe*, eds Blades, J. C., Turnshek, D. and Norman, C. A. (Cambridge: Cambridge University Press), pp. 127–143.

Bersanelli, M., Bouchet, F. R., Efstathiou, G. *et al.* (1995). *Phase A Study for the Cobras/Samba Mission* (Paris: European Space Agency), D/SCI(96)3.

Bertola, F. and Capaccioli, M. (1975). Dynamics of early type galaxies. I – The rotation curve of the elliptical galaxy NGC 4697, *Astrophysical Journal*, **200**, 439–445.

Bertola, F. and Galletta, G. (1979). Ellipticity and twisting of isophotes in elliptical galaxies, *Astronomy and Astrophysics*, **77**, 363–365.

Bertola, F., Bettoni, D., Danziger, J., Sadler, E., Spark, L. and de Zeeuw, T. (1991). Testing the gravitational field in elliptical galaxies: NGC 5077, *Astrophysical Journal*, **373**, 369–390.

Bessel, F. W. (1839). Bestimmung der Entfernung des 61sten Sterne des Schwans (Determination of the distance of 61-Cygni), *Astronomische Nachrichten*, **16**, 64–96.

Best, P. N., Longair, M. S. and Röttgering, H. J. A. (1996). Evolution of the aligned structures in $z \sim 1$ radio galaxies, *Monthly Notices of the Royal Astronomical Society*, **280**, L9–L12.

(1998). HST, radio and infrared observations of 28 3CR radio galaxies at redshift $z \sim 1$. II – Old stellar populations in central cluster galaxies, *Monthly Notices of the Royal Astronomical Society*, **295**, 549–567.

(2000). Ionization, shocks and evolution of the emission-line gas of distant 3CR radio galaxies, *Monthly Notices of the Royal Astronomical Society*, **311**, 23–36.

Bethe, H. A. (1939). Energy production in stars, *Physical Review*, **55**, 434–456.

Bethe, H. A. and Critchfield, C. L. (1938). The formation of deuterons by proton combination, *Physical Review*, **54**, 248–254.

Binney, J. (1978). On the rotation of elliptical galaxies, *Monthly Notices of the Royal Astronomical Society*, **183**, 501–514.

Birkinshaw, M. (1990). Observations of the Sunyaev–Zeldovich effect, in *The Cosmic Microwave Background: 25 Years Later*, eds Mandolesi, N. and Vittorio, N. (Dordrecht: Kluwer Academic Publishers), pp. 77–94.

Blackett, P. M. S. (1925). The ejection of protons from nitrogen nuclei, photographed by the Wilson method, *Proceedings of the Royal Society of London*, **A107**, 349–360.

Blackett, P. M. S. (1948). A possible contribution to the light of the night sky from the Cherenkov radiation emitted by cosmic rays, in *The Emission Spectra of the Night Sky and Aurorae, Gassiot Committee Report* (London: Physical Society of London), pp. 34–35.

Blackett, P. M. S. and Occhialini, G. P. S. (1933). Some photographs of the tracks of penetrating radiation, *Proceedings of the Royal Society of London*, **A139**, 699–722.

Blain, A. W. and Longair, M. S. (1993). Sub-millimetre cosmology, *Monthly Notices of the Royal Astronomical Society*, **264**, 509–521.

Blandford, R. D. and Ostriker, J. P. (1978). Particle acceleration by astrophysical shocks, *Astrophysical Journal*, **221**, L29–L32.

Blandford, R. D. and Znajek, R. L. (1977). Electromagnetic extraction of energy from Kerr black holes, *Monthly Notices of the Royal Astronomical Society*, **179**, 433–456.

Blewett, J. P. (1946). Radiation losses in the induction electron accelerator, *Physical Review*, **69**, 87–95.

Bludman, S. A. and Ruderman, M. A. (1977). Induced cosmological constant expected above the phase transition restoring the broken symmetry, *Physical Review Letters*, **38**, 255–257.

Blumenthal, G. R., Faber, S. M., Primack, J. R. and Rees, M. J. (1984). Formation of galaxies and large-scale structure with cold dark matter, *Nature*, **311**, 517–525.

Bohr, N. (1913a). On the constitution of atoms and molecules: Part 1, On the constitution of atoms and molecules, *Philosophical Magazine, Sixth Series*, **23**, 1–25.

(1913b). On the constitution of atoms and molecules: Part 2, Systems containing only a single nucleus, *Philosophical Magazine, Sixth Series*, **23**, 476–502.

(1913c). On the constitution of atoms and molecules: Part 3, Systems containing several nuclei, *Philosophical Magazine, Sixth Series*, **23**, 857–875.

Böhringer, H. (1994). Clusters of galaxies, in *Frontiers of Space and Ground-based Astronomy*, eds Wamsteker, W., Longair, M. S. and Kondo, Y. (Dordrecht: Kluwer Academic Publishers), pp. 359–368.

Boksenberg, A. (1997). Quasar absorption lines: reflections and views, in *The Hubble Space Telescope and the High Redshift Universe*, eds Tanvir, N. R., Aragón-Salamanca, A. and Wall, J. V. (Singapore: World Scientific Publishing Company), pp. 283–294.

Bolte, M. (1997). Globular clusters: old, in *Critical Dialogues in Cosmology*, ed. Turok, N. (Singapore: World Scientific), pp. 156–168.

Bolton, C. T. (1972). Identifications of Cyg X-1 with HDE 226868, *Nature*, **235**, 271–273.

Bolton, J. and Stanley, G. J. (1949). The position and probable identification of the source of galactic radio-frequency radiation Taurus A, *Australian Journal of Scientific Research*, **A2**, 139–148.

Bolton, J. G., Stanley, G. J. and Slee, O. B. (1949). Positions of three discrete sources of galactic radio-frequency radiation, *Nature*, **164**, 101–102.

Bolyai, J. (1832). Appendix: Scientiam spatii absolute veritam exhibens (Appendix explaining the absolutely true science of space). Published as an appendix to the essay by his father, F. Bolyai, *An Attempt to Introduce Studious Youth to the Elements of Pure Mathematics* (Maros Vásárhely, Transylvania).

Bondi, H. and Gold, T. (1948). The steady-state theory of the expanding Universe, *Monthly Notices of the Royal Astronomical Society*, **108**, 252–270.

Bosma, A. (1981). 21-cm line studies of spiral galaxies II. The distribution and kinematics of neutral hydrogen in spiral galaxies of various morphological types, *Astronomical Journal*, **86**, 1825–1846.

Boss, B. (1918). Real stellar motions (abstract), *Popular Astronomy*, **26**, 686.

Bothe, W. and Kolhörster, W. (1929). The nature of high-altitude radiation, *Zeitschrift für Physik*, **56**, 751–777.

Bowen, I. S. (1927). The origin of the chief nebular lines, *Publications of the Astronomical Society of the Pacific*, **39**, 295–297.

   (1928). The origin of the nebular lines and the structure of the planetary nebulae, *Astrophysical Journal*, **67**, 1–15.

Bowyer, C. S., Field, G. B. and Mack, J. E. (1968). Detection of an anisotropic soft X-ray background, *Nature*, **217**, 32–34.

Bowyer, S., Byram, E. T., Chubb, T. A. and Friedman, H. (1964). Lunar occulation of X-ray emission from the Crab Nebula, *Science*, **146**, 912–917.

Boyle, B. J., Jones, L. R., Shanks, T., Marano, B., Zitelli, V. and Zamorani, G. (1991). QSO evolution and clustering at $z < 2.9$, *Proceedings of the Workshop on The Space Distribution of Quasars: Astronomical Society of the Pacific Conference Series*, **21**, 191–201.

Boyle, B. J., Shanks, T., Croom, S. M. *et al.* (2000). The 2dF QSO redshift survey – I. The optical luminosity function of quasi-stellar objects, *Monthly Notices of the Royal Astronomical Society*, **317**, 1014–1022.

Boyle, W. S. and Smith, G. E. (1970). Charge coupled semiconductor devices, *Bell System Technical Journal*, **49**, 587–593.

Braccesi, A., Formiggini, L. and Gandolfi, E. (1970). Magnitudes, colours and coordinates of 175 ultraviolet excess objects in the field 13^{h}, $+36°$, *Astronomy and Astrophysics*, **5**, 264–279.

Bracewell, R. N., ed. (1959). *Paris Symposium on Radio Astronomy* (Stanford: Stanford University Press).

Bracewell, R. N. and Roberts, J. A. (1954). Aerial smoothing in radio astronomy, *Australian Journal of Physics*, **7**, 615–640.

Bradley, J. (1728). An account of a new discovered motion of the fixed stars, *Philosophical Transactions of the Royal Society*, **35**, 637–661.

Bradt, H. L. and Peters, B. (1950). Abundance of lithium, beryllium, boron and other light nuclei in the primary cosmic radiation and the problem of cosmic-ray origin, *Physical Review*, **80**, 943–953.

Braes, L. L. E. and Miley, G. K. (1971). Radio emission from Scorpius X-1 at 21.2 cm, *Astronomy and Astrophysics*, **14**, 160–163.

Branch, D. and Patchett, B. (1973). Type I supernovae, *Monthly Notices of the Royal Astronomical Society*, **161**, 71–83.

Branch, D. and Tammann, G. A. (1992). Type I supernovae as standard candles, *Annual Review of Astronomy and Astrophysics*, **30**, 359–389.

Brans, C. and Dicke, R. H. (1961). Mach's principle and a relativistic theory of gravitation, *Physical Review*, **124**, 925–935.

Brooks, J. R., Isaak, G. R. and van der Raay, H. B. (1976). Observatons of free oscillations of the Sun, *Nature*, **259**, 92–95.

Broten, N. W., Legg, T. H., Locke, J. L. *et al.* (1967). Radio interferometry with a baseline of 3074 km, *Astrophysical Journal*, **72**, 787–800.

Broten, N. W., Oka, T., Avery, L. W., Macleod, J. M. and Kroto, H. W. (1978). The detection of HC_9N in interstellar space, *Astrophysical Journal*, **223**, L105–L107.

Brown, T. M., Stebbins, R. T. and Hill, H. A. (1976). Observed oscillations of the apparent solar diameter, in *Solar and Stellar Pulsation Conference*, eds Cox, A. N. and Dupree, R. G. (Los Alamos: Los Alamos Scientific Laboratory), pp. 1–6.

Bruzual, G. and Charlot, S. (2003). Stellar population synthesis at the resolution of 2003, *Monthly Notices of the Royal Astronomical Society*, **344**, 1000–1028.

Burbidge, E. M., Burbidge, G. R., Fowler, W. A. and Hoyle, F. (1957). Synthesis of the elements in stars, *Reviews of Modern Physics*, **29**, 547–650.

Burbidge, E. M., Burbidge, G. R. and Prendergast, K. H. (1960). The rotation, mass distribution and mass of NGC 5866, *Astrophysical Journal*, **131**, 282–292.

(1965). The rotation and mass of the SA Galaxy NGC 681, *Astrophysical Journal*, **142**, 154–159.

Burbidge, E. M., Burbidge, G. R. and Sandage, A. R. (1963). Evidence for the occurence of violent events in the nuclei of galaxies, *Reviews of Modern Physics*, **35**, 947–972.

Burbidge, E. M., Lynds, C. R. and Burbidge, G. R. (1966). On the measurement and interpretation of absorption features in the spectrum of the quasi-stellar object 3C 191, *Astrophysical Journal*, **144**, 447–451.

Burbidge, E. M., Lynds, C. R. and Stockton, A. N. (1968). Further observations of quasi-stellar objects with absorption-line spectra: Ton 1530, PKS 0237-23, and PHL 938, *Astrophysical Journal*, **152**, 1077–1093.

Burbidge, G. R. (1959). Estimates of the total energy in particles and magnetic field in the non-thermal radio sources, *Astrophysical Journal*, **129**, 849–851.

Burbidge, G. R. and Burbidge, E. M. (1967). *Quasi-stellar Objects* (New York: Freeman and Company).

Butcher, H. and Oemler, Jr., A. (1978). The evolution of galaxies in clusters. I – ISIT photometry of Cl 0024+1654 and 3C 295, *Astrophysical Journal*, **219**, 18–30.

(1984). The evolution of galaxies in clusters. V – A study of populations since $z \sim 0.5$, *Astrophysical Journal*, **285**, 426–438.

Cameron, A. G. W. (1957). Nuclear reactions in stars and nucleogenesis, *Publications of the Astronomical Society and the Pacific*, **69**, 201–222.

(1958). Nuclear astrophysics, *Annual Review of Nuclear Science*, **8**, 299–326.

Campbell, W. W. (1901). A preliminary determination of the motion of the Solar System, *Astrophysical Journal*, **13**, 80–89.

(1910). Some peculiarities in the motions of the stars, *Lick Observatory Bulletin 6*, No. 196, 125–133.

Campbell, W. W. and Moore, J. H. (1928). Radial velocities of stars brighter than visual magnitude 5.51 as determined at Mount Hamilton and Santiago, *Publications of the Lick Observatory*, **16**, 1–399.

Cannon, A. J. and Pickering, E. C. (1901). Spectra of bright southern stars photographed with the 13-inch Boyden telescope as part of the Henry Draper Memorial, *Annals of the Harvard College Observatory (Part II)*, **28**, 131–263.

Carlstrom, J. E., Joy, M. K., Grego, L. *et al.* (2000). Imaging the Sunyaev–Zel'dovich effect, in *Particle Physics and the Universe: Proceedings of Nobel Symposium 198*, eds Bergström, L., Carlson, P. and Fransson, C. (Stockholm: Physica Scripta), pp. 148–155.

Carroll, S. M., Press, W. H. and Turner, E. L. (1992). The cosmological constant, *Annual Review of Astronomy and Astrophysics*, **30**, 499–542.

Carter, B. (1971). Axisymmetric black hole has only two degrees of freedom, *Physical Review Letters*, **26**, 331–333.

  (1974). Large number coincidences and the anthropic principle in cosmology, in *Confrontation of Cosmological Theories with Observational Data*, IAU Symposium no. 63, ed. Longair, M. S. (Dordrecht: D. Reidel Publishing Company), pp. 291–298.

Chaboyer, B. (1998). The age of the Universe, *Physics Reports*, **307**, 23–30.

Chadwick, J. (1932). Possible existence of a neutron, *Nature*, **129**, 312.

Chambers, K. C., Miley, G. K. and van Breugel, W. J. M. (1987). Alignment of radio and optical orientations in high-redshift radio galaxies, *Nature*, **329**, 604–606.

Chandrasekhar, S. (1931). The maximum mass of ideal white dwarfs, *Astrophysical Journal*, **74**, 81–82.

  (1939). *An Introduction to the Study of Stellar Structure* (Chicago: Chicago University Press).

  (1943a). Dynamical friction I. General considerations: the coefficient of dynamical friction, *Astrophysical Journal*, **97**, 255–262.

  (1943b). Dynamical friction II. The rate of escape of stars from clusters and the evidence for the operation of dynamical friction, *Astrophysical Journal*, **97**, 263–273.

  (1943c). Dynamical friction III. A more exact theory of the rate of escape of stars from clusters, *Astrophysical Journal*, **98**, 54–60.

  (1946). On the radiative equilibrium of a stellar atmosphere. X., *Astrophysical Journal*, **103**, 351–370.

  (1980). The role of general relativity in astronomy: retrospect and prospect, *Highlights of Astronomy*, vol. 5, ed. Wayman, P. A. (Dordrecht: D. Reidel Publishing Company), pp. 45–61.

  (1981). *Hydrodynamic and Hydromagnetic Stability* (New York: Dover Publications).

Chandrasekhar, S. and Henrich, L. R. (1942). An attempt to interpret the relative abundances of the elements and their isotopes, *Astrophysical Journal*, **95**, 288–298.

Charbonneau, D., Brown, T. M., Latham, D. W. and Mayor, M. (2000). Detection of planetary transits across a Sun-like star, *Astrophysical Journal*, **529**, L45–L48.

Charbonneau, D., Brown, T. M., Noyes, R. W. and Gilliland, R. L. (2002). Detection of an extrasolar planet atmosphere, *Astrophysical Journal*, **568**, 377–384.

Charles, P. (1998). Black holes in our Galaxy: observations, in *Theory of Black Hole Accretion Disks*, eds Abramowicz, M. A., Björnsson, G. and Pringle, J. E. (Cambridge: Cambridge University Press), pp. 1–20.

Cheung, A. C., Rank, D. M., Townes, C. H., Thornton, D. D. and Welch, W. J. (1968). Detection of NH₃ molecules in the interstellar medium by their microwave emission, *Physical Review Letters*, **221**, 1701–1705.

(1969). Detection of water in interstellar regions by its microwave radiation, *Nature*, **221**, 626–628.

Chevalier, R. A. (1992). Supernova 1987A at five years of age, *Nature*, **355**, 691–696.

Christensen-Dalsgaard, J. and Gough, D. O. (1980). Is the Sun helium deficient?, *Nature*, **288**, 544–547.

Christy, R. F. (1964). The calculation of stellar pulsation, *Reviews of Modern Physics*, **36**, 555–571.

(1968). The theory of Cepheid variables, *Quarterly Journal of the Royal Astronomical Society*, **9**, 13–39.

Clark, G. W., Garmire, G. P. and Kraushaar, W. L. (1968). Observation of high-energy cosmic gamma rays, *Astrophysical Journal Letters*, **153**, L203–L207.

Clavel, J., Reichert, G. A., Alloin, D. *et al.* (1991). Steps toward determination of the size and structure of the broad-line region in active galactic nuclei. I – An 8 month campaign of monitoring NGC 5548 with IUE, *Astrophysical Journal*, **366**, 64–81.

Cockcroft, J. D. and Walton, E. T. S. (1932). Disintegration of lithium by swift protons, *Nature*, **129**, 649.

Cocke, W. J., Disney, M. J. and Taylor, D. J. (1969). Discovery of optical signals from pulsar NP 0532, *Nature*, **221**, 525–527.

Cohen, M. H., Cannon, W., Purcell, G. H. *et al.* (1971). The small-scale structure of radio galaxies and quasar-stellar sources at 3.8 centimetres, *Astrophysical Journal*, **170**, 207–218.

Coles, P. and Lucchin, F. (1995). *Cosmology – The Origin and Evolution of Cosmic Structure* (Chichester: John Wiley & Sons).

Colgate, S. A. and McKee, C. (1969). Early supernova luminosity, *Astrophysical Journal*, **157**, 623–643.

Colless, M., Dalton, G., Maddox, S. *et al.* (2001). The 2dF galaxy redshift survey: spectra and redshifts, *Monthly Notices of the Royal Astronomical Society*, **328**, 1039–1063.

Comstock, G. C. (1897). On the application of interference methods to the determination of the effective wavelength of starlight, *Astrophysical Journal*, **5**, 26–35.

Connolly, A. J., Szalay, A. S., Dickinson, M., SubbaRao, M. U. and Brunner, R. J. (1997). The evolution of the global star formation history as measured from the Hubble Deep Field, *Astrophysical Journal*, **486**, L11–L14.

Costa, E., Frontera, F., Heise, J. *et al.* (1997). Discovery of an X-ray afterglow associated with the gamma-ray burst of 28 February 1997, *Nature*, **387**, 783–785.

Cowan, J. J., Thielemann, F.-K. and Truran, J. W. (1991). Radioactive dating of the elements, *Annual Reviews of Astronomy and Astrophysics*, **29**, 447–497.

Cowan Jr, C. L., Reines, F., Harrison, F. B., Kruse, H. W. and McGuire, A. D. (1956). Detection of the free neutrino: a confirmation, *Science*, **124**, 103–104.

Cowie, L. L., Hu, E. M. and Songaila, A. (1995). Detection of massive forming galaxies at redshifts $z > 1$, *Nature*, **377**, 603–605.

Cowie, L. L., Lilly, S. J., Gardner, J. and McLean, I. S. (1988). A cosmologically significant population of galaxies dominated by very young star formation, *Astrophysical Journal*, **332**, L29–L32.

Cowie, L. L., Songaila, A., Hu, E. M. and Cohen, J. D. (1996). New insight on galaxy formation and evolution from Keck spectroscopy of the Hawaii deep fields, *Astronomical Journal*, **112**, 839–864.

Cowley, A. P. (1992). Evidence for black holes in stellar binary systems, *Annual Reviews of Astronomy and Astrophysics*, **30**, 287–310.

Cowling, T. G. (1935). The stability of gaseous spheres, *Monthly Notices of the Royal Astronomical Society*, **96**, 42–60.

Cox, A. N. (1965). Stellar absorption coefficients and opacities, in *Stellar Structure – Stars and Stellar Systems: Compendium of Astronomy and Astrophysics, Vol. VIII*, eds Aller, L. H. and McLaughlin, D. B. (Chicago: University of Chicago Press), pp. 195–268.

Cox, D. P. and Smith, B. W. (1974). Large-scale effects of supernova remnants on the Galaxy: generation and maintenance of a hot network of tunnels, *Astrophysical Journal Letters*, **189**, L105–L108.

Crittenden, R., Bond, R., Davis, R. L., Efstathiou, G. and Steinhardt, P. J. (1993). Imprint of gravitational waves on the cosmic microwave background, *Physical Review Letters*, **71**, 324–327.

Curie, M. P. and Sklodowska-Curie, M. (1898). On a new radioactive substance contained in pitchblende, *Comptes Rendus*, **127**, 175–178.

Curie, M. P., Sklodowska-Curie, M. and Bémont, G. (1898). On a new, strongly radioactive substance, contained in pitchblende, *Comptes Rendus*, **127**, 1215–1217.

Curtis, H. D. (1917). New stars in spiral nebulae, *Publications of the Astronomical Society of the Pacific*, **29**, 180–181.

(1918a). Descriptions of 762 nebulae and clusters photographed with the Crossley reflector, *Publications of the Lick Observatory*, **13**, 11–42. The reference to 'a curious straight ray' in NGC 4486 (M87) appears on page 31.

(1918b). A study of occulting matter in the spiral nebulae, *Publications of the Lick Observatory*, **13**, 45–54.

(1921). The scale of the Universe, *Bulletin of the National Research Council*, **2**, 194–217.

Dashevsky, V. M. and Zeldovich, Y. B. (1964). Propagation of light in a nonhomogeneous non-flat Universe II, *Astronomicheskii Zhurnal*, **41**, 1071–1074. Translation in *Soviet Astronomy*, **8**, 1965, 854–856.

Davidson, W. (1962). The cosmological implications of the recent counts of radio sources, I. Analysis of the results and their immediate interpretation, *Monthly Notices of the Royal Astronomical Society*, **123**, 425–435.

Davidson, W. and Davies, M. (1964). Interpretation of the counts of radio sources in terms of a 4-parameter family of evolutionary universes, *Monthly Notices of the Royal Astronomical Society*, **127**, 241–255.

Davies, R. L., Efstathiou, G., Fall, S. M., Illingworth, G. and Schechter, P. L. (1983). The kinematic properties of faint elliptical galaxies, *Astrophysical Journal*, **266**, 41–57.

Davis Jr., L. and Greenstein, J. L. (1951). The polarization of starlight by aligned dust grains, *Astrophysical Journal*, **114**, 206–240.

Davis, M. and Peebles, P. J. E. (1983). A survey of galaxy redshifts. V – The two-point position and velocity correlations, *Astrophysical Journal*, **267**, 465–482.

Davis, M., Efstathiou, G., Frenk, C. and White, S. D. M. (1985). The evolution of large-scale structure in a universe dominated by cold dark matter, *Astrophysical Journal*, **292**, 371–394.

(1992a). The end of cold dark matter?, *Nature*, **356**, 489–494.

Davis, M., Geller, M. J. and Huchra, J. (1978). The local mean mass density of the Universe – new methods for studying Galaxy clustering, *Astrophysical Journal*, **221**, 1–18.

Davis, R. (1955). Attempt to detect the antineutrinos from a nuclear reactor by the $Cl^{37}(\nu^-, e^-)Ar^{37}$ reaction, *Physical Review*, **97**, 766–769.

Davis, R. L., Hodges, H. M., Smoot, G. F., Steinhardt, P. J. and Turner, M. S. (1992b). Cosmic microwave background probes models of inflation, *Physical Review Letters*, **69**, 1856–1859.

de Bernardis, P., Ade, P. A. R., Bock, J. J. *et al.* (2000). A flat Universe from high-resolution maps of the cosmic microwave background radiation, *Nature*, **404**, 955–959.

de Sitter, W. (1917a). On Einstein's theory of gravitation and its astronomical consequences, *Monthly Notices of the Royal Astronomical Society*, **78**, 3–28.

(1917b). On the relativity of inertia. Remarks concerning Einstein's latest hypothesis, *Proceedings of the Royal Academy of Amsterdam*, **19**, 1217–1225.

de Vaucouleurs, G. (1971). The large-scale distribution of galaxies and clusters of galaxies, *Publications of the Astronomical Society of the Pacific*, **83**, 113–143.

(1974). Structure, dynamics and statistical properties of galaxies (invited paper), in *The Formation and Dynamics of Galaxies, IAU Symposium 58*, ed. Shakeshaft, J. R. (Dordrecht: D. Reidel Publishing Company), pp. 1–53.

de Vaucouleurs, G., de Vaucouleurs, A., Corwin Jr, H. G., Buta, R. J., Paturel, G. and Fouque, P. (1991). *Third Reference Catalogue of Bright Galaxies: Containing Information on 23,024 Galaxies With Reference to Papers Published Between 1913 and 1988* (Berlin: Springer-Verlag).

Dekel, A. (1986). Biased galaxy formation, *Comments on Astrophysics*, **11**, 235–256.

Dekel, A. and Rees, M. J. (1987). Physical mechanisms for biased galaxy formation, *Nature*, **326**, 455–462.

Dekel, A., Burstein, D. and White, S. D. M. (1997). Measuring Ω, in *Critical Dialogues in Cosmology*, ed. Turok, N. (Singapore: World Scientific), pp. 175–192.

Dennise, J.-F., Le Roux, E. and Steinberg, J. C. (1957). Novelles observations du rayonnement du ciel sur la longeur d'onde 33 cm (New observations of the background radiation at a wavelength of 33 cm), *Comptes Rendus*, **244**, 3030–3033.

Deubner, F.-L. (1975). Observations of low wavenumber nonradial eigenmodes of the Sun, *Astronomy and Astrophysics*, **44**, 371–375.

Dicke, R. H. (1961). Dirac's cosmology and Mach's principle, *Nature*, **192**, 440–441.

Dicke, R. H. and Peebles, P. J. E. (1979). Big Bang cosmology – enigmas and nostrums, in *General Relativity: An Einstein Centenary Survey*, eds Hawking, S. W. and Israel, W. (Cambridge: Cambridge University Press), pp. 504–517.

Dicke, R. H., Peebles, P. J. E., Roll, P. G. and Wilkinson, D. T. (1965). Cosmic black-body radiation, *Astrophysical Journal*, **142**, 414–419.

Dingle, H. (1937). Modern Aristotelianism, *Nature*, **139**, 784–786.

Dirac, P. A. M. (1928a). The quantum theory of the electron, *Proceedings of the Royal Society of London*, **A117**, 610–624.

(1928b). The quantum theory of the electron II, *Proceedings of the Royal Society of London*, **A118**, 351–361.

(1931). Quantum singularities in the electromagnetic field, *Proceedings of the Royal Society of London*, **A133**, 60–72.

(1937). The cosmical constants, *Nature*, **139**, 323.

Djorgovski, S. G. and Davis, M. (1987). Fundamental properties of elliptical galaxies, *Astrophysical Journal*, **313**, 59–68.

Dombrovski, V. A. (1954). On the nature of the radiation from the Crab Nebula, *Dokladi Akademiya Nauk SSSR*, **94**, 1021–1024.

Doroshkevich, A. G. and Novikov, I. D. (1964). Mean density of radiation in the metagalaxy and certain problems in relativistic cosmology, *Dokladi Akademiya Nauk SSSR*, **154**, 809–811. Translation in *Soviet Physics Doklady*, **9**, 1964, 111–113.

Doroshkevich, A. G., Novikov, I. D., Sunyaev, R. A. and Zeldovich, Y. B. (1971). Helium production in the different cosmological models, in *Highlights of Astronomy*, vol. 2, ed. de Jager, C. (Dordrecht: D. Reidel Publishing Company), pp. 313–327.

Doroshkevich, A. G., Sunyaev, R. A. and Zeldovich, Y. B. (1974). The formation of galaxies in Friedmannian universes, in *Confrontation of Cosmological Theories with Observational Data*, IAU Symposium no. 63, ed. Longair, M. S. (Dordrecht: D. Reidel Publishing Company), pp. 213–225.

Doroshkevich, A. G., Zeldovich, Y. B., Sunyaev, R. A. and Khlopov, M. Y. (1980a). Astrophysical implications of the neutrino rest mass – Part II. The density-perturbation spectrum and small-scale fluctuations in the microwave background, *Pis'ma v Astronomicheskii Zhurnal*, **6**, 457–464.

(1980b). Astrophysical implications of the neutrino rest mass – Part III. The non-linear growth of perturbations and hidden mass, *Pis'ma v Astronomicheskii Zhurnal*, **6**, 465–469.

Draine, B. T. (2003). Interstellar dust grains, *Annual Reviews of Astronomy and Astrophysics*, **41**, 241–289.

Draper, H. (1879). On photographing the spectra of the stars and planets, *American Journal of Science and Arts*, **18**, 419–425.

Dressler, A. (1980). Galaxy morphology in rich clusters – implications for the formation and evolution of galaxies, *Astrophysical Journal*, **236**, 351–365.

(1984). The evolution of galaxies in clusters, *Annual Review of Astronomy and Astrophysics*, **22**, 185–222.

Dressler, A., Lynden-Bell, D., Burstein, D. *et al.* (1987). Spectroscopy and photometry of elliptical galaxies. I – A new distance estimator, *Astrophysical Journal*, **313**, 42–58.

Dreyer, J. L. E. (1888). New general catalogue of nebulae and clusters of stars, *Memoirs of the Royal Astronomical Society*, **49**, 1–237.

(1895). Index catalogue of nebulae, *Memoirs of the Royal Astronomical Society*, **51**, 185–228.

(1908). Index catalogue of nebulae and cluster stars, *Memoirs of the Royal Astronomical Society*, **59**, 105–198.

Dunbar, D. N. F., Pixley, R. E., Wenzel, W. A. and Whaling, W. (1953). The 7.68-MeV state of C^{12}, *Physical Review*, **92**, 649–650.

Dunlop, J. S. (1994). The cosmological evolution of active galaxies, in *Frontiers of Space and Ground-Based Astronomy: The Astrophysics of the 21st Century*, eds Wamsteker, W., Longair, M. S. and Kondo, Y. (Dordrecht: Kluwer Academic Publishers), pp. 395–407.

Dunlop, J. S. and Peacock, J. A. (1990). The redshift cut-off in the luminosity function of radio galaxies and quasars, *Monthly Notices of the Royal Astronomical Society*, **247**, 19–42.

Dunlop, J. S., McLure, R. J., Yamada, T. *et al.* (2004). Discovery of the galaxy counterpart of HDF 850.1, the brightest submillimetre source in the Hubble Deep Field, *Monthly Notices of the Royal Astronomical Society*, **350**, 769–784.

Dwek, E., Arendt, R. G., Hauser, M. G. *et al.* (1995). Morphology, near-infrared luminosity, and mass of the Galactic bulge from COBE DIRBE observations, *Astrophysical Journal*, **445**, 716–730.

Dyer, C. C. and Roeder, R. C. (1972). The distance–redshift relation for universes with no intergalactic medium, *Astrophysical Journal*, **174**, L115–L117.

Dyson, F. W., Eddington, A. S. and Davidson, C. (1920). A determination of the deflection of light by the Sun's gravitational field, from observations made at the total eclipse of May 29, 1919, *Philosophical Transactions of the Royal Society*, **220**, 291–333.

Eales, S., Rawlings, S., Law-Green, D., Cotter, G. and Lacy, M. (1997). A first sample of faint radio sources with virtually complete redshifts. I – Infrared images, the Hubble diagram and the alignment effect, *Monthly Notices of the Royal Astronomical Society*, **291**, 593–615.

Eastman, R. G. and Kirshner, R. P. (1989). Model atmospheres for SN 1987A and the distance to the Large Magellanic Cloud, *Astrophysical Journal*, **347**, 771–793.

Eddington, A. S. (1908). On the mathematical theory of two star drifts, and on the systematic motions of zodiacal stars, *Monthly Notices of the Royal Astronomical Society*, **68**, 588–605.

(1916a). The kinetic energy of a star cluster, *Monthly Notices of the Royal Astronomical Society*, **76**, 525–528.

(1916b). On the radiative equilibrium of the stars, *Monthly Notices of the Royal Astronomical Society*, **77**, 16–35.

(1917). Further notes on the radiative equilibrium of the stars, *Monthly Notices of the Royal Astronomical Society*, **77**, 596–612.

(1920). The internal constitution of the stars, *Observatory*, **43**, 341–358.

(1924). On the relation between the masses and luminosities of the stars, *Monthly Notices of the Royal Astronomical Society*, **84**, 308–332.

Eddington, A. S. (1926a). Diffuse matter in interstellar space, *Proceedings of the Royal Society*, **A111**, 424–456.

(1926b). *The Internal Constitution of the Stars* (Cambridge: Cambridge University Press). Reprinted 1988.

(1930). On the stability of Einstein's spherical world, *Monthly Notices of the Royal Astronomical Society*, **90**, 668–678.

(1935). Remark in the paper 'Relativistic degeneracy' read to the meeting of the Royal Astronomical Society on January 11 1935, *Observatory*, **58**, 37–39.

(1941). Discussion of Sir Arthur Eddington's contribution, in *Novae and White Dwarfs*, ed. Shaler, A. J. (Paris: Herrmann et Cie), pp. 262–267.

(1946). *Fundamental Theory* (Cambridge: Cambridge University Press).

Edge, D. O., Shakeshaft, J. R., McAdam, W. B., Baldwin, J. E. and Archer, S. (1959). A survey of radio sources at a frequency of 159 Mc/s, *Memoirs of the Royal Astronomical Society*, **68**, 37–60.

Efstathiou, G. (1990). Cosmological perturbations, in *Physics of the Early Universe*, eds Peacock, J. A., Heavens, A. F. and Davies, A. T. (Edinburgh: SUSSP Publications), pp. 361–463.

Eggert, J. (1919). Über den Dissoziationszustand der Fixsterngase (On the dissociation state of gases in the fixed stars), *Physikalische Zeitschrift*, **20**, 570–574.

Eguchi, K., Enomoto, S., Furuno, K. *et al.* (2003). First results from KamLAND: evidence for reactor anti-neutrino disappearance, *Physical Review Letters*, **90**, id. 021802(1–6).

Einasto, J. (2001). Dark matter and large scale structure, in *Historical Development of Modern Cosmology*, ASP Conference Series vol. 252, eds Martinez, V. J., Trimble, V. and Pons-Bordeia, M. J. (San Francisco: ASP), pp. 85–107.

Einstein, A. (1907). Über das Relativitätsprinzip und die aus demselben gezogenen Folgerungen (On the relativity principle and the conclusions drawn from it), *Jahrbuch der Radioaktivität und Elektronik*, **4**, 411–462.

(1911). Über den Einfluss der Schwerkraft auf die Ausbreitung des Lichtes (On the influence of gravitation on the propagation of light), *Annalen der Physik*, **35**, 898–908.

(1912). Relativität und Gravitation. Erwiderung auf eine Bemerkung von M. Abraham (Relativity and gravitation. Reply to a comment by M. Abraham), *Annalen der Physik*, **38**, 1059–1064.

(1915). Die Feldgleichung der Gravitation (The field equations of gravitation), *Sitzungsberichte, Königlich Preussische Akademie der Wissenschaften (Berlin)*, **II** (1915), 844–847.

(1916a). *Die Grundlage der Allgemeinen Relativitätstheorie (The Foundation of the General Theory of Relativity)*. (Leipzig: J. A. Barth). Acknowledgement to the contribution of Marcel Grossmann is printed on p. 6.

(1916b). Die Grundlage der Allgemeinen Relativitätstheorie (The foundation of the general theory of relativity), *Annalen der Physik*, **49**, 769–822.

(1917). Kosmologische Betrachtungen zur Allgemeinen Relativitätstheorie (Cosmological considerations in the general theory of relativity), *Sitzungsberichte, Königlich Preussische Akademie der Wissenschaften (Berlin)*, **I** (1917), 142–152.

Einstein, A. (1919). Spielen Gravitationsfelder im Aufbau der materiellen Elementarteilchen eine wesentliche Rolle? (Do gravitational fields play a significant role for the structure of elementary particles?), *Sitzungsberichte, Königlich Preussische Akademie der Wissenschaften*, **I** (1919), 349–356.

(1922a). Bemerkung zu der Arbeit von A. Friedmann 'Ueber die Kruemmung des Raumes' (Remark on the work of A. Friedmann 'On the curvature of space'), *Zeitschrift für Physik*, **11**, 326.

(1922b). Kyoto address of December 1922, in *Einstein Köen-Roku*, ed. Ishiwara, J. (Tokyo: Tokyo-Tosho, 1977).

(1923). Notiz zu der Arbeit von A. Friedmann 'Über die Krümmung des Raumes' (A note on the work of A. Friedmann 'On the curvature of space'), *Zeitschrift für Physik*, **16**, 228.

Einstein, A. and de Sitter, W. (1932). On the relation between the expansion and the mean density of the Universe, *Proceedings of the National Academy of Sciences*, **18**, 213–214.

Einstein, A. and Grossmann, M. (1913a). *Entwurf einer verallgemeinerten Relativitätstheorie und einer Theorie der Gravitation* (*Outline of a Generalised Theory of Relativity and of a Theory of Gravitation*) (Leipzig: Teubner).

(1913b). Entwurf einer verallgemeinerten Relativitätstheorie und einer Theorie der Gravitation (Outline of a generalised theory of relativity and of a theory of gravitation), *Zeitschrift für Mathematik und Physik*, **62**, 225–259.

Elder, F. R., Gurewitsch, A. M., Langmuir, R. V. and Pollock, H. C. (1947). Radiation from electrons in a synchrotron, *Physical Review*, **71**, 829–830.

Ellis, R. (1987). Galaxy surveys at high redshift – past, present and future, in *High Redshift and Primaeval Galaxies*, eds Bergeron, J., Kunth, D., Rocca-Volmerange, B. and Tran Thanh Van, J. (Gif sur Yvette: Edition Frontières), pp. 3–16.

Ellis, R. G. (1997). Faint blue galaxies, *Annual Review of Astronomy and Astrophysics*, **35**, 389–443.

Elsasser, W. M. (1933). Sur le principe de Pauli dans les noyaux (On Pauli's principle for nuclei), *Journal de Physique et le Radium, Series 7*, **4**, 549–556.

Emden, R. (1907). *Gaskugeln* (Leipzig and Berlin: B. G. Teubner).

Epstein, I. (1950). A note on energy generation, *Astrophysical Journal*, **112**, 207–210.

Ewen, H. I. and Purcell, E. M. (1951). Radiation from Galactic hydrogen at 1420 MHz, *Nature*, **168**, 356.

Faber, S. M. and Jackson, R. E. (1976). Velocity dispersions and mass-to-light ratios for elliptical galaxies, *Astrophysical Journal*, **204**, 668–683.

Fabian, A. C. (1994). Cooling flows in clusters of galaxies, *Annual Review of Astronomy and Astrophysics*, **32**, 277–318.

Fabricant, D. G., Lecar, M. and Gorenstein, P. (1980). X-ray measurements of the mass of M87, *Astrophysical Journal*, **241**, 552–560.

Falk, S. W. and Arnett, W. D. (1973). A theoretical model for Type II supernovae, *Astrophysical Journal*, **180**, L65–L68.

Fall, S. M. (1997). A global perspective on star formation, in *The Hubble Space Telescope and the High Redshift Universe*, eds Tanvir, N. R., Aragón-Salamanca, A. and Wall, J. V. (Singapore: World Scientific Publishing Company), pp. 303–308.

Fanaroff, B. L. and Riley, J. M. (1974). The morphology of extragalactic radio sources of high and low luminosity, *Monthly Notices of the Royal Astronomical Society*, **167**, 31P–36P.

Faulkner, J. (2001). Low-mass red giants as binary stars without angular momentum, in *Evolution of Binary and Multiple Star Systems; A Meeting in Celebration of Peter Eggleton's 60th Birthday*, ASP Conference Series vol. 229, eds Podsiadlowski, P., Rappaport, S., King, A. R., D'Antona, F. and Burder, L. (San Francisco: ASP), pp. 3–14.

Feast, M. W. and Catchpole, R. M. (1997). The Cepheid period-luminosity zero-point from Hipparcos trigonometrical parallaxes, *Monthly Notices of the Royal Astronomical Society*, **286**, L1–L5.

Felten, J. (1977). Study of the luminosity function for field galaxies, *Astronomical Journal*, **82**, 861–878.

Fermi, E. (1926). Sulla quantizzazione del gas perfecto monoatomica (On the quantisation of a perfect monatomic gas), *Rendiconti della Accademia Nazionale dei Lincei*, **3**, 145–149.

   (1934a). Tentativo di una teoria dei raggi β (An attempt at a theory of β-rays), *Nuovo Cimento*, **11**, 1–19.

   (1934b). Versuch einer Theorie der β(-Strahlen) (An attempt at a theory of β-rays), *Zeitschrift für Physik*, **88**, 161–177.

   (1949). On the origin of the cosmic radiation, *Physical Review*, **75**, 1169–1174.

Fich, M. and Tremaine, S. (1991). The mass of the Galaxy, *Annual Review of Astronomy and Astrophysics*, **29**, 409–445.

Fichtel, C. E., Simpson, G. A. and Thompson, D. J. (1978). Diffuse gamma radiation, *Astrophysical Journal*, **222**, 833–849.

Field, G. B. (1965). Thermal instability, *Astrophysical Journal*, **142**, 531–567.

   (1974). Interstellar abundances: gas and dust, *Astrophysical Journal*, **187**, 453–469.

Field, G. B., Goldsmith, D. W. and Habing, H. J. (1969). Cosmic-ray heating of the interstellar gas, *Astrophysical Journal Letters*, **55**, L149–L154.

Fitch, W. S., Pacholczyk, A. G. and Weymann, R. J. (1967). Light variations of the Seyfert galaxy NGC 4151, *Astrophysical Journal*, **150**, L67–L70.

Fixsen, D. J., Cheng, E. S., Gales, J. M., Mather, J. C., Shafer, R. A. and Wright, E. L. (1996). The cosmic microwave background spectrum from the full COBE FIRAS data set, *Astrophysical Journal*, **473**, 576–587.

Fizeau, H. and Foucault, L. (1847). Recherches sur les interférences des rayons calorifiques (Researches on the interference of heat rays), *Comptes Rendus de l'Academie des Sciences*, **25**, 447–450.

Ford, H. C., Harms, R. J., Tsvetanov, Z. I. *et al.* (1994). Narrowband HST images of M87: evidence for a disk of ionized gas around a massive black hole, *Astrophysical Journal Letters*, **435**, L27–L30.

Forman, W., Jones, C., Cominsky, L. *et al.* (1978). The fourth UHURU catalog of X-ray sources, *Astrophysical Journal Supplement Series*, **38**, 357–412.

Foucault, L. (1849). Lumière électrique (Electric light), *L'Institut, Journal Universel des Sciences*, **17**, 44–46.

Fowler, R. H. (1926). On dense matter, *Monthly Notices of the Royal Astronomical Society*, **87**, 114–122.

Fowler, R. H. and Milne, E. A. (1923). The intensities of absorption lines in stellar spectra, and the temperatures and pressures in the reversing layers of stars, *Monthly Notices of the Royal Astronomical Society*, **83**, 403–424.

(1924). The maxima of absorption lines in stellar spectra (second paper), *Monthly Notices of the Royal Astronomical Society*, **84**, 499–515.

Fowler, W. A. (1958). Completion of the proton–proton reaction chain and the possibility of energetic neutrino emission by hot stars, *Astrophysical Journal*, **127**, 551–556.

Fraunhofer, J. (1817a). Bestimmung des Brechungs- und Farbenzerstreuungs-Vermögens Verschiedener Glasarten, in Bezug auf die Vervollkommnung Achromatischer Fernröhre (On the refractive and dispersive power of different species of glass in reference to the improvement of achromatic telescopes, with an account of the lines or streaks which cross the spectrum), *Denkschriften der königlichen Akademie der Wissenschaften zu München*, **5**, 193–226. Translation: *Edinburgh Philosophical Journal*, **9**, 288–299 (1823); **10**, 26-40 (1824).

(1817b). Bestimmung des Brechungs- und Farbenzerstreuungs-Vermögens Verschiedener Glasarten, in Bezug auf die Vervollkommnung Achromatischer Fernröhre (On the refractive and dispersive power of different species of glass in reference to the improvement of achromatic telescopes, with an account of the lines or streaks which cross the spectrum), *Gilberts Annalen der Physik*, **56**, 264–313.

(1821). Neue Modifikation des Lichtes durch Gegenseitige Einwirkung und Beugung der Strahlen, und Gesetze Derselben (New modifications of light through interactions and diffraction of rays and its laws), *Denkschriften der königlichen Akademie der Wissenschaften zu München*, **8**, 1–76.

(1823). Kurzer Bericht von den Resultaten neuerer Versuche über die Gesetze des Lichtes, und die Theorie Derselben (A short account of the results of recent experiments upon the laws of light and its theory), *Gilberts Annalen der Physik*, **74**, 337–378. Translation: *Edinburgh Journal of Science*, **7**, 101–113, 251–262 (1827); **8**, 7–10 (1828).

Frazier, E. N. (1968). A spatio-temporal analysis of velocity fields in the solar photosphere, *Zeitschrift für Astrophysik*, **68**, 345–356.

Freedman, W. L., Madore, B. F., Gibson, B. K. *et al.* (2001). Final results from the Hubble Space Telescope Key Project to measure the Hubble constant, *Astrophysical Journal*, **533**, 47–72.

Freeman, K. C. (1970). On the disks of spiral and S0 galaxies, *Astrophysical Journal*, **160**, 811–830.

Freier, P., Lofgren, E. J., Ney, E. P., Oppenheimer, F., Bradt, H. L. and Peters, B. (1948). Evidence for heavy nuclei in the primary cosmic radiation, *Physical Review*, **74**, 213–217.

Frenk, C. (1986). Galaxy clustering and the dark-matter problem, *Philosophical Transactions of the Royal Astronomical Society*, **A320**, 517–541.

Fried, D. L. (1965). Statistics of a geometric representation of wavefront distortion, *Journal of the Optical Society of America*, **55**, 1427–1435.

Friedman, A. A. (1922). On the curvature of space, *Zeitschrift für Physik*, **10**, 377–386.

Friedman, A. A. (1923). *The World as Space and Time* (Petrograd: Academia).

(1924). On the possibility of a world with constant negative curvature, *Zeitschrift für Physik*, **12**, 326–332.

Friedman, H., Lichtman, S. W. and Byram, E. T. (1951). Photon counter measurements of solar X-rays and extreme ultraviolet light, *Physical Review*, **83**, 1025–1030.

Friedrich, W., Knipping, P. and Laue, M. von (1912). Interferenz-Erscheinungen bei Röntgenstrahlen (Interference effects with Röntgen rays), *Sitzberichte der Königlich Bayerischen Akademie der Wissenschaften*, pp. 303–322.

Fritz, G., Henry, R. C., Meekins, J. F., Chubb, T. A. and Friedman, H. (1969). X-ray pulsar in the Crab Nebula, *Science*, **164**, 709–712.

Frost, E. B. (1909). Spectroscopic notes, *Astrophysical Journal*, **29**, 233–239.

Fukuda, S., Fukuda, Y., Ishitsuka, *et al.* (2001). Solar ^{8}B and hep neutrino measurements from 1258 days of super-Kamiokande data, *Physical Review Letters*, **86**, 5651–5655.

Fukuda, Y., Hayakawa, T., Inoue, K. *et al.* (1996). Solar neutrino data covering solar cycle 22, *Physical Review Letters*, **77**, 1683–1686.

Galbraith, W. and Jelley, J. V. (1953). Light pulses from the night sky associated with cosmic rays, *Nature*, **171**, 349–350.

(1955). Light-pulses from the night sky and Cherenkov radiation, Part 1, *Journal of Atmospheric and Terrestrial Physics*, **6**, 250–262.

Gallego, J., Zamorano, J., Aragón-Salamanca, A. and Rego, M. (1995). The current star formation rate of the local Universe, *Astrophysical Journal*, **455**, L1–L4.

Gamow, G. (1928). Zur Quantentheorie der Atomzertrümmerung (On the quantum theory of atomic destruction), *Zeitschrift für Physik*, **52**, 510–515.

(1937). *Atomic Nuclei and Nuclear Transformations* (Oxford: Oxford University Press).

(1939). Physical possibilities of stellar evolution, *Physical Review*, **55**, 718–725.

(1946). Expanding Universe and the origin of elements, *Physical Review*, **70**, 572–573.

(1970). *My World Line*, (New York: Viking Press). The reference to Einstein's admission of 'the greatest blunder of my life' is on p. 44.

Garcia-Munoz, M., Mason, G. M. and Simpson, J. A. (1977). The age of the Galactic cosmic rays derived from the abundances of ^{10}Be, *Astrophysical Journal*, **217**, 857–877.

Garnavich, P. M., Kirshner, R. P., Challis, P. *et al.* (1998). Constraints on cosmological models from Hubble Space Telescope observations of high-z supernovae, *Astrophysical Journal Letters*, **493**, L53–L58.

Geiger, H. and Müller, W. (1928). Das Electronenzählrohr (The electron-counting tube), *Physicalische Zeitschrift*, **29**, 839–841.

(1929). Technische Bemerkungen zum Electronenzählrohr (Technical remarks on the electron-counting tube), *Physicalische Zeitschrift*, **30**, 489–493.

Geller, M. J. and Huchra, J. P. (1989). Mapping the Universe, *Science*, **246**, 897–903.

Genzel, R., Schödel, R., Ott, T. *et al.* (2003). Near-infrared flares from accreting gas around the supermassive black hole at the Galactic centre, *Nature*, **425**, 934–937.

Gershtein, S. S. and Zeldovich, Y. B. (1966). Rest mass of a muonic neutrino and cosmology, *Pisma v Zhurnal Eksperimentalnoi i Teoreticheskoi Fiziki*, **4**, 174–177.

Ghez, A. M., Morris, M., Becklin, E. E., Tanner, A. and Krememek, T. (2000). The accelerations of stars orbiting the Milky Way's central black hole, *Nature*, **407**, 349–351.

Giacconi, R., Gursky, H., Paolini, F. R. and Rossi, B. B. (1962). Evidence for X rays from sources outside the solar system, *Physical Review Letters*, **9**, 439–443.

Giacconi, R., Gursky, H., Kellogg, E., Schreier, E. and Tananbaum, H. (1971a). Discovery of periodic X-ray pulsations in Centaurus X-3 from UHURU, *Astrophysical Journal*, **167**, L67–L73.

Giacconi, R., Kellogg, E., Gorenstein, P., Gursky, H. and Tananbaum, H. (1971b). An X-ray scan of the Galactic plane from UHURU, *Astrophysical Journal*, **165**, L27–L35.

Giacconi, R., Bechtold, J., Branduardi, G. *et al.* (1979). A high-sensitivity X-ray survey using the Einstein Observatory and the discrete source contribution to the extragalactic X-ray background, *Astrophysical Journal Letters*, **234**, L1–L7.

Giavalisco, M., Livio, M., Bohlin, R. C., Macchetto, F. D. and Stecher, T. P. (1996). On the morphology of the HST faint galaxies, *Astronomical Journal*, **112**, 369–377.

Gibbons, G. W. and Hawking, S. W. (1977). Cosmological event horizons, thermodynamics, and particle creation, *Physical Review*, **D15**, 2738–2751.

Gibbons, G. W., Hawking, S. W. and Siklos, S. T. C., eds (1983). *The Very Early Universe: Proceedings of the Nuffield Workshop, Cambridge, UK, June 21–July 9, 1982* (Cambridge: Cambridge University Press).

Gibbons, G. W., Shellard, E. P. S. and Rankin, S. J., eds (2003). *The Future of Theoretical Physics and Cosmology* (Cambridge: Cambridge University Press).

Ginzburg, V. L. (1951). Cosmic rays as a source of Galactic radio-radiation, *Doklady Akademiya Nauk SSSR*, **76**, 377–380.

Ginzburg, V. L. and Kirzhnits, D. A. (1964). On the superfluidity of neutron stars, *Zhurnal Experimentalnoi i Teoretichseskikh Fizica*, **47**, 2006–2007. Translation in *Soviet Physics JETP*, **20**, 1965, 1346–1348.

Glazebrook, K., Ellis, R. S., Colless, M., Broadhurst, T. J., Allington-Smith, J. R. and Tanvir, N. R. (1995). The morphological identification of the rapidly evolving population of faint galaxies, *Monthly Notices of the Royal Astronomical Society*, **275**, L19–L22.

Gold, T. (1968). Rotating neutron stars as the origin of pulsating radio sources, *Nature*, **218**, 731–732.

(1969). Rotating neutron stars and the nature of pulsars, *Nature*, **221**, 25–27.

Goldhaber, G., Boyle, B., Bunclark, P. *et al.* (1996). Cosmological time dilation using Type Ia supernovae as clocks, *Nuclear Physics B Proceedings Supplements*, **51**, 123–127.

Goldreich, P. and Julian, W. H. (1969). Pulsar electrodynamics, *Astrophysical Journal*, **157**, 869–880.

Goobar, A. and Perlmutter, S. (1995). Feasibility of measuring the cosmological constant lambda and mass density omega using Type IA supernovae, *Astrophysical Journal*, **450**, 14–18.

Gott, J. R., Melott, A. L. and Dickinson, M. (1986). The sponge-like topology of large-scale structure in the Universe, *Astrophysical Journal*, **306**, 341–357.

Gough, D. O. (1977). Random remarks on solar hydrodynamics, in *The Energy Balance and Hydrodynamics of the Solar Chromosphere and Corona*, eds Bonnet, R. M. and Delache, P. (Clermont-Ferrand: G. de Bussac), pp. 3–36.

Gower, J. F. R. (1966). The source counts from the 4C survey, *Memoirs of the Royal Astronomical Society*, **133**, 151–161.

Gower, J. F. R., Scott, P. F. and Wills, D. (1967). A survey of radio sources in the declination ranges -07 to 20 and 40 to 80, *Monthly Notices of the Royal Astronomical Society*, **71**, 49–144.

Graham Smith, F. (1951). An accurate determination of the positions of four radio stars, *Nature*, **168**, 555.

Greenstein, J. L. and Matthews, T. A. (1963). Red-shift of the unusual radio source 3C 48, *Nature*, **197**, 1041–1042.

Greenstein, J. L. and Schmidt, M. (1964). Red-shifts of the radio sources 3C 48 and 3C 273, *Astrophysical Journal*, **140**, 1–43.

Gregory, J. (1668). *Geometriae Pars Universalis* (Padua: Published by the heirs of Paolo Frambotti), p. 148.

Gribben, J. and Rees, M. J. (1989). *Dark Matter, Mankind and Anthropic Cosmology* (New York: Bantam Books).

Gull, S. F. (1975). The X-ray, optical and radio properties of young supernova remnants, *Monthly Notices of the Royal Astronomical Society*, **171**, 263–278.

Gunn, J. E. (1978). The Friedmann models and optical observations in cosmology, in *Observational Cosmology: 8th Advanced Course, Swiss Society of Astronomy and Astrophysics, Saas-Fee 1978*, eds Maeder, A., Martinet, L. and Tammann, G. (Geneva: Geneva Observatory Publications), pp. 1–121.

Gunn, J. E. and Peterson, B. A. (1965). On the density of neutral hydrogen in intergalactic space, *Astrophysical Journal*, **142**, 1633–1636.

Gursky, H., Giacconi, R., Paolini, F. R. and Rossi, B. B. (1963). Further evidence for the existence of galactic X-rays, *Physical Review Letters*, **11**, 530–535.

Gursky, H., Kellogg, E. M., Murray, S., Leong, C., Tananbaum, H. and Giacconi, R. (1971). A strong X-ray source in the Coma cluster observed by UHURU, *Astrophysical Journal Letters*, **167**, L81–L84.

Guth, A. (1981). Inflationary Universe: a possible solution to the horizon and flatness problems, *Physical Review*, **D23**, 347–356.

(1997). *The Inflationary Universe: The Quest for a New Theory of Cosmic Origins* (Reading, Massachusetts: Addison-Wesley).

(2003). Inflation and cosmological perturbations, in *The Future of Theoretical Physics and Cosmology*, eds Gibbons, G. W., Shellard, E. P. S. and Rankin, S. J. (Cambridge: Cambridge University Press), pp. 725–754.

Hale, G. E. (1928). The possibilities of large telescopes, *Harper's Magazine*, **156**, 639–646.

Hall, J. S. (1949). Observations of the polarized light from stars, *Science*, **109**, 166–167.

Halley, E. (1718). Considerations on the change of the latitudes of some of the principal fixt stars, *Philosophical Transactions of the Royal Society*, **30**, 736–738.

Halm, J. (1911). Further considerations relating to the systematic motions of the stars, *Monthly Notices of the Royal Astronomical Society*, **71**, 610–639.

Halverson, N. W., Leitch, E. M., Pryke, C. *et al.* (2002). Degree angular scale interferometer first results: a measurement of the cosmic microwave background angular power spectrum, *Astrophysical Journal*, **568**, 38–45.

Hamilton, A. J. S., Kumar, P., Lu, E. and Matthews, A. (1991). Reconstructing the primordial spectrum of fluctuations of the Universe from the observed nonlinear clustering of galaxies, *Astrophysical Journal*, **374**, L1–L4.

Hampel, W., Handt, J., Heusser, G. *et al.* (1999). GALLEX solar neutrino observations: results for GALLEX IV, *Physics Letters B*, **447**, 127–133.

Hanany, S., Ade, P., Balbi, A. *et al.* (2000). MAXIMA-1: a measurement of the cosmic microwave background anisotropy on angular scales of $10' - 5°$, *Astrophysical Journal*, **545**, L5–L9.

Hardcastle, M. J., Alexander, P., Pooley, G. G. and Riley, J. M. (1996). The Jets in 3C 66B, *Monthly Notices of the Royal Astronomical Society*, **278**, 273–284.

Harms, R. J., Ford, H. C., Tsvetanov, Z. I. *et al.* (1994). HST FOS spectroscopy of M87: evidence for a disk of ionized gas around a massive black hole, *Astrophysical Journal Letters*, **435**, L35–L38.

Harrison, B. K., Wakano, M. and Wheeler, J. A. (1958). Matter-energy at high density; end-point of thermonuclear evolution, in *Onzième Conseil de Physique Solvay, La Structure et l'evolution de l'univers* (Brussels: Editions Stoops), pp. 124–146.

Harrison, B. K., Thorne, K. S., Wakano, M. and Wheeler, J. A. (1965). *Gravitational Theory and Gravitational Collapse* (Chicago: University of Chicago Press).

Harrison, E. R. (1970). Fluctuations at the threshold of classical cosmology, *Physical Review*, **D1**, 2726–2730.

Hartmann, J. F. (1904). Investigations of the spectrum and orbit of Delta Orionis, *Astrophysical Journal*, **19**, 268–286.

Hasinger, G., Burg, R., Giacconi, R. *et al.* (1993). A deep X-ray survey in the Lockman Hole and the soft X-ray log N–log S, *Astronomy and Astrophysics*, **275**, 1–15.

Hausman, M. A. and Ostriker, J. P. (1977). Cannibalism among galaxies – dynamically produced evolution of cluster luminosity functions, *Astrophysical Journal*, **217**, L125–L129.

Hawking, S. W. (1972). Black holes in general relativity, *Communications in Mathematical Physics*, **25**, 152–166.

Hawking, S. W. and Ellis, G. R. (1973). *The Large Scale Structure of Space-Time* (Cambridge: Cambridge University Press).

Hawking, S. W. and Penrose, R. (1969). The singularities of gravitational collapse and cosmology, *Proceedings of the Royal Society*, **A314**, 529–548.

Hayakawa, S. and Matsuoka, M. (1964). Part V. Origin of cosmic X-rays, *Supplement of Progress of Theoretical Physics (Japan)*, **30**, 204–228.

Hayashi, C. (1950). Proton-neutron concentration ratio in the expanding Universe at the stages preceding the formation of the elements, *Progress of Theoretical Physics (Japan)*, **5**, 224–235.

(1961). Stellar evolution in early phases of gravitational contraction, *Publications of the Astronomical Society of Japan*, **13**, 450–452.

Hazard, C., Mackey, M. B. and Shimmins, A. J. (1963). Investigation of the radio source 3C 273 by the method of lunar occultations, *Nature*, **197**, 1037–1039.

Helmholtz, H. von (1854). On the interaction of natural forces, *Philosophical Magazine (Series 4)*, **11**, 489–518. Lecture delivered at Königsberg, 7 February 1854.

Helou, G., Soifer, B. T. and Rowan-Robinson, M. (1985). Thermal infrared and nonthermal radio – remarkable correlation in disks of galaxies, *Astrophysical Journal*, **298**, L7–L11.

Henderson, T. (1840). On the parallax of α Centauri, *Memoirs of the Royal Astronomical Society*, **11**, 61–68.

Henyey, L. G. and Keenan, P. C. (1940). Interstellar radiation from free electrons and hydrogen atoms, *Astrophysical Journal*, **91**, 625–630.

Herbig, G. H. (1952). Emission-line stars in Galactic nebulosities, *Journal of the Royal Astronomical Society of Canada*, **46**, 222–233.

Hernquist, L., Katz, N., Weinberg, D. H. and Miralda-Escudé, J. (1996). The Lyman-alpha forest in the cold dark matter model, *Astrophysical Journal*, **457**, L51–L55.

Herschel, J. F. W. (1840). On the chemical action of the rays of the solar spectrum on preparations of silver and other substances, both metallic and non-metallic, and on some photographic processes, *Philosophical Transactions of the Royal Society of London*, **130**, 1–59.

(1864). General catalogue of nebulae and clusters of stars, *Philosophical Transactions of the Royal Society*, **154**, 1–137.

Herschel, W. (1783). On the proper motion of the Sun and Solar System; with an account of several changes that have happened among the fixed stars since the time of Mr. Flamstead, *Philosophical Transactions of the Royal Society*, **73**, 247–283.

(1785). On the construction of the heavens, *Philosophical Transactions of the Royal Society*, **75**, 213–268.

(1800a). Experiments on the refrangibility of the invisible rays of the Sun, *Philosophical Transactions of the Royal Society*, **90**, 284–292.

(1800b). Experiments on the solar, and on the terrestrial rays that occasion heat; with a comparative view of the laws to which light and heat, or rather the rays which occasion them, are subject, in order to determine whether they are the same, or different. Part I, *Philosophical Transactions of the Royal Society*, **90**, 293–326.

(1800c). Experiments on the solar, and on the terrestrial rays that occasion heat; with a comparative view of the laws to which light and heat, or rather the rays which occasion them, are subject, in order to determine whether they are the same, or different. Part II, *Philosophical Transactions of the Royal Society*, **90**, 437–538.

(1800d). Investigation of the powers of the prismatic colours to heat and illuminate objects; with remarks that prove the different refrangibility of radiant heat. To which is added, an inquiry into the method of viewing the Sun advantageously, with telescopes of large apertures and high magnifying powers, *Philosophical Transactions of the Royal Society*, **90**, 255–283.

(1802). Catalogue of 500 new nebulae, nebulous stars, planetary nebulae, and clusters of stars; with remarks on the construction of the heavens, *Philosophical Transactions of the Royal Society*, **92**, 477–528.

Hertzsprung, E. (1905). Zur Strahlung der Sterne I (On the radiation of stars I), *Zeitschrift für Wissenschaftliche Photographie*, **3**, 429–442.

(1906). Ueber die Optische Stärke der Strahlung des Schwartzen Körpers und das Minimale Lichtäquivalent (On the optical intensity of black-body radiation and the equivalent minimum light emission), *Zeitschrift für Wissenschaftliche Photographie*, **4**, 43–54.

Hertzsprung, E. (1907). Zur Strahlung der Sterne II (On the radiation of stars II), *Zeitschrift für Wissenschaftliche Photographie*, **5**, 86–107.

(1911). Über die Verwendung Photographischer Effectiver Wellenlängen zur Bestimmung von Farbenäquivalenten (On the use of photographic effective wavelengths for the determination of equivalent colours), *Publikationen des Astrophysikalischen Observatoriums zu Potsdam*, **22**, 1–40.

(1913). Über die Räumliche Verteilung der Veränderlichen vom δ Cephei-Typus (On the spatial distribution of the variables of δ-Cephei type), *Astronomische Nachrichten*, **196**, 201–209.

(1919). Bermerkungen zur Statistik der Sternparallaxen (Remarks on the statisics of stellar parallaxes), *Astronomische Nachrichten*, **208**, 89–96.

Hess, V. F. (1912). Über Beobachtungen der durchdringenden Strahlung bei sieben Freiballonfahrten (Concerning observations of penetrating radiation on seven free balloon flights), *Physikalische Zeitschrift*, **13**, 1084–1091.

Hesser, J. E., Harris, W. E., VandenBerg, D. A., Allwright, J. W. B., Shott, P. and Stetson, P. (1987). A CCD color-magnitude study of 47 Tucanae, *Publications of the Astronomical Society of the Pacific*, **99**, 739–808.

Hewish, A., Bell, S. J., Pilkington, J. D. H., Scott, P. F. and Collins, R. A. (1968). Observations of a rapidly pulsating radio source, *Nature*, **217**, 709–713.

Hey, J. S. (1946). Solar radiations in the 4–6 metre radio wave-length band, *Nature*, **157**, 47–48.

Hey, J. S., Parsons, S. J. and Phillips, J. W. (1946). Fluctuations in cosmic radiation at radio-frequencies, *Nature*, **158**, 234.

Higgs, P. W. (1964). Broken symmetries, massless particles and gauge fields, *Physics Letters*, **12**, 132–133.

Hiltner, W. A. (1949). Polarization of light from distant stars by the interstellar medium, *Science*, **109**, 165.

Hirata, K. S., Inoue, K., Kajita, T., Kifune, T. and Kihara, K. (1990). Results from one thousand days of real-time, directional solar-neutrino data, *Physical Review Letters*, **65**, 1297–1300.

Hjellming, R. and Wade, C. (1971). Further radio observations of Scorpius X-1, *Astrophysical Journal*, **170**, 523–528.

Hjorth, J., Sollerman, J., Møller, P. *et al.* (2003). A very energetic supernova associated with the γ-ray burst of 29 March 2003, *Nature*, **423**, 847–850.

Hoag, A. A. and Smith, M. G. (1977). Faint emission-line quasi-stellar object candidates, *Astrophysical Journal*, **217**, 362–381.

Hoyle, F. (1946). The chemical composition of the stars, *Monthly Notices of the Royal Astronomical Society*, **106**, 225–259.

(1948). A new model for the expanding Universe, *Monthly Notices of the Royal Astronomical Society*, **108**, 372–382.

(1954). On nuclear reactions occurring in very hot stars. I. The synthesis of elements from carbon to nickel, *Astrophysical Journal Supplement*, **1**, 121–146.

Hoyle, F. and Fowler, W. A. (1963a). On the nature of strong radio sources, *Monthly Notices of the Royal Astronomical Society*, **125**, 169–176.

Hoyle, F. and Fowler W. A. (1963b). Nature of strong radio sources, *Nature*, **197**, 533–535.

Hoyle, F. and Lyttleton, R. A. (1942). On the internal constitution of the stars, *Monthly Notices of the Royal Astronomical Society*, **102**, 177–193.

Hoyle, F. and Schwarzschild, M. (1955). On the evolution of type II stars, *Astrophysical Journal Supplement*, **2**, 1–40.

Hoyle, F. and Tayler, R. J. (1964). The mystery of the cosmic helium abundance, *Nature*, **203**, 1108–1110.

Hoyle, F., Burbidge, G. R. and Sargent, W. L. W. (1966). On the nature of the quasi-stellar sources, *Nature*, **209**, 751–753.

Hu, W. (1996). Concepts in CMB anisotropy formation, in *The Universe at High-z, Large-Scale Structure and the Cosmic Microwave Background*, eds Martinez-Gonzales, E. and Sanz, J. L. (Berlin: Springer-Verlag), pp. 207–240.

Hu, W. and Sugiyama, N. (1995). Anisotropies in the cosmic microwave background: an analytic approach, *Astrophysical Journal*, **444**, 489–506.

Hu, W., Sugiyama, N. and Silk, J. (1997). The physics of microwave background anisotropies, *Nature*, **386**, 37–43.

Hubble, E. P. (1925). Cepheids in spiral nebulae, *Publications of the American Astronomical Society*, **5**, 261–264.

  (1926). Extra-galactic nebulae, *Astrophysical Journal*, **64**, 321–369.

  (1929). A relation between distance and radial velocity among extra-galactic nebulae, *Proceedings of the National Academy of Sciences*, **15**, 168–173.

  (1935). Angular rotations of spiral nebulae, *Astrophysical Journal*, **81**, 334–335.

  (1936). *The Realm of the Nebulae* (New Haven: Yale University Press).

Hubble, E. P. and Humason, M. (1934). The velocity–distance relation among extra-galactic nebulae, *Astrophysical Journal*, **74**, 43–80.

Huchra, J. and Brodie, J. (1987). The M87 Globular Cluster System. I – Dynamics, *Astronomical Journal*, **93**, 779–784.

Hudson, M. J., Dekel, A., Courteau, S., Faber, S. M. and Willick, J. A. (1995). Ω and biasing from optical galaxies versus POTENT mass, *Monthly Notices of the Royal Astronomical Society*, **274**, 305–316.

Huggins, W. (1868). Further observations on the spectra of some of the stars and nebulae, with an attempt to determine therefrom whether these bodies are moving towards or from the Earth, also observations on the spectra of the Sun and of the Comet II, *Philosophical Transactions of the Royal Society of London*, **158**, 529–564.

  (1869). Note on the heat of the stars, *Proceedings of the Royal Society of London*, **17**, 309–312.

Huggins, W. and Miller, W. A. (1864a). On the spectra of some of the fixed stars, *Philosophical Transactions of the Royal Society of London*, **154**, 413–435.

  (1864b). On the spectra of some of the nebulae; a supplement to the paper 'On the spectra of some fixed stars', *Philosophical Transactions of the Royal Society of London*, **154**, 437–444.

Hughes, D. H., Serjeant, S., Dunlop, J. *et al.* (1998). High-redshift star formation in the Hubble Deep Field revealed by a submillimetre-wavelength survey, *Nature*, **394**, 241–247.

Hulse, R. A. and Taylor, J. H. (1975). Discovery of a pulsar in a binary system, *Astrophysical Journal Letters*, **195**, L51–L53.

Hulst, H. C. van de (1945). Radio waves from space: origin of radiowaves, *Nederlands Tijdschrift voor Natuurkunde*, **11**, 210–221.

(1949a). Interstellar polarization and magneto-hydrodynamic waves, in *Problems of Cosmical Aerodynamics: Proceedings of IUTAM–IAU Symposium on Cosmical Gas Dynamics*, eds Burgers, J. M. and van de Hulst, H. C. (Dayton, Ohio: Central Air Documents Office), pp. 45–58.

(1949b). The solid particles of interstellar space, *Recherches Astronomiques de l'Observatoire d'Utrecht*, no. 11, Part 2, 1–50.

Humason, M. L., Mayall, N. U. and Sandage, A. R. (1956). Redshifts and magnitudes of extra-galactic nebulae, *Astronomical Journal*, **61**, 97–162.

Hummer, D. G. and Mihalas, D. (1988). The equation of state for stellar envelopes. I – An occupation probability formalism for the truncation of internal partition functions, *Astrophysical Journal*, **331**, 794–814.

Illingworth, G. (1977). Rotation (?) in 13 elliptical galaxies, *Astrophysical Journal Letters*, **218**, L43–L47.

Inskip, K. J., Best, P. N., Longair, M. S. and MacKay, D. J. C. (2002). Infrared magnitude-redshift relations for luminous radio galaxies, *Monthly Notices of the Royal Astronomical Society*, **329**, 277–289.

Irwin, M., McMahon, R. G. and Hazard, C. (1991). APM optical surveys for high redshift quasars, *Proceedings of the Workshop on The Space Distribution of Quasars: Astronomical Society of the Pacific Conference Series*, **21**, 117–126.

Ivanenko, D. and Pomeranchuk, I. (1944). On the maximal energy attainable in a betatron, *Physical Review*, **65**, 343.

Jansky, K. G. (1933). Electrical disturbances apparently of extraterrestrial origin, *Proceedings of the Institution of Radio Engineers*, **21**, 1387–1398.

Jeans, J. H. (1902). The stability of a spherical nebula, *Philosophical Transactions of the Royal Society of London*, **199**, 1–53.

(1917). Remark in discussion of the Royal Astronomical Society of 8 December 1916, *Observatory*, **40**, 43.

(1926). Diffuse matter in interstellar space: Letter to the Editor, *Observatory*, **49**, 333–335. See the final paragraph.

Jennison, R. C. and Das Gupta, M. K. (1953). Fine structure of the extra-terrestrial radio source Cygnus 1, *Nature*, **172**, 996–997.

Jõeveer, M. and Einasto, J. (1978). Has the Universe the cell structure?, in *The Large Scale Structure of the Universe*, eds Longair, M. S. and Einasto, J. (Dordrecht: D. Reidel Publishing Company), pp. 241–251.

Johnson, H. L. (1962). Infrared stellar photometry, *Astrophysical Journal*, **135**, 69–77.

(1965). Interstellar extinction in the Galaxy, *Astrophysical Journal*, **141**, 923–942.

Johnson, H. L. and Morgan, W. W. (1953). Fundamental stellar photometry for standards of spectral type in the revised system of the Yerkes spectral atlas, *Astrophysical Journal*, **117**, 313–352.

Johnson, H. L., Mitchell, R. I., Iriate, B. and Wisniewski, W. Z. (1966). UBVRIJKL photometry of the bright stars, *Communications of the Lunar Planetary Laboratory*, **4**, 99–110.

Johnson, W. N. III and Haymes, R. C. (1973). Detection of a gamma-ray spectral line from the Galactic-center region, *Astrophysical Journal*, **184**, 103–126.

Jones, M., Saunders, R., Alexander, P. *et al.* (1998). An image of the Sunyaev–Zel'dovich effect, *Nature*, **365**, 320–323.

Joy, A. H. (1939). Rotational effects, interstellar absorption, and certain dynamical constants of the Galaxy determined from Cepheid variables, *Astrophysical Journal*, **89**, 356–376.

(1945). T Tauri variable stars, *Astrophysical Journal*, **102**, 168–195.

Kaiser, C. R. and Alexander, P. (1997). A self-similar model for extragalactic radio sources, *Monthly Notices of the Royal Astronomical Society*, **286**, 215–222.

Kaiser, N. (1986). Evolution and clustering of rich clusters, *Monthly Notices of the Royal Astronomical Society*, **222**, 323–345.

Kapahi, V. K. (1987). The angular size-redshift relation as a cosmological tool, in *Observational Cosmology*, eds Hewitt, A., Burbidge, G. and Fang, L.-Z. (Dordrecht: D. Reidel Publishing Co.), pp. 251–265.

Kapteyn, J. C. (1892). To what stellar system does our Sun belong?, *Publications of the Astronomical Society of the Pacific*, **4**, 259–260.

(1905). Star streaming, *Report of the British Association for the Advancement of Science*, **257**, pp. 237–265.

(1906). *Plan of Selected Areas* (Groningen: Astronomical Laboratory).

(1922). First attempt at a theory of the arrangement and motion of the sidereal system, *Astrophysical Journal*, **55**, 302–328.

Kapteyn, J. C. and Rhijn, P. J. van (1920). On the distribution of the stars in space especially in the high galactic latitudes, *Astrophysical Journal*, **52**, 23–38.

Katz, N., Weinberg, D. H., Hernquist, L. and Miranda-Escudé, J. (1996). Damped Lyman-alpha and Lyman-limit absorbers in the cold dark matter model, *Astrophysical Journal*, **457**, L57–L60.

Kauffmann, G., Colberg, J. M., Diaferio, A. and White, S. D. M. (1999). Clustering of galaxies in a hierarchical Universe: I. Methods and results at $z = 0$, *Monthly Notices of the Royal Astronomical Society*, **303**, 188–206.

Kaufmann, W. (1902). Die Elektromagnetische Masse des Elektrons (On the electromagnetic mass of the electron), *Physikalische Zeitschift*, **4**, 54–56.

Kellermann, K. I. (1993). The cosmological deceleration parameter estimated from the angular-size/redshift relation for compact radio sources, *Nature*, **361**, 134–136.

Kennefick, J. D., Djorgovski, S. G. and de Carvalo, R. R. (1995). The luminosity function of $z > 4$ quasars from the second Palomar Sky Survey, *Astronomical Journal*, **110**, 2553–2565.

Kennicutt, R. C. (1989). The star formation law in galactic discs, *Astrophysical Journal*, **344**, 685–703.

Kerr, R. P. (1963). Gravitational field of a spinning mass as an example of algebraically special metrics, *Physical Review Letters*, **11**, 237–238.

Khachikian, E. Y. and Weedman, D. W. (1971). A spectroscopic study of luminous galactic nuclei, *Astrofizika*, **7**, 389–406.

    (1974). An atlas of Seyfert galaxies, *Astrophysical Journal*, **192**, 581–589.

Kiang, T. and Saslaw, W. C. (1969). The distribution in space of clusters of galaxies, *Monthly Notices of the Royal Astronomical Society*, **143**, 129–138.

Kibble, T. W. B. (1976). Topology of cosmic domains and strings, *Journal of Physics A: Mathematical and General*, **9**, 1387–1398.

Kiepenheuer, K. O. (1950). Cosmic rays as the source of general Galactic radio emission, *Physical Review*, **79**, 738–739.

Kirchhoff, G. (1859). Ueber den Zusammenhang zwischen Emission und Absorption von Licht und Wärme (On the connection between emission and absorption of light and heat), *Berlin Monatsberichte*, pp. 783–787.

    (1861). Untersuchungen über das Sonnenspektrum und die Spectren der Chemischen Elemente (Investigations of the solar spectrum and the spectra of the chemical elements), Part 1, *Abhandlungen der königlich Preussischen Akademie der Wissenschaften zu Berlin*, pp. 63–95.

    (1862). Untersuchungen über das Sonnenspektrum und die Spectren der Chemischen Elemente (Investigations of the solar spectrum and the spectra of the chemical elements), Part 1 (continued), *Abhandlungen der königlich Preussischen Akademie der Wissenschaften zu Berlin*, pp. 227–240.

    (1863). Untersuchungen über das Sonnenspektrum und die Spectren der Chemischen Elemente (Investigations of the solar spectrum and the spectra of the chemical elements), Part 2, *Abhandlungen der königlich Preussischen Akademie der Wissenschaften zu Berlin*, pp. 225–240.

Kirshner, R. and Kwan, J. (1974). Distances to extragalactic supernovae, *Astrophysical Journal*, **193**, 27–36.

Kirshner, R. P. and Oke, B. (1975). Supernova 1972e in NGC 5253, *Astrophysical Journal*, **200**, 574–581.

Klebesadel, R. W., Strong, I. B. and Olson, R. A. (1973). Observations of gamma-ray bursts of cosmic origin, *Astrophysical Journal Letters*, **182**, L85–L88.

Kleinmann, D. E. and Low, F. J. (1967). Discovery of an infrared nebula in Orion, *Astrophysical Journal*, **149**, L1–L4.

Kniffen, D. A., Chipman, E. and Gehrels, N. (1994). The gamma-ray sky according to Compton: a new window to the Universe, in *Frontiers of Space and Ground-Based Astronomy*, eds Wamsteker, W., Longair, M. S. and Kondo, Y. (Dordrecht: Kluwer Academic Publishers), pp. 5–16.

Knop, R. A., Aldering, G., Amanullah, R. *et al.* (2003). New constraints on Ω_M, Ω_Λ, and w from an independent set of 11 high-redshift supernovae observed with the Hubble Space Telescope, *Astrophysical Journal*, **598**, 102–137.

Kobold, H. A. (1895). Untersuchungen des Eigenbewegung des Auwers-Bradley Catalogs nach Bessel'schen Methode (Investigation of the proper motions from the Auwers–Bradley Catalogue using Bessel's method), *Abhandlungen der Kaiserlicher Leopoldinisch-Carolinschen Deutschen Akademie der Naturforscher*, **64**, 213–365.

Kogut, A., Banday, A. J., Bennett, C. L. *et al.* (1996). Tests for non-Gaussian statistics in the DMR four-year sky maps, *Astrophysical Journal*, **464**, L29–L33.

Kolb, E. W. and Turner, M. S. (1990). *The Early Universe* (Redwood City, California: Addison–Wesley Publishing Co.).

Kolhörster, W. (1913). Messungen der Durchdringenden Strahlung im Freiballon in Grösseren Höhen (Measurements of penetrating radiation in free balloon flights at great altitudes), *Physikalische Zeitschrift*, **14**, 1153–1156.

Kompaneets, A. (1956). The establishment of thermal equilibrium between quanta and electrons, *Zhurnal Eksperimentalnoi i Teoreticheskoi Fiziki*, **31**, 876–885. Translation in *Soviet Physics*, **4**, 1957, 730–737.

Koo, D. C. and Kron, R. (1982). QSO counts – a complete survey of stellar objects to $B = 23$, *Astronomy and Astrophysics*, **105**, 107–119.

Kormendy, J. (1982). Observations of galaxy structure and dynamics, in *Morphology and Dynamics of Galaxies: Twelfth Advanced Course of the Swiss Society of Astronomy and Astrophysics*, eds Martinet, L. and Mayor, M. (Sauverny, Switzerland: Geneva Observatory), pp. 113–288.

Kormendy, J. and Richstone, D. O. (1995). Inward bound – the search for supermassive black holes in galactic nuclei, *Annual Review of Astronomy and Astrophysics*, **33**, 581–624.

Kramers, H. A. (1923). On the theory of X-ray absorption and of the continuous X-ray spectrum, *Philosophical Magazine*, **46**, 836–871.

Kraushaar, W. L., Clark, G. W., Garmire, G. P., Borken, R., Higbie, P. and Agogino, M. (1965). Explorer XI experiment on cosmic gamma rays, *Astrophysical Journal*, **141**, 845–863.

Kroto, H. W., Heath, J. R., O'Brien, S. C., Curl, R. F. and Smalley, R. E. (1985). C(60): buckminsterfullerene, *Nature*, **318**, 162–163.

Kroto, H. W., Kirby, C., Walton, D. R. M. *et al.* (1978). The detection of cyanohexatriyne, $H(C\equiv C)_3CN$, in Heiles's Cloud 2, *Astrophysical Journal*, **219**, L133–L137.

Kroto, H. W., Heath, J. R., O'Brien, S. C., Curl, R. F. and Smalley, R. E. (1987). Long carbon chain molecules in circumstellar shells, *Astrophysical Journal*, **314**, 352–355.

Kruskal, M. D. (1960). Maximal extension of Schwarzschild metric, *Physical Review*, **119**, 1743–1745.

Krymsky, G. F. (1977). A regular mechanism for the acceleration of charged particles on the front of a shock wave, *Doklady Akademiya Nauk SSSR*, **234**, 1306–1308.

Krzeminski, W. (1973). *International Astronomical Union Circular no. 2612*.

    (1974). The identification and UBV photometry of the visible component of the Centaurus X-3 binary system, *Astrophysical Journal Letters*, **192**, L135–L138.

Kuiper, G. P., Wilson, W. and Cashman, R. J. (1947). An infrared stellar spectrometer, *Astrophysical Journal*, **106**, 243–250.

Kundic, T., Turner, E. L., Colley, W. N. *et al.* (1997). A robust determination of the time delay in 0957+561A, B and a measurement of the global value of Hubble's constant, *Astrophysical Journal*, **482**, 75–82.

Labeyrie, A. (1975). Interference fringes obtained on Vega with two optical telescopes, *Astrophysical Journal*, **196**, L71–L75.

Lagage, P. O. and Cesarsky, C. J. (1983). The maximum energy of cosmic rays accelerated by supernova shocks, *Astronomy and Astrophysics*, **125**, 249–257.

Lamb, H. (1932). *Hydrodynamics*, 6th edn (Cambridge: Cambridge University Press).

Lanczos, K. (1922). Bemerkung zur de Sitterschen Welt (Remarks on de Sitter's world model), *Physikalische Zeitschrift*, **23**, 539–543.

Landau, L. D. (1932). On the theory of stars, *Physicalische Zeitschrift der Sowjetunion*, **1**, 285–288.

(1938). Origin of stellar energy, *Nature*, **141**, 333–334.

Lane, J. H. (1870). On the theoretical temperature of the Sun; under the hypothesis of a gaseous mass maintaining its volume by its internal heat, and depending on the laws of gases as known by terrestrial experiment, *American Journal of Science and Arts, 2nd Series*, **50**, 57–74.

Langley, S. P. (1886). On hitherto unrecognised wave-lengths, *American Journal of Science*, **32**, 83–106.

(1900). The absorption lines in the infra-red spectrum of the Sun, *Annals of the Smithsonian Astrophysical Observatory*, **1**, 5–21.

Lanzetta, K. M., Wolfe, A. M. and Turnshek, D. A. (1995). The IUE Survey for damped Lyman-α and Lyman-limit absorption systems, *Astrophysical Journal*, **440**, 435–457.

Large, M. I., Vaughan, A. E. and Mills, B. Y. (1968). A pulsar supernova association?, *Nature*, **220**, 340–341.

Larson, R. B. (1969a). The emitted spectrum of a proto-star, *Monthly Notices of the Royal Astronomical Society*, **145**, 297–308.

(1969b). Numerical calculations of the dynamics of collapsing proto-star, *Monthly Notices of the Royal Astronomical Society*, **145**, 271–295.

Lattes, C. M. G., Occhialini, G. P. S. and Powell, C. F. (1947). Observations on the tracks of slow mesons in photographic emulsions, *Nature*, **160**, 453–456.

Laue, M. von (1912). Eine quantative Prüfung der Theorie für die Interferenzerscheinungen bei Röntgenstrahlung (A quantitative test of the theory of X-ray interference phenomena), *Sitzberichte der Königlich Bayerischen Akademie der Wissenschaften*, pp. 363–373.

Le Verrier, U. J. J. (1859). Sur la théorie de Mercure et sur le mouvement du périhélie de cette planète (On the theory of Mercury and the movement of the perihelion of this planet), *Comptes Rendus*, **49**, 379–383.

Leavitt, H. S. (1912). Periods of 25 variable stars in the Small Magellanic Cloud, *Harvard College Observatory Circular*, No. 173, 1–2.

Leger, A. and Puget, J. L. (1984). Identification of the 'unidentified' IR emission features of interstellar dust?, *Astronomy and Astrophysics*, **137**, L5–L8.

Leibacher, J. W. and Stein, R. F. (1971). A new description of the solar five-minute oscillation, *Astrophysical Letters*, **7**, 191–192.

Leighton, R. B. (1960). (In Discussion on) Considerations on local velocity fields in stellar atmospheres: Prototype – the solar atmosphere, in *Aerodynamic Phenomena in Stellar Atmospheres*, ed. Thomas, R. N. (Bologna: Nicola Zanichelli), pp. 321–327.

Leighton, R. B., Noyes, R. W. and Simon, G. W. (1962). Velocity fields in the solar atmosphere. I. Preliminary report, *Astrophysical Journal*, **135**, 474–499.

Lemaître, G. (1927). A homogeneous Universe of constant mass and increasing radius, accounting for the radial velocity of extra-galactic nebulae, *Annales de la Société Scientifique de Bruxelles*, **A47**, 29–39. Translation in *Monthly Notices of the Royal Astronomical Society*, **91**, (1931), 483–490.

Lemaître, G. (1931a). The beginning of the world from the point of view of quantum theory, *Nature*, **127**, 706.

(1931b). The expanding Universe, *Monthly Notices of the Royal Astronomical Society*, **91**, 490–501.

(1933). Spherical condensations in the expanding Universe, *Comptes Rendus de L'Academie des Sciences de Paris*, **196**, 903–904.

Leventhal, M., MacCallum, C. J. and Stang, P. D. (1978). Detection of 511 keV positron annihilation radiation from the Galactic center direction, *Astrophysical Journal Letters*, **225**, L11–L14.

Liddle, A. R. and Lyth, D. (2000). *Cosmological Inflation and Large-Scale Structure* (Cambridge: Cambridge University Press).

Lifshitz, E. (1946). On the gravitational stability of the expanding Universe, *Journal of Physics, Academy of Sciences of the USSR*, **10**, 116–129.

Lightman, A. P. and Schechter, P. L. (1990). The omega dependence of peculiar velocities induced by spherical density perturbations, *Astrophysical Journal Supplement Series*, **74**, 831–832.

Lilly, S. J. and Cowie, L. L. (1987). Deep infrared surveys, in *Infrared Astronomy with Arrays*, eds Wynn-Williams, C. G. and Becklin, E. E. (Honolulu: Institute for Astronomy, University of Hawaii Publications), pp. 473–482.

Lilly, S. J. and Longair, M. S. (1984). Stellar populations in distant radio galaxies, *Monthly Notices of the Royal Astronomical Society*, **211**, 833–855.

Lilly, S. J., Tresse, L., Hammer, F., Crampton, D. and LeFevre, O. (1995). The Canada–France redshift survey. VI. Evolution of the galaxy luminosity function to $z \sim 1$, *Astrophysical Journal*, **455**, 108–124.

Lin, H., Kirshner, R. P., Shectman, S. A., Landy, S. D., Oemler, A. and Tucker, D. L. (1996). The power spectrum of galaxy clustering in the Las Campanas Redshift Survey, *Astrophysical Journal*, **471**, 617–635.

Lindblad, B. (1925). Star-streaming and the structure of the stellar system, *Arkiv för Matematik, Astronomi och Fysik*, **19A** (21), 1–8.

(1927). On the cause of the ellipsoidal distribution of stellar velocities, *Arkiv för Matematik, Astronomi och Fysik*, **20A** (17), 1–7.

Linde, A. D. (1974). Is the Lee constant a cosmological constant?, *Zhurnal Experimentalnoi i Teoretichseskikh Fizica (JETP) Letters*, **19**, 183–184.

(1982). A new inflationary Universe scenario: a possible solution of the horizon, flatness, homogeneity, isotropy and primordial monopole problems, *Physics Letters*, **108B**, 389–393.

(1983). Chaotic inflation, *Physics Letters*, **129B**, 177–181.

Lineweaver, C. H. (2005). Inflation and the cosmic microwave background, in *The New Cosmology: Proceedings of the 16th International Physics Summer School, Canberra*, ed. Colless, M. (Singapore: World Scientific).

Lobachevsky, N. I. (1829). On the principles of geometry, *Kazanski Vestnik*) (*Kazan Messenger*).

(1830). On the principles of geometry, *Kazanski Vestnik (Kazan Messenger*).

Lockyer, J. N. (1900). *Inorganic Evolution* (London: Macmillan and Co.).

(1914). Notes on stellar classification II, *Nature*, **94**, 618–619.

Longair, M. S. (1966). On the interpretation of radio source counts, *Monthly Notices of the Royal Astronomical Society*, **133**, 421–436.

(1971). Observational cosmology, *Reports of Progress in Physics*, **34**, 1125–1248.

(1978). Radio astronomy and cosmology, in *Observational Cosmology: 8th Advanced Course, Swiss Society of Astronomy and Astrophysics, Saas-Fee 1978*, eds Maeder, A., Martinet, L. and Tammann, G. (Geneva: Geneva Observatory Publications), pp. 125–257.

(1997). The Friedman Robertson–Walker models: on bias, errors and acts of faith, in *Critical Dialogues in Cosmology*, ed. Turok, N. (Singapore: World Scientific), pp. 285–308.

Longair, M. S., Ryle, M. and Scheuer, P. A. G. (1973). Models of extended radio sources, *Monthly Notices of the Royal Astronomical Society*, **164**, 253–270.

Lovelace, R. V. E. and Romanova, M. M. (2003). Relativistic Poynting jets from accretion disks, *Astrophysical Journal*, **596**, L159–L162.

Lovell, A. C. B. (1987). The emergence of radio astronomy in the UK after World War II, *Quarterly Journal of the Royal Astronomical Society*, **28**, 1–9.

Low, F. J. (1961). Low-temperature germanium bolometer, *Journal of the Optical Society of America*, **51**, 1300–1304.

(1966). The infrared brightness temperature of Uranus, *Astrophysical Journal*, **146**, 326–328.

Low, F. J. and Aumann, H. H. (1970). Observations of Galactic and extragalactic sources between 50 and 300 microns, *Astrophysical Journal Letters*, **162**, L79–L85.

Low, F. J. and Johnson, H. L. (1964). Stellar photometry at 10 μm, *Astrophysical Journal*, **139**, 1130–1134.

Low, F. J., Aumann, H. H. and Gillespie, C. M. (1970). Closing astronomy's last frontier – far infrared, *Astronautics and Aeronautics*, **8**, 26–30.

Lund, N. (1984). Cosmic ray abundances, elemental and isotopic, in *Cosmic Radiation in Contemporary Astrophysics*, ed. Shapiro, M. M. (Dordrecht: D. Reidel Publishing Company), pp. 1–26.

Lundmark, K. (1920). The relations of the globular clusters and spiral nebulae to the stellar system, *Küngliga Svenska Vetenskaps-Akademiens Handlingar*, **60**, no. 8.

(1921). The Spiral Nebula Messier 33, *Publications of the Astronomical Society of the Pacific*, **33**, 324–327.

Lynden-Bell, D. (1967). Statistical mechanics of violent relaxation in stellar systems, *Monthly Notices of the Royal Astronomical Society*, **136**, 101–121.

(1969). Galactic nuclei as collapsed old quasars, *Nature*, **223**, 690–694.

Lynden-Bell, D., Faber, S. M., Burstein, D. *et al.* (1988). Spectroscopy and photometry of elliptical galaxies, *Astrophysical Journal*, **326**, 19–49.

Lyubimov, V. A., Novikov, E. G., Nozik, V. Z., Tretyakov, E. F. and Kozik, V. S. (1980). An estimate of the ν_e mass from the β-spectrum of tritium in the valine molecule, *Physics Letters*, **138**, 30–56.

Maanen, A. van (1916). Preliminary evidence of internal motion in the spiral nebula Messier 101, *Astrophysical Journal*, **44**, 210–228.

(1921). Internal motion in four spiral nebulae, *Publications of the Astronomical Society of the Pacific*, **33**, 200–202.

(1935). Internal motion in spiral nebulae, *Astrophysical Journal*, **81**, 336–337.

McCarthy, P. J., van Breugel, W. J. M., Spinrad, H. and Djorgovski, G. (1987). A correlation between the radio and optical morphologies of distant 3CR radio galaxies, *Astrophysical Journal*, **321**, L29–L33.

Macchetto, F. D. and Dickinson, M. (1997). Galaxies in the young Universe, *Scientific American*, **276**, 66–73.

McClintock, J. E. (1992). Black holes in the Galaxy, in *Proceedings of the Texas ESO/CERN Symposium on Relativistic Astrophysics, Cosmology and Fundamental Particles*, eds Barrow, J. D., Mestel, L. and Thomas, P. A. (New York: New York Academy of Sciences), pp. 495–502.

McCray, R. (1993). Supernova SN 1987A revisited, *Annual Review of Astronomy and Astrophysics*, **31**, 175–216.

McCrea, W. H. (1929). The hydrogen chromosphere, *Monthly Notices of the Royal Astronomical Society*, **89**, 483–497.

(1951). Relativity theory and the creation of matter, *Proceedings of the Royal Society of London*, **206**, 562–575.

(1970). A philosophy for Big Bang cosmology, *Nature*, **228**, 21–24.

McKee, C. F. and Ostriker, J. P. (1977). A theory of the interstellar medium – three components regulated by supernova explosions in an inhomogeneous substrate, *Astrophysical Journal*, **218**, 148–169.

McKellar, A. (1941). Molecular lines from the lowest states of the atomic molecules composed of atoms probably present in interstellar space, *Publications of the Dominion Astrophysical Observatory (Victoria)*, **7**, 251–272.

McLeod, J. M. and Andrew, B. H. (1968). The radio source VRO 42.22.01, *Astrophysical Letters*, **1**, 243.

Madau, P., Ferguson, H. C., Dickinson, M. E., Giavalisco, M., Steidel, C. C. and Fruchter, A. (1996). High-redshift galaxies in the Hubble Deep Field: colour selection and star formation history to $z \sim 4$, *Monthly Notices of the Royal Astronomical Society*, **283**, 1388–1404.

Madau, P., Pozzetti, L. and Dickinson, M. (1998). The star formation history of field galaxies, *Astrophysical Journal*, **242**, 106–116.

Maddox, S. J., Efstathiou, G., Sutherland, W. G. and Loveday, J. (1990). Galaxy correlations on large scales, *Monthly Notices of the Royal Astronomical Society*, **242**, 43P–47P.

Maeder, A. (1994). A selection of 10 most topical stellar problems, in *Frontiers of Space and Ground-Based Astronomy*, eds Wamsteker, W., Longair, M. S. and Kondo, Y. (Dordrecht: Kluwer Academic Publishers), pp. 177–186.

Mahoney, W. A., Ling, J. C., Wheaton, W. A. and Jacobson, A. S. (1984). HEAO 3 discovery of Al-26 in the interstellar medium, *Astrophysical Journal*, **286**, 578–585.

Malkan, M. and Sargent, W. L. (1982). The ultraviolet excess of Seyfert 1 galaxies and quasars, *Astrophysical Journal*, **254**, 22–37.

Malmquist, K. G. (1920). A study of stars of spectral type A, *Meddelanden från Lunds Astronomiska Observatorium, Series II*, no. 22 (Lund: Scientia Publishers), pp. 1–69.

Margon, B. and Ostriker, J. P. (1973). The luminosity function of Galactic X-ray sources: a cut-off and a 'standard candle'?, *Astrophysical Journal*, **186**, 91–96.

Markarian, B. E. (1967). Galaxies with an ultraviolet continuum, *Astrofizica*, **3**, 24–38.

Markarian, B. E., Lipovetsky, V. A. and Stepanian, D. A. (1981). Galaxies with ultraviolet continuum XV, *Astrofizica*, **17**, 619–627. Translation in *Astrophysics*, **17**, 1982, 321–332.

Marx, G. and Szalay, A. S. (1972). Cosmological limit on neutretto mass, in *Neutrino '72*, vol. 1 (Budapest: Technoinform), pp. 191–195.

Mather, J. C., Cheng, E. S., Eplee Jr, R. E. *et al.* (1990). A preliminary measurement of the cosmic microwave background spectrum by the Cosmic Background Explorer (COBE) Satellite, *Astrophysical Journal*, **354**, L37–L40.

Matt, G., Fabian, A. C. and Reynolds, C. S. (1997). Geometrical and chemical dependence of K-shell X-ray features, *Monthly Notices of the Royal Astronomical Society*, **289**, 175–184.

Matthews, T. A. and Sandage, A. R. (1963). Optical identification of 3C 48, 3C 196 and 3C 286 with stellar objects, *Astrophysical Journal*, **138**, 30–56.

Matthews, T. A., Morgan, W. W. and Schmidt, M. (1964). A discussion of galaxies identified with radio sources, *Astrophysical Journal*, **140**, 35–49.

Matthewson, D. S. and Ford, V. L. (1970). Polarization observations of 1800 stars, *Memoirs of the Royal Astronomical Society*, **74**, 139–182.

Matz, S. M., Share, G. H., Leising, M. D., Chupp, E. L. and Vestrand, W. T. (1988). Gamma-ray line emission from SN 1987A, *Nature*, **331**, 416–418.

Maury, A. C. and Pickering, E. C. (1897). Spectra of bright stars photographed with the 11-inch Draper telescope as part of the Henry Draper Memorial, *Annals of the Harvard College Observatory (Part I)*, **28**, 1–128.

Mayer-Hasselwander, H. A., Kanbach, G., Bennett, K. *et al.* (1982). Large-scale distribution of Galactic gamma radiation observed by COS-B, *Astronomy and Astrophysics*, **105**, 164–175.

Mayor, M. and Queloz, D. (1995). A Jupiter-mass companion to a solar-type star, *Nature*, **378**, 355–359.

Mazzarella, J. M. and Balzano, V. A. (1986). A catalog of Markarian galaxies, *Astrophysical Journal Supplement Series*, **62**, 751–819.

Menzel, D. H. (1926). The planetary nebulae, *Publications of the Astronomical Society of the Pacific*, **38**, 295–312.

(1931). The general theory of absorption and emission lines, *Publications of the Lick Observatory*, **17**, 213–243.

Messier, C. (1784). Catalogue de Nébuleuses et des Amas d'Étoiles (Catalogue of nebulae and clusters of stars), in *Connaissance des Temps, ou Connaisance des Mouvemans Céleste, Pour l'année bissextile 1784* (Paris: L'Académie Royal des Sciences), pp. 117–269.

Mészáros, P. (1974). The behaviour of point masses in an expanding cosmological substratum, *Astronomy and Astrophysics*, **37**, 225–228.

Mészáros, P. (2002). Theories of gamma-ray bursts, *Annual Review of Astronomy and Astrophysics*, **40**, 137–169.

Mészáros, P. and Rees, M. J. (1993). Gamma-ray bursts: multiwaveband spectral predictions for blast wave models, *Astrophysical Journal*, **418**, L59–L62.

Metcalfe, N., Shanks, T., Campos, A., Fong, R. and Gardner, J. P. (1996). Galaxy formation at high redshifts, *Nature*, **383**, 236–237.

Michell, J. (1767). An inquiry into the probable parallax, and magnitude of the fixed stars, from the quantity of light which they afford us, and the particular circumstances of their situation, *Philosophical Transactions of the Royal Society*, **57**, 234–264.

(1784). On the means of discovering the distance, magnitude, etc. of the fixed stars, in consequence of the diminution of the velocity of their light, in case such a diminution should be found to take place in any of them, and such other data should be procured from observations, as would be farther necessary for that purpose, *Philosophical Transactions of the Royal Society*, **74**, 35–57.

Michelson, A. A. (1890). On the application of interference methods to astronomical measurements, *Philosophical Magazine*, **30**, 1–21.

Michelson, A. A. and Pease, F. G. (1921). Measurement of the diameter of Alpha Orionis with the interferometer, *Astrophysical Journal*, **53**, 249–259.

Migdal, A. B. (1959). Superfluidity and the moments of inertia of nuclei, *Zhurnal Experimentalnoi i Teoretichseskikh Fizica*, **37**, 249–263. Translation in *Soviet Physics JETP*, **10**, 1960, 176–185.

Mihara, T., Makashima, K., Ohashi, T., Sakao, T. and Tashiro, M. (1990). New observations of the cyclotron absorption feature in Hercules X-1, *Nature*, **346**, 250–252.

Mikheyev, S. P. and Smirnov, A. Y. (1985). Resonance enhancement of oscillations in matter and solar neutrino spectroscopy, *Soviet Journal of Nuclear Physics*, **42**, 913–917.

Miley, G. K. (1968). Variation of the angular sizes of quasars with red-shift, *Nature*, **218**, 933–934.

(1971). The radio structure of quasars – a statistical investigation, *Monthly Notices of the Royal Astronomical Society*, **152**, 477–490.

Miley, G. K., Perola, G. C., van der Kruit, P. and van der Laan, H. (1972). Active galaxies with radio trails in clusters, *Nature*, **237**, 269–272.

Miller, G. E. and Scalo, J. M. (1979). The initial mass function and stellar birthrate in the solar neighborhood, *Astrophysical Journal Supplement Series*, **41**, 513–547.

Mills, B. Y. and Slee, O. B. (1957). A preliminary survey of radio sources in a limited region of the sky at a wavelength of 3.5 m, *Australian Journal of Physics*, **10**, 162–194.

Milne, E. (1948). *Kinematic Relativity* (Oxford: Clarendon Press).

Milne, E. A. and McCrea, W. H. (1934a). Newtonian expanding Universe, *Quarterly Journal of Mathematics*, **5**, 64–72.

(1934b). Newtonian Universes and the curvature of space, *Quarterly Journal of Mathematics*, **5**, 73–80.

Minkowski, R. (1941). Spectra of supernovae, *Publications of the Astronomical Society of the Pacific*, **53**, 224–225.

(1942). The Crab Nebula, *Astrophysical Journal*, **96**, 199–213.

Minkowski, R. (1960a). International cooperative efforts directed toward optical identification of radio sources, *Proceedings of the National Academy of Sciences of the United States of America*, **46**, 13–19.

(1960b). A new distant cluster of galaxies, *Astrophysical Journal*, **132**, 908–910.

Minnaert, M. and Mulders, G. (1930). Intensity measurement of the Fraunhofer lines in the wavelength region 5150 to 5270 Å, *Zeitschrift für Astrophysik*, **1**, 192–199.

Mirabel, I. F. and Rodrigues, L. F. (1994). A superluminal source in the Galaxy, *Nature*, **371**, 46–48.

(1998). Microquasars in our Galaxy, *Nature*, **392**, 673–676.

Miranda-Escudé, J., Cen, R., Ostriker, J. P. and Rauch, M. (1996). The Lyman alpha forest from gravitational collapse in the CDM + lambda model, *Astrophysical Journal*, **471**, 582–616.

Mitchell, R. J., Culhane, J. L., Davison, P. J. N. and Ives, J. C. (1976). Ariel 5 observations of the X-ray spectrum of the Perseus Cluster, *Monthly Notices of the Royal Astronomical Society*, **175**, 29P–34P.

Miyoshi, M., Moran, J., Herrnstein, J. *et al.* (1995). Evidence for a black-hole from high rotation velocities in a sub-parsec region of NGC4258, *Nature*, **373**, 127–129.

Møller, P. and Jakobsen, P. (1990). The Lyman continuum opacity at high redshifts – through the Lyman forest and beyond the Lyman valley, *Astronomy and Astrophysics*, **228**, 299–309.

Monck, W. H. S. (1895). The spectra and colours of the stars, *Journal of the British Astronomical Association*, **5**, 416–419.

Moran, J. M., Crowther, P. P., Burke, B. F. *et al.* (1967). Spectral line interferometry with independent time standards at stations separated by 845 kilometers, *Science*, **157**, 676–677.

Morgan, W. W., Keenan, P. C. and Kellman, E. (1943). *Atlas of Stellar Spectra, with an outline of Spectral Classification*. (Chicago: University of Chicago Press).

Morgan, W. W., Sharpless, S. and Osterbrock, D. (1951). Some features of Galactic structure in the neighbourhood of the Sun (abstract), *Astronomical Journal*, **57**, 3.

Morgan, W. W., Whitford, A. E. and Code, A. D. (1953). Studies in Galactic structure. I. A preliminary determination of the space distribution of the blue giants, *Astrophysical Journal*, **118**, 318–322.

Muller, C. A. and Oort, J. H. (1951). The interstellar hydrogen line at 1420 MHz and an estimate of galactic rotation, *Nature*, **168**, 356–358.

Myers, S. T., Baker, J. E., Readhead, A. C. S., Leitch, E. M. and Herbig, T. (1997). Measurements of the Sunyaev–Zeldovich effect in the nearby clusters A478, A2142, and A2256, *Astrophysical Journal*, **485**, 1–21.

Nakajima, T., Oppenheimer, B. R., Kulkarni, S. R., Golimowski, D. A., Matthews, K. and Durrance, S. T. (1995). Discovery of a cool brown dwarf, *Nature*, **378**, 463–465.

Nemiroff, R. J. (1994). A century of gamma ray burst models, *Comments on Astrophysics*, **17**, 189–205.

Neugebauer, G. and Leighton, R. B. (1969). *Two-micron Sky Survey: A Preliminary Catalogue* (Washington: NASA SP-3047).

Newman, E. T., Couch, K., Chinnapared, K., Exton, A., Prakash, A. and Torrence, R. (1965). Metric of a rotating charged mass, *Journal of Mathematical Physics*, **6**, 918–919.

Neyman, J., Scott, E. L. and Shane, C. D. (1954). The index of clumpiness of the distribution of images of galaxies, *Astrophysical Journal Supplement*, **1**, 269–293.

Novikov, I. D. (1964). On the possibility of [the] appearance of large scale inhomogeneities in the expanding Universe, *Journal of Experimental and Theoretical Physics*, **46**, 686–689.

Oda, M., Gorenstein, P., Gursky, H., Kellogg, E., Schreier, E., Tananbaum, H. and Giacconi, R. (1971). X-ray pulsations from Cygnus X-1 observed from UHURU, *Astrophysical Journal*, **166**, L1–L7.

O'Dell, C. R., Peimbert, M. and Kinman, T. D. (1964). The planetary nebulae in the globular cluster M15, *Astrophysical Journal*, **140**, 119–129.

Ohm, E. A. (1961). Project Echo: receiving system, *Bell System Technical Journal*, **40**, 1065–1094.

Oke, J. B. (1950). A theoretical Hertzsprung–Russell diagram for red dwarf stars, *Journal of the Royal Astronomical Society of Canada*, **44**, 135–148.

Oliver, S. J., Rowan-Robinson, M. and Saunders, W. (1992). Infrared background constraints on the evolution of IRAS galaxies, *Monthly Notices of the Royal Astronomical Society*, **256**, 15P–22P.

Omnes, R. (1969). Possibility of matter–antimatter separation at high temperatures, *Physical Review Letters*, **23**, 38–40.

Oort, J. H. (1927). Observational evidence confirming Lindblad's hypothesis of a rotation of the Galactic system, *Bulletin of the Astronomical Institutes of the Netherlands*, **3**, 275–282.

    (1932). The force exerted by the stellar system in the direction perpendicular to the Galactic plane and some related problems, *Bulletin of the Astronomical Institutes of the Netherlands*, **6**, 249–287.

    (1958). Distribution of galaxies and density in the Universe, in *Solvay Conference on The Structure and Evolution of the Universe* (Brussels: Institut International de Physique Solvay), pp. 163–181.

Oort, J. H. and Walraven, T. (1956). Polarization and composition of the Crab Nebula, *Bulletin of the Astronomical Institutes of the Netherlands*, **12**, 285–311.

Oort, J. H., Kerr, F. J. and Westerhout, G. (1958). The Galactic system as a spiral nebula, *Monthly Notices of the Royal Astronomical Society*, **118**, 379–389.

Öpik, E. (1922). An estimate of the distance of the Andromeda Nebula, *Astrophysical Journal*, **55**, 406–410.

    (1938). Stellar structure, source of energy, and evolution, *Publications of the Astronomical Observatory of the University of Tartu*, **30** (3), 1–115.

    (1951). Stellar models with variable compositions. II Sequences of models with energy generation proportional to the fifteenth power of temperature, *Proceedings of the Royal Irish Academy*, **54**, 49–77.

Oppenheimer, B. R., Kulkarni, S. R., Matthews, K. and Nakajima, T. (1995). Infrared spectrum of the cool brown dwarf GL229B, *Science*, **270**, 1478–1479.

Oppenheimer, J. R. and Snyder, H. (1939). On continued gravitational contraction, *Physical Review*, **56**, 455–459.

Oppenheimer, J. R. and Volkoff, G. M. (1939). On massive neutron cores, *Physical Review*, **55**, 374–381.

Osmer, P. S. (1982). Evidence for a decrease in the space density of quasars at z more than about 3.5, *Astrophysical Journal*, **253**, 28–37.

Osterbrock, D. E. and Rogerson, J. B. (1961). The helium and heavy-element content of gaseous nebulae and the Sun, *Publications of the Astronomical Society of the Pacific*, **73**, 129–134.

Ostriker, J. P. and Gunn, J. E. (1969). On the nature of pulsars. I. Theory, *Astrophysical Journal*, **157**, 1395–1417.

Ostriker, J. P. and Peebles, P. J. E. (1973). A numerical study of the stability of flattened galaxies: or, can cold galaxies survive?, *Astrophysical Journal*, **186**, 467–480.

Ostriker, J. P. and Tremaine, S. D. (1975). Another evolutionary correction to the luminosity of giant galaxies, *Astrophysical Journal*, **202**, L113–L117.

Oswalt, T. D., Smith, J. A., Wood, M. A. and Hintzen, P. (1996). A lower limit of 9.5 gyr on the age of the galactic disk from the oldest white dwarf stars, *Nature*, **382**, 692–694.

Owen, F. N. and Ledlow, M. J. (1994). The FR I/II break and the bivariate luminosity function in Abell clusters of galaxies, in *First Stromlo Symposium: Physics of Active Galactic Nuclei*, ASP Conference Series, vol. 34, eds Bicknell, G. V., Dopita, M. A. and Quinn, P. J. (San Francisco: ASP), pp. 319–323.

Pacini, F. (1967). Energy emission from a neutron star, *Nature*, **216**, 567–568.

(1968). Rotating neutron stars, pulsars and supernova remnants, *Nature*, **219**, 145–146.

Padmanabhan, T. (1993). *Structure Formation in the Universe* (Cambridge: Cambridge University Press).

(1996). *Cosmology and Astrophysics Through Problems* (Cambridge: Cambridge University Press), p. 437–440.

Page, L. (1997). Review of observations of the CMB, in *Critical Dialogues in Cosmology*, ed. Turok, N. (Singapore: World Scientific), pp. 343–362.

Pagel, B. E. J. (1997). *Nucleosynthesis and Chemical Evolution of Galaxies* (Cambridge: Cambridge University Press).

Panagia, N., Gilmozzi, R., Macchetto, F., Adorf, H.-M. and Kirshner, R. P. (1991). Properties of the SN 1987A circumstellar ring and the distance to the Large Magellanic Cloud, *Astrophysical Journal*, **380**, L23–L26.

Pankey Jr., T. (1962). *Possible Thermonuclear Activities in Natural Terrestrial Minerals*, Ph.D. thesis, Howard University.

Papaloizou, J. C. B. and Pringle, J. E. (1984). The dynamical stability of differentially rotating discs with constant specific angular momentum, *Monthly Notices of the Royal Astronomical Society*, **208**, 721–750.

Partridge, R. B. (1980a). Flucutations in the cosmic microwave background radiation at small angular scales, *Physica Scripta*, **21**, 624–629.

(1980b). New limits on small-scale angular fluctuations in the cosmic microwave background, *Astrophysical Journal*, **235**, 681–687.

Partridge, R. B. (1999). Current status of the cosmic microwave background radiation, in *Cosmological Parameters and the Evolution of the Universe*, IAU Symposium no. 183, ed. Sato, K. (Dordrecht: Kluwer Academic Publishers), pp. 74–87.

Pauli, W. (1925). Über den Zusammenhang des Abschlusses der Elektronengruppen im Atom mit der Komplexstruktur der Spektrum (On the connection of filled shell phenomena in atoms with complex structure of atomic spectra), *Zeitschrift für Physik*, **31**, 765–783.

Payne, C. H. (1925). *Stellar Atmospheres: Harvard College Observatory Monographs, No. 1* (Cambridge, Massachusetts: Harvard University Press).

Peacock, J. A. and Dodds, S. J. (1994). Reconstructing the linear power spectrum of cosmological mass fluctuations, *Monthly Notices of the Royal Astronomical Society*, **267**, 1020–1034.

Peacock, J. A. and Heavens, A. F. (1985). The statistics of maxima in primordial density perturbations, *Monthly Notices of the Royal Astronomical Society*, **217**, 805–820.

Peacock, J. A., Cole, S., Norberg, P. *et al.* (2001). A measurement of the cosmological mass density from clustering in the 2df galaxy redshift survey, *Nature*, **410**, 169–173.

Pearson, T. J. and Readhead, A. C. S. (1984). Image formation by self-calibration in radio astronomy, *Annual Reviews of Astronomy and Astrophysics*, **22**, 97–130.

Pearson, T. J., Unwin, S. C., Cohen, M. H. *et al.* (1981). Superluminal expansion of quasar 3C273, *Nature*, **290**, 365–368.

Pearson, T. J., Unwin, S. C., Cohen, M. H. *et al.* (1982). Superluminal expansion of 3C273, in *Extragalactic Radio Sources*, eds Heeschen, D. S. and Wade, C. M. (Dordrecht: D. Reidel Publishing Company), pp. 355–356.

Pearson, T. J., Mason, B. S., Readhead, A. C. S. *et al.* (2003). The anisotropy of the microwave background to $l = 3500$: mosaic observations with the cosmic background imager, *Astrophysical Journal*, **591**, 556–574.

Pease, F. G. (1921). The angular diameter of α Bootis by the interferometer, *Publications of the Astronomical Society of the Pacific*, **33**, 171–173.

(1931). Interferometer methods in astronomy, *Ergebnisse der Exakten Naturwissenschaften*, **10**, 84–96.

Peebles, P. J. E. (1966a). Primeval helium abundance and the primeval fireball, *Physical Review Letters*, **16**, 410–413.

(1966b). Primordial helium abundance and the primordial fireball II, *Astrophysical Journal*, **146**, 542–552.

(1968). Recombination of the primeval plasma, *Astrophysical Journal*, **153**, 1–11.

(1976). A cosmic virial theorem, *Astrophysics and Space Science*, **45**, 3–19.

(1980). *The Large-Scale Structure of the Universe* (Princeton: Princeton University Press).

(1982). Large-scale background temperature fluctuations due to scale-invariant primaeval perturbations, *Astrophysical Journal*, **263**, L1–L5.

(1993). *Principles of Physical Cosmology* (Princeton: Princeton University Press).

Peebles, P. J. E. and Yu, J. T. (1970). Primeval adiabatic perturbation in an expanding Universe, *Astrophysical Journal*, **162**, 815–836.

Pei, Y. C. and Fall, S. M. (1995). Cosmic chemical evolution, *Astrophysical Journal*, **454**, 69–76.

Penrose, R. (1965). Gravitational collapse and space-time singularities, *Physical Review Letters*, **14**, 57–59.

(1969). Gravitational collapse: the role of general relativity, *Rivista Nuovo Cimento*, **1**, 252–276.

Penzias, A. A. and Wilson, R. W. (1965). A measurement of excess antenna temperature at 4080 MHz, *Astrophysical Journal*, **142**, 419–421.

Perley, R. A., Dreher, J. W. and Cowan, J. J. (1984). The jet and filaments in Cygnus A, *Astrophysical Journal*, **285**, L35–L38.

Perlmutter, S., Boyle, B., Bunclark, P. *et al.* (1996). High-redshift supernova discoveries on demand: first results from a new tool for cosmology and bounds on q_0, *Nuclear Physics B*, **51**, 20–29.

Perlmutter, S., Gabi, S., Goldhaber, G. *et al.* (1997). Measurements of the cosmological parameters omega and lambda from the first seven supernovae at $z > 0.35$, *Astrophysical Journal*, **483**, 565–581.

Perlmutter, S., Aldering, G., della Valle, M. *et al.* (1998). Discovery of a supernova explosion at half the age of the Universe, *Nature*, **391**, 51–54.

Permutter, S., Aldering, G., Goldhaber, G., *et al.* (1999). Measurements of Ω and Λ from 42 high–redshift supernovae, *Astrophysical Journal*, **517**, (2), 565–586.

Peterson, B. M., Balonek, T. J., Barker, E. S. *et al.* (1991). Steps toward determination of the size and structure of the broad-line region in active galactic nuclei. II – An intensive study of NGC 5548 at optical wavelengths, *Astrophysical Journal*, **368**, 119–137.

Pettini, M., King, D. L., Smith, L. J. and Hunstead, R. W. (1997). The metallicity of high-redshift galaxies: the abundance of zinc in 34 damped Ly-α systems from $z = 0.7$ to 3.4, *Astrophysical Journal*, **486**, 665–680.

Pettit, E. and Nicholson, S. B. (1928). Stellar radiation measurements, *Astrophysical Journal*, **68**, 279–308.

Phillips, M. M. (1993). The absolute magnitudes of Type IA supernovae, *Astrophysical Journal*, **413**, L105–L108.

Pickering, E. C. (1890). The Draper catalogue of stellar spectra photographed with the 8-inch Bache telescope as part of the Henry Draper Memorial, *Annals of the Harvard College Observatory*, **27**, 1–388.

(1896). Stars having peculiar spectra. New variable stars in Crux and Cygnus, *Astrophysical Journal*, **4**, 369–370. This note is also Harvard College Observatory Circular no. 12.

(1897). Stars having peculiar spectra, *Astrophysical Journal*, **5**, 92–94. This note is also Harvard College Observatory Circular no. 16.

(1908). Revised Harvard photometry, *Annals of the Harvard College Observatory*, **50**, 1–252.

(1912). Distribution of stellar spectra, *Annals of the Harvard College Observatory*, **56** (1), 1–26.

Pilkington, J. D. H. and Scott, P. F. (1965). A survey of radio sources between declinations 20 and 40, *Monthly Notices of the Royal Astronomical Society*, **69**, 183–224.

Planck, M. (1900). On the theory of the laws of the energy distribution in the normal spectrum, *Verhandlungen der Deutschen Physikalische Gesellschaft*, **2**, 237–245.

Plaskett, J. S. (1923). The H and K lines of calcium in O-type stars, *Monthly Notices of the Royal Astronomical Society*, **84**, 80–93.

Plaskett, J. S. and Pearce, J. A. (1933). The problems of diffuse matter in the Galaxy, *Publications of the Dominion Astrophysical Observatory*, **5**, 167–237.

Pouillet, C.-S. (1838). Mémoire sur la chaleur solaire, sur les pouvoirs rayonnants et absorbants de l'air atmosphérique, et sur la temperature de l'espace (Memoir on the heat of the Sun, on the radiative and absorptive powers of atmospheric air and on the temperature of space), *Comptes Rendus de l'Academie des Sciences*, **7**, 24–65.

Pozdnyakov, L. A., Sobol, I. M. and Sunyaev, R. A. (1983). Comptonization and the shaping of X-ray source spectra: Monte Carlo calculations, *Astrophysics and Space Science Reviews*, **2**, 189–331.

Prendergast, K. H. and Burbidge, G. R. (1968). On the nature of some Galactic X-ray sources, *Astrophysical Journal*, **151**, L83–L88.

Press, W. H. and Schechter, P. (1974). Formation of galaxies and clusters of galaxies by self-similar gravitational condensation, *Astrophysical Journal*, **187**, 425–438.

Pskovskii, Y. P. (1977). Light curves, color curves, and expansion velocity of Type I supernovae as functions of the rate of brightness decline, *Astronomicheskii Zhurnal*, **54**, 1188–1201. Translation in *Soviet Astronomy*, **21**, 1977, 675–682.

    (1984). Photometric classification and basic parameters of Type I supernovae, *Astronomicheskii Zhurnal*, **61**, 1125–1136. Translation in *Soviet Astronomy*, **28**, 1984, 658–664.

Radhakrishnan, V. and Cooke, D. J. (1969). Magnetic poles and the polarisation structure of pulsar radiation, *Astrophysics Letters*, **3**, 225–229.

Radhakrishnan, V. and Manchester, R. N. (1969). Detection of a change of state in the pulsar PSR 0833−45, *Nature*, **222**, 228–229.

Radhakrishnan, V., Cooke, D. J., Komesaroff, M. M. and Morris, D. (1969). Evidence in support of a rotational model for the pulsar PSR 0833-45, *Nature*, **221**, 443–446.

Rappaport, S., Doxsey, R. and Zaumen, W. (1971). A search for X-ray pulsations from Cygnus X-1, *Astrophysical Journal*, **168**, L43–L47.

Rappaport, S. A. and Joss, P. C. (1983). X-ray pulsars in massive binary systems, in *Accretion Driven Stellar X-ray Sources*, eds Lewin, W. H. G. and van den Heuvel, E. P. J. (Cambridge: Cambridge University Press), pp. 1–39.

Razin, V. A. (1958). The polarization of cosmic radio radiation at wavelengths of 1.45 and 3.3 meters, *Astronomicheskii Zhurnal*, **35**, 241–252. Translation in *Soviet Astronomy*, **2**, 1958, 216–225.

Reber, G. (1940). Cosmic static, *Astrophysical Journal*, **91**, 621–624.

    (1944). Cosmic static, *Astrophysical Journal*, **100**, 279–287.

Rees, M. J. (1966). Appearance of relativistically expanding radio sources, *Nature*, **211**, 468–470.

    (1967). Studies in radio source structure – I. A relativistically expanding model for variable quasi-stellar radio sources, *Monthly Notices of the Royal Astronomical Society*, **135**, 345–360.

Rees, M. J. (1971). New interpretation of extragalactic radio sources, *Nature*, **229**, 312–317.
   (1976). Beam models for double sources and the nature of the primary energy source, in *The Physics of Non-thermal Radio Sources*, ed. Setti, G. (Dordrecht: D. Reidel Publishing Company), pp. 107–120.
Rees, M. J. and Ostriker, J. P. (1977). Cooling, dynamics and fragmentation of massive gas clouds – clues to the masses and radii of galaxies and clusters, *Monthly Notices of the Royal Astronomical Society*, **179**, 541–559.
Rees, M. J. and Sciama, D. W. (1968). Large-scale density inhomogeneities in the Universe, *Nature*, **217**, 511–516.
Rees, M. J., Phinney, E. S., Begelman, M. C. and Blandford, R. D. (1982). Ion-supported tori and the origin of radio jets, *Nature*, **295**, 17–21.
Reichley, P. E. and Downs, G. S. (1969). Observed decrease in the periods of pulsar PSR 0833−45, *Nature*, **222**, 229–230.
Reines, F. and Cowan Jr, C. L. (1956). The neutrino, *Nature*, **178**, 446–449.
Richards, D. W. and Comella, J. M. (1969). The period of pulsar NP 0532, *Nature*, **222**, 551–552.
Riemann, B. (1854). *Über die Hypothesen welche der Geometrie zu Grunde liegen (On the Hypotheses that Lie at the Foundations of Geometry)* (Göttingen: University of Göttingen). Habilitationschrift.
Riess, A. G., Press, W. H. and Kirshner, R. P. (1995). Using Type IA supernova light curve shapes to measure the Hubble constant, *Astrophysical Journal*, **438**, L17–L20.
Rindler, W. (1956). Visual horizons in world models, *Monthly Notices of the Royal Astronomical Society*, **116**, 662–677.
Ritchey, G. W. (1917). Novae in spiral nebulae, *Publications of the Astronomical Society of the Pacific*, **29**, 210–212.
Ritter, A. (1883a). Untersuchungen über die Constitution Gasförmiger Weltkörper (Researches on the Constitution of Gaseous Celestial Bodies), *Wiedemanns Annalen*, **20**, 897–927.
   (1883b). Untersuchungen über die Constitution Gasförmiger Weltkörper (Researches on the Constitution of Gaseous Celestial Bodies), *Wiedemanns Annalen*, **20**, 137–160.
   (1898). On the constitution of gaseous celestial bodies, *Astrophysical Journal*, **8**, 293–315.
Robertson, H. P. (1928). On relativistic cosmology, *Philosophical Magazine*, **5**, 835–848.
   (1935). Kinematics and world structure, *Astrophysical Journal*, **82**, 284–301.
Rochester, G. D. and Butler, C. C. (1947). Evidence for the existence of new unstable elementary particles, *Nature*, **160**, 855–857.
Rogers, A. E. E., Hinteregger, H. F., Whitney, A. R. *et al.* (1974). The structure of radio sources 3C 273B and 3C 84 deduced from the 'closure' phases and visibility amplitudes observed with three-element interferometers, *Astrophysical Journal*, **193**, 293–301.
Rogers, F. J. and Iglesias, C. A. (1994). Astrophysical opacity, *Science*, **263**, 50–55.
Rogerson, J. B. and York, D. G. (1973). Interstellar deuterium abundance in the direction of Beta Centauri, *Astrophysical Journal*, **186**, L95–L98.

Rogerson, J. B., Spitzer, L., Drake, J. F. *et al.* (1973a). Spectrophotometric results from the Copernicus satellite. I. Instrumentation and performance, *Astrophysical Journal*, **181**, L97–L102.

Rogerson, J. B., York, D. G., Drake, J. F., Jenkins, E. B., Morton, D. C. and Spitzer, L. (1973b). Spectrophotometric results from the Copernicus satellite. III. Ionization and composition of the intercloud medium, *Astrophysical Journal Letters*, **181**, L110–L115.

Roll, P. G. and Wilkinson, D. T. (1966). Cosmic background radiation at 3.2 cm – support for cosmic black-body radiation, *Physical Review Letters*, **16**, 405–407.

Röntgen, W. C. (1895). Über eine neue Art von Strahlen (On a new type of ray. Preliminary communication), *Erste Mittheilung: Sitzungsberichte der Physikalisch-Medizinische Gesellschaft, Würzburg*, **137**, 132–141.

Rosseland, S. (1924). Note on the absorption of radiation within a star, *Monthly Notices of the Royal Astronomical Society*, **84**, 525–528.

Rossi, B. (1970). An X-ray pulsar in the Crab Nebula, in *Non-Solar X- and Gamma-Ray Astronomy*, IAU Symposium No. 37, ed. Gratton, L. (Dordrecht: D. Reidel Publishing Company), pp. 183–184.

Rougoor, G. W. and Oort, J. H. (1960). Distribution and motion of interstellar hydrogen in the Galactic system with particular reference to the region within 3 kiloparsecs of the center, *Proceedings of the National Academy of Sciences of the United States of America*, **46**, 1–13.

Rowan-Robinson, M. (1968). The determination of the evolutionary properties of quasars by means of the luminosity-volume test, *Monthly Notices of the Royal Astronomical Society*, **141**, 445–458.

Rowan-Robinson, M., Benn, C. R., Lawrence, A., McMahon, R. G. and Broadhurst, T. J. (1993). The evolution of faint radio sources, *Monthly Notices of the Royal Astronomical Society*, **263**, 123–130.

Rubin, V. C., Thonnard, N. and Ford, W. K. (1980). Rotational properties of 21 Sc galaxies with a large range of luminosities and radii from NGC 4605 ($R = 4$ kpc) to UGC2885 ($R = 122$ kpc), *Astrophysical Journal*, **238**, 471–487.

Ruderman, M. (1969). *Neutron-Starquakes and Pulsar Periods* (New York: Department of Physics, New York University).

Russell, H. N. (1912a). On the determination of the orbital elements of eclipsing variable stars I, *Astrophysical Journal*, **35**, 315–340.

(1912b). On the determination of the orbital elements of eclipsing variable stars II, *Astrophysical Journal*, **36**, 54–74.

(1914a). Relations between the spectra and other characteristics of stars, *Popular Astronomy*, **22**, 275–294.

(1914b). Relations between the spectra and other characteristics of stars, *Popular Astronomy*, **22**, 331–351.

(1914c). Relations between the spectra and other characteristics of the stars, I. Historical, *Nature*, **93**, 227–230.

(1914d). Relations between the the spectra and other characteristics of the stars, II. Brightness and spectral class, *Nature*, **93**, 252–258.

Russell, H. N. (1914e). Relations between the the spectra and other characteristics of the stars, III., *Nature*, **93**, 281–286.

(1921). Response at the meeting of the Royal Astronomical Society on Friday 11 February 1921 on receipt of the Gold Medal of the Society, *Observatory*, **44**, 71–2.

(1922). The theory of ionization and the sun-spot spectrum, *Astrophysical Journal*, **55**, 119–144.

(1925). The problem of stellar evolution, *Nature*, **116**, 209–212.

(1929). On the composition of the Sun's atmosphere, *Astrophysical Journal*, **70**, 11–82.

Russell, H. N. and Saunders, F. A. (1925). New regularities in the spectra of the alkaline earths, *Astrophysical Journal*, **61**, 38–69.

Russell, H. N. and Shapley, H. (1912a). On darkening at the limb in eclipsing variables I, *Astrophysical Journal*, **36**, 239–254.

(1912b). On darkening at the limb in eclipsing variables II, *Astrophysical Journal*, **36**, 385–408.

Russell, H. N., Adams, W. S. and Moore, C. E. (1928). A calibration of Rowland's scale of intensities for solar lines, *Astrophysical Journal*, **68**, 1–8.

Rutherford, E. (1899). Uranium radiation and the electrical conduction produced by it, *Philosophical Magazine, Series 5*, **47**, 109–163.

(1907). Some cosmical aspects of radioactivity, *Journal of the Royal Astronomical Society of Canada*, **1**, 145–165.

(1919). Collisions of α particles with light atoms, IV. An anomalous effect in nitrogen, *Philosophical Magazine, Series 6*, **37**, 581–587.

(1920). Nuclear constitution of atoms, *Proceedings of the Royal Society of London*, **A97**, 374–400.

Rutherford, E. and Andrade, E. N. da C. (1913). The reflection of γ-rays from crystals, *Nature*, **92**, 267.

Rutherford, E. and Chadwick, J. (1921). The artificial disintegration of light elements, *Philosophical Magazine, Series 6*, **42**, 809–825.

Rutherford, E. and Royds, T. (1909). The nature of the α particle from radioactive substances, *Philosophical Magazine, Series 6*, **15**, 281–286.

Ryle, M. (1955). Radio stars and their cosmological significance, *The Observatory*, **75**, 137–147.

Ryle, M. and Graham Smith, F. (1948). A new intense source of radio-frequency radiation in the constellation of Cassiopeia, *Nature*, **162**, 462–463.

Ryle, M. and Neville, A. C. (1962). A radio survey of the north polar region with a 4.5 minutes of arc pencil-beam system, *Monthly Notices of the Royal Astronomical Society*, **125**, 39–56.

Ryle, M., Elsmore, B. and Neville, A. C. (1965). High resolution observations of the radio sources in Cygnus and Cassiopeia, *Nature*, **205**, 1259–1262.

Sachs, R. K. and Wolfe, A. M. (1967). Perturbations of a cosmological model and angular variations in the microwave background, *Astrophysical Journal*, **147**, 73–90.

Saha, M. N. (1920). Ionization in the solar chromosphere, *Philosophical Magazine*, **40**, 479–488.

Saha, M. N. (1921). On the physical theory of stellar spectra, *Proceedings of the Royal Society of London*, **99A**, 135–153.

Sahu, K. C., Livio, M., Petro, L. *et al.* (1997). The optical counterpart to gamma-ray burst GRB 970228 observed using the Hubble Space Telescope, *Nature*, **387**, 476–478.

Sakharov, A. D. (1965). The initial stage of an expanding Universe and the appearance of a nonuniform distribution of matter, *Zhurnal Eksperimentalnoi i Teoreticheskoi Fiziki*, **49**, 345–358. Translation in *Soviet Physics JETP*, **22**, 1966, 241–249.

  (1967). Violation of CP invariance, C asymmetry, and baryon asymmetry of the Universe, *Zhurnal Experimentalnoi i Teoretichseskikh Fizica (JETP) Letters*, **5**, 32–35.

Salpeter, E. E. (1952). Nuclear reactions in stars without hydrogen, *Astrophysical Journal*, **115**, 326–328.

  (1955). The luminosity function and stellar evolution, *Astrophysical Journal*, **121**, 161–167.

  (1964). Accretion of interstellar matter by massive objects, *Astrophysical Journal*, **140**, 796–800.

Sampson, R. A. (1895). On the rotation and mechanical state of the Sun, *Memoirs of the Royal Astronomical Society*, **51**, 123–183.

Sandage, A. R. (1958). Current problems in the extragalactic distance scale, *Astrophysical Journal*, **127**, 513–526.

  (1961a). The ability of the 200-inch telescope to discriminate between selected world models, *Astrophysical Journal*, **133**, 355–392.

  (1961b). *The Hubble Atlas of Galaxies* (Washington D.C.: Carnegie Institution of Washington); Publication 618.

  (1965). The existence of a major new constituent of the Universe: the quasistellar galaxies, *Astrophysical Journal*, **141**, 1560–1578.

  (1968). Observational cosmology, *The Observatory*, **88**, 91–106.

  (1970). Cosmology – the search for two numbers, *Physics Today*, **23**, 34–41.

  (1995). Practical cosmology: inventing the past, in *The Deep Universe*, by Sandage, A. R., Kron, R. G. and Longair, M. S., eds Binggeli, B. and Buser, R. (Berlin: Springer-Verlag), pp. 1–232.

Sandage, A. R. and Hardy, E. (1973). The redshift-distance relation. VII. Absolute magnitudes of the first three ranked cluster galaxies as a function of cluster richness and Bautz–Morgan cluster type: the effect on q_0, *Astrophysical Journal*, **183**, 743–758.

Sandage, A. R. and Schwarzschild, M. (1952). Inhomogeneous stellar models II. Models with exhausted cores in gravitational contraction, *Astrophysical Journal*, **116**, 463–476.

Sandage, A. R., Osmer, P., Giacconi, R. *et al.* (1966). On the optical identification of Sco X-1, *Astrophysical Journal*, **146**, 316–321.

Sanders, W. T., Kraushaar, W. L., Nousek, J. A. and Fried, P. M. (1977). Soft diffuse X-rays in the southern Galactic hemisphere, *Astrophysical Journal Letters*, **217**, L87–L91.

Sargent, W. L. W., Young, P. J., Lynds, C. R., Boksenberg, A., Shortridge, K. and Hartwick, F. D. A. (1978). Dynamical evidence for a central mass concentration in the galaxy M87, *Astrophysical Journal*, **221**, 731–744.

Saunders, W., Rowan-Robinson, M., Lawrence, A. *et al.* (1990). The 60 μm and far-infrared luminosity functions of IRAS Galaxies, *Monthly Notices of the Royal Astronomical Society*, **242**, 318–337.

Schade, D., Lilly, S. J., Crampton, D., Hammer, F., LeFevre, O. and Tresse, L. (1995). Canada–France redshift survey: Hubble Space Telescope imaging of high-redshift field galaxies, *Astrophysical Journal Letters*, **451**, L1–L4.

Schalén, C. (1936). Über Probleme der Interstellaren Absoption (On problems of interstellar absorption), *Nova Acta Regiae Societatis Scientiarum Upsaliensis*, Series IV, Vol. 10, no. 1 (also *Uppsala Astronomiska Observatoriums Meddelanden*, no. 64).

Schechter, P. (1976). An analytic expression for the luminosity function of galaxies, *Astrophysical Journal*, **203**, 297–306.

Scheiner, J. (1899). On the spectrum of the Great Nebula in Andromeda, *Astrophysical Journal*, **9**, 149–150.

Scheuer, P. A. G. (1957). A statistical method for analysing observations of faint radio stars, *Proceedings of the Cambridge Philosophical Society*, **53**, 764–773.

   (1965). A sensitive test for the presence of atomic hydrogen in intergalactic space, *Nature*, **207**, 963.

   (1974). Models of extragalactic radio sources with a continuous energy supply from a central object, *Monthly Notices of the Royal Astronomical Society*, **166**, 513–528.

   (1975). Radio astronomy and cosmology, in *Stars and Stellar Systems*, vol. 9, eds Sandage, A. R., Sandage, M. and Kristian, J. (Chicago: University of Chicago Press), pp. 725–760.

Schmidt, B. P., Kirshner, R. P. and Eastman, R. G. (1992). Expanding photospheres of Type II supernovae and the extragalactic distance scale, *Astrophysical Journal*, **395**, 366–386.

Schmidt, B. V. (1931). Ein lichtstarkes komafreies Spiegelsystem (A wide-field coma-free mirror system), *Zentralzeitung für Optik und Mechanik*, **52**, 25–26.

Schmidt, G. C. (1898). Ueber die von den Thorvebindungen und einigen anderen Substanzen ausgehende Strahlung (On the emitted radiation from thorium compounds and several other substances), *Annalen der Physik und Chemie (Wiedemanns Annalen)*, **65**, 141–151.

Schmidt, M. (1963). 3C 273: a star-like object with large red-shift, *Nature*, **197**, 1040.

   (1965). Large redshifts of five quasi-stellar sources, *Astrophysical Journal*, **141**, 1295–1300.

   (1968). Space distribution and luminosity functions of quasi-stellar sources, *Astrophysical Journal*, **151**, 393–409.

Schmidt, M. and Green, R. F. (1983). Quasar evolution derived from the Palomar bright quasar survey and other complete quasar surveys, *Astrophysical Journal*, **269**, 352–374.

Schmidt, M. and Matthews, T. A. (1964). Redshift of the quasi-stellar radio sources 3C 47 and 3C 147, *Astrophysical Journal*, **139**, 781–785.

Schmidt, M., Schneider, D. P. and Gunn, J. E. (1986). Spectroscopic CCD surveys for quasars at large redshift. II – A PFUEI transit survey, *Astrophysical Journal*, **310**, 518–533.

Schmidt , M., Schneider, D. P. and Gunn, J. E. (1995). Spectrscopic CCD surveys for quasars at large redshift. IV. Evolution of the luminosity function from quasars detected by their Lyman-alpha emission, *Astronomical Journal*, **110**, 68–77.

Schneider, D. P., Schmidt, M. and Gunn, J. E. (1994). Spectroscopic CCD surveys for quasars at large redshift. 3: The Palomar transit GRISM survey catalog, *Astronomical Journal*, **107**, 1245–1269.

Schödel, R., Ott, T., Genzel, R. *et al* (2002). A star in a 15.2-year orbit around the super-massive black hole at the centre of the Milky Way, *Nature*, **419**, 694–696.

Schönberg, M. and Chandrasekhar, S. (1942). On the evolution of the main-sequence stars, *Astrophysical Journal*, **96**, 161–171.

Schott, G. A. (1912). *Electromagnetic Radiation* (Cambridge: Cambridge University Press).

Schramm, D. N. (1997). The age of the Universe, in *Critical Dialogues in Cosmology*, ed. Turok, N. (Singapore: World Scientific), pp. 81–91.

Schramm, D. N. and Wasserburg, G. J. (1970). Nucleochronologies and the mean age of the elements, *Astrophysical Journal*, **162**, 57–69.

Schreier, E., Levinson, R., Gursky, H., Kellogg, E., Tananbaum, H. and Giacconi, R. (1972). Evidence for the binary nature of Centaurus X-3 from UHURU X-ray observations, *Astrophysical Journal*, **172**, L79–L89.

Schuster, A. (1902). The solar atmosphere, *Astrophysical Journal*, **16**, 320–327.

    (1905). Radiation through a foggy atmosphere, *Astrophysical Journal*, **21**, 1–22.

Schwarzschild, K. (1900a). Über das zulässige Krümmungsmass des Raumes (On an upper limit to the curvature of space), *Vierteljahrsschrift der Astronomischen Gesellschaft*, **35**, 337–347.

    (1900b). Über die Photographische Vergleichung der Helligkeit Verschiedenfarbiger Sterne (On the photographic comparison of the brightness of different coloured stars), *Sitzungsberichte der Kaiserlichen Akademie der Wissenschaften in Wien, Mathematisch-naturwissenschaft, Klasse 2a*, **109**, 1127–1134.

    (1906). Über das Gleichgewicht der Sonnenatmosphäre (On the equilibrium of the solar atmosphere), *Nachrichten von der Königlichen Gesselschaft der Wissenschaften zu Göttingen, Mathematisch-physikalische Klasse*, **41**, pp. 1–24.

    (1907). Über die Eigenbewegung der Fixsterne (On the proper motion of the fixed stars), *Nachrichten von der Gesellschaft der Wissenschaften zu Göttingen*, pp. 614–631.

    (1916). Über das Gravitationsfeld einis Massenpunktes nach der Einsteinschen Theorie (On the gravitational field of a point mass according to Einsteinian theory), *Sitzungsberichte der Königlich Preussischen Akademie der Wissenschaften zu Berlin*, **1**, 189–196.

Schwarzschild, M. (1958). *Structure and Evolution of the Stars* (Princeton: Princeton University Press).

    (1979). A numerical model for a triaxial stellar system in dynamical equilibrium, *Astrophysical Journal*, **232**, 236–247.

Schwinger, J. (1946). Electron radiation in high energy accelerators, *Physical Review*, **70**, 798.

    (1949). On the classical radiation of accelerated electrons, *Physical Review*, **75**, 1912–1925.

Scott, P. F., Carreira, P., Cleary, K. *et al* (2003). First results from the Very Small Array – III. The cosmic microwave background power spectrum, *Monthly Notices of the Royal Astronomical Society*, **341**, 1076–1083.

Seaton, M. J., Yan, Y., Mihalas, D. and Pradhan, A. K. (1994). Opacities for stellar envelopes, *Monthly Notices of the Royal Astronomical Society*, **266**, 805–828.

Secchi, A. (1866). Nouvelles Recherches sur l'Analyse Spectrale de la Lumière des Étoiles (New researches on the spectral analysis of the light of the stars), *Comptes Rendus*, **63**, 621–628.

(1868). Sur les Spectres Stellaires (On the spectra of the stars), *Comptes Rendus*, **66**, 124–126.

Seldner, M., Siebars, B., Groth, E. J. and Peebles, P. J. E. (1977). New reduction of the Lick catalog of galaxies, *Astronomical Journal*, **82**, 249–256.

Sellgren, K. (1984). The near-infrared continuum emission of visual reflection nebulae, *Astrophysical Journal*, **277**, 623–633.

Serjeant, S., Dunlop, J. S., Mann, R. G. *et al*. (2003). Submillimetre observations of the Hubble Deep Field and flanking fields, *Monthly Notices of the Royal Astronomical Society*, **344**, 887–904.

Severny, A. B., Kotov, V. A. and Tsap, T. T. (1976). Observations of solar pulsations, *Nature*, **259**, 87–89.

Seyfert, C. K. (1943). Nuclear emission in spiral nebulae, *Astrophysical Journal*, **97**, 28–40.

Shakeshaft, J. R., Ryle, M., Baldwin, J. E., Elsmore, B. and Thomson, J. H. (1955). A radio survey of radio sources between declinations −38 and +83, *Memoirs of the Royal Astronomical Society*, **67**, 106–154.

Shakura, N. and Sunyaev, R. A. (1973). Black holes in binary systems. Observational appearance, *Astronomy and Astrophysics*, **24**, 337–355.

Shane, C. D. and Wirtanen, C. A. (1957). The distribution of galaxies, *Publications of the Lick Observatory*, **22**, 1–60.

Shapiro, I. I. (1964). Fourth test of general relativity, *Physical Review Letters*, **13**, 789–791.

Shapiro, M. M. (1991). A brief introduction to the cosmic radiation, in *Cosmic Rays, Supernovae and the Interstellar Medium*, eds Shapiro, M. M., Silberberg, R. and Wefel, J. P. (Dordrecht: Kluwer Academic Publishers), pp. 1–28.

Shapley, H. (1915). Orbits of eighty seven eclipsing binaries – a summary, *Astrophysical Journal*, **38**, 158–174.

(1917). Note on the magnitude of novae in spiral nebulae, *Publications of the Astronomical Society of the Pacific*, **29**, 213–217.

(1918). Studies based on the colors and magnitudes in stellar clusters. VII. The distances, distribution in space, and dimensions of 60 globular clusters, *Astrophysical Journal*, **48**, 154–181.

(1921). The scale of the Universe, *Bulletin of the National Research Council*, **2**, 171–193.

Shellard, P. (2003). The future of cosmology: observational and computational prospects, in *The Future of Theoretical Physics and Cosmology*, eds Gibbons, G. W., Shellard, E. P. S. and Rankin, S. J. (Cambridge: Cambridge University Press), pp. 755–780.

Shklovsky, I. S. (1953). On the nature of the radiation from the Crab Nebula, *Dokladi Akademiya Nauk SSSR*, **90**, 983–986.

Shklovsky, I. S. (1967). The nature of the X-ray source Sco X-1, *Astronomicheskii Zhurnal*, **44**, 930–938. Translation in *Soviet Astronomy*, **11**, 1967, 749–755.

Shmoanov, T. (1957). A method for measuring the absolute effective radiation temperature of radio emission at low equivalent temperatures, *Pribory i Tekhnika Experimenta (Instruments and Experimental Methods)*, **1**, 83–86.

Shu, F. H., Adams, F. C. and Lizano, S. (1987). Star formation in molecular clouds – observation and theory, *Annual Reviews of Astronomy and Astrophysics*, **25**, 23–81.

Silk, J. (1968). Cosmic black-body radiation and galaxy formation, *Astrophysical Journal*, **151**, 459–471.

Silk, J. and Wyse, R. F. G. (1993). Galaxy formation and Hubble sequence, *Physics Reports*, **231**, 293–365.

Simpson, J. A. (1983). Elemental and isotopic composition of Galactic cosmic rays, *Annual Reviews of Nuclear and Particle Science*, **33**, 323–381.

Skobeltsyn, D. (1929). Über eine neue Art sehr schneller β-strahlen (On a new type of very fast β-ray), *Zeitschrift für Physik*, **54**, 686–702.

Slipher, V. M. (1909). Peculiar star spectra suggestive of selective absorption of light in space, *Bulletins of the Lowell Observatory*, **2**, 1–2.

(1917). A spectrographic investigation of spiral nebulae, *Proceedings of the American Philosophical Society*, **56**, 403–409.

Smail, I., Ivison, R. J. and Blain, A. W. (1997). A deep sub-millimeter survey of lensing clusters: a new window on galaxy formation and evolution, *Astrophysical Journal Letters*, **490**, L5–L8.

Smith, H. J. and Hoffleit, D. (1963). Light variations in the superluminous radio galaxy 3C 273, *Nature*, **198**, 650–651.

Smoot, G. F., Bennett, C. L., Kogut, A. *et al.* (1992). Structure in the COBE differential microwave radiometer first-year maps, *Astrophysical Journal*, **396**, L1–L5.

Sneden, C., McWilliam, A., Preston, G. W., Cowan, J. J., Burris, D. L. and Armosky, B. J. (1992). The ultra-metal-poor, neutron-capture-rich giant star CS 22892-052, *Astrophysical Journal*, **467**, 819–840.

Snell, R. L., Loren, R. B. and Plambeck, R. L. (1980). Observations of CO in L1551 – evidence for stellar driven wind shocks, *Astrophysical Journal Letters*, **239**, L17–L22.

Snyder, L. E., Buhl, D., Zuckerman, B. and Palmer, P. (1969). Microwave detection of interstellar formaldehyde, *Physical Review Letters*, **22**, 679–681.

Soldner, J. G. von (1804). On the deflection of a light ray from its straight motion due to the attraction of a world body which it passes closely, *Astronomisches Jahrbuch für das Jahr 1804* (Berlin: Späthen), pp. 161–172.

Spergel, D. N., Verde, L., Peiris, H. V. *et al.* (2003). First-year Wilkinson Microwave Anisotropy Probe (WMAP) observations: determination of cosmological parameters, *Astrophysical Journal Supplement Series*, **148**, 175–194.

Spite, F. and Spite, M. (1982). Abundance of lithium in unevolved halo stars and old disk stars – interpretation and consequences, *Astronomy and Astrophysics*, **115**, 357–366.

Spitzer, L. and Savedoff, M. P. (1950). The temperature of interstellar matter, *Astrophysical Journal*, **111**, 593–608.

Staelin, D. H. and Reifenstein III, E. C. (1968). Pulsating radio sources near the Crab Nebula, *Science*, **162**, 1481–1483.

Stahler, S. W., Shu, F. J. and Taam, R. E. (1980). The evolution of protostars I – Global formation and results, *Astrophysical Journal*, **241**, 637–654.

Starobinsky, A. A. (1985). Cosmic background anisotropy induced by isotropic flat-spectrum gravitational-wave perturbations, *Soviet Astronomy Letters*, **11**, 133–137. In Russian: *Pis'ma k Astronomicheskii Zhurnal*, **11**, 1985, 323–330.

Stebbins, J., Hufford, C. M. and Whitford, A. E. (1940). The mean coefficient of selective absorption in the Galaxy, *Astrophysical Journal*, **92**, 193–199.

Steidel, C. C. (1998). Galaxy evolution: has the 'epoch of galaxy formation' been found?, in *Eighteenth Texas Symposium on Relativistic Astrophysics and Cosmology*, eds Olinto, A. V., Frieman, J. A. and Schramm, D. N. (River Edge, N.J.: World Scientific Publishing Company), pp. 124–135.

Steidel, C. C. and Hamilton, D. (1992). Deep imaging of high redshift QSO fields below the Lyman limit. I – The field of Q0000−263 and galaxies at $z = 3.4$, *Astronomical Journal*, **104**, 941–949.

Stobie, R. S. (1969). Cepheid pulsation-III. Models fitted to a new mass-luminosity relation, *Monthly Notices of the Royal Astronomical Society*, **144**, 511–535.

Stockton, A. N. and Lynds, C. R. (1966). The remarkable absorption spectrum of 3C 191, *Astrophysical Journal*, **144**, 451–453.

Stoner, E. C. (1929). The limiting density in white dwarf stars, *Philosophical Magazine*, **7**, 63–70.

Storrie-Lombardi, L. J., McMahon, R. G. and Irwin, M. J. (1996). Evolution of neutral gas at high redshift: implications for the epoch of galaxy formation, *Monthly Notices of the Royal Astronomical Society*, **283**, L79–L83.

Stoughton, D., Lupton, R. H., Bernardi, M. *et al.* (2002). Sloan Digital Sky Survey: early data release, *Astronomical Journal*, **123**, 485–548.

Strömberg, G. (1924). The asymmetry in stellar motions and the existence of a velocity-restriction in space, *Astrophysical Journal*, **59**, 228–251.

Strömgren, B. (1932). The opacity of stellar matter and the hydrogen content of the stars, *Zeitschrift für Astrophysik*, **4**, 118–152.

(1933). On the interpretation of the Hertzsprung–Russell Diagram, *Zeitschrift für Astrophysik*, **7**, 222–238.

(1939). The physical state of interstellar hydrogen, *Astrophysical Journal*, **89**, 526–547.

Struve, F. G. W. (1840). Über die Parallaxe des Sterns α Lyrae (On the parallax of the star α Lyrae), *Astronomische Nachrichten*, **396**, 177–180. The page numbers refer to the columns of the journal.

Suess, H. E. and Urey, H. C. (1956). Abundances of the elements, *Reviews of Modern Physics*, **28**, 53–74.

Sunyaev, R. A. and Zeldovich, Y. B. (1970a). Interaction of matter and radiation in the hot model of the Universe, *Astrophysics and Space Science*, **7**, 21–30.

(1970b). Small-scale fluctuations of relic radiation, *Astrophysics and Space Science*, **7**, 3–19.

Sunyaev, R. A. and Zeldovich, Y. B. (1980). Microwave background radiation as a probe of the contemporary structure and history of the Universe, *Annual Review of Astronomy and Astrophysics*, **18**, 537–560.

Szalay, A. S. and Marx, G. (1976). Neutrino rest mass from cosmology, *Astronomy and Astrophysics*, **49**, 437–441.

Tanaka, Y., Nandra, K., Fabian, A. C. *et al.* (1995). Gravitationally redshifted emission implying an accretion disk and massive black-hole in the active galaxy MCG:-6-30-15, *Nature*, **375**, 659–661.

Tananbaum, H., Gursky, H., Kellogg, E. M., Levinson, R., Schreier, E. and Giacconi, R. (1972). Discovery of a periodic binary X-ray source in Hercules from UHURU, *Astrophysical Journal*, **174**, L144–L149.

Taylor, J. H. (1992). Pulsar timing and relativistic gravity, *Philosophical Transactions of the Royal Society*, **341**, 117–134.

Tegmark, M., Strauss, M. A., Blanton, M. R. *et al.* (2004). Cosmological parameters from SDSS and WMAP, *Physical Review D*, **69**, 103501 (1–28).

Thomson, W. (1854a). On the mechanical energies of the Solar System, *British Association Report, Part II*.

(1854b). On the mechanical energies of the Solar System, *Philosophical Magazine*, **8**, 409–430.

Thorne, K. S., Price, R. H. and Macdonald, D. A. (1986). *Black Holes: The Membrane Paradigm* (New Haven: Yale University Press).

Tinsley, B. M. (1980). Evolution of the stars and gas in galaxies, *Fundamentals of Cosmic Physics*, **5**, 287–388.

Tolman, R. C. (1934). Effect of inhomogeneity on cosmological models, *Proceedings of the National Academy of Sciences*, **20**, 169–176.

Tonks, L. and Langmuir, I. (1929). Oscillations in ionized gases, *Physical Review*, **33**, 195–210.

Tonry, J. L., Schmidt, B. P., Barris, B. *et al.* (2003). Cosmological results from high-z supernovae, *Astrophysical Journal*, **594**, 1–24.

Toomre, A. and Toomre, J. (1972). Galactic bridges and tails, *Astrophysical Journal*, **178**, 623–666.

Toutain, T. and Frölich, C. (1992). Characteristics of solar p-modes – results from the IPHIR experiment, *Astronomy and Astrophysics*, **257**, 287–297.

Tremaine, S. and Gunn, J. E. (1979). Dynamical role of light neutral leptons in cosmology, *Physical Review Letters*, **42**, 407–410.

Tremaine, S. and Richstone, D. O. (1977). A test of a statistical model for the luminosities of bright cluster galaxies, *Astrophysical Journal*, **212**, 311–316.

Trimble, V. L. and Thorne, K. S. (1969). Spectroscopic binaries and collapsed stars, *Astrophysical Journal*, **156**, 1013–1019.

Trümper, J., Pietsch, W., Reppin, C., Voges, W., Steinbert, R. and Kendziorra, E. (1978). Evidence for strong cyclotron line emission in the hard X-ray spectrum of Hercules X-1, *Astrophysical Journal Letters*, **219**, L105–L110.

Trumpler, R. J. (1930). Preliminary results on the distances, dimensions, and space distribution of open star clusters, *Lick Observatory Bulletin*, **14**, 154–188.

Tully, R. B. and Fisher, J. R. (1977). A new method of determining distances to galaxies, *Astronomy and Astrophysics*, **54**, 661–673.

Turok, N., ed. (1997). *Critical Dialogues in Cosmology* (Singapore: World Scientific).

Turtle, A. J. (1963). The spectrum of the Galactic radio emission, II, *Monthly Notices of the Royal Astronomical Society*, **126**, 405–417.

Turtle, A. J., Pugh, J. F., Kenderdine, S. and Pauliny-Toth, I. I. K. (1962). The spectrum of the Galactic radio emission, I. Observations of low resolving power, *Monthly Notices of the Royal Astronomical Society*, **124**, 297–312.

Tyson, A. (1990). Spectrum and origin of the extragalactic optical background radiation, in *Galactic and Extragalactic Background Radiation*, eds Bowyer, S. and Leinert, C. (Dordrecht: Kluwer Academic Publishers), pp. 245–255.

Ulrich, M. H., Boksenberg, A., Bromage, G. E. *et al.* (1984). Detailed observations of NGC 4151 with IUE – III. Variability of the strong emission lines from 1978 February to 1980 May, *Monthly Notices of the Royal Astronomical Society*, **206**, 221–238.

Ulrich, R. K. (1970). The five-minute oscillations of the solar surface, *Astrophysical Journal*, **162**, 993–1002.

Unsöld, A. (1928). Über die Struktur der Fraunhoferschen Linien und die Quantitative Spektralanalyse der Sonnenatmosphäre (On the structure of the Fraunhofer lines and a quantitative spectral analysis of the solar atmosphere), *Zeitschrift für Physik*, **46**, 765–781.

Urey, H., Brickwedde, F. G. and Murphy, G. M. (1932). A hydrogen isotope of mass 2, *Physical Review*, **39**, 164–165.

Urry, C. M. and Padovani, P. (1994). Unification of BL Lac objects and FR1 radio galaxies, in *First Stromlo Symposium: Physics of Active Galactic Nuclei*, ASP Conference Series, vol. 34, eds Bicknell, G. V., Dopita, M. A. and Quinn, P. J. (San Francisco: ASP), pp. 215–226.

Vacanti, G., Cawley, M. F., Colombo, E. *et al.* (1991). Gamma-ray observations of the Crab Nebula at TeV energies, *Astrophysical Journal*, **377**, 469–479.

Vashakidze, M. A. (1954). On the degree of polarization of the light near extragalactic nebulae and the Crab Nebula, *Astronomicheskikh Tsirkular*, no. 147, 11–13.

Velikhov, E. P. (1959). Stability of an ideally conducting liquid flowing between cylinders rotating in a magnetic field, *Zhurnal Eksperimentalnoi i Teoreticheskoi Fiziki*, **36**, 1398–1404. Translation in *Soviet Physics – JETP*, **9**, 1959, 995–998.

Vidal-Madjar, A., Laurent, C., Bonnet, R. M. and York, D. G. (1977). The ratio of deuterium to hydrogen in interstellar space. III – The lines of sight to Zeta Puppis and Gamma Cassiopaeia, *Astrophysical Journal*, **211**, 91–107.

Villard, P. (1900a). Sur la réflection et la réfraction des rayons cathodique et les rayons déviables de radium (On the reflection and refraction of cathode rays and the deviable rays of radium), *Comptes Rendus de L'Academie des Sciences*, **130**, 1010–1012.

(1900b). Sur le rayonnement du radium (On the radiation of radium), *Comptes Rendus de L'Academie des Sciences*, **130**, 1178–1179.

Vogel, H. C. (1874). Spectralanalytische Mitteilungen (Communications on spectral analysis), *Astronomische Nachrichten*, **84**, 113–124.

Vorontsov-Velyaminov, B. A. (1959). *Atlas and Catalogue of Interacting Galaxies, Part I* (Moscow: Sternberg Institute, Moscow State University).

(1977). Atlas of interacting galaxies, part II, and the concept of fragmentation of galaxies, *Astronomy and Astrophysics Supplement Series*, **28**, 1–117.

Wagoner, R. V. (1973). Big-Bang nucleosynthesis revisited, *Astrophysical Journal*, **179**, 343–360.

Wagoner, R. V., Fowler, W. A. and Hoyle, F. (1967). On the synthesis of elements at very high temperatures, *Astrophysical Journal*, **148**, 3–49.

Walker, A. G. (1936). On Milne's theory of world structure, *Proceedings of the London Mathematical Society, Series 2*, **42**, 90–127.

Walker, M. F. (1956). Studies of extremely young clusters, *Astrophysical Journal Supplement*, **2**, 365–387.

Walker, R. G. and Price, S. D. (1975). *Air Force Cambridge Research Laboratories Infrared Sky Survey* (Cambridge, Massachusetts: AFCRL TR-0373).

Wall, J. V. (1996). Space distribution of radio source populations, in *Extragalactic Radio Sources, IAU Symposium no. 175*, eds Ekers, R., Fanti, C. and Padrielli, L. (Dordrecht: Kluwer Academic Publishers), pp. 547–552.

Walsh, D., Carswell, R. F. and Weymann, R. J. (1979). 0957+561A, B – twin quasistellar objects or gravitational lens?, *Nature*, **279**, 381–384.

Wandel, A. and Mushotzky, R. F. (1986). Observational determination of the masses of active galactic nuclei, *Astrophysical Journal*, **306**, L61–L66.

Warren, S. J., Hewett, P. C., Irwin, M. J., McMahon, R. G. and Bridgeland, M. T. (1987). First observation of a quasar with a redshift of 4, *Nature*, **325**, 131–133.

Warren, S. J., Hewett, P. C. and Osmer, P. S. (1994). A wide-field multicolor survey for high-redshift quasars, $z \geq 2.2$. III: The luminosity function, *Astrophysical Journal*, **421**, 412–433.

Weaver, H., Williams, D. R. W., Dieter, N. H. and Lum, W. T. (1965). Observations of a strong unidentified microwave line and of emission from the OH molecule, *Nature*, **208**, 29–31.

Webber, W. R. (1983). Cosmic ray electrons and positrons – a review of current measurements and some implications, in *Composition and Origin of Cosmic Rays*, ed. Shapiro, M. M. (Dordrecht: D. Reidel Publishing Company), pp. 83–100.

Weber, J. (1961). *General Relativity and Gravitational Waves*, Interscience Tracts on Physics and Astronomy (New York: Interscience).

(1966). Observation of the thermal fluctuations of a gravitational-wave detector, *Physical Review Letters*, **17**, 1228–1230.

(1969). Evidence for discovery of gravitational radiation, *Physical Review Letters*, **22**, 1320–1324.

(1970). Anisotropy and polarization in the gravitational-radiation experiments, *Physical Review Letters*, **25**, 180–184.

Webster, B. L. and Murdin, P. (1972). Cygnus X-1: a spectroscopic binary with a heavy companion?, *Nature*, **235**, 37–38.

Weedman, D. (1994). Starburst galaxies at high redshift, in *First Stromlo Symposium: Physics of Active Galactic Nuclei*, ASP Conference Series, vol. 34, eds Bicknell, G. V., Dopita, M. A. and Quinn, P. J. (San Francisco: ASP), pp. 409–415.

Weekes, T. C., Cawley, M. F., Fegan, D. J. *et al.* (1989). Observation of TeV gamma rays from the Crab Nebula using the atmospheric Cerenkov imaging technique, *Astrophysical Journal*, **342**, 379–395.

Weinberg, S. (1989). The cosmological constant problem, *Reviews of Modern Physics*, **61**, 1–23.

(1997). Theories of the cosmological constant, in *Critical Dialogues in Cosmology*, ed. Turok, N. (Singapore: World Scientific), pp. 195–203.

Weinheimer, C. (2001). Neutrino mass from tritium β-decay, in *Dark Matter in Astro- and Particle Physics, Proceedings of the International Conference DARK 2000*, ed. Klapdor-Kleingrothaus, H. V. (Berlin: Springer-Verlag), pp. 513–519.

Weinreb, S., Barrett, A. H., Meeks, M. L. and Henry, J. C. (1963). Radio observations of OH in the interstellar medium, *Nature*, **200**, 829–831.

Weinreb, S., Meeks, M. L., Carter, J. C., Barrett, A. H. and Rogers, A. E. E. (1965). Observations of polarized OH emission, *Nature*, **208**, 440–441.

Weizsäcker, C. F. von (1937). Element transformation inside stars. I, *Physikalische Zeitschrift*, **38**, 176–191.

(1938). Element transformation inside stars. II, *Physikalische Zeitschrift*, **39**, 633–646.

Wesselink, A. J. (1947). The observations of brightness, colour and radial velocity of δ-Cephei and the pulsation hypothesis, *Bulletin of the Astronomical Institutes of the Netherlands*, **10**, 91–99. Errata, **10**, 258 and 310.

Westerhout, G., Seeger, C. L., Brouw, W. N. and Tinbergen, J. (1962). Polarization of the Galactic 75-cm radiation, *Bulletin of the Astronomical Institutes of the Netherlands*, **16**, 187–212.

Weyl, H. (1923). Zur allgemeinen Relativitätstheorie (On the theory of general relativity), *Physikalische Zeitschrift*, **29**, 230–232.

Weymann, R. J. (1966). The energy spectrum of radiation in the expanding Universe, *Astrophysical Journal*, **145**, 560–571.

Wheeler, J. A. (1968). Our Universe: the known and the unknown, *American Scientist*, **56**, 1–20.

(1977). Genesis and observership, in *Foundational Problems in the Special Science*, eds Butts, R. E. and Hintikka, J. (Dordrecht: D. Reidel Publishing Company), pp. 3–33.

White, S. D. (1989). Observable signatures of young galaxies, in *The Epoch of Galaxy Formation*, eds Frenk, C. S., Ellis, R. S., Shanks, T., Heavens, A. F. and Peacock, J. A. (Dordrecht: Kluwer Academic Publishers), pp. 15–30.

Whitford, A. E. (1948). An extension of the interstellar absorption-curve, *Astrophysical Journal*, **107**, 102–105.

Whitney, A. R., Shapiro, I. I., Rogers, A. E. E. *et al.* (1971). Quasars revisited: rapid time variations observed via very-long-baseline interferometry, *Science*, **173**, 225–230.

Will, C. M. (2001). The confrontation between general relativity and experiment, *Living Review in Relativity*, **4**. Online article: cited on 15 August 2001 http://www.livingreviews.org/Articles/Volume4/2001-4will/.

Wilson, C. T. R. (1901). On the ionisation of atmospheric air, *Proceedings of the Royal Society of London*, **68**, 151–161.

Wilson, R. W., Jefferts, K. B. and Penzias, A. A. (1970). Carbon monoxide in the Orion Nebula, *Astrophysical Journal Letters*, **161**, L43–L44.

Windhorst, R. A., Dressler, A. and Koo, D. A. (1987). Ultradeep optical identifications and spectroscopy of faint radio sources, in *Observational Cosmology*, eds Hewitt, A., Burbidge, G. and Fang, L.-Z. (Dordrecht: D. Reidel Publishing Co.), pp. 573–576.

Windhorst, R. A., Fomalont, E. B., Kellermann, K. I. *et al.* (1995). Identification of faint radio sources with optically luminous interacting disk galaxies, *Nature*, **375**, 471–474.

Wirtz, C. W. (1922). Einiges zur Statistik der Radialgeschwindigkeiten von Spiralnebeln und Kugelsternhaufen (Some remarks on the statistics of the radial velocities of spiral nebulae and star clusters), *Astronomische Nachrichten*, **215**, 349–354.

Wolf, M. (1923). On the dark nebula NGC 6960, *Astronomische Nachrichten*, **219**, 109–116.

Wolfe, A. M. (1988). Damped Ly-α absorption systems, in *QSO Absorption Lines: Probing the Universe*, eds Blades, J. C., Turnshek, D. and Norman, C. A. (Cambridge: Cambridge University Press), pp. 306–317.

Wolfenstein, L. (1978). Neutrino oscillations in matter, *Physical Review D*, **17**, 2369–2374.

Wollaston, W. H. (1802). A method of examining refractive and dispersive powers, by prismatic reflection, *Philosophical Transactions of the Royal Society*, **92**, 365–380.

Wolszczan, A. and Frail, D. (1992). A planetary system around the millisecond pulsar PSR 1257+12, *Nature*, **255**, 145–147.

Woodard, M. F. and Libbrecht, K. G. (1988). On the measurement of solar rotation using high-degree *p*-mode oscillations, in *Seismology of the Sun and Sun-Like Stars*, ed. Rolfe, E. J. (Noorwijk: ESA Publications), pp. 67–71.

Wright, W. W., Palmer, H. K., Albrecht, S. and Campbell, W. W. (1911). Radial velocities of 150 stars south of declination $-20°$ determined by the D. O. Mills expedition period 1903–1906, *Publications of the Lick Observatory*, **9** (Part 4), 71–347.

Young, P. J., Westphal, J. A., Kristian, J., Wilson, C. P. and Landauer, F. P. (1978). Evidence for a supermassive object in the nucleus of the galaxy M87 from SIT and CCD area photometry, *Astrophysical Journal*, **221**, 721–730.

Young, T. (1802). On the theory of light and colours, *Philosophical Transactions of the Royal Society*, **92**, 12–48.

Yukawa, H. (1935). On the interaction of elementary particles. I, *Proceedings of the Physical-Mathematical Society of Japan*, **17**, 48–57.

Zanstra, H. (1926). An application of the quantum theory to the luminosity of diffuse nebulae, *Physical Review*, **27**, 644.

(1927). An application of the quantum theory to the luminosity of diffuse nebulae, *Astrophysical Journal*, **65**, 50–70.

Zeldovich, Y. B. (1964a). The fate of a star and the evolution of gravitational energy upon accretion, *Soviet Physics Doklady*, **9**, 195–197.

(1964b). Observations in a Universe homogeneous in the mean, *Astronomicheskii Zhurnal*, **41**, 19–24. Translation in *Soviet Astronomy*, **8**, 1964, 13–16.

(1965). Survey of modern cosmology, *Advances of Astronomy and Astrophysics*, **3**, 241–379.

(1968). The cosmological constant and the theory of elementary particles, *Uspekhi Fizisheskikh Nauk*, **95**, 209–230.

(1970). Gravitational instability: an approximate theory for large density perturbations, *Astronomy and Astrophysics*, **5**, 84–89.

Zeldovich, Y. B. (1972). A hypothesis, unifying the structure and the entropy of the Universe, *Monthly Notices of the Royal Astronomical Society*, **160**, 1P–3P.

Zeldovich, Y. B. and Guseynov, O. H. (1966). Collapsed stars in binaries, *Astrophysical Journal*, **144**, 840–841.

Zeldovich, Y. B. and Novikov, I. D. (1964). Mass of quasi-stellar objects, *Soviet Physics Doklady*, **9**, 834–837.

Zeldovich, Y. B. and Sunyaev, R. A. (1969). The interaction of matter and radiation in a hot-model Universe, *Astrophysics and Space Science*, **4**, 301–316.

(1980). Astrophysical implications of the neutrino rest mass. I – The Universe, *Pis'ma v Astronomicheskii Zhurnal*, **6**, 451–456.

Zeldovich, Y. B., Kurt, D. and Sunyaev, R. A. (1968). Recombination of hydrogen in the hot model of the Universe, *Zhurnal Eksperimentalnoi i Teoreticheskoi Fiziki*, **55**, 278–286. Translation in *Soviet Physics – JETP*, **28**, 1969, 146–150.

Zhevakin, S. A. (1953). On the theory of Cepheids. I., *Astronomisheskii Zhurnal*, **30**, 161–179.

Zuckerman, B., Turner, B. E., Johnson, D. R. *et al.* (1975). Detection of interstellar trans-ethyl alcohol, *Astrophysical Journal*, **196**, L99–L102.

Zwicky, F. (1933). Rotverschiebung von Extragalaktischen Nebeln (The redshift of extra-galactic nebulae), *Helvetica Physica Acta*, **6**, 110–118.

(1937). On the masses of nebulae and of clusters of nebulae, *Astrophysical Journal*, **86**, 217–246.

(1942). On the large scale distribution of matter in the Universe, *Physical Review*, **61**, 489–503.

(1968). *Catalogue of Selected Compact Galaxies and of Post-eruptive Galaxies* (Guemlingen, Switzerland: F. Zwicky).

Name index

Object index

Subject index

The principal references to broad topics of major significance are highlighted in bold type. Topics described in more detail in the explanatory supplements to the chapters are also printed in bold type. References to the books mentioned in the text, notes and explanatory supplements are shown in italic type with the author's name in brackets. For books with three or more authors, *et al.* is used.

519